WIND ENERGY COMES OF AGE

PAUL GIPE

JOHN WILEY & SONS, INC.
New York • Chichester • Toronto • Brisbane • Singapore

This text is printed on acid-free paper.

Library of Congress Cataloging in Publication Data:

Gipe, Paul.
Wind energy comes of age / by Paul Gipe.
p. cm.
Includes index.
ISBN 0-471-10924-X
1. Wind power. I. Title.
TJ820.G56 1995
333.792—dc20 94-36564

Printed in the United States of America

10 9 8 7 6 5 4 3 2 1

WIND ENERGY COMES OF AGE

The Wiley Series in Sustainable Design

The Wiley Series in Sustainable Design has been created for professionals responsible for, and individuals interested in, the design and construction of the built environment. The series is dedicated to the advancement of knowledge in design and construction that serves to sustain the natural environment. Titles in the series cover innovative and emerging topics related to the design and development of the human community, within the context of preserving and enhancing the quality of the natural world. Consistent with their content, books in the series are produced with care taken in the selection of recycled and non-polluting materials.

Gray World, Green Heart
Technology, Nature and the Sustainable Landscape
Robert J. Thayer, University of California, Davis

Regenerative Design for Sustainable Development
John T. Lyle, California State Polytechnic University, Pomona

Audubon House
Building the Environmentally Responsible, Energy-Efficient Office
National Audubon Society
Croxton Collaborative, Architects

Design with Nature
Ian L. McHarg

Wind Energy Comes of Age
Paul Gipe

To the men and women who breathed life into a dormant, though never dead, technology.

"A fascinating story, not only of technology, but of the heart and soul behind it. A must for anyone who really wants to understand how wind power developed and for all those who want to understand what it takes to move sustainable technology toward the marketplace."

Carl J. Weinberg, President
Weinberg Associates

"*Wind Energy Comes of Age* is the most comprehensive and stimulating account I have read of wind power's promise in the diversified, competitive, and environmentally sustainable energy future on which our collective futures depend."

William Grant, Director
Midwest Energy Efficiency Program
Izaak Walton League of America

"With a clear environmental imperative to generate cleaner energy, this book presents a compelling case that by harnessing the secure and inexhaustible force of wind power we can begin to move towards a more sustainable energy future."

Fiona Weightman
Friends of the Earth

"Paul Gipe's *Wind Energy Comes of Age* is an essential 'a to z' encyclopedia on wind technology. Mr. Gipe's long history with the wind industry in California shows in the breadth of his presentation—he offers both a basic introduction to wind energy, as well as detailed data on the engineering, economic, and environmental issues associated with wind energy. I have no doubt that wind energy will continue to play an important role in California's energy future and that this book will be a key handbook for future planners and developers around the world."

Charles R. Imbrecht, Chairman
California Energy Commission

Contents

Foreword

BY CHRISTOPHER FLAVIN

The mid-nineties would seem to be a poor time to introduce a new energy technology. Oil prices are languishing at half the levels reached a decade earlier, and public concern about the environment has plummeted from its peak in the early nineties. Yet wind power installations in 1994 appear to have reached a record high—well above the earlier peak achieved in 1985—and the next few years show every sign of being even more bullish as "wind energy comes of age."

Unlike the wind industry of the eighties, which was almost entirely concentrated in California and Denmark, today's wind power boom is a truly global phenomenon, extending from the Great Plains of Minnesota to the grasslands of Argentina, and from the northern plains of Germany to the southern plains of India. By the end of 1994, some 25,000 wind turbines were in operation worldwide, providing a peak generating capacity of roughly 3,500 megawatts. During the next few years, that capacity is likely to rise by as much as 20 percent annually.

During the 13 years that I have followed and written about the wind power industry, I have seen its total output grow by a factor of 200. In that time, it has gone from being a glitzy tax shelter for wealthy dentists and Hollywood moguls to a serious global industry, with 1994 revenues of more than $1 billion. Even more dramatic changes are ahead, however, for wind energy is starting to figure prominently in many national energy plans, and could become a major source of income and employment in some countries.

At a time when climate scientists are increasingly concerned about carbon dioxide emissions from coal plants, and the nuclear industry is dying, the significance of a major new source of electricity is clear. In the U.S., the wind potential of just three states—North and South Dakota and Texas—could supply all the country's electricity. At the global level, the total wind potential is estimated to be roughly five times current electricity use.

Each new industry needs a bard—to sing its praises, describe its potential, and warn of problems before they become debilitating. In Paul Gipe, the wind industry has found that figure. His nearly two decades working in and with wind energy have given him a unique perspective, and his frank, "tell-it-like-it-is" style, allow Paul to provide the hard-headed assessment that wind power needs as it makes the transition to global prominence.

In *Wind Energy Comes of Age,* Paul Gipe has done this and more—assembling an ambitious, compelling, and at times critical description of the state of wind power in the mid-nineties. He charts the rise of wind power over the past decade, describing the ups and downs of the California and Danish markets, then goes on to assess the economics and reliability of today's wind power technology.

Rather than taking industry claims at their word, Paul digs beneath the surface, demonstrating that while wind power is not quite as economical as some promoters claim, it is now decidedly less expensive than coal and nuclear power in many parts of the world—a big change from the situation just a decade ago. His perspective is unique: Paul has lived for over a decade virtually among the Tehachapi wind farms in southern California, and he has also traveled extensively in Europe where he has met most of the continent's wind power leaders.

To its credit, *Wind Energy Comes of Age* tackles even the most nettlesome issues plaguing the wind industry, including the problem of bird kills, often referred to euphemistically as "avian mortality." Although the magnitude of the problem is not yet fully clear, Paul raises important warning flags about the dangers of not taking it and other environmental issues seriously. Unless the industry heeds Paul's warnings, it will lose the environmental high-ground that helped get it where it is today.

Another delicate issue tackled in *Wind Energy Comes of Age* is the role of government-sponsored research and development. Since the mid-seventies, the governments of several countries—encouraged by environmentalists and others—have invested more than $1 billion in the development of wind energy technology. Paul argues that in the United States and Germany, which have had some of the largest wind R&D budgets, most of the money has been spent on high-tech aerospace technology that has failed to perform in the field.

He is particularly critical of the engineers who continue to advocate huge multi-megawatt wind machines that have proven highly unreliable and uneconomical in the past. He notes that most of today's commercial wind turbines result from more incremental improvements, deriving from the efforts of the farm machine industry and other less sophisticated technologies.

Paul heaps praise on Denmark's wind power program, which he believes has done a better job of promoting incremental improvements in wind technology, while avoiding more sophisticated but less reliable approaches. This he believes has given Danish wind companies a reliability record that is the envy of the rest of the industry. He also points with praise to tax credits and other government incentives that have spurred wind energy development in many countries, and have encouraged private companies to invest in R&D.

Whether one agrees with such critiques or not, *Wind Energy Comes of Age* provides a valuable guide to the global wind industry as it unfolds in the decade ahead. Even those who feel stung by his criticisms would do well to remember the fate of the nuclear power industry, and others that chose to ignore early problems. As wind power makes the transition to major global industry during the next decade, clear thinking and careful analysis will be more important than ever.

Christopher Flavin is Vice-President for Research at the Worldwatch Institute in Washington, D.C. He is the co-author, with Nicholas Lenssen, of *Power Surge: Guide to the Coming Energy Revolution* (W.W. Norton, 1994). He serves on the boards of the American Wind Energy Association and the Business Council for a Sustainable Energy Future.

Preface

This book draws on personal experiences from nearly two decades of writing about wind energy and from a decade of observing California's wind industry. It also includes observations gleaned during the early 1990s from several tours of European wind projects, where I spoke with manufacturers, trade associations, and consultants about the future of wind energy. As part of the research for this book, I gazed at, listened to, walked among, and photographed wind turbines in Denmark, Germany, the Netherlands, Belgium, France, and Britain.

As both an observer of the wind industry and a participant in it, I have had the rare opportunity to see a dream materialize—to see wind energy come of age. As an observer, I have chronicled the industry's remarkable progress in trade publications as well as two previous books. As a participant, I have installed small wind turbines in Pennsylvania, testified before legislators about the benefits of wind energy, stood before angry crowds who were fearful that the wind turbines I advocated would take their firstborn, and answered so many questions from a public fascinated by the wind turbines in the Tehachapi Pass that I Installed a low-power radio station to do it automatically.

As the reader will quickly grasp, I am not at arm's length from my subject. For much of my professional life, I have argued that wind energy makes economic and environmental sense, and I still believe that it does. Now, more than ever, wind energy promises a safe and reliable source of electricity for today and for tomorrow.

This book is partly an attempt to document wind energy's achievements. Part I is an examination of where the technology stands today. In it I trace how far we have come in improving reliability and performance and explain how medium-sized wind turbines slew the giant progeny of the aerospace industry. I also warn against following a technological path littered with previous failures and suggest that despite its falling cost, wind energy should never become "too cheap to meter."

Like other technologies, wind energy has both its beauty marks and its blemishes. During the early 1990s, I became alarmed that the wind industry in the United States seemed headed toward repeating some of the environmental mistakes we made in the 1980s. Fortunately, I have found that American wind developers—rather than the malicious marauders depicted by some critics—are, for the most part, managers with good intentions. Often they are simply ignorant of community or environmental values that others, myself included, take for granted. For this reason, I look closely at wind's aesthetic impact in Part II and suggest ways it can be minimized. I also discuss wind energy's broader impacts on people, land, and wildlife. How successful we in the wind industry are at working with our neighbors, siting our projects with care, and building aesthetically pleasing wind turbines will determine the extent to which wind energy ultimately fulfills its potential. But I also quantify the benefits of this relatively benign technology, and call on the environmental community to take a more active role in fostering its development. Together we can ensure that wind technology remains the environmentally-sound source of energy that it promised to be.

I explore the integration of wind turbines with utilities in Part III, and consider whether wind is a suitable mate for methane. I also look at the technology's potential contribution to our energy mix and urge the sustained orderly development of wind energy to avoid the boom and bust cycle seen in California during the 1980s.

Wind Energy Comes of Age offers a discourse useful to wind's proponents while providing suggestions on how the industry can treat its blemishes. We have come so far during the past decade that now is an appropriate time for us to reflect on where we have been and on where we should be going.

PAUL GIPE
Tehachapi, CA
August 1994

Acknowledgments

Writing a book of this scope has required the assistance of numerous colleagues. I am especially grateful for the generous assistance of Risø National Laboratory in Denmark, ECN in the Netherlands, the Deutsches Windenergie Institut in Germany, the Energy Technology Support Unit in Great Britain, and Battelle Pacific Northwest Laboratory, Sandia National Laboratories, and the National Renewable Energy Laboratory in the United States.

I would like to thank individually the many people who contributed information or comments: Birger Madsen and Per Krogsgaard, BTM Consult; Jos Beurskens and Nico van der Borg, ECN; Armin Keuper and Jens-Peter Molly, DEWI; Peter Edwards, Wind Electric; John Armstrong and Peter Simpson, Wind Energy Group; Ian Mays, Chong Tan, Jeremy Bass, and Michael Anderson, Renewable Energy Systems; David Lindley and Peter Musgrove, National Wind Power; Keith Pitcher, Yorkshire Water; Henning Holst, Winkra; Finn Godtfredsen, Per Lundsager, and Neil Mortensen, Risø; Per Lading, Windgineering; Finn Hansen, Ken Karas, and Kevin Cousineau, Zond Systems; Dan Ancona, DOE; David Spera, DASCON Engineering; Robert Thresher, Neil Kelley, and Jim Tangler, NREL; Randy Swisher, AWEA; Ian Page, ETSU; Johannes Poulsen and Oscar Holst Jensen, Vestas DWT; Aloys Wobben, Enercon; Vagn Trend Poulsen, Nordtank; Andrew Garrad and Peter Jamieson, Garrad Hassan; Robert Poore and Bob Lynette, R. Lynette Associates; Don Smith, D.R. Smith Consulting; Mary Ilyin, PG&E; Peter Karnøe, Institute of Organization and Industrial Sociology; Matthias Heymann, Universität Stuttgart; Jamie Chapman, OEM Development; Eric Miller, Joanie Stewart, and Dick Curry, KENETECH Windpower; Dick Widseth, Phoenix Industries; Henning Kruse and Henrik Stiesdal, Bonus; John Twidell, De Montfort University; Ruud De Bruijne, NOVEM; Bill Vachon, Vachon & Assoc.; Meade Gougeon, Gougeon Bros., Apostolos Fragoulis, CRES; Dirk Kooman, Nedwind; Pramod Kulkarni, CEC; Jan Olesen, Difko; Martin Pasqualetti, Arizona State University; Lothar Schultze and Uwe Carstensen, DGW; Bernard Chabot and Guy Simonot, ADEME; Jean-

Michel Germa, Cabinet Germa; Bill Canter, WindStats; Yves-Bruno Civel, *Systémes Solaires;* Neal Emmerton, DWEA; Peggy Friis, Elsamprojekt; Jens Vesterdal, ELSAM; Poul Nielsen, DEFU; Troels Thomsen, LM Glasfiber; Hans Ohlsson, Nutek; Catherine Mitchell, University of Sussex; Roger Kelly, Centre for Alternative Technology; Phyllis Bosley, Towson State University; Tony Robotham, Coventry University; Carol White, Graphic Masters; Robert Thayer, Department of Environmental Design at U.C.–Davis; Helen Colijn; Mark Haller; Liz Ewald, C.S.U.–Bakersfield; Harry Halloran, Renewable Energy Development Corporation, for his support; and Nancy Nies, my wife, for her encouragement and patient perusal of the manuscript.

Conventions

Monetary values are expressed in the originating currency and converted to approximate U.S. dollars. If a dimension is given in the metric system, it is generally also given in English units. When the rotor diameter is used as a shorthand for the size of the wind turbine, the convention is to use the metric system.

The results of calculations are frequently rounded to the nearest significant figure, and approximations are frequently used. Like any other technology dependent on a natural resource—farming, mining, or drilling for oil—mathematical calculations are only an approximation of what can be expected in the real world.

The engineering priesthood has coined numerous terms for kinematic devices that convert the energy in the wind into a usable form. During the boom in research and development spending during the 1970s, WECS, an acronym for "wind energy conversion systems," was in vogue. Later, attorneys and brokers in California adopted WTG, for "wind turbine generator," to imply a level of sophistication that was often lacking. Both terms are often used in technical literature as substitutes for more comprehensible terms. The convention here is to use the terms *wind turbine* and *wind machine.* The term *wind turbine* refers primarily to electricity-generating wind machines. *Windmill* refers to the American farm windmill and the "Dutch" or European windmill.

Although comprehensive, *Wind Energy Comes of Age* is not exhaustive. Readers interested specifically in small wind turbines should see *Wind Power for Home & Business.* Those wishing more technical information should see Appendix C.

Wind Energy Comes of Age

Visquem de l'aire del cel. Catalan proverb, "Let us live from the air of the sky."

Late in the afternoon, Per Krogsgaard taps on his keyboard *"Danske Vindmøllepark 1."* The modem in his office on the west coast of Denmark dials, and in a few moments numbers begin flashing onto the computer screen from a world away. "Ja," Per mumbles, "the wind's been good. We haven't seen Palm Springs produce like this in years."

On the other side of the world, Kevin Cousineau reaches his aerie in California's Tehachapi Mountains. Like Krogsgaard, he keys in arcane commands on his desktop computer. Via a geosynchronous satellite somewhere over the Pacific, data from wind turbines feeding a diesel-powered network on the island of Maui begin flooding onto Cousineau's screen. Meanwhile, down on the first floor, windsmiths are listening to a briefing on work safety before starting their day. Across the Tehachapi valley, Darrell Bone begins monitoring the ebb and flow of electricity onto the lines of Southern California Edison from the more than 5000 wind turbines lacing the nearby hills.

Not since the wind was used to sail the world's seas and pump water from the lowlands of northern Europe has wind energy been used on such a grand scale as is now found in California and Denmark. No longer will wind energy be seen as the domain of a disheveled miller with corn flour in his hair, furling the cloth sails on his wooden windmill. This archaic image has given way to that of trained professionals tending their sleek aeroelectric generators by computer.

From the deserts of California to the shores of the North Sea, wind energy has come of age as a commercial generating technology. Wind turbines now provide commercial bulk power in California, Hawaii, Minnesota, Alberta, Denmark, Germany, Great Britain, Spain, Italy, Greece, the Netherlands, India, and China, and the list continues to grow.

Wind energy has come of age not only for customers of electric utilities but also for those who live beyond the end of utility lines. Wind now works for those like Ed Wulf, who installed a small wind turbine instead of extending the utility line to his home. Wind also works for the women of Ain Tolba, who no longer have to fetch water several kilometers from their village in Morocco now that modern wind turbines do the work for them. Today, wind energy is improving the quality of life for people around the globe.

The numbers are telling. Production of wind-generated electricity has risen from practically zero in the early 1980s to more then 6 terawatt-hours (TWh) in 1994, enough to meet the residential consumption equivalent to that of Copenhagen, Amsterdam, Dublin, and Zurich combined. Generation will increase to 7 TWh by 1995, and could top 30 TWh per year in Europe and North America early in the next century. Wind generating capacity worldwide reached 3500 megawatts (MW) in 1994 from more than 25,000 wind turbines and was increasing at the rate of 400 MW to 500 MW per year.

Today, wind turbines worldwide generate as much electricity as that produced by a conventional power plant fueled with either coal or uranium, and wind energy has done so with a fraction of the financial incentives heaped on those technologies. More than half the generating capacity in California has been installed since the federal energy tax credits that launched the industry in the early 1980s expired in 1985, and more than one-third of the capacity in Denmark has been installed since 1989, when the Danish subsidy program ceased.

Wind energy has made its most significant contribution in California, where the modern wind industry was reborn and where many had prematurely written its obituary. With 12% of the U.S. population and the world's seventh largest economy, California is both a major producer and a major consumer of energy. California serves as an example of both the best and the worst in energy policy. The Golden State produces more wind-generated electricity than anywhere else on earth—nearly half of worldwide production in 1994—and more geothermal, biomass, and solar energy as well. At the same time, California continues to emit almost as much carbon dioxide as the entire nation of France, and California consumes as much electricity as Great Britain with only half the population.

While not without its problems, wind energy represents a remarkable success story, says Charles Imbrecht, chairman of the California Energy Commission.[1] Wind energy was one of the few commercial successes to emerge from the scramble to develop alternative energy after the oil embargoes of the 1970s. George Maneatis, president of Pacific Gas & Electric during the late 1980s, went so far as to describe California's wind power plants as *the* most successful alternative source of energy developed in the state.[2]

But the way in which this success was achieved surprised advocates and critics alike. Most had envisioned using wind energy either with small wind turbines installed on farms and at homes scattered across the countryside, or with giant machines erected by electric utilities that used blades stretching 100 meters (m) [300 feet (ft)] or more in length. Neither approach succeeded in North America, although Denmark launched an industry successfully by serving a dispersed rural market. Instead, private developers, primarily in California, installed medium-sized wind turbines by the hundreds in large arrays, and now operate the turbines collectively as wind-driven power plants.

Wind energy is no longer the sole domain of political activists. It is now a worldwide industry with billions of dollars at stake. Dale Osborn, a marketing executive for California wind turbine manufacturer U.S. Windpower, told the *San Francisco Chronicle* that wind energy had come of age and was "not a bunch of people with ponytails anymore."[3] Investors poured more than $3 billion (thousand million) into California's wind plants. From 1985 to 1993, three American companies (U.S. Windpower, SeaWest, and Zond Systems) alone raised more than $1 billion to build wind plants after the expiration of energy tax credits. To meet the American Wind Energy Association's goal of installing 10,000 MW by the year 2000 and the European Community's goal of installing 8000 MW by the year 2005, the industry will need to raise another $15 billion, says Ken Karas, chief executive of Zond Systems.

The rapid growth of wind energy pushed the technology beyond a financial milestone in 1993, when worldwide sales exceeded $1 billion, nearly half of that from the sale of wind-generated electricity, the remainder from the sale of wind turbines, mostly in Europe. The industry had not seen revenue this great since the height of California wind development in 1985, a time when revenues were due almost entirely to equipment sales.

Wind energy has matured not only in California but in Europe as well. The steady growth in European use of wind energy will lead European generation to overshadow that of California in 1995. Germany's *Windenergie Tage* (Wind Energy Days), a biennial exhibition of wind turbines, drew 5000 people in 1991. By 1993 the small port city of Husum was overrun by a crowd of 8600 eager for the latest information on the technology.

The litany of plaudits showered on wind energy for its hard-won credibility comes from technicians on the staff of the California Energy Commission (CEC) as well as utility executives. The CEC's Sam Rashkin, after reviewing six years of data from California's Wind Project Performance Reporting System, concludes that "the facts are irrefutable; the technology has proven itself." Rashkin, not one given to effusive praise, observes that "the rapid growth in installed wind capacity in California has been accompanied by an even greater growth in production and performance." According to the CEC's analysis, the productivity of California's wind turbines improved over 50% from 1985 to 1990. "This performance improvement is even greater," says Rashkin, "if turbines installed since 1985 are evaluated separately." The latter represent the more reliable and productive technology available today. Consequently, California wind projects now perform within the range initially established by the CEC for the technology. "While performance has improved over 50%," says Rashkin, "costs have been almost halved, from approximately $1,900 per kilowatt in 1985 to $1000 per kilowatt in 1990. Thus wind technology now produces more at a lower capital cost."[4] Greater productivity at lower cost led the CEC to decide that "among all commercially available fossil-fuel and renewable energy technologies evaluated for utility power plants, wind technology has the lowest levelized costs" for use in California.[5]

The stance of Southern California Edison toward wind energy has run the gamut from outright hostility to cautious partnership. So it is illuminating to see how SCE, which buys more wind-generated electricity than any other utility in the world, now looks at the technology. According to Michael Peevey, SCE president at the time, "The wind power industry has come a long way in the past decade, perhaps further than many of us thought possible." Peevey

(a)

(c)

(b)

Figure I.1. Wind turbines in juxtaposition to conventional power plants. (a) Bonus's Combi turbines at the Avedøre gas-fired power plant within Copenhagen's city limits. (Courtesy Bonus Energy, Brande, Denmark.) (b) Nordtank turbines at a former peat mine in Ireland. (Courtesy of Nordtank Energy Group, Balle, Denmark.) (c) WindMaster turbines behind a dike at a gas-fired power plant near Lelystad, the Netherlands.

found that wind turbines on SCE's system were operating "at a capacity factor [a measure of productivity] of 25 to 30%—more than twice our original estimate," and concluded that "wind-generated electricity will have a place in our future."[6]

If success can be measured by the power of one's enemies, wind energy has indeed come of age when the president of the West Virginia Coal Association feels compelled to condemn it on public radio, and when former lobbyists for the British nuclear industry find it necessary to form an organization whose sole purpose is to stop wind energy in its tracks. They have reason to be concerned. Wind turbines now generate electricity majestically on mined-out peat lands owned by Bord Na Mona (Ireland's state peat mining company) near Bellacorick, alongside gas-fired power plants in Denmark and the Netherlands, and within the shadow of a nuclear plant in Wales (Figure I.1). Wind energy is certainly no immediate threat to King Coal, but the irony of the Bonneville Power Administration's buying wind-generated electricity from Carbon County, Wyoming, or that of the Irish Electricity Board's purchasing wind energy from Arigna Fuels atop an abandoned strip mine, is not lost on the fossil fuel industry. Nor is Hydro Quebec's decision during the mid-1990s to consider substituting wind turbines for a proposed hydroelectric plant on the Saint Lawrence River.

Wind energy's success, "forged in the crucible of California's deserts" as Union of Concerned Scientists's Don Aitken puts it, has pushed the technology beyond that of merely another "alternative." This success demonstrates that the technology works, and that wind energy can produce sizable amounts of electricity. Although wind energy suffered severe growing pains and struggled through a stormy adolescence during the 1980s, it has matured. Wind energy is now ready to take its place alongside fossil and nuclear fuels as a conventional source of electricity.

From California and the Pacific northwest to the heartland states and New England, from the mountaintops of Wales and the dikes of the Netherlands to the coast of India and the steppes of Mongolia, wind energy has come of age and is on the road to making a valuable contribution toward a sustainable electricity supply.

Introduction Endnotes

1. Charles (Chuck) Imbrecht, chairman of the California Energy Commission, banquet address before the annual conference of the American Wind Energy Association, San Francisco, September 1989.
2. George Maneatis, president, Pacific Gas & Electric Co., statement at RETSIE '89, Santa Clara, CA, June 21, 1989.
3. U.S. Windpower changed its name to KENETECH Windpower in 1993. Because more people recognize the firm's original name rather than its new identify, "U.S. Windpower" is used throughout.

4. Sam Rashkin, "California Wind Project Performance: A Review of Wind Performance Results from 1985 to 1990," Windpower '91, annual conference of the American Wind Energy Association, Palm Springs, CA, September 1991.
5. Ibid.
6. Michael Peevey, president, Southern California Edison Co., address before American Wind Energy Association's annual conference, Windpower '91, Palm Springs, CA, September 24, 1991.

I

WHERE THE TECHNOLOGY STANDS TODAY

In this book we examine wind energy's newfound respectability and the problems it faces. In Part I we learn where wind energy is used today and the technology employed. We also trace the sometimes tortuous path of wind energy's maturation from the darling of researchers and alternative energy advocates of the 1970s to the commercial reality of today's investment bankers. In Part II we explore wind energy's environmental impacts as well as its benefits. In Part III we look toward the future.

1

Overview

Wind Werkt! Dutch for "Wind works!"

Analysts use two measures to describe wind's role in the energy mix: units of capacity (MW), the amount of power wind turbines are capable of producing; and units of energy in kilowatt-hour(s) (kWh), the amount of electricity they actually generate. Capacity is an effective shorthand for describing the relative position of wind energy in nations around the globe. Yet it is just that, a shorthand. Capacity is never a substitute for the kilowatt-hours of electricity generated by wind turbines, the bottom line in any discussion of wind energy. The goal of those who consider wind energy a timely, economical, and environmentally sensible way to meet society's energy needs, after all, is the production of wind-generated electricity; it is useful to keep that perspective in view. It is not the number of wind turbines, or even their size, that is important; it is the amount of energy they generate.

Worldwide Status

Wind turbines generated more than 6 TWh in 1994. California wind plants produced 47% of that amount. Denmark and Germany each produced 17% (Figure 1.1). California accounted for the bulk of wind generating capacity in North America through 1994. There were small wind power plants operating in several other states, including Hawaii, Oregon, Montana, and Massachusetts. Larger pilot wind plants were also operating in Minnesota and Alberta (Figure 1.2). By the mid-1990s several major projects were under way in Iowa, Oregon, Washington, Texas, and Maine, and projects in other states were under negotiation.

In 1993, Europe passed an important milestone when total installed capacity exceeded 1000 MW. Denmark and Germany, followed by Great Britain and

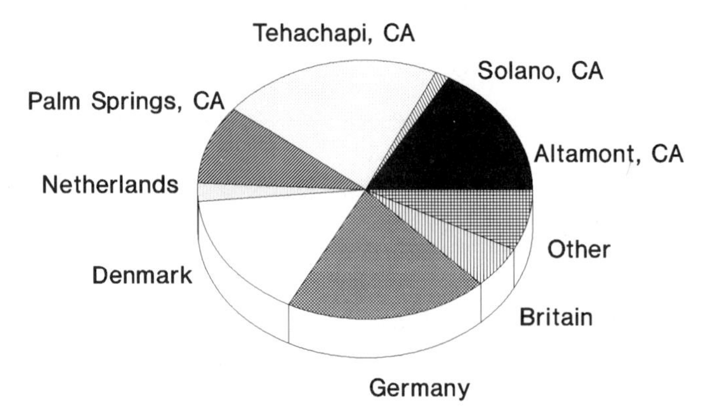

Figure 1.1. World wind generation. In 1994, California produced half the world's wind-generated electricity; Denmark, one-fifth.

the Netherlands, lead Europe in installed capacity (Figure 1.3). Sites on the west coast of Denmark and in the mountains of Wales and Cornwall, which lie transverse to the prevailing westerlies, are the windiest in Europe and rival those of California's famed mountain passes. By the end of 1995 the installed capacity in western Europe will exceed not only that of California but also that of North America as well.

California's Tehachapi-Mojave wind resource area leads the world in wind-generated electricity (Figure 1.4). The Tehachapi area took the top spot in 1992 after a decade-long reign by California's Altamont Pass. Denmark's generation surpassed Altamont's for the first time in 1993. By 1994 Germany began rivaling Denmark as Europe's foremost producer of wind-generated electricity.

In the developing world China had installed some 30 MW of medium-sized

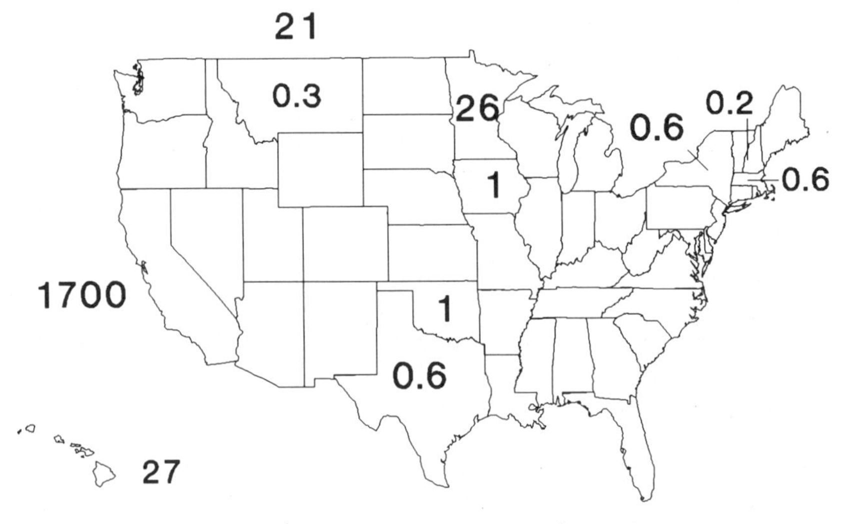

Figure 1.2. North American wind capacity.

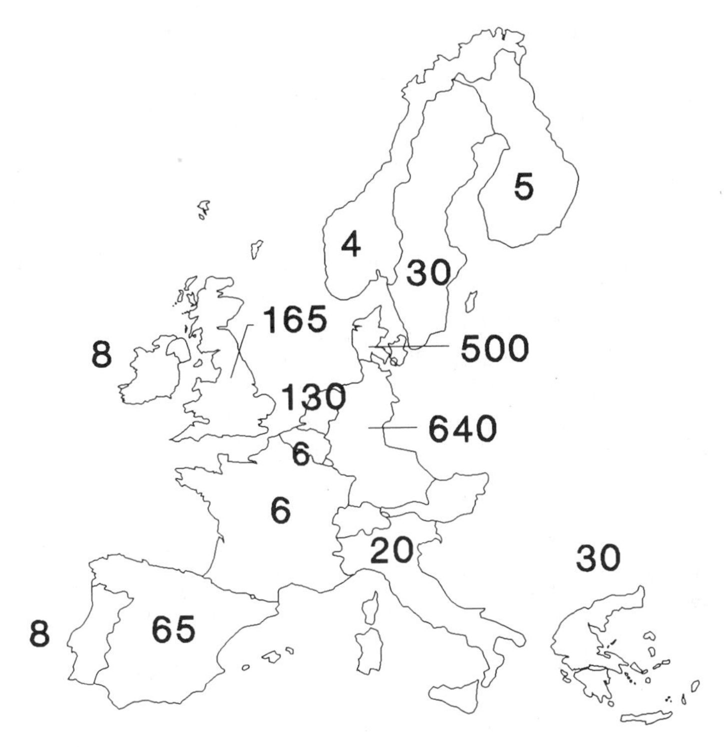

Figure 1.3. 1994 European wind capacity. Denmark and Germany lead Europe in installed wind generating capacity.

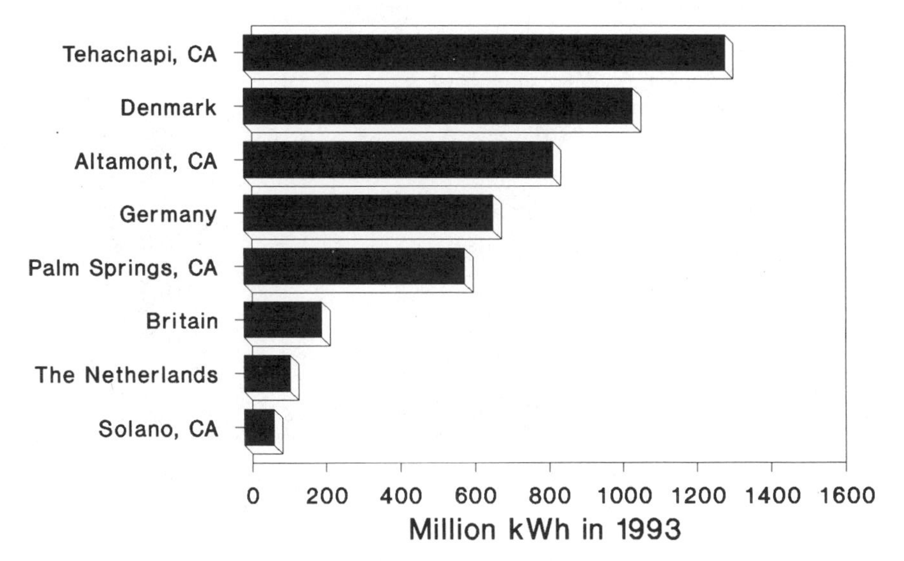

Figure 1.4. Major centers of wind generation.

wind turbines by the mid-1990s, and India had installed more than three times that number. Small wind turbines, machines with rotors less than 10 m (33 ft) in diameter, meanwhile, contribute a modest but not insignificant amount of generating capacity worldwide, including 10 MW in North America, and another 10 MW in Asia. Surprisingly, multiblade farm windmills, an often overlooked form of wind technology, deliver more than 250 MW of pumping power worldwide.

Wind development in North America has been sporadic. Development peaked during the mid-1980s at the height of the tax credit–inspired boom in California. According to the California Energy Commission, nearly 400 MW was installed in 1985 and another 300 MW in 1986. Development in the European Community has progressed more consistently than it has in North America (Figure 1.5). During the late 1980s, Europe surpassed North America in the rate of new installations, and by the mid-1990s, European wind development equaled that of California during its peak.

With a steadily increasing installed base, it was only a matter of time before European production of wind-generated electricity began to exceed that in North America (Figure 1.6). California's percentage of worldwide generation fell below 50% at the end of 1994 for the first time in a decade.

Models of Wind Development

Historically, wind turbines have been used to pump water or provide power at remote sites. This is still an important role even today for those in the

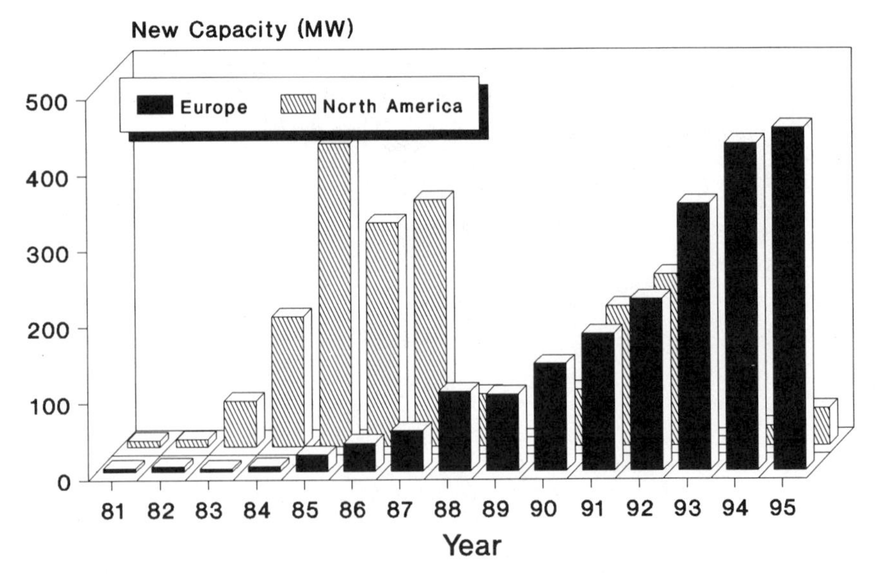

Figure 1.5. New wind generating installations.

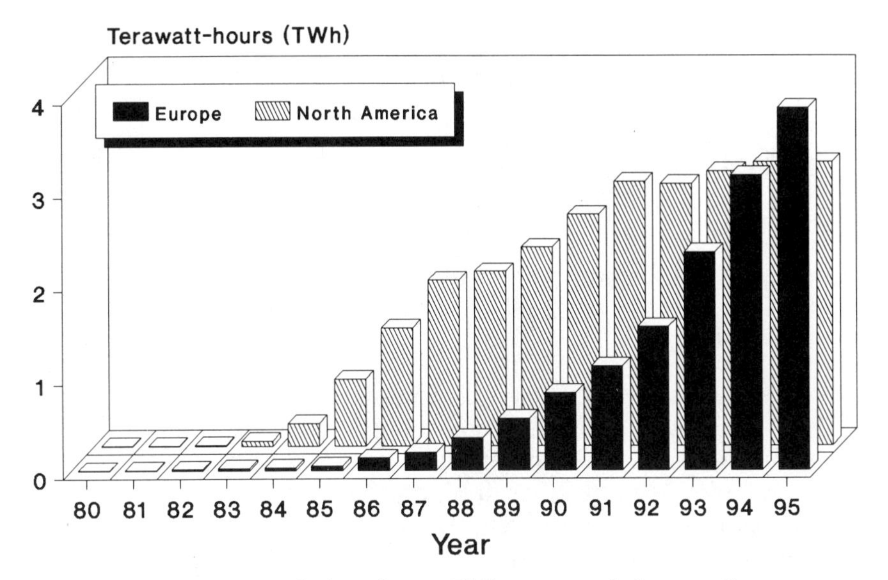

Figure 1.6. North American and European wind generation.

developing world and for those in developed countries who live *off-the-grid,* beyond the reach of power lines. In such applications, wind turbines are typically distributed as single units across the landscape. Wind turbines generating utility-compatible electricity can also be used individually, or in small clusters, or concentrated in large arrays to generate bulk electricity much like any other power plant (Figure 1.7).

Wind Power Plants

Before the 1980s, wind energy development focused on the individual wind turbine. By the late 1970s, this perspective began to change as attention shifted to maximizing collective generation from an array of many wind turbines. From today's vantage point, this idea seems logically consistent with all prior utility experience: Power plants are composed of several generating units.

The idea has historical precedents. At Kinderdijk and elsewhere in the Netherlands, the Dutch ganged clusters of machines in linear arrays along dikes and canals as needed. However, prior to the 1980s, the concept, as applied to modern wind turbines, seemed revolutionary: The wind industry was in the business of building power plants and generating electricity with wind energy, not simply in the business of building wind turbines.

Reflecting the concept's newness are the many terms that have arisen to describe it: *wind farms, wind parks, wind power electrical generating facilities, wind-driven generating plants, wind power plants,* and the related *wind power*

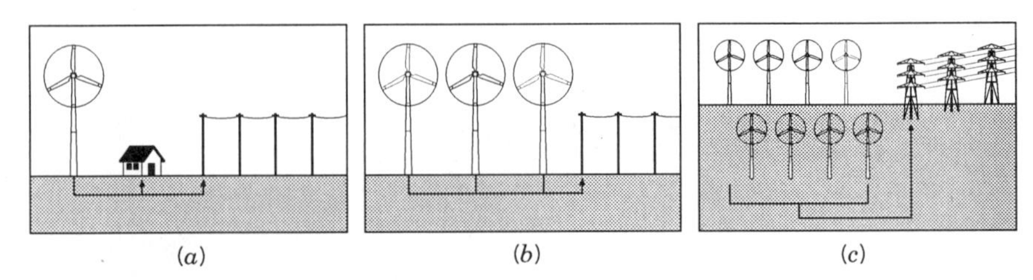

Figure 1.7. Interconnected applications. Wind turbines can be interconnected with the utility network singly (a) at a home or farm, in small clusters (b), or in large arrays (c).

stations. Early on, finding the best nomenclature created a dilemma. On the one hand, advocates wanted a term connoting wind's technological success and its coming of age as a conventional source of electricity as in the term *wind power plant.* On the other hand, proponents also wanted to preserve the association with the enlightened land use—the stewardship—that the term *wind farm* implies.

Wind farms is an expression that still finds adherents. Wind cognoscenti adopted the term in the late 1970s because wind generation and farming depend on seasonal cycles, the turbines are planted in rows like fields of corn (maize), and there is a literary association between the rural areas where turbines are sited and with harvesting a renewable crop. Yet the term's agrarian overtones disturbed some. Financiers preferred a more sophisticated term for their well-heeled clientele; thus *wind power electrical generating facility* gained currency briefly in southern California. Fortunately, it died a quick death. But the financiers did point out a need for a more accurate description of wind projects.

The term *wind parks* grew out of the razzle-dazzle world of California real-estate development, in which groups of commercial buildings become "industrial parks." The term is now pervasive throughout the world. Even the normally sober Danes have adopted it to describe some projects as in *Vindmøllepark.* Southern California Edison officially endorses *wind parks* even though the utility labels its own power plants as *generating stations.* The word *parks,* however, carries with it connotations of sylvan landscapes or natural preserves protected from commercial use. Large assemblages of wind turbines can in no way be construed as parks. Critics might even charge that wind energy's proponents deliberately choose to continue using the term parks for its positive connotations rather than the term plants, with its utilitarian implications.

Eventually, it became evident to utility planners and engineers alike that these were indeed power plants, differing from conventional plants only in that they were wind driven. From the utility sector hence came the expression *wind-driven generating plant. Wind power stations* is an equally descriptive term. The Electric Power Research Institute in Palo Alto, California, for example, has formally adopted the term *wind power stations.* Subsequently, these expressions were shortened to *wind plant* or *wind station.*

What, then, is a wind power plant? Generally, it is any cluster of wind turbines used for the bulk generation of electricity. A wind plant contrasts with a single wind turbine used to meet on-site needs that characterized wind energy before development began in California. Wind plants vary widely in size. In California, project size has grown with the industry. Today, wind plants range in size from 40 to 400 turbines. In Europe, arrays are much smaller, averaging in the tens of machines.

In contrast to North America, in Europe the line between a cluster of wind turbines and a wind power plant is less distinct (Figure 1.8). In Great Britain, where there are few individual wind turbines, most wind turbines are installed commercially in wind plants. But British wind plants, some comprising only 10 turbines, are more akin to clusters than they are the wind plants found in California.

Massive California-style development is not necessarily the future of wind energy—any more so than the scenes of derrick stacked upon derrick on Signal Hill in Los Angeles at the turn of the century represent today's oil industry. It is but one model of wind development that has proved successful (Figure 1.9). Commercial development of wind energy in the midwestern United States and on the American Great Plains will probably use rectangular arrays similar to those near Palm Springs, California, although the spacing among turbines will be much greater.

Wind turbines can also be sited to advantage on terrain with strong linear features, such as along dikes and breakwaters, along coastlines or along field boundaries and fence lines on tilled farmland (Figure 1.10).

Despite the thousands of turbines on the broad floor of the San Gorgonio Pass near Palm Springs, a few sites in the Altamont Pass, and a 1000-turbine

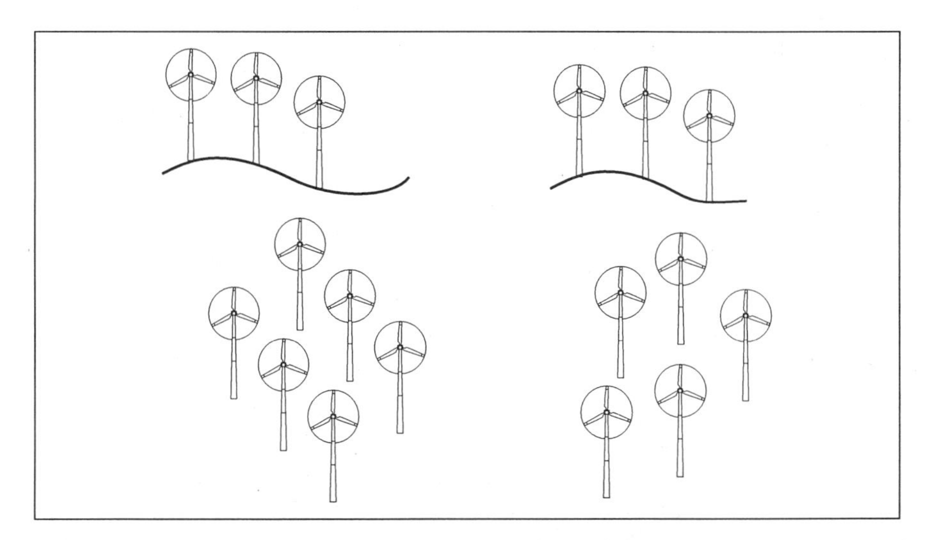

Figure 1.8. Clusters. Wind turbines can be grouped in small clusters on either hilly or level terrain, as in Wales and Cornwall.

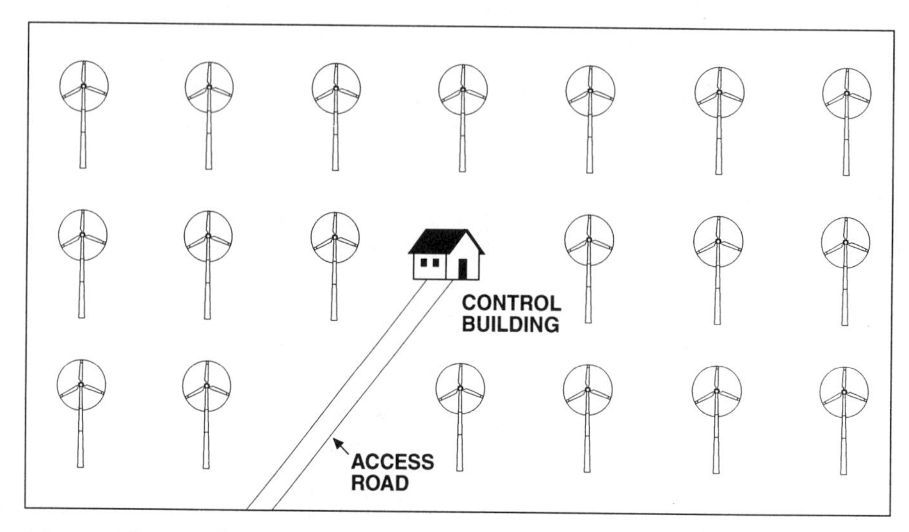

Figure 1.9. Rectilinear arrays. On level or gently rolling terrain, wind turbines are deployed in symmetrical phalanxes as at Tændpibe–Velling Mærsk in Denmark or on the Whitewater River floodplain near Palm Springs, California.

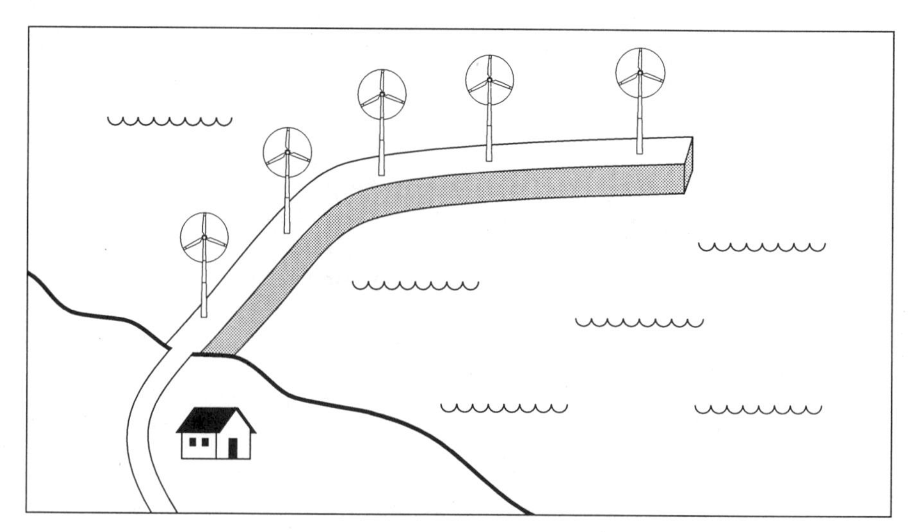

Figure 1.10. Linear arrays. Wind turbines can be strung along linear features like pearls on a necklace, as on the harbor breakwaters at Ebeltoft, Denmark and Zeebrugge, Belgium; alongside the dikes of the Noordoost polder in the Netherlands and the Friedrich-Wilheim-Lübke koog in northern Germany; or between field boundaries, as found in Denmark.

array near Mojave, most wind turbines in California have been installed in long, undulating arrays on ridgetops transverse to the prevailing wind (Figure 1.11). One ridge in the Tehachapi Pass is so steep that the developer cut multiple benches for successive rows of turbines and erected them on towers of different heights to use fully all the available space. This stepped array or "wind wall" acts much like a dam in a river.

Distributed Wind Turbines

Wind energy in the United States has followed both development patterns: wind plants and individual turbines (Figure 1.12). The 17,000 wind turbines in California's large arrays generate the most electricity and most of the attention. Yet there are another 5500 small wind turbines operating unheralded in distributed applications across the country, most interconnected with the local utility like their bigger brethren in California (Figure 1.13). In contrast, most of the wind-generated electricity in Denmark is produced by single medium-sized wind turbines serving rural homes and farmsteads. Only a portion is generated by wind plants.

Wind turbines are also used elsewhere around the world in a variety of applications less conspicuous than wind power plants. Small wind machines, for example, are often used in isolated or remote areas to pump water mechanically or to charge batteries in stand-alone power systems (Figure 1.14). Today, three-fourths of all small wind turbines built are destined for stand-alone power systems at remote sites. Some find their way to homesteads in Canada and Alaska far from the nearest village. Others serve mountaintop telecommunications sites where utility power could seldom be economically justified. An increasing number are being put to use by homeowners in the United States who are determined to produce their own power, even though they could buy electricity from the local utility if they desired.

Extremely small "micro" wind turbines have revolutionized life off-the-

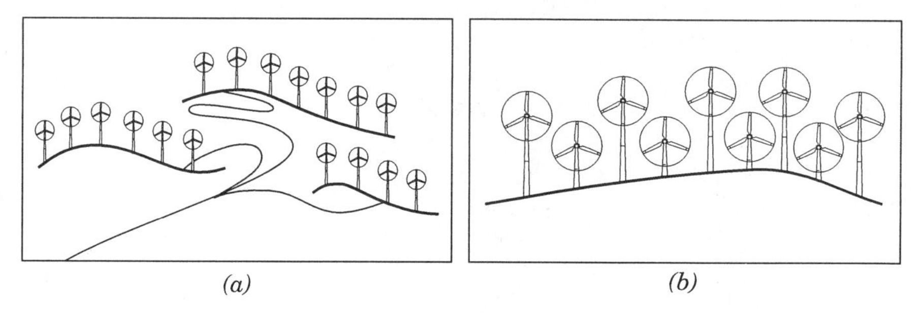

Figure 1.11. Ridgetop arrays. In hilly or mountainous terrain, wind turbines are arrayed on the ridgetops as in the Altamont Pass (a), or stacked atop towers of different heights on the ridgeline as in a wind wall near Tehachapi, California (b).

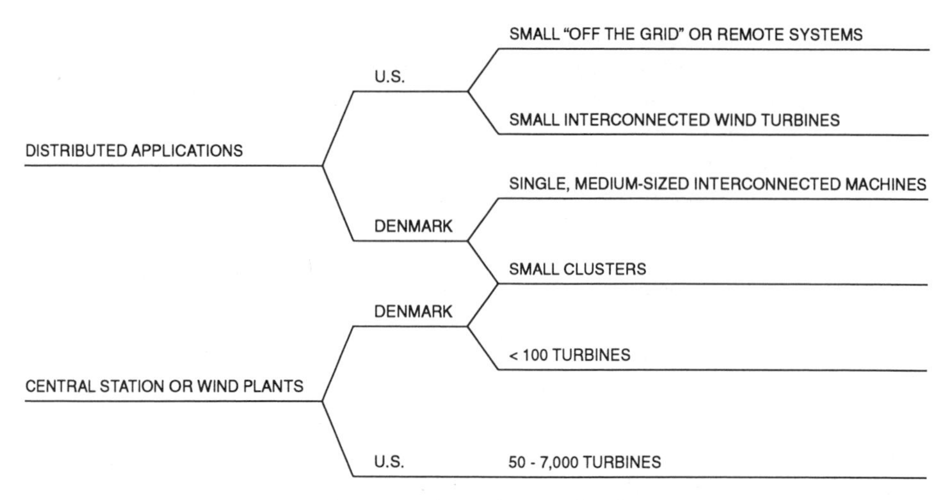

Figure 1.12. Models of wind development.

grid. These are machines so small they can be carried on horseback. At less than 3 m (10 ft) in diameter, microturbines are mere gnats alongside the big machines in California and Denmark. Yet they deliver a valuable service, providing cathodic protection of pipelines in North America and power to nomadic herdsmen on the Mongolian steppe, Asia's Great Plains (Figure 1.15). With the advent of microturbines and advancements in photovoltaic cells, Pacific Gas & Electric found the number of stand-alone power systems in northern California mushrooming during the early 1990s. They expect the market to continue expanding as urbanites move to increasingly remote areas not served by utility power.

Some utilities are going so far as to provide stand-alone power systems themselves instead of building new power lines. The government of New South Wales now subsidizes stand-alone power systems for remote cattle stations in Australia's outback in lieu of extending a power line from the provincial utility. Even the conservative Électricité de France has found that it makes economic sense to install wind turbines in rural areas of France and its overseas territories rather than extending their lines to any and all. Stand-alone power systems, however, seldom use wind energy exclusively. In many areas of the world, wind and solar resources complement each other: winter's winds balancing summer's sun. Wind and solar hybrids capitalize on each technology's assets, enabling designers to reduce the size, and the cost, of each component (Figure 1.16). Hybrids perform even better when coupled with small backup generators to reduce the amount of battery storage needed.

Hybrids contribute even more value in developing countries, where one-third of the world's people live without electricity. Many developing nations are scrambling to expand their power systems to meet the demand for rural electrification. But extending utility service from the cities to remote villages

Figure 1.13. Small wind turbine. Enertech 1800 generating utility-compatible power for a home near Palm Springs, California. This 13-ft (4-m)-diameter wind turbine has been in use since the early 1980s.

is a seldom affordable luxury. Thus hybrid systems, even though they generate little power in comparison to central power plants, can meet the modest needs of Third World villages.

Low per capita consumption magnifies a hybrid system's benefits because so little electricity is needed to raise the quality of life. One kilowatt hour of electricity provides 10 times more services in India than it does in Indiana. Two small wind turbines, which would supply only two homes with electric heat in the United States, can pump safe drinking water for 4000 people in Morocco. Consequently, developing countries are turning to wind and solar energy as a cost-effective way to meet the electrical needs of rural areas.

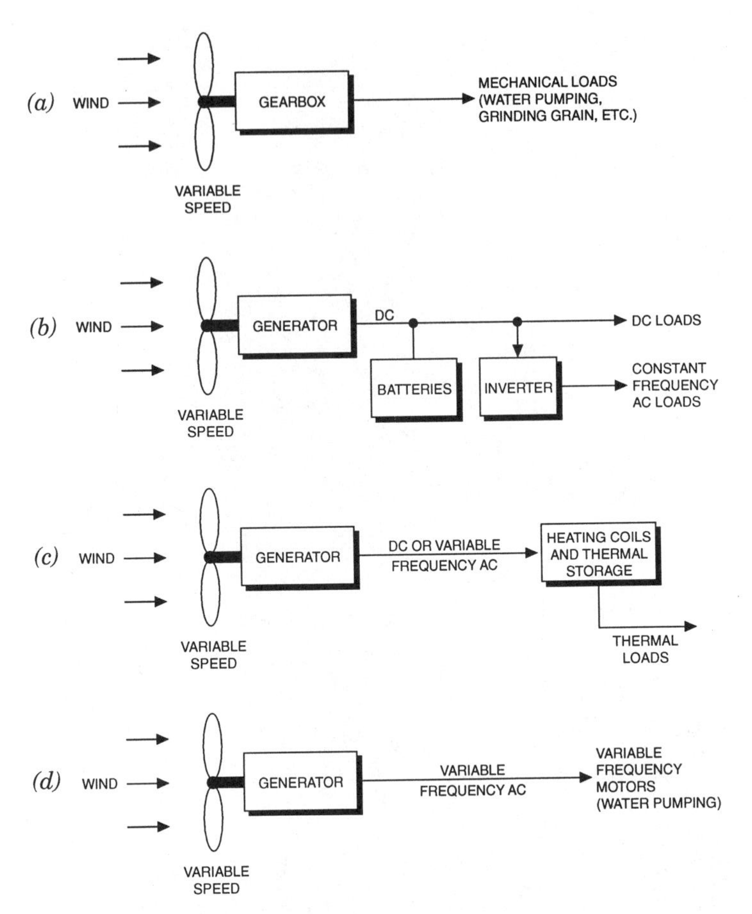

*Figure 1.14. Stand-alone wind machines. Historically, wind machines have been used to mechanically grind grain or pump water (*a*) and to charge batteries in remote power systems (*b*). During the 1970s and early 1980s, several attempts were made to use wind turbines for heating (*c*). Today, electricity-generating wind turbines can be used to power well pumps directly (*d*). (Reprinted from* Wind Power for Home & Business, *courtesy of Chelsea Green Publishing Company.)*

When one wind turbine is insufficient, the modularity of small wind turbines permits tailoring the hybrid system to the size needed. Bergey Windpower installed an array of their small wind turbines at the village of Xcalac on the Yucatan peninsula in Mexico's first "wind farm." The six Bergey Excels, each capable of generating 10 kilowatts (kW) were combined with 12 kW of photovoltaic cells and a 35-kW diesel generator. This hybrid power system offset the construction of a proposed $3.2 million power line at a fraction of the cost.

Whether in developing or developed countries, wind turbines, both large and small, can prove useful to utilities by providing a nonconventional means for meeting rural demand (Figure 1.17). Pacific Gas & Electric, for example,

Figure 1.15. Nomadic microturbine. Marlec wind turbine used by Mongolian nomads. The turbine is so small that it can be carried on horseback. (Courtesy of Peter Fraenkel, IT Power, Eversley, Hants, England.)

has found that by reducing the loads at the end of heavily used lines, they can avoid constructing costly new transmission capacity. They can reduce loads by boosting conservation or by installing modular sources of generation, such as photovoltaic panels and small wind turbines, close to the point of

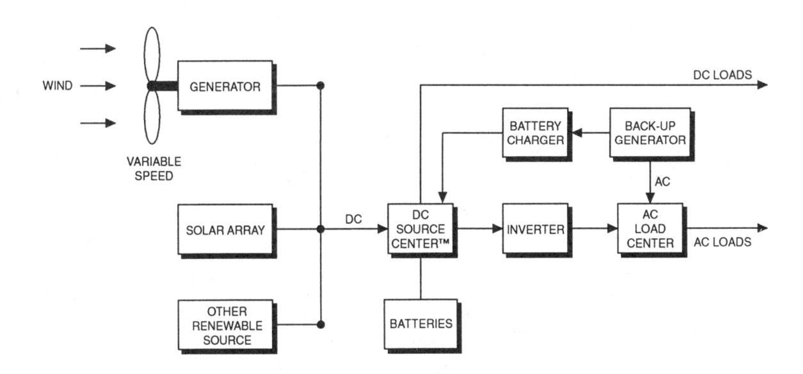

Figure 1.16. Hybrid power system. Combining solar and wind energy in a stand-alone power system offers the best of each technology for those living off-the-grid. (DC Source Center is a registered Trademark of Photron. Courtesy of Photron, Willits CA.)

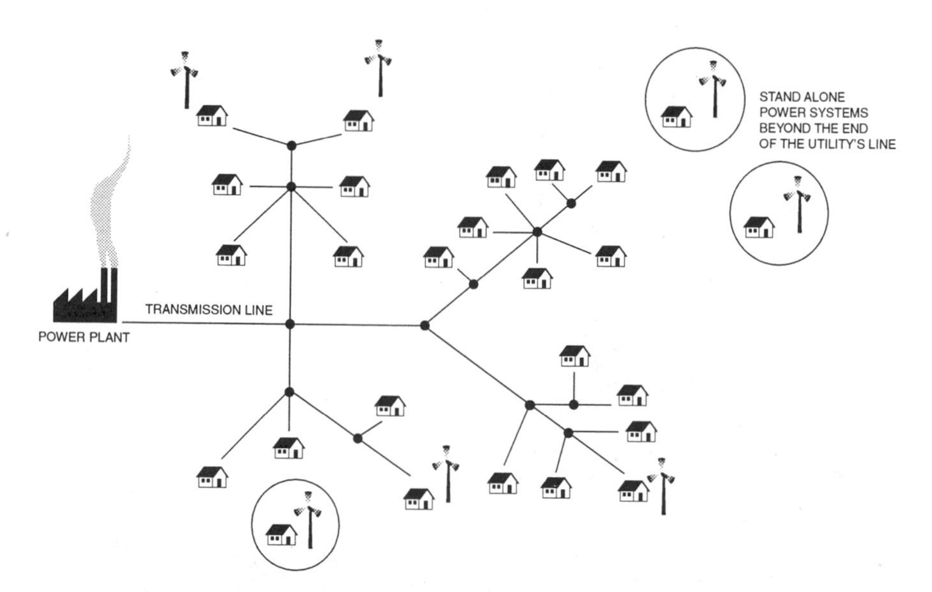

Figure 1.17. End of the line. Wind turbines at the end of the utility's distribution system can boost weak transmission lines. Combined with photovoltaics in a hybrid system, wind turbines can also provide power for those living off-the-grid. (Reprinted from Wind Power for Home & Business, *courtesy of Chelsea Green Publishing Company.)*

demand. The Electric Power Research Institute even suggests that small utility-supplied loads in some areas of the Great Plains could be better met by disconnecting the line and using "distributed sources of generation" instead. For example, one lightly loaded power line in Kansas serves remote wells for stock watering. Farm windmills once pumped these wells and could easily do so again.

Today more than 1 million of the classic or "American" water-pumping windmills (Figure 1.18) are still in use worldwide. Some on the American Great Plains have been in quasi-continuous use for 80 years. The multiblade farm windmill performs its function well, yet it suffers from a major limitation: The windmill must be located at the well, often not the best wind location.

Modern wind turbines, in contrast, can be sited to best advantage and drive the well pump electrically from a distance. With contemporary electronics, batteries and inverters are no longer needed, and the wind turbine can be wired directly to the pump motor. In Laredo, Texas, a community college installed four 10-kW Bergey wind turbines to pump irrigation water independent of the local utility. Medium-sized wind turbines can also be used to offload conventional irrigation well motors electrically or mechanically via an overrunning clutch (Figure 1.19).

Because the need for heating is often greatest when winter winds are howling, several attempts have been made on both sides of the Atlantic to

Figure 1.18. Pumping water: one of wind energy's most important historical uses. Aermotor water-pumping windmill in foreground, Enertech E-44s in background, at Altamont Pass.

use the wind for home heating (Figure 1.20). The University of Massachusetts pioneered the "wind furnace" concept in the United States by using electric resistance heating, as did S.J. Windpower in Denmark. At Cornell University and in Ireland, researchers experimented with water churns.

All were to no avail. Wind turbines are capital intensive, and these researchers thought that by eliminating the need for generating utility-compatible power, they could build cheap wind turbines that would produce a low-grade form of energy: heat. They failed to reckon with low-cost, mass-produced induction generators. It costs no more to generate utility-compatible electricity

Figure 1.19. Wind-assisted irrigation. Experimental Darrieus turbine at the U.S. Department of Agriculture's experiment station near Bushland, Texas in the late 1970s.

with an induction generator than it does to produce low-grade heat. Thus no savings materialized.

Today, wind turbines can be used for home heating. But these machines generate utility-compatible electricity that can be used for a multitude of other, more valued purposes. Often, in these applications the wind turbine meets all other electrical loads first, then is used for heating to avoid selling any excess generation back to the utility at less than the retail rate.

One of the potentially large applications for wind energy may be powering electric vehicles (Figure 1.21). Conceptually, this is simply an extension of the existing role of wind power plants. The arrays would serve an increase in demand on a utility system by the advent of electric vehicles, for example, in

Figure 1.20. Wind furnace: attempt by the University of Massachusetts in the late 1970s to harness the wind for home heating.

the Los Angeles basin. But cars represent only one form of transportation. Don Smith, an engineer formerly with Pacific Gas & Electric, envisions thousands of wind turbines powering newly electrified railroads as they cross the Great Plains. Unlike Europe, most railroads in North America are diesel powered. Smith suggests electrifying the transcontinental routes and installing wind turbines along the right-of-way in a linear array of continental proportions.

Elements Essential for Wind Development

From the foregoing it is reasonable to wonder why California and Denmark have thousands of wind turbines and France has few, and why European wind development is surpassing that in North America. A simplistic explanation is

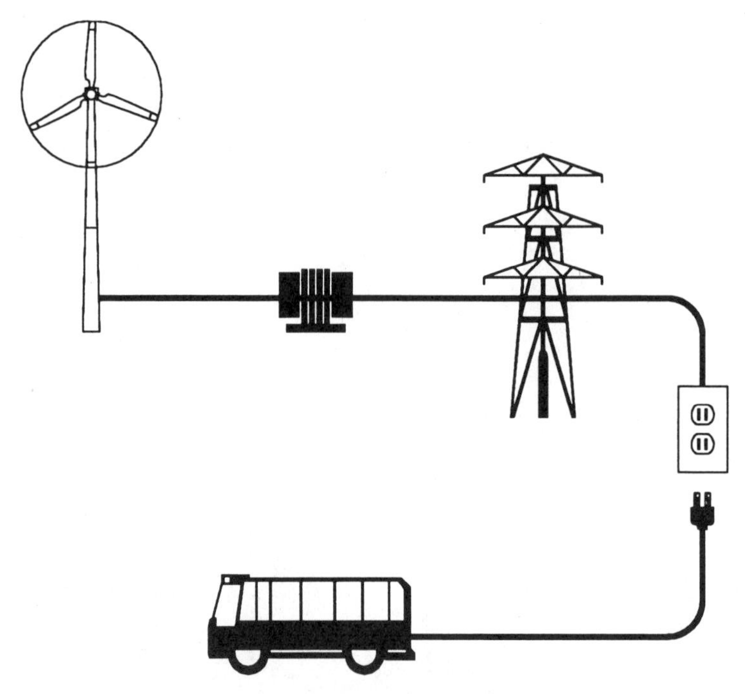

Figure 1.21. Wind-powered transportation. Wind-generated electricity can be used to power electric vehicles.

that California and Denmark are willing to pay a fair price for wind-generated electricity, France is not, and Europeans as a whole are more willing than North Americans to pay for renewable resources. The truth, of course, is far more complex.

There are several elements essential for using wind energy successfully. Often a key ingredient is missing and wind energy remains merely a wistful vision. Even where all elements are present, they reside in delicate balance; any disruption upsets the equilibrium.

With recent advancements in the technology the use of wind energy is primarily a political decision. There are many places in North America outside California and in France where ample resources make wind energy competitive with other fuels. The adage "where there's a will, there's a way" is apt. But the drama of whether or not to use wind energy is played out in numerous private decisions, be it Minnesota farmers installing wind turbines in their backyards, Danish families investing in cooperatives, or Credit Suisse financing a large California wind power plant.

On the surface, the choices may not always appear rational. For example, it may make more economic sense for a midwestern farmer to install a small wind turbine even though a medium-sized machine is far more cost-effective. A Danish farmer with the same wind resource may, and usually does, reach exactly the opposite conclusion under Danish economic and political conditions.

From a financial perspective the elements that determine wind energy development are cost, revenue, and the desired return on investment. Other factors may be important, such as national energy policies, but these are often implemented through incentives such as tax credits, or disincentives such as taxes, that are reflected in the financial equation.

Costs include both the cost of installation and the cost of operation. The installed cost is not only the cost of the turbine and its installation, but also the cost of necessary infrastructure, for example, the roads and transmission lines to a wind power plant.

Revenue is determined by the wind resource, the turbines' performance, and the value of the energy produced. For a wind plant, the value of the electricity is determined by the purchase power rate or buyback tariff. For a homeowner living off-the-grid, the value is the cost of the electricity they would otherwise buy from the utility, plus the cost of building the power line to reach their homestead.

The desired rate of return is influenced by the investment community's perception of the project's risk, the return from competing investments of similar risk, the type of investment, tax laws, and the political climate. During the early 1980s, for example, wind projects in the United States were perceived as risky ventures (they were) and demanded high rates of return. Congress offered federal tax credits to compensate and to stimulate investment in a technology that was deemed desirable.

Incentives in the United States during the 1980s, in the form of state and federal tax credits, effectively reduced wind energy's capital cost. Lowering the cost side of the equation boosted the potential rate of return on wind investments, assuming that the revenue side remained constant.

For wind energy to work, a potential user needs ample wind, a place to put a wind turbine, a market for the energy it will produce, and some assurance that the product, electricity, will reach the market and fetch the price necessary for a sufficiently long period to make all the effort worthwhile. A wind turbine without wind is like a dam without water. As will be explained, the power in the wind is an exponential function of wind speed. Slight changes in the average speed produce significant changes in the amount of energy available. Finding a site with an ample wind resource is no simple matter except in Denmark, a nation with the best understood wind resource in the world.

How much wind is necessary is a function of how much wind energy is worth. Less wind is needed when the value of wind energy is high. For someone living off-the-grid or for a remote telecommunications site, the value of wind energy is extremely high because of the difficulty and expense of bringing in fuel or building a power line. Many have discovered that anyone more than 1 kilometer (km) (0.6 mile) from a utility line will find wind energy easily competitive at average wind speeds of 4 meters per second (m/s) (9 mph) or greater. On the other hand, if a utility has excess generating capacity and is selling electricity at a distressed price of $0.035 per kilowatt-hour to stimulate consumption, it may be difficult for wind energy to compete even at sites with average speeds of 10 m/s (22 mph), where there is 15 times more energy available.

One reason for California's early prominence was a program begun by the California Energy Commission in 1977 to map the state's wind resources. The subsequent report identified several areas that looked promising. Three of the mountain passes found in the survey have some of the best wind resources in the world and have since seen extensive wind development. Some of these sites have average wind speeds in excess of 8 m/s (18 mph). However, meteorologists underestimated the difficulty of characterizing wind speeds over hilly terrain from a few widely spaced measurements and overestimated the average speed at many sites. Some sites with operating turbines have average speeds of only 6 m/s (13 mph).

Because of the high value placed on wind-generated electricity in northern Europe, wind speeds of 5 to 6.5 m/s (11 to 15 mph) are often sufficient. In contrast, average wind speeds in excess of 7 m/s (16 mph) are necessary in Great Britain because of the risk entailed by tariffs with an unusully short period of fixed prices.

If there is sufficient wind, wind turbines need a place to call home. Wind machines are not abstractions. They are physical objects rooted in the landscape. Often considered benign relative to conventional energy sources, wind turbines do raise objections about visual pollution, noise, and other land-use impacts. Siting can pose a major obstacle. During the early 1980s, land with good wind resources in California was inexpensive, and siting authority was obtained easily except for an area near wealthy Palm Springs, where development was delayed until 1983. Siting turbines in densely populated Europe is typically more difficult than in North America. Even so, wind turbines are an accepted rural land use in Denmark.

Like any other enterprise, no wind turbine can be successful without a market for its product. This is straightforward for a homeowner living off-the-grid; they will use all the electricity the wind turbine can produce. Midwestern farmers with medium-sized turbines, on the other hand, may need to sell some of the wind turbines' excess generation back to the local utility. They will need an agreement with the utility to do so, and to make it worthwhile the utility will have to pay a fair price for the wind-generated electricity.

In the United States, federal law requires utilities to buy nonutility generation. The law allows each utility to determine the price it pays. Midwestern farmers may be permitted to sell their excess generation to the utility, but the utility may pay only a fraction of its retail rate. This may be insufficient to justify installing a wind turbine that generates more electricity than can be consumed on site. Consequently, farmers in the United States may be forced to buy smaller, less cost-effective wind turbines than would a comparable farmer in Denmark.

Like their American counterparts, Danish farmers are permitted to sell excess generation to the local utility. But Danish utilities pay more than North American utilities for wind-generated electricity, enabling Danish farmers to buy the most cost-effective turbines available.

The same relationships apply to California. Eventually, there were willing buyers, an attractive purchase price for energy and capacity, 30-year contracts, and reasonable transmission and interconnection costs. By the early 1980s the state's two major utilities had succumbed to political and financial pressure and began buying generation from nonutility sources. At the time California's Public Utility Commission began to foster a series of contracts, with standard terms, between the utilities and independent energy producers. These standard contracts provided a framework for bankers to estimate revenues from wind turbines using the wind resources identified in earlier state studies.

As in most business transactions, for the arrangement to be successful, both parties must benefit. Whether a utility benefits from wind generation could well determine the ultimate success of wind energy. Where these conditions are absent or change, relations with the buyer (the utility) become tenuous even with a legally binding contract. In California the political climate changed dramatically in the late 1980s when the price of oil collapsed, removing any political imperative for renewable energy development. Subsequently, the utilities saw little to gain from independent generation and followed these trends with efforts to limit further wind industry expansion.

Next to maximizing reward, a financier's most important objective is reducing risk. Wind resource risk can be controlled only by siting turbines with care. Yet there is sufficient risk from the annual variation in the wind to make most bankers extremely nervous. Developers try to control technological risk by choosing wind turbines with good track records and then maintaining them properly. Yet machines do occasionally fail.

The one factor strictly under human control, the purchase power rate or tariff for sales to the electric utility, is subject to political uncertainty. To minimize this risk, bankers prefer to finance projects that have contracts with fixed payments for a specified number of years. This assures them that if all else works properly, they will recoup their investment when expected.

Contracts are not absolutely essential. Some countries, such as Denmark, are so politically stable and the role of wind energy so widely acknowledged that investors and bankers alike have faith that the government will honor its policies. Yet elsewhere, as in the United States, the political winds can change direction quickly, leaving the owner of a valuable investment at risk without the protection of a contract.

To attract capital and to ensure a revenue stream, operators have found that fixed-price contracts are preferable to floating-price contracts even if the fixed price is lower than the average floating price. Lenders have seen the international price of energy swing too widely too often to gamble on future energy prices. Both investors and operators rest more easily with a firm price even if the return is less than that possible in the short term under a floating-price contract.

Contract length, particularly that of fixed-price contracts, may also determine a wind plant's ability to attract capital. For example, a 10-year contract may not be sufficient to convince financiers that they can safely recover their investment while the contract is in effect. Contracts in California remain in force 30 years, but the fixed-price portion covers only the first 10 years.

Even when all other factors are favorable, costly interconnection requirements can make a project unfeasible. Disagreement between Danish utilities and wind turbine owners over interconnection costs hindered new installations during the mid-1990s.

In California, operators must pay for all interconnection costs that benefit only the wind company. Where transmission capacity is adequate to carry the expected wind plant's generation, costs are limited to the transmission lines linking the wind plant with the grid. But where lines of limited capacity are the sole means to market (the utility load), the operator could be forced to build additional transmission lines of considerable length. Just such a situation developed in southern California's Tehachapi Mountains during the late 1980s.

The California Experience

With a land area nearly twice that of the United Kingdom, and an economy rivaling that of Italy, California ia a nation within a nation.[1] Since the gold rush of 1849 the state's economy has profoundly affected the rest of the United States. Commercial success in trend-setting California often produced success elsewhere. The automobile may have been born in Detroit, but it found a home in California. Wind turbines are no different from other revolutionary technologies, such as microcomputers, that found fertile soil in California's golden hills.

Wind turbines began to sprout like mushrooms on California hillsides because several factors converged simultaneously to lift wind companies out of the doldrums of government-sponsored research. California grew to dominate worldwide wind development during the early 1980s because the state has some of the most energetic winds in North America, and where these occur, low-cost land was abundant; at the time California had the most favorable purchase power rates and the most cooperative utilities in the nation; it had an abundance of wealth; it had a favorable investment climate (new ideas are more readily accepted in California than anywhere else in the United States); and California offered lucrative incentives to match those of the federal government. But most important, in 1978 Congress passed the National Energy Act to encourage conservation and development of the nation's indigenous energy resources. The act included provisions that opened the market to nonutility generators (independent energy producers), permitting the private generation of electricity. The act also offered tax incentives to stimulate development.

The Public Utility Regulatory Policies Act (PURPA), one of the energy act's many parts, guaranteed a market for the electricity generated by independent producers. It required utilities to buy electricity at a fair price and to sell backup power to independent producers at nondiscriminatory rates. Through PURPA, Congress attempted to eliminate many of the institutional barriers that had blocked wind and solar energy at the utility level.[2]

Federal energy tax credits (10%) created by the act took effect immediately. But the tax credits were of little immediate help because PURPA's implementation, and the market it opened, was delayed until the early 1980s as individual states sought to interpret federal regulations. Then in 1980, with the energy situation even more pressing, Congress passed the Crude Oil Windfall Profits Tax Act, which boosted the business energy tax credits to 15% and extended their applicability through the end of 1985.[3]

Congress had previously authorized a 10% investment tax credit to spur economic recovery. This tax credit applied to all capital investment regardless of whether it was used for a robot at an auto plant in Indiana, a giant shovel at a strip mine in Ohio, or a wind turbine in California. Both purchase of a wind turbine for a farm or business, and investment in a California wind plant qualified for credits against federal taxes totaling 25%: the investment credit plus the business energy credit.

To encourage alternative sources of energy, California offered an additional 25% state tax credit. The state and federal credits were not simply additive, but the total reduction in tax liability neared 50%, causing an avalanche of funds to flow into California solar energy investments, including wind and solar power plants as well as solar water heaters.

Together the federal and state incentives were far higher than subsidies in Denmark (30%) and encouraged many fledgling companies to rush into business. The tax credits were so lucrative that they attracted those who knew more about constructing a deal than they knew about building wind turbines. An incident at Oak Creek Energy Systems captures the frenetic pace of the period. Oak Creek, a wind developer near Tehachapi, was making money so fast during the early 1980s that they misplaced a check for $500,000. Auditors eventually found the check—still negotiable—nearly a decade later during bankruptcy proceedings.

The tax credits led to a boom-and-bust development cycle, but they succeeded. The credits created an industry almost overnight, an industry that continues to produce wind-generated electricity. By compensating for the risk of investing in untested new technology, the credits attracted the interest of up to 50,000 individual investors and raised $2 billion for California's burgeoning wind plants. Few made as much money as they were promised, and the risk was probably greater than many were led to believe. Most investors have earned a reasonable return—but not all. The credits created a nightmare for some.

Helmut Stich and his wife Karen bought a wind turbine in 1985 for $163,000. Only years later did they learn that their accountant, who they believed was advising them in good faith, received a hefty commission on the transaction. In this, the accountant was no different than many other "financial advisers" in California at the time who directed their clients into wind energy investments. Stich, a German émigré, eventually lost all $163,000 at the hands of the very government that encouraged him to make the investment in the first place. (The credits were, after all, an incentive by Congress to encourage investment in alternative sources of energy.)

Following a change in national policy with the election of Ronald Reagan, an agency of Stich's adopted country, the Internal Revenue Service (IRS), denied the tax credits on Stich's wind investment along with that of many others.[4] Despite clear congressional intent, the IRS conducted an aggressive campaign to challenge, retroactively, investments in wind and solar energy. Under the Internal Revenue Code, the taxpayer is guilty until proven innocent. Stich, for various personal reasons, was unable to fight the IRS. Others were and won their cases on appeal to tax court in Washington, DC. It was sometimes a Pyrrhic victory. In one case the litigation expense exceeded the cost of the wind turbine.[5] Fortunately, the IRS was grossly incompetent or they might have won more cases. They succeeded, as with the Stichs, principally by intimidating taxpayers from appealing IRS decisions.

For other investors their turbines seldom worked properly or the companies managing them siphoned off much of the meager earnings. Yet despite the horror stories, many investors did profit from California's windy passes and its entrepreneurial climate.

Both PURPA and the tax credits were necessary for wind development. But by themselves they were insufficient to launch an industry. Even with PURPA, California utilities were reluctant to negotiate power sales contracts or interconnection agreements with either homeowners who wished to use a wind turbine for supplemental power, or third parties who wished to build wind power plants and generate electricity commercially.

All that changed in California just before year end 1979 when the state's Public Utility Commission (PUC) fined Pacific Gas & Electric Co. (one of the largest investor-owned utilities in the United States) $15 million for not considering conservation and alternative energy in their future generating mix.[6] Attitudes among the state's utilities suddenly softened.

The following year, California's ambitious governor, Jerry Brown, organized a conference to attract financial interest in commercial wind development. The governor sought development in areas of California where state-funded studies had identified "excellent" wind resources. One year later, independent energy producers installed the first wind turbines in two of the state's three windiest passes.

Despite this progress, development proceeded slowly—so slowly that in 1983 the PUC fined Southern California Edison $8 million for failing to follow orders to accelerate development of cogeneration and other alternatives. Wide-scale development became possible only after the PUC established standard contracts, including one with forecast prices through the year 2000. By the early 1990s these contracts were fueling a heated regulatory debate in California. Utilities wanted to abrogate the contracts, and the PUC's Division of Ratepayers Advocates were helping them find every loophole possible.

During the 1970s California's demand for electricity grew by 7% per year. In 1974, analysts expected utilities to build a nuclear plant every 40 miles along California's long coastline to meet the demand. When the legislature banned new nuclear power plants (until a means was found for disposing of radioactive wastes), utilities had few choices. After 1973 oil was no longer an option and the National Energy Act prohibited utilities from burning natural gas. Utilities were forced to a traditional fuel: coal. By the end of the 1970s California had four giant coal plants on the drawing boards.

According to Dave Branchcomb of Henwood Energy Services, California faced a dramatic capacity shortfall in the early 1980s that utility-owned coal plants would be unable to meet. Utilities met their obligation to serve by turning to power plants (facilities) built by independent companies that qualified under PURPA. In the spring of 1983, budding independent energy producers concluded a five-week-long negotiating session with the PUC and the state's two major private utilities: Southern California Edison and Pacific Gas & Electric. The negotiations resulted in an Interim Standard Offer Number 4 (ISO4) contract as an expedient for determining the long-run costs avoided by not building the coal-fired plants then envisioned. The price of PG&E's proposed coal plant was $0.08 per kilowatt-hour in 1992 dollars, nearly that of the ISO4 agreement. Actual costs of building the plant could have been higher, not counting any social or environmental costs.

Nonutility generators also isolated ratepayers from the risk of building and operating the plant, including the risk of future pollution controls on, for example, carbon dioxide emissions. Unlike utilities, "qualifying facilities" under PURPA are unable to pass cost increases along to ratepayers. If a wind power plant performs less than expected, investors suffer, not ratepayers. Thus the state's ratepayers could have been saddled with coal-fired power plants and their attendant pollution had California not pursued qualifying facilities, such as wind power plants.

By definition the ISO4 contracts were based on estimated long-run costs. Short-run costs, in contrast, fluctuate with changes in the price of fuels. During the early 1980s, short-run energy costs were $0.07 per kilowatt-hour. "SCE was with us at the altar," says Dan Richards on the development of the interim standard contracts after more than two years of PUC hearings. Richards, an attorney specializing in PURPA, says it was SCE's own price forecast that the PUC adopted for the fixed-price portion of the ISO4 contracts.[7]

In 1983, the utilities' forecasted ISO4 price was $0.06 per kilowatt-hour, equivalent at the time to Pacific Gas & Electric's avoided cost. In 1984, PG&E's avoided cost moved slightly ahead of the ISO4 price. As the utilities switched from oil to natural gas, the avoided cost plummeted. A decade later the avoided cost hovered around $0.035 per kilowatt-hour, less than the cost 10 years earlier despite inflation. In the meantime, the ISO4 forecasted price reached

$0.10 per kilowatt-hour in the early 1990s and had increased to $0.14 per kilowatt-hour by the mid-1990s, when the fixed-price portion of many contracts were scheduled to expire.

The ISO4 contracts were in two parts: a 10-year fixed-price period, and a 20-year period of floating prices. A wind turbine installed in 1985 when the federal energy credits expired would exhaust its fixed-price portion of the 30-year contract in 1995. After the fixed-price portion expires, there is a floor price derived from the capacity value of the wind turbine. Analysts expect that the capacity payment, equivalent to about $0.02 per kilowatt-hour, plus the avoided energy cost would result in a payment of about $0.055 per kilowatt-hour.

Branchcomb and investors in California wind turbines see little difference between their payments under the ISO4 contracts and PG&E's payment for its Diablo Canyon nuclear plant. During the early 1990s, PG&E was earning $0.11 per kilowatt-hour from Diablo Canyon generation, the "ultimate SO4" contract says Branchcomb. Farther south, SCE was earning less, $0.09 per kilowatt-hour, from its San Onofre nuclear station, but still substantially more than the widely touted avoided energy cost.[8]

Hindsight is always 20/20, and in light of low prices for fossil fuel in the mid-1990s, California consumers could save money by burning natural gas instead of using wind energy—or nuclear power. The CEC worries, though, that California would put all its generating eggs in the natural gas basket, creating a situation in the late 1990s much like it faced in the late 1970s. When the Iranian oil embargo struck, sending shock waves through the state's economy, 90% of the state's generation was dependent on one fuel: oil. Wind, solar, and geothermal all provide the state with generating diversity that can dampen supply disruptions or wild swings in the cost of any one fuel. The CEC has often ruled that this diversity has value.

With standard contracts that provided an assured revenue stream available simply for the asking, and more than ample investment incentives, development in California began apace. Development focused on three windy passes: Altamont in the north, and Tehachapi and San Gorgonio in the south. A limited number of turbines have also been installed in the Pacheco Pass near the San Luis Reservoir and in Solano County at the confluence of the Sacramento and San Joaquin rivers. The Altamont Pass east of San Francisco accounts for about one-third of the state's wind generation, while the Tehachapi Pass northeast of Los Angeles produces half; the San Gorgonio Pass near Palm Springs generates much of the remainder.

By the mid-1990s a dozen privately held firms were operating 1700 MW of wind-generating capacity and delivering nearly 3 TWh annually to the state's electric utilities. Three firms account for more than half of California's installed capacity: U.S. Windpower, SeaWest, and Zond Systems.

Generation grew rapidly after the first turbines were installed in 1981: quadrupling from 1983 to 1984, tripling from 1984 to 1985, and nearly doubling from 1985 to 1986. Increases in generation have moderated somewhat since 1986 because new turbines represent an increasingly smaller proportion of the total installed, and because the major gains in productivity—by improved reliability—have already been achieved.

The total number of turbines installed and the capacity they represent declined in 1988, reflecting the removal of inoperative turbines, many of which were hastily installed during the industry's rapid growth (Figure 1.22). These machines contributed little to generation and detracted from the industry's overall performance.

All three areas are sparsely populated where cattle grazing was the dominant land use before development began. The value of the land for grazing decreases from the Altamont Pass in the north to the San Gorgonio Pass in the south. All three areas are classified as either semiarid or desert, but the Altamont receives relatively more moisture than do the passes farther south. The Tehachapi Pass is a gateway to the high desert from the Great Valley, whereas the San Gorgonio Pass is the gateway from the Los Angeles coastal basin to the Sonoran or low deserts of the great southwest.

Even in such a large state as California, land is an important commodity. The availability of land and its low cost enabled the industry to gain a foothold with what was, at that time, a marginal technology.

The federal energy tax credits that fueled California's growth expired at the end of 1985 and were not renewed. The 10% federal investment tax credit also expired at the end of 1985 by the 1986 tax reform act. California's state tax credit was reduced to 15% in 1986, the year in which it expired. As a result of the changes in the tax code, investments in wind energy became less

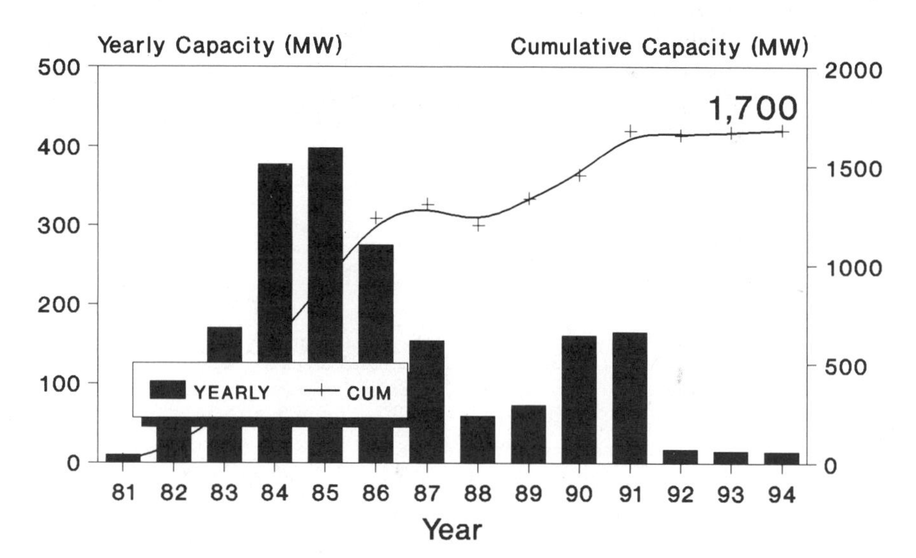

Figure 1.22. California wind capacity. Despite the loss of federal energy tax credits in 1985, wind turbines continued to be installed in California, although at a reduced rate.

attractive. Although still an economic investment, the loss of the credits reduced the flow of capital to the industry during the late 1980s.

Financiers now raise capital from conventional sources, such as banks and leasing companies. More than half of California's wind capacity has been installed since the federal energy tax credits expired by using conventional financing.

The European Experience

California became a proving ground not only of wind energy but of European wind turbines in particular. Nearly 50% of California's generating capacity was of European origin by the mid 1990s (Table 1.1). California demonstrated conclusively that thousands of wind turbines could produce significant amounts of wind-generated electricity—that the technology could be made to work. The California market also pumped millions of dollars into the coffers of European, especially Danish, manufacturers, financing advances in European wind technology.[9]

The boom in sales to California produced a thriving Danish export business. According to a study by the Electric Power Research Institute, the Danish wind industry was earning about $200 million per year during the early 1990s.[10] Much of the earnings came from export sales. For example, Denmark installed some 40 MW domestically in 1992 and exported another 120 MW. While domestic sales remained stagnant in 1993, exports grew to 180 MW. With the slack California market in the mid-1990s, most Danish turbines were exported to other European countries, notably Germany and Britain. During the early 1990s, European wind installations were growing at a rate of 200 to 400 MW per year.

Table 1.1
Wind Turbines in California

	Turbines		*Capacity*	
Country of Origin	Number	Percent	MW	Percent
United States	7,786	49	725	44
Denmark	6,778	43	676	41
Japan	660	4	165	10
Belgium	174	1	37	2
Great Britain	112	1	34	2
Germany	283	2	11	1
The Netherlands	63	0	8	0
1992 Total	15,856		1,655	

Source: California Energy Commission, *1992 WPPRS Annual Report,* June 1993.

European incentive programs began in the late 1980s and early 1990s in Germany, the Netherlands, Italy, Great Britain, and Sweden. In 1992 the European Community issued its Alterner program, which for the first time set a target for the development of wind energy in Europe: 8000 MW by the year 2005.[11]

Germany

Despite the soaring costs of reunification Germany became the world's hotbed of wind energy development during the mid-1990s. Germans were installing new turbines at the rate of 200 MW per year. By the end of 1994 there were 500 MW of wind capacity operating in Germany, most in the two northern *länder* of Niedersachsen (Lower Saxony) and Schleswig-Holstein. Reunification added another province with a long coastline, Mecklinburg-Vorpomern, where wind energy was also growing rapidly. By the end of 1995, Germany will surpass 700 MW. (Figure 1.23). The competitive German market has spawned several new manufacturers and led major Danish suppliers, such as Bonus, Vestas, and Nordtank, to expand their German affiliates. German manufacturer Enercon continues to dominate the market, but Danish manufacturers are in hot pursuit (Figure 1.24).

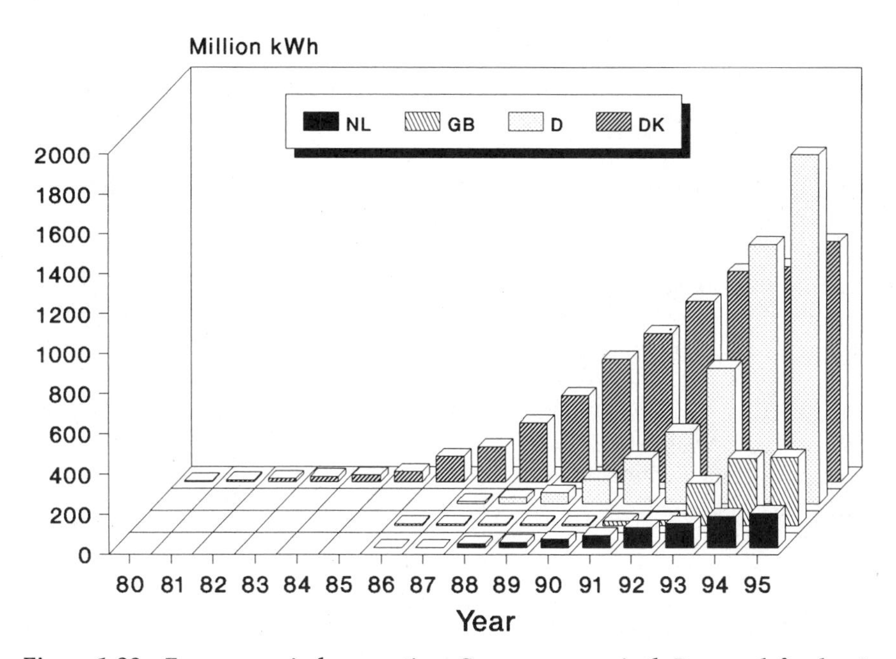

Figure 1.23. European wind generation. Germany now rivals Denmark for dominance of European wind generation.

Figure 1.24. German wind turbines. Enercon E-40s emerge from morning mist on the North German Plain.

Three factors have contributed to the boom in German wind development: A subsidy for research and development, a buyback policy similar to that in Denmark, and high utility rates. The high cost of electricity made wind an attractive option once interconnection between privately owned wind turbines and the local utility was permitted by the federal government. But development grew dramatically in 1989 when the German ministry for technological development, *Bündesministerium für Forschung und Technik* (BMFT), began paying 8 pfennings ($0.048) per kilowatt-hour for wind generation from turbines accepted into the agency's research program. (The subsidy is good for 10 years.) The popular program was limited to 100 MW and was quickly oversubscribed. In 1991 the BMFT expanded the program to 250 MW but reduced the payment to 6 pfennings ($0.036) per kilowatt-hour.

The ministry's 250-MW program could result in the installation of as much as 320 MW of wind generating capacity because the BMFT rates wind turbine capacity conservatively at 10 m/s (about 22 mph). Most manufacturers, however, rate their turbines at much higher wind speeds. Thus a 250-kW wind turbine could qualify as 200 kW under the BMFT program.

Subsequently, the federal government passed a law similar to PURPA in the United States, requiring utilities to buy renewably generated electricity at 90% of the utilities' average retail rate for about DM 0.166 ($0.10) per kilowatt-hour nationwide. The new tariffs, coupled with payments under the BMFT's research program, offered a potential gold mine to wind developers (Table 1.2).

Table 1.2
Payment for Wind-Generated Electricity in Germany

	DM/kWh	Dollars/kWh
90% of retail rate	0.166	0.0996
BMFT R&D tariff	0.06	0.036
	0.226	0.1356

Note: Exchange rate: 1 DM = $0.6.
Source: Jamie Chapman, EPRI, 1993.

Independent of the federal government, the northern states of Niedersachsen and Schleswig-Holstein launched their own wind programs that dwarfs any elsewhere in Europe or North America. Each state has committed to building 1000 MW of wind generating capacity. According to Uwe Carstensen of the German Wind Energy Association, *Deutsche Gesellschaft für Windenergie* (DGW), Niedersachsen is slightly more ambitious, targeting completion in the year 2000, while Schleswig-Holstein is looking at the more modest date of 2010, when it expects to generate 10% of the state's electricity with wind energy. The move by the two German states puts each on a par with the nations of Denmark and the Netherlands.

The level of commitment to wind energy in Niedersachsen is evident in its creation of the German Wind Energy Institute, *Deutsche Windenergie Institut* (DEWI), at Wilhelmshaven in 1990. The state-supported center, which studies problems of integrating wind energy into the utility network and wind energy's environmental impact as well as more conventional wind energy research topics, is more productive than many national wind energy programs.

Each state within Germany's federal system may also offer their own incentives. Subsidies vary from 14 to 17% of installed costs in Schleswig-Holstein and Niedersachsen to 30 to 50% in Bavaria and other interior states. The utility serving Hamburg, HEW, pays an incentive of DM 0.10 ($0.06) per kilowatt-hour on top of the 90% retail tariff, bringing the total tariff to DM 0.266 ($0.156) per kilowatt-hour in the city-state. However, the premium is available only until 1996.[12]

During the heyday of government-sponsored research and development, says DGW's Carstensen, subsidies went directly to aerospace manufacturers, that is, big companies building big turbines. Now, he says, the subsidy goes to the end user, the farmer or landowner who buys wind turbines, and aerospace companies have fled the highly competitive market.

Carstensen notes that the BMFT subsidy became a means of protecting German manufacturers from foreign (Danish) competition. Two-thirds of BMFT's support went to German firms. This has given small German companies a start. The Bundestag, the lower house of Germany's parliament, wants to maintain some market protection, but the goal has changed from the development of a new industry, says Carstensen, to the development of renewables for their own sake.

In 1991 all German installations received the BMFT subsidy. But by 1993 the program was so successful that only one-third of the turbines being installed were participants. The demand was so great that BMFT planners could target recipients who were buying wind turbines that produced the most desirable noise and performance characteristics. Buyers, such as farmers, invested 10 to 20% in equity and financed 75% as debt available from banks at a special interest rate of 7% for renewable energy projects. Most of the debt is guaranteed by the value of the wind turbine.

By the end of 1993 the consulting firm of Winkra had become Germany's largest wind developer, with a portfolio of 100 machines representing 30 MW. According to Winkra's managing director Henning Holst, there is a trend toward larger installations in Germany. Until now most wind turbines had been installed in single units or small clusters. But increasing difficulties siting turbines is forcing developers to bigger turbines and larger projects to concentrate as much generation on one site as possible.

The Netherlands

In 1992 wind energy in the Netherlands passed a major benchmark: Modern wind turbines were, for the first time, officially incorporated into *Nationale Mollendag,* National Windmill Day, held the second Saturday of May. It took the cautious Dutch more than a decade to link their country's historic use of the wind with modern wind technology. The Dutch passed another landmark in 1992, when total installed capacity in the Netherlands reached 100 MW.

Although well short of their schedule to install 1000 MW by the year 2000, the Dutch have installed wind turbines in more varied applications than in any other nation. They have also done as much as the Danes to integrate modern wind turbines into the fabric of the nation. Wind turbines can now be seen in all of the Netherlands' windy provinces, from Zeeland in the south to Friesland and Groningen in the north. There are working wind turbines singly or in small clusters in the parking lots of shopping centers; at restaurants, locks, pumping plants, harbors, farms, and homes; and along the country's famed dikes. The Dutch have even begun to incorporate these modern windmills into their well-oiled tourism industry. Brochures, maps, and booklets propose modern wind turbines as stops along itineraries that include traditional windmills, museums, and tulip beds.[13]

Wind energy has developed more slowly in the Netherlands than in Germany or Denmark for several reasons. The densely populated country has a slow and time-consuming permitting process to ensure that the Dutch can live harmoniously with each other. Further, early Dutch wind development had been hindered by a misdirected emphasis on big wind turbines, then by a faulty utility-administered program.

As in Germany, Britain, and the United States, the Dutch invested initially in research and development. They were somewhat more successful. Unlike the experimental machines built elsewhere, one large wind turbine, the NEWECS 45 near Medemblik, and one medium-sized turbine, NEWECS 25 in Curaçao, that they did build are still operating. When the NEWECS 45, which cost NLG 4.6 million ($2.7 million) in 1985, is operating properly, it is capable of generating 1.8 million kilowatt-hours per year. At a cost of NLG 50,000 ($30,000) per year to maintain the 45-m turbine, the cost of operating the experimental machine is $0.017 per kilowatt-hour, nearly twice that of contemporary designs.[14]

In 1991 the government, the utilities, and the seven windiest provinces reached agreement on how to meet the nation's goal of 1000 MW by the year 2000, the first phase of an ambitious target of 2000 MW by the year 2010. Each province agreed to accept a portion of the total capacity, and the utilities agreed to install 250 MW by 1995 as part of their Environmental Action Plan (MAP in Dutch) for reducing global warming emissions.

The utilities subsequently created an unwieldy program dubbed Windplan, which required Dutch manufacturers to meet unusual technical requirements not found in other wind energy programs. After domestic manufacturers adapted their designs to meet the utilities' demands, Windplan faltered and reneged on their orders for Dutch wind turbines, leaving manufacturers with designs ill-suited for the international marketplace. This nearly killed the fledgling Dutch wind industry.

The Dutch began offering financial incentives for wind development in the mid-1980s. These incentives, in the form of a capital subsidy, were substantial, NLG 700 ($400) per kilowatt of capacity, equivalent to 30% of installed cost.[15] The capital subsidy led Dutch manufacturers to boost the kilowatt rating of their turbines artificially to maximize the subsidy. The government set a limit to prevent abuse, but it was ineffective. As a result, Dutch wind turbines had at one time some of the highest specific ratings (kilowatts of rated capacity per square meter of rotor swept area) of the wind industry. The Dutch soon realized their mistake and in 1991 began applying the capital subsidy on the swept area of the wind turbine's rotor, a more direct measure than generator size of how much electricity a wind turbine will produce.

Historically, like tax credits in the United States during the 1980s, capital subsidies encourage the construction of equipment. The goal, however, is the production of wind-generated electricity, not necessarily the installation of machinery. Capital subsidies may or may not deliver the result sought because the wind turbines that qualify may stand idle or perform poorly. To function effectively, incentives must be targeted directly at the action desired. When the objective is to generate electricity, the incentives must be based on kilowatt-hours of production. Denmark and Germany have been successful because the tariff for wind-generated electricity in kilowatt-hours is sufficiently high to justify wind investments.

This is the route that the Association of Private Wind Energy Installations,

De Vereniging van Particuliere Windenergie Exploitanten (PAWEX), wants the Dutch program to take. PAWEX represents cooperatives and private wind turbine owners, which together account for 20% of Dutch capacity. During the mid-1990s they were negotiating with the government and with EnergieNed, the Dutch association of distribution companies, to boost the buyback tariff.

In 1994 Dutch utilities were paying a tariff based on a marginal cost of NLG 0.07 ($0.042) per kilowatt-hour. They were also paying an incentive for renewables and cogeneration as part of their environmental action plan of NLG 0.03 to 0.08 ($0.018 to $0.048) per kilowatt-hour, depending on the utility. Thus some Dutch utilities were paying a total of NLG 0.15 ($0.09) per kilowatt-hour. PAWEX was seeking to standardize the incentive payment at the high end of the range across all Dutch utilities so that all wind turbine owners would receive NLG 0.15 per kilowatt-hour. Coupled with the capital subsidy from NOVEM (*Nederlandse Onderneming voor Energie en Milieu*), the Dutch agency for energy and the environment, equivalent to about 25% of installed cost, PAWEX was optimistic that the growth of Dutch wind energy would be limited only by constraints on siting.

In rural areas some siting conflicts have eased since a Dutch bank that finances farmers has discovered wind's potential. As the European Community (EC) begins to reduce farm supports, Dutch farmers, like those in France, Denmark, and Germany will become increasingly hard pressed. Rabo Bank has begun to "suggest" that farmers consider wind energy as a new source of revenue that will enable them to stay on the land. Facing tough times, Dutch farmers are turning increasingly to a resource on which their ancestors once relied.

Great Britain

The British Isles are renowned for some of the best wind sites in Europe. Andrew Garrad of Garrad Hassan estimates that Britain has 28 times the wind resources of Denmark. Yet by the early 1990s little had been achieved because of the difficulty interconnecting wind turbines with England's state-owned utility, the Central Electricity Generating Board.

The situation changed dramatically in 1989 when the Electricity Act privatized electricity generation. To the chagrin of Thatcherites, the "City" (Britain"s financial center) refused to invest in privatized nuclear utilities. The government was stuck with the nuclear industry on its books and no assurance of a market for its costly electricity after privatization. The Thatcher government conceived the Non-Fossil Fuel Obligation (NFFO) to prop up Britain's ailing nuclear industry, which generates 20% of England's electricity, by guaranteeing it a market at premium prices. The NFFO requires (obligates) the regional electricity companies (RECs), which buy and distribute electricity, to provide a portion of their supply from nonfossil sources, that is, from nuclear power.

The RECs buy this portion at a premium price and pass the cost along to their customers. The NFFO applies only to English and Welsh RECs, with special provisions for Scotland.

However, the government overlooked a subtle detail: Wind energy is also a nonfossil resource. Accordingly, a special tranche of NFFO was created for renewables, with a goal of 1500 MW by the year 2000.[16] Of the 11% NFFO levy on utility bills in 1993 1% went to renewables, the rest to nuclear. In the initial awards, landfill gas, biogas, and incineration took 80% of nonnuclear capacity.

Unlike incentives in Denmark and Germany, the NFFO process is legalistic and cumbersome. Developers bid against each other on the price they believe necessary to build specific projects. The first two tranches of NFFO capacity that were issued in the early 1990s guarantee premium payments only through 1998. The contracts continue thereafter at the then avoided cost. Because the post-1998 avoided cost, or "pool price," is uncertain, so are future revenues. Thus a project completed in 1992 must both earn enough in six years to recover its costs and earn a profit or the project was unfinanceable.

This is an unusually short amortization period for such a long-lived investment as a wind plant and entails a high degree of risk. Any mechanical or legal problems preventing the turbines from operating could quickly spell bankruptcy. As a consequence, the NFFO premium price for wind energy in the first two tranches is the highest in the world: 9 pence ($0.144) per kilowatt-hour in the first tranche, and a "strike price" of 11 pence ($0.176) per kilowatt-hour in the second tranche. It was initially difficult to finance British wind projects, even with NFFO, due to the perceived novelty of the technology, the variability of the wind resource, and the high transaction costs associated with relatively small projects.

Peter Edwards' wind plant in Cornwall, Britain's first, illustrates how the NFFO works. Edwards installed ten wind turbines on his farm near Delabole in late 1991 under the first NFFO tranche. He financed two-thirds of the project's cost with a bank loan and a small grant from the European Community. Edwards raised the equity portion from a combination of sources, including an investment by the local REC. The turbines will generate 8.6 million kilowatt-hours per year at the 7.5-m/s (17-mph) site, earning about £774,000 ($1.24 million) per year. At that rate, the £3.4 million ($5.4 million) project should pay for itself within five years, well before expiration of the fixed-price tariff in 1998. Subsequently, the tariff falls to the then prevailing wholesale price: possibly as low as 2 to 2.5 pence ($0.032 to $0.04) per kilowatt-hour. This will cut revenues three-fourths.

British analysts believe that the low pool price discourages wind development independent of NFFO. They argue that the Thatcher government, in a free-market frenzy, sold off the CEGB at less than its fair market value. Thus the capital or equipment cost of the two new privatized utilities is artificially low. The government, in effect, issued the two new utilities a huge one-time subsidy. With below-market costs for its power plants, the post-1998 pool price will also be lower than the market would have otherwise set. Many of England's existing nuclear plants also received favorable government financing (30 to 40-year loans at 5% interest) unavailable to the private sector.

Through the early 1990s only two rounds of contracts had been awarded. The third tranche for England and Wales and the first Scottish Renewable Order were announced in late 1994. These new contracts will add 400 MW of wind-generating capacity to the nearly 165 MW already operating in England and Wales, and result in Scotland's first 100 MW. Unlike previous NFFO tranches, the fixed price awarded under these contracts does not expire in 1998. The new contracts' 15-year fixed-price period nearly halved the price for wind energy.

With the intermittency of the first three tranches, British wind development has followed the California model of boom (contract awards and feverish activity) and bust (no contracts and no construction) only on a much smaller scale. The NFFO, with its requirement for costly financing and legions of attorneys, has also discouraged dispersed or locally owned applications. There are very few individually owned wind turbines in Britain.

Italy

Despite a cumbersome regulatory process and an emphasis on domestic designs, the Italian wind program finally began to bear fruit in the mid-1990s. With the installation of a 1.5-MW experimental turbine at its Alta Nurra test site on Sardinia and completion of several small wind plants, installed capacity reached nearly 20 MW by the end of 1994. For many years the only operating turbines in Italy were found at Alta Nurra.

Because of planning obstacles, the government worries that Italy may miss its target of 60 MW by 1995. To speed up the process, the Italian government has directed each region to identify areas suitable for wind and other renewable projects. Meanwhile, ENEL has begun surveying wind resources in promising regions in hopes of accelerating development.

ENEL, the recently privatized national utility, had run afoul of permitting delays at several of its sites. To meet planning restrictions, ENEL limited road construction to the absolute minimum necessary for access. And as developers are now doing in Cornwall and Wales, they used local building stone for the façades of control buildings and transformer sheds. Because tourists flock to the western part of Sardinia, where ENEL plans one 10-MW project, the utility made a special effort to harmonize the wind plant with the terrain. The Italians hope to avoid constructing any new roads. Wherever possible they will use existing paths and open areas and have adjusted the layout of the turbines accordingly. As a result, the Monte Arci plant will occupy more land than first envisioned, but ENEL's cooperation has won the reluctant support of local planners. ENEL even went to the extra length of designing the control building to resemble a typical Sardinian farmhouse.

Wind plants over 3 MW qualify for 30% capital subsidy; state-supported demonstration projects qualify for 50%. Projects also qualify for a buyback rate of 171 lira ($0.136) per kilowatt-hour for five years. Payment is reduced if the project takes advantage of the capital subsidy. But even together, the five-year fixed-price period may be insufficient to assure bankers that the initial investment will be recouped, thus limiting wind development to state-supported projects. Italy's reoccurring economic and political crises could throw the Italian wind program into jeopardy just as it begins to get off the ground.

Spain

Spain is unlikely to become the "next California" as its energy agency is fond of saying. At best, the Spanish wind program will result in modest development through the mid-1990s. Spanish wind development is hampered by a lack of financeable utility contracts and by a weak industrial infrastructure.

There is also an overt reliance on direct government subsidy that slows expansion to the pace of Spanish bureaucracy and sometimes that of the distant European Community bureaucracy in Brussels. Because Spain is one of the least developed countries in the European Community, developers have discovered a gold mine in the EC's valorem fund, a program designed to spur regional development. Most Spanish wind projects are subsidized by the valorem program.

Even with the EC's valorem funds, Spain's wind program has gotten off to a slow start. By 1988 wind turbines were producing only one-ninth of that expected by *Instituto para la Diversificación y Ahorro de la Energía* (IDAE). Development has consistently fallen far short of IDAE's goals and will remain a long way from providing the 4% of Spain's primary energy needs that they once envisioned.

Spain expects to operate a total of 83 MW of wind capacity by the end of 1995. Several demonstration projects have been completed. The most significant is southwest of Gibraltar between Algeciras and Tarifa. If not Spain, at least "Tarifa is the European California," says Emillio Zurutuza, vice president of Campania Sevillana de Electricidad, a utility serving 8 million customers in southern Spain. In 1985 the utility started a project near Tarifa at a 8.5-m/s (19-mph) site using an array of different machines. By 1994 30 MW was being generated by 250 turbines at Tarifa, making it one of Europe's largest concentrations of wind turbines. With 20 MW, Plantas Eólica del Sur (PESUR) is the largest single wind plant at Tarifa. PESUR is composed of 150 turbines built by a Abengoa–U.S. Windpower joint venture and 34 machines built by Endesa, a large Spanish conglomerate.

Sweden

Sweden expects that total capacity will reach 100 MW in 1996 as a result of private development using an incentive program introduced in the early 1990s. Initially, the five-year program paid 25% of the installed cost of turbines larger than 60 kW. This was later increased to 35%. Coupled with a buyback rate of $0.06 per kilowatt-hour available from some Swedish utilities, wind energy is finally moving forward after nearly two decades of research. However, most Swedish utilities pay only 25 øre ($0.045) per kilowatt-hour, forcing potential users to find highly energetic sites, even with the capital subsidies.

Finland

Of unknown potential only a few years ago, the Finnish market for wind turbines could be large. Finland has the highest per capita consumption of electricity in Europe, and demand is rising. Although expected efficiency gains will be substantial, conservation is not considered the sole solution to demands for new capacity. Further development of nuclear and fossil plants is constrained by environmental considerations. Consequently, Finland is developing an interest in wind energy.

Results from initial resource assessments suggests that development of coastal sites alone could produce 5 to 7 TWh of wind generation, 10% of the nation's 60-TWh consumption. Even greater potential may have been found on the fjelds of Lapland. "We've discovered much more wind than anyone ever expected," says Markhu Autti of Kemijoki Oy, a hydropower company in northern Finland. Finnish meteorologists believe that the fjelds—smoothly contoured mountains with bald summits about 500 m above sea level—pierce the ground inversions common in cold climates, exposing them to high winds. Because development may be hindered by extreme cold and severe icing, the Finns are searching for wind technology that can perform well in the harsh environment.

In early 1993 the Finnish Ministry of Trade and Industry announced a modest program to build 100 MW of wind capacity by the year 2005. The government will pay about FIM 0.35 ($0.875) per kilowatt-hour, which includes 40% of the retail price plus a 20% value-added tax refund.

Greece

Like other Mediterranean countries, Greece has been slow to tap its abundant wind resource. By 1995 Greece had installed 30 MW of capacity, mostly as part of wind-diesel systems, despite studies showing that the country could generate 14% of its electricity, 6.4 TWh per year, from the wind. In 1992 the Minister for Energy set a goal of 400 MW by the year 2000, 150 MW of which would be built by the state-owned Public Power Corporation (PPC).[17] The principal obstacle to expansion has been the reluctance of PPC to interconnect with private producers and to pay a tariff comparable to its costs of generation.

France

The situation is similar in France, where the state-owned utility, Électricité de France (EDF), controls entry to the market. After an often feeble and nearly always disastrous attempt to develop an indigenous wind industry during the 1980s, France relaunched a small wind program during the early 1990s. The move comes at a time when France faces increasing pressure from the European Community to open its electricity market to competition and new questions are being raised about the country's dependence on nuclear power.

The program by the French Agency for Energy and the Environment, *Agence de l'Environnement et de la Maîtrise de l'Énergie* (ADEME), concentrates on small wind-diesel systems for its overseas territories, off-the-grid applications in remote areas of continental France, and two small wind plants interconnected with EDF's lines.

Only 2.5 MW had been installed by 1994, mostly in the country's first wind plant. Located at Port-la-Nouvelle in southern France, the 7.2-m/s (16-mph) site in the foothills of the Pyrenees overlooking the Golfe du Lion is raked by a fierce northwest wind akin to the infamous Mistral of the Rhône valley: the Tramontana. Cabinet Germa, the Parisian consulting firm that installed the turbines, financed the project partly with grants from the EC (25%), ADEME and the regional government (10%), and the landowner, a nearby cement plant. The small plant of five turbines sells its 5.1 million kilowatt-hours of annual production to EDF for an average weighted price of FF 0.295 ($0.059) per kilowatt-hour. The tariff includes a paltry FF 0.126 ($0.025) per kilowatt-hour during the summer months when EDF is awash in excess nuclear capacity, and FF 0.518 franc ($0.104) per kilowatt-hour during the winter.

Following the successful installation of a single turbine on Dunkerque's public beach in the early 1990s, a second project will be completed in 1995 in the north of France near the Belgian border. The regional government of Nord-Pas-de-Calais, a heavily industrialized region that includes Dunkerque and Calais, in cooperation with Espace Eolien Developpement, will install nine turbines on the digue or breakwater protecting Dunkerque's inner harbor. The 3-MW project is similar to the 23 turbines on the breakwater at Zeebrugge farther north along the Belgian coast and could eventually be expanded to 17 turbines capable of generating 13.2 million kilowatt-hours per year at the 6.4-m/s (14.3-mph) site.

The principal obstacle to further wind plant development in France is EDF's low tariff. But cracks are developing in EDF's monopoly. Several of the utility's reactors will soon begin approaching the end of their useful lives, and unless

EDF begins to diversify its generating mix, it will have to start another round of costly and controversial nuclear construction.

EDF's long campaign to electrify the French economy has succeeded all too well. France doubled electricity consumption from 1974 to 1984, twice the growth rate of other European countries, locking France into dependence on the atom. In 1974 EDF's investments in nuclear energy accounted for 9% of France's gross domestic product, equivalent to the construction of 16,000 km (nearly 10,000 miles) of autoroutes. By 1980 nuclear power accounted for 10% of the French industrial work force, according to French energy specialists.[18]

Unsurprisingly for a nation that meets 75% of its electricity with nuclear power, the French ministry for industry and commerce found nuclear energy the least costly source of new generation in a mid-1990s study, but acknowledged, for the first time, that wind energy is on the threshold of competitiveness, albeit after the year 2000. The government has stated that the encouraging results justify continued "experimentation" with wind technology.

EDF is most interested in the ability of wind energy to cut the cost of serving remote sites. The utility has opted to install hybrid wind and solar systems at several locations in southern France instead of extending their transmission lines. They are also considering wind energy for reducing the cost of serving French territories in the Caribbean.

Because the French Antilles are administered as part of continental France, EDF charges only FF 0.66 ($0.132) per kilowatt-hour even though it costs five times that much to generate the electricity. To cut the cost of serving La Desirade, a small island east of Guadeloupe, EDF contracted with French wind turbine manufacturer Vergnet S.A. to operate a small wind plant for less than half of EDF's generating cost. The 144 kW provided by the 12 small wind turbines, the French West Indies first wind plant, now supplies 70% of the island's electricity.

In many ways, Denmark lies at the opposite extreme from France. Wind energy is abundant, and valued; and long ago, the land of Niels Bohr, one of the founding fathers of nuclear power, turned away from the atom.

Chapter 1 Endnotes

1. California is the world's seventh largest economy. Its rank depends on the value of the dollar, which fluctuates.
2. Thomas Starrs, "Legislative Incentives and Energy Technologies: Government's Role in the Development of the California Wind Energy Industry," Ecology Law Quarterly, Boalt School of Law, University of California, Berkeley, CA, 1988.
3. Ibid.
4. Ronald Reagan removed solar panels from the White House roof in a statement of his view toward renewable energy. Later in his transmittal letter to Congress on the 1986

Tax Reform Act, President Reagan specifically cited wind turbines in California as an example of why the tax code needed "reform."

5. Helmut Stich, letter to Vice President Al Gore, August 24, 1993. Stich's letter refers to separate tax court decisions concerning Victor E. Duzer and Gordon Tanner.
6. David Roe, *Dynamos and Virgins* (New York: Random House, 1984). An exciting account of leaked documents and power politics.
7. Dan Richards, oral comments at meeting of qualifying facility representatives at Southern California Edison, Rosemead, CA, March 5, 1992.
8. Dave Branchcomb, Henwood Energy Systems, oral comments before the Independent Energy Producers Association's "All Hands" meeting, Sacramento, CA, February 23, 1993.
9. Jamie Chapman, "European Wind Technology," Electric Power Research Institute, Palo Alto, CA, March 1993, pp. 2–5.
10. Ibid., pp. 1–3.
11. Birger Madsen, "Overview of the European Wind Energy Market," paper presented at Windpower '92, the American Wind Energy Association's annual conference, Seattle, WA, October 19–23, 1992.
12. Sara Knight, "High Payments and Subsidies," *Windpower Monthly,* 8:10, October, p. 15.
13. See the 1992 Holland Map issued by the Netherlands Board of Tourism, Leideschendam. The suggested tour of Friesland includes a stop at SEP's Sexbierum wind plant. See also Helen Colijn, *The Backroads of Holland: Scenic Excursions by Bicycle, Car, Train, or Boat* (San Francisco: Bicycle Books, 1992), which includes sights of both traditional and modern wind turbines, and Chris Westra, *A Closer Look at Wind Energy* (Amsterdam: Chris Westra Produktie, 1993), which features cycling routes to wind turbines.
14. Westra, *A Closer Look at Wind Energy,* pp. 44–45.
15. Birger Madsen, "Evaluation of the Stimulation of Wind Energy in Europe," European Strategy Document on Wind Energy utilization, Status Report B, The Association of Danish Windmill Manufacturers, Herning, Denmark, September 1990, pp. 17–18
16. The 1500 MW goal is in units of declared net capacity (DNC). This is the equivalent amount of conventional base-load capacity that each technology would offset. For example, if British nuclear plants operate at 70% capacity factor and wind turbines operate at 30% capacity factor, 1 MW DNC of wind plants is equal to 2.3 MW of installed wind capacity. Only a portion of the goal will be met by wind energy.
17. Apostolos Fragoulis, "Wind Energy in Greece: Development, Perspectives, CO_2 Production Potential," Centre for Renewable Energy Sources, Pikermi, Greece, undated paper.
18. Jean-Claude Debeir, Jean-Paul Deléage, and Daniel Hémery, *In the Servitude of Power: Energy and Civilization Through the Ages,* trans. John Barzman (London: Zed Books, 1991), pp. 199–200.

2

The Vikings Are Coming

God Vind! An expression of Danish sailors on parting, literally "good wind." The equivalent phrase can be found in most European languages.

The blond-haired young man leaned against the wall of the bar. His blue eyes searched over the Saturday night crowd at the Mountain Inn, Tehachapi's sole night spot. Once a haunt of cowboys who came into town for its country-western music, the inn now had a new clientele. They were lean, lithe, and clad in Levis so tight that even the cowboys winced. Their fair skin was rosy from working in the unaccustomed sun of southern California's desert. Heads turned that night when an unattached woman sidled up to the stranger and they headed for the dance floor. The locals could tell he wasn't one of them. Instead of pointy-toed boots, he wore wooden clogs. Later, when asked where he was from, he replied in a heavy British accent, "Denmark." Lest his questioner had never heard of the small Scandinavian country, he added, "where the Vikings came from." He was the vanguard of what later became nearly an invasion. Soon everyone in Tehachapi would hear of the Vikings from Denmark.

Europeans of several nationalities came to the United States during the early 1980s seeking their fortunes in wind energy. Bob Jans is Belgian, Flemish to be exact. Jans, equally capable of making a transaction in English, French, German, or Spanish, imported the first Danish wind turbines to the United States in the early 1980s. Jim Dehlsen, founder of Zond Systems, one of California's first wind farm developers, quickly followed suit. Fittingly, Dehlsen's surname is Danish, and after he visited Denmark in 1981 the trickle of Danish machines entering California soon turned into a flood as 40-ft shipping containers snaked their way across the Atlantic. At the peak of the wind boom in California, Europeans were shipping 2000 wind turbines to the United States each year. It was a veritable invasion force, with Denmark's Mærsk Lines leading the way—not with fearsome dragon-prowed long boats, but with loaded container ships.

Denmark and the Danish Experience

The development of wind energy in Denmark has followed a far different path than it has in North America. Unlike American wind turbine design, which was concentrated in the hands of the aerospace industry, Danish wind technology grew out of the agricultural sector as a natural by-product of the Danish economy, where farming and metalworking play a major role.

Denmark is a modern nation whose farms and industry provide its citizens with one of the world's highest standards of living. A small country about the size of Indiana, Denmark's landscape is characterized primarily by small landholders. With three-fourths of its total land area devoted to farming, Denmark has been likened to a large agricultural factory. Although only one-tenth of Denmark's 5 million inhabitants work the land, agriculture accounts for about 40% of its exports. The country, one of the few in Europe that feeds itself, exports both food (Danish butter, bacon, ham, and cookies can be found in stores throughout North America) and the agricultural tools that produce it. Less well known, perhaps, is Danish skill in metal fabrication: It ranks among the top five shipbuilding nations in the world.

Denmark has only two major cities, Copenhagen and Århus. Most of its population is distributed uniformly across a relatively flat glaciated countryside. Its position jutting into the stormy North Sea, its 5000 miles of indented coastline, and its open farmland also endow it with a significant wind resource, not unlike that of the American Great Plains. This is fortuitous, for Denmark is a net energy importer.

Danish Energy Supply

In 1973, imported oil met 95% of Denmark's power needs. Following the oil embargo, interest in alternative energy mounted. By 1975 the Danish Academy of Technical Sciences concluded that 10% of the nation's electricity could be met by wind power without disturbing the existing utility system.[1] Denmark has never looked back.

Danes are culturally predisposed toward the wind. It powered their early conquests and carried Norsemen as far as the New World. The red and white Dannebrog pennants can be seen fluttering from countless flagpoles across the country, and European or "Dutch" windmills are still a common sight on the landscape. With such an ample indigenous resource and a memory of how wind turbines had come to the country's rescue during two world wars, the Danish parliament, the Folketing, naturally turned to wind energy as one means of weaning the country from imported oil. They began by establishing the Test Station for Wind Turbines, *Provestationen for Mindre Vindmøller,* at Risø National Laboratory in 1978 (Figure 2.1), and followed with a 30% investment subsidy in 1979.

Figure 2.1. Wind turbine under test at Denmark's Risø National Laboratory. Most testing of complete turbines is now done in the field rather than at test centers. Research reactor in background is Denmark's only nuclear plant. Denmark has chosen wind energy over nuclear power.

By 1980 manufacturers and experimenters were turning out their first products, and the Danish wind industry was born. Riisager, WindMatic, Bonus, Vestas, and a host of others introduced wind turbines about 10 m (33 ft) in diameter which powered 30-kW generators. Believing that the industry had become established, parliament began phasing out the incentive program in 1985, gradually reducing the installation subsidy to 10% in 1989, and dropping it altogether by 1990.

In 1981, Denmark launched its first energy plan, which included a goal of 1000 MW of wind power by the year 2000. A decade later, Denmark, which had become self-sufficient in oil following its discoveries in the North Sea, responded to the Toronto agreement for a 20% reduction in CO_2 emissions in the year 2005 by announcing its Energy 2000 plan. The plan directs the country to meet 10% of its electricity supply from wind energy. Despite Danish conservation efforts, consumption of electricity increased from 19 TWh in 1975 to 31 TWh in 1990. Thus, the plan requires an estimated 3 TWh per year from wind generation by 2005 or as much as that produced annually in California during the mid-1990s.

To reach this objective, Denmark will have to add 1000 MW to the 500 MW operating in 1994. This will require installing 2000 to 4000 of today's 250- to 500-kW turbines. There were already 3500 wind turbines operating in 1994.

Historical Development

The Danes are not newcomers to wind energy, nor are wind turbines new to Denmark. Poul la Cour, the "Danish Edison," began experimenting with wind-generated electricity in 1891. At the time, the 3000 "Dutch" windmills then in use were providing the equivalent of half again as much as all the animal power then supporting Danish agriculture. As early as 1903, the Danish Wind Power Society was fostering wind-generated electricity. At the same time, la Cour had refined the *klapsejlsmølle* (clap-sail windmill), following turn-of-the-century tests in his wind tunnel.

Danes commercialized La Cour's design during World War I. By the end of the war more than one-fourth of all rural power stations in Denmark were using wind turbines. During the long wartime blockade, the 3 MW provided by these crude wind machines and the widespread use of small farm windmills typically 5 m (16 ft) in diameter were a valuable source of power to an impoverished rural population. Although most windmills were used for mechanical power, there was the wind power equivalent of 120 to 150 MW operating in Denmark by 1920[2] (Table 2.1).

Denmark again turned to the wind during World War II, when nearly 90 turbines were installed, including the 30-kW Lykkegaard wind turbine, patterned after la Cour's *klapsejlsmølle,* and F.L. Smidth's more modern Aeromotors. The estimated 3 million kilowatt-hours per year that these machines generated were a welcome addition at Danish power stations short of fuel.

F.L. Smidth became one of the world's first firms to marry the rapidly advancing field of aerodynamics to wind turbine manufacturing. The diversified company designed machinery for working with concrete and used this technology to build concrete silos and chimneys. They also built small airplanes. Smidth's wind turbines borrowed from both fields by using modern airfoils upwind of a concrete tower. The laminated wood blades were swept back, or

Table 2.1
Danish Historical Development

Year	Type	Units	kW	Total Equivalent MW
1900	"Dutch" windmill	3,000	20	60
1918	La Cour	120	25	3
1920	Farm windmill, 5 m	30,000	2	60
				123
1943	Smidth 17.5 m	12	60	0.7
	Smidth 24 m	7	70	0.5
	Lykkegaard	60	30	1.8
				3.0

Source: Bent Rasmussen and Flemming Øster, "Power from the Wind," in *Wind Energy in Denmark: Research and Technological Development,* F. Øster and H. M. Andersen, eds. (Copenhagen: Danish Energy Agency, 1990), pp. 7–11.

coned, toward the tower, a feature found decades later on Micon, Nordtank, and other Danish wind turbines.[3]

After the war, interest in wind energy again waned. Johannes Juul's interest, however, did not. In 1950, Juul began testing an 8-m prototype for Danish utility SEAS. He subsequently modified a Smidth turbine used on the island of Bogø. With the experience gained from these two machines, Juul began work on his crowning achievement, the mill at Gedser, which was to become the forerunner of all later Danish wind turbines (Figure 2.2).

The three-bladed stall-regulated upwind rotor at Gedser spanned 24 m (79 ft). For overspeed protection, Juul devised a simple system for pitching the tips of each blade. Installed in 1956, the Gedser mill operated in regular service from 1959 through 1967. During its lifetime, it generated 2.2 million kilowatt-hours and was capable of producing 350,000 kWh annually, yielding a respectable 800 kWh per square meter of rotor swept area at the windy site. When Denmark looked again to wind energy for help in meeting one more energy crisis, the country had a working model still standing at Gedser 150 km (90 miles) south of Copenhagen.

Engineers at Risø scaled up Juul's Gedser design nearly threefold in 1980, even replicating the struts and stays, for one of what were to become the twin turbines at Nibe. Like Gedser, Nibe A used a stall-regulated fixed-pitch rotor upwind of a concrete tower. To hedge their bets and test their ability to use variable pitch on a rotor 40 m (130 ft) in diameter, they built a second turbine, Nibe B.

Juul's stayed rotor was difficult to adapt to a turbine of this size and the rotor on Nibe A was eventually replaced. Nibe B on the other hand, has been in regular service since 1984. Yet wind energy came of age in Denmark not at Nibe, and not at Risø, but at farms and at homes throughout the countryside.

Figure 2.2. Gedser mill. The forerunner of modern Danish wind turbines. The Gedser mill used three blades upwind of the tower and pitchable blade tips to control overspeed. All are hallmarks of successful Danish wind turbine design. (Courtesy of U.S. Department of Energy.)

Right Product at the Right Place

Rising electricity prices after the oil embargo and the Folketing's capital incentives for renewables spurred the interest of Danish farmers in an old standby: wind energy. They began clamoring for a new generation of wind turbines. Their timing was right.

During the late 1970s the European market for Danish farm equipment slackened dangerously, forcing manufacturers to seek new products for their rural customers. They quickly adapted their surplus capacity to a new market: wind turbines. The broad distribution of people across the Danish landscape demanding modern wind turbines, a good wind resource, and a manufacturing sector accustomed to building heavy machinery for a demanding rural market were the essential ingredients for launching a successful domestic wind industry.

Danish success was also due in part to "geographical proximity." Because the country is so small, manufacturers could service their own turbines, often directly from the factory. More than half of the Vestas turbines in Denmark, for example, are serviced directly by the manufacturer. This enabled companies to learn quickly from their mistakes and to keep their turbines in operation as physical proof for potential buyers that the company's machines were a good investment.

Two sites illustrate geographical proximity at work: U.S. Windpower's Altamont Pass wind plant and Vestas' plant at Tændpibe–Velling Mærsk. Both are located close to assembly plants and both are serviced directly by the manufacturer. The 4400 wind turbines operating at U.S. Windpower's Altamont site are merely an extension of the firm's assembly operation a few miles away in Livermore. Similarly, Vestas' plant at Lem is only a few kilometers away from Tændpibe–Velling Mærsk. This provides hands-on knowledge of how the turbines perform in the field and quick feedback to designers about problems as they develop, accelerating refinements. For any major repairs, the turbines can simply be removed and taken to the shop for "factory" service. Turbines at greater distances must be serviced in the field.

Armin Keuper of the German Wind Energy Institute, DEWI, saw the same effect at work in Germany during the early 1990s. In a survey of German wind turbines, Keuper discovered that various brands of wind turbines were concentrated near their point of manufacture. The densest concentration of Enercon turbines, for example, were found in Lower Saxony's Ostfriesland province, where the turbines are built. This simplifies service and provides immediate feedback for refining the design in much the same way as is seen in Denmark.

These conditions contrasted markedly with those in the United States. Since the United States is a world leader in aerospace, early U.S. programs poured research money into the aerospace industry, under the assumption (an incorrect one, as it turns out) that aerospace was the industrial sector most capable of building wind turbines. After all, wind turbines do look much like aircraft. Fortunately for the future of wind energy, Denmark has no aerospace industry.

Only in superficial ways do wind turbines resemble aircraft. Wind turbines must produce cost-effective electricity. To do so they must perform like other power plant machinery; that is, they must work reliably over long periods with little maintenance. Whereas a wind turbine must operate many hours on only a few hours of service, an aircraft flies only a few hours relative to many hours of skilled maintenance.

Wind turbines must operate far more hours than automobiles operate, and with less maintenance. Most wind turbines will operate more hours per year than an automobile will during its entire lifetime. For example, the 100,000-mile life of a typical American car is equivalent to 2000 hours of operation at 50 mph. Wind turbines in California or Denmark operate about 6000 hours per year.

As a consequence, early U.S. designs resulting from government research programs were marvels of the aerospace arts: highly efficient turbines operating at high speeds. Unfortunately, these designs proved noisy and unreliable. By the mid-1990s no aerospace company was building wind turbines in the United States and no designs from that period remained on the market. The one successful American manufacturer of medium-sized wind turbines, U.S. Windpower, designed and built its own turbine independent of the official U.S. wind program.

Unlike U.S. Windpower, which built its reputation on the ability to raise financing, all other U.S. manufacturers outside the aerospace industry were small, undercapitalized firms with little staying power. For comparison, the medium-sized Danish manufacturers that entered the market were moderately capitalized and experienced at marketing to a rural clientele. Still, there were several noteworthy financial failures of Danish manufacturers, including two of the top three, Vestas and Nordtank; their early capitalization and know-how did permit them, however, to establish a solid manufacturing base that remains a major force today.

Further, the huge geographic size of the U.S. market, as well as the vast difference in terrain and wind speeds from one region to the next, also hindered the development of a successful domestic industry. As in Denmark, nearly all early marketing efforts in the United States focused on rural applications for homeowners and farmers. In the United States, this resulted in the distribution of small wind turbines across the vast expanse of the continent. Many turbines were installed in low-wind areas, where they produced little electricity. The small U.S. firms and their distributors found it impossible to service such widely scattered turbines. The machines that failed often remained inoperative for months, if not years, discouraging potential buyers. When these small firms encountered technical or financial problems, they were often unable to correct them and quickly failed.

Dependable Home Market

One of the keys to Denmark's success has been a consistent national policy resulting in a strong domestic market for wind energy. During the 1980s Denmark installed 30 to 50 MW of wind capacity per year. Thus there was a relatively reliable demand for manufacturers to meet. This assurred them of a market sufficient to finance continued development of new, more cost-effective turbines.

Although regulatory uncertainty in the mid-1990s hobbled domestic sales, wind turbines generated 3.5% of Danish electricity in 1993. Altogether Danish manufacturers had built 12,500 wind turbines during the 1980s and early 1990s, 9000 (or 1000 MW) of which were exported, most to California. Half of the wind capacity in Europe was built in Denmark.[4]

In contrast to California turbines, most wind machines in Denmark serve homes, farms, or small businesses, and most installations comprise only one turbine. There are few wind plants, and these are small by U.S. standards. Most Danish wind power plants are small enough to be viewed as clusters of distributed or individual turbines. The average size of Danish wind plants varies from 10 to 50 turbines each. There are only 100 units at Tændpibe–Velling Mærsk, the site with the greatest number of turbines in Denmark. In contrast, there are 1060 turbines at SeaWest's site near Mojave, California.

Government planners and Danish utilities had hoped they could corral Maverick individual wind turbines into more easily managed wind plants before they stampeded across the Danish landscape. They argued that wind farms would reduce installation and maintenance costs. But these economies have proven elusive, because the wind plants permitted by planners, typically 5 to 10 MW, are too small to gain economies of scale. By 1993 three-fourths of all wind turbines in Denmark were installed singly or in small clusters.

In 1986 the Danish government reached an accord with the country's utilities for the installation of 100 MW in utility-owned wind plants by 1990. ELSAM, the utility serving Jutland and Funen, was responsible for 55 MW, and ELKRAFT, whose service area includes Copenhagen, was responsible for the remainder. They met their goal in 1992. Until 1988 the contribution by utility-constructed wind power plants was negligible (Figure 2.3). By the end of 1993 utilities accounted for one-fourth of total Danish wind generation.

The utilities reached another agreement with a parliamentary committee on a second 100-MW program in 1990. Their approval was contingent on planning agencies assigning sufficient sites at the regional and local levels. They had reason to hesitate. The utilities had encountered unexpected opposition to their proposals for the first 100 MW.

Banding Together

Danes pride themselves on their unusual ability to act individually while also working together cooperatively. Nearly all farms are owned by their operators, yet nearly all Danish farmers are members of farm cooperatives. It is these cooperatives that process the Danish foods found on the shelves of stores in North America. "Cooperative societies have played a major role in the development of [Danish] agricultural prosperity," says geographer Jesse Wheeler.[5]

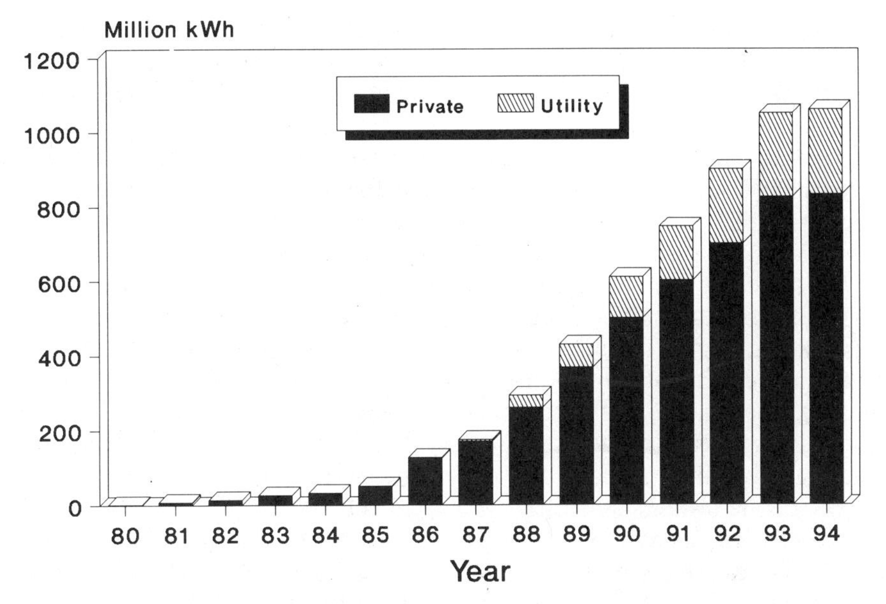

Figure 2.3. Danish wind generation. Individually or cooperatively owned wind turbines account for three-fourths of Denmark's wind-generated electricity.

Danish wind turbine cooperatives and an association of wind turbine owners have had a profound effect on the development of wind energy in the country. Prior to the entry of the utilities into the market during the late 1980s, nearly all wind turbines in Denmark were installed individually or by cooperatives. Nearly all the early wind plants are owned cooperatively, and most of the remainder are municipally owned. Some 100,000 Danish households or 250,000 people (5% of the population) own a stake in a windmill guild or cooperative.[6]

An early champion of cooperatively developed wind energy is Torgny Møller, publisher of *Naturlig Energi,* a pioneering wind power magazine. With Flemming Tranæs and others, Møller formed *Danske Vindkraftværker* (literally, "Danish wind power plants"), the Danish Windmill Owners' Association, in 1978. The group grew from a series of informal quarterly meetings of backyard experimenters, hobbyists, and environmentalists, which began in the mid-1970s.

Problems with the early turbines that had been sold to idealistic buyers led the owners' association to demand minimum design standards. The most important for the future of wind energy was the requirement for a fail-safe, redundant braking system, such as the tip brakes invented by Juul in the 1950s. This single provision probably did more than any other to further Danish wind technology, because it attempted to ensure the survival of the wind turbine when something went wrong—and it occasionally did. As a part of their role, the association also regularly reports on the reliability of Danish turbines and publishes a survey of owner satisfaction. They have also compelled Danish manufacturers to fulfill their production guarantees where the turbines have underperformed.

Tranæs, who bought two 55-kW turbines together with 12 other families, pays 3500 DKK (about $600) per year to the association. The association represents 80% of Denmark's installed wind capacity. The association protects owners' interests and advances wind energy in Denmark. From Tranæs's view in the trenches, the influence of the owners' association is absolutely essential in the give and take of the political arena. In a victory for Tranæs, the owners' association pushed for inclusion of single turbine installations in an early 1990s agreement between the government and utilities on the future of wind energy. The utilities and government planners had wanted to direct turbines into wind farms and clusters, eliminating the individually owned single turbine that had been the backbone of Danish wind development.

Often, the owners' association works in conjunction with *Foreningen af Danske Vindmøllefabrikanter,* the Association of Danish Windmill Manufacturers, Denmark's industry trade group. Together, the two have formed an effective alliance for using thousands of Danish wind turbine owners to win support in a politically fragmented parliament. By their sheer number, wind turbine owners now represent an influential voice in Danish politics. Collective action by Danish wind turbine owners has been so successful that the German wind energy association, DGW, plans to replicate the model south of the border.

Payment and Performance

Wind energy grew rapidly in Denmark not only because of the early capital subsidy, but also because Danish utilities pay a fair price for wind-generated electricity. By agreement with the government, utilities pay 85% of the retail rate when buying electricity from privately owned wind turbines, or more than $0.05 per kilowatt-hour (Table 2.2). Through the early 1990s, payment was limited to cooperatives or to owners of single wind turbines under 150 kW. For owners of larger turbines, and for cooperative members living outside the district where the cooperative's turbines were installed, payment was limited to 70% of the retail rate. The size limit for individually owned turbines was eventually raised from 150 kW to 250 kW and then eliminated in the mid-1990s.

During the 1980s, Denmark increased electricity taxes to 0.31 DKK ($0.05), including a 0.10 DKK ($0.02) per kilowatt-hour tax on carbon dioxide. Of these taxes, private owners of wind turbines are exempted from 0.27 DKK ($0.045) per kilowatt-hour. (Since early 1992, utilities have also been exempted 0.10 DKK per kilowatt-hour in carbon taxes for wind energy that they generate

Table 2.2
Payment for Wind-Generated Electricity in Denmark

	1993 DKK	Dollars
85% of pretax retail rate	0.31	0.052
Electricity tax offset	0.17	0.028
Carbon dioxide tax offset	0.1	0.017
	0.58	0.097
For comparison		
Average retail rate, including taxes	1	0.167

Note: Exchange rate: 6 DKK = $1.

themselves.) Coupled with the credit against energy taxes, utilities were paying the equivalent of $0.10 per kilowatt-hour for wind-generated electricity in the mid-1990s, in comparison to retail rates (including taxes) of nearly $0.17 per kilowatt-hour.[7]

The value of wind-generated electricity in offsetting taxes is substantial. For those belonging to a windmill cooperative (*fællesmølle;* shared turbine) the revenue from sales of wind-generated electricity is tax free. As in all other countries, a unit of tax-free income is worth more than a unit of taxable income. A savings of 1 DKK is worth 1.47 DKK of taxable income at Danish tax rates.[8]

Until 1994, parliament limited participation in cooperatives to those living within the district where the turbines were located, or an adjoining district. The government also limited the shares available to each cooperative subscriber to 1.5 times the member's domestic *electricity* consumption, equivalent to about 9000 kWh per year. In early 1994, parliament debated easing the restriction requiring that cooperative members live near their turbines, and raising the shares they may own to 20,000 kWh per year or the equivalent of domestic *energy* consumption. The new policy would require that only half the cooperative's owners live near the wind turbines, permitting, for the first time, urban dwellers to participate in wind projects in rural areas.

These provisions enable Danes to invest in wind energy in one of several ways. Farmers who install and operate wind turbines on their farms offset farm electricity consumption. As in the United States, the cost of energy consumption is a tax-deductible business expense. Danish farmers are also exempted from the value-added tax on electricity. These provisions reduce the after-tax cost of electricity, cutting the net benefit of self-generation. Even so, at windy Danish sites the attractive buyback rate makes it economically feasible for farmers to install the most cost-effective size wind turbine available, whether or not it generates more electricity than can be consumed on site.

In contrast, the low tariffs for wind-generated electricity in the United States discourage farmers from installing the most cost-effective turbines, because their excess generation will be sold to the utility at unprofitable rates. American farmers are thus forced to buy smaller-than-optimum wind turbines to avoid generating a surplus.

Danes may also cooperatively buy a wind turbine located elsewhere in their district or an adjoining district. They will receive $0.10 per kilowatt-hour tax free for sales up to 9000 kWh per year from their cooperative-owned turbine. The $900 they receive in payment is worth the equivalent of $1300 in gross income.[9]

One oft-overlooked aspect of Danish success with small cooperatives and single turbine installations is the backing of Danish financial institutions. Like buying a house or other costly capital investment, buyers seldom pay cash for the entire amount of the wind turbine. They finance a portion of the cost. To the envy of many commercial wind plant developers elsewhere in the world, Danish banks and finance societies willingly provide 10- to 12-year loans at attractive rates for 60 to 80% of the installed cost.[10]

Critics who examine Danish electricity prices and those of the rest of Europe are quick to note that the prices are much higher than in North America; they attribute this to mismanagement or, in the case of Denmark, the contribution of wind-generated electricity. Yet Europeans are just as competent at generating electricity as Americans. Like the cost of energy in other forms, half the cost of electricity in European countries is tax, added in part to discourage consumption. Denmark buys some of the cheapest fossil fuel in Europe and produces some of the lowest-cost electricity in the European Community. According to Risø's Finn Godtfredsen, modern Danish wind turbines at coastal sites will generate electricity competitive with that of new coal-fired power plants in Denmark for DKK 0.35 to 0.40 ($0.059 to $0.067) per kilowatt-hour.[11]

As have other countries, Denmark has accomplished this feat by improving the performance and lowering the cost of wind turbines. The peak efficiency of Danish wind machines has increased from 35 to 40% in the early 1980s to 45 to 48% in the early-1990s, says Godtfredsen, and availability has risen to a stable 98%. Overall performance has risen as well. The average specific yield of turbines in Denmark doubled, from 400 to 450 at the industry's inception to 750 to 850 kWh/m^2 of rotor swept area in 1992. Risø expects to squeeze another 10% increase in specific yield out of the turbines during the mid-1990s.[12]

The installed cost of Danish wind turbines has decreased 40% relative to swept area since the early 1980s. Most of the savings have come from the wind turbine and tower, which account for 70% of total costs. Because of stiff domestic competition, prices for Danish wind turbines are very homogeneous when expressed in costs per annual energy output (Figure 2.4). This reflects what economists call *market transparency.* Through the educational efforts

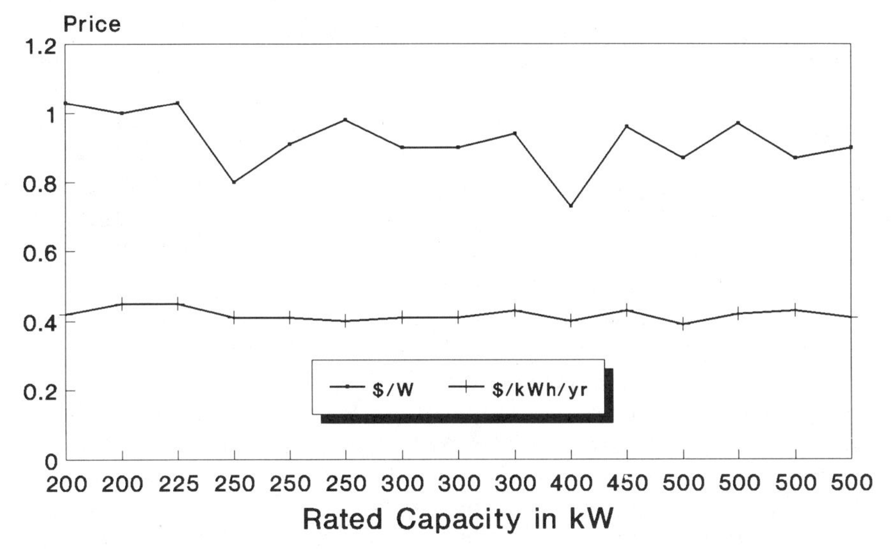

Figure 2.4. 1993 Specific price of Danish wind turbines. Although the specific price in terms of dollars per watt varies from one manufacturer to the next because of inconsistent rating practices, the specific cost in dollars per kilowatt-hour generated per year is nearly constant.

of Denmark's famed folkschools, publications such as *Naturlig Energi,* and the windmill owners' association, Danish consumers are not easily misled by prices in dollars per kilowatt of capacity. Instead, they look at price relative to expected generation. At coastal sites, the price of Danish wind turbines hovers around $0.40 to $0.45 per kilowatt-hour of annual generation, despite differences in the price per kilowatt of rated capacity from $800 to $1000.[13]

Nationwide Land-Use Planning

Although thousands of wind turbines had been installed in Denmark without difficulty, Danish utilities, to their chagrin, encountered opposition to their proposed wind plants. Legally, Danish utilities are cooperatives and are owned by their customers. In practice, however, they were often viewed as outsiders, and their future neighbors treated proposed wind farms no differently than they would have treated any other type of power plant.

At the utilities' request, the government appointed a special wind turbine siting committee in 1991 to find sites for the second 100-MW utility program and to recommend general siting rules for all other turbines. Planning has become necessary because wind turbines have increased in size by an order of magnitude since the early 1980s, and their numbers continue to proliferate.[14] The committee estimated that the Danish landscape could absorb 1000 to 2800 MW of wind capacity, taking into account local objections and the preservation of scenic areas.

Every five years local governments review their land-use plans. Because of the siting committee, most Danish counties have now included wind energy in their regional plans. These plans designate zones where wind turbines are prohibited, where single turbines may be installed, where wind plants are permitted, and where clusters or single turbines may be erected with special dispensation by the county government.

Many municipalities have also begun to include wind turbines in local land-use plans. One of the first to do so was Thisted, where there are already more wind turbines than anywhere else in Denmark. Because a large portion of the Thisted district is protected coastline, large swaths of the district were declared off-limits to wind energy. Only 30 MW of permissible sites was eventually identified. Even in Denmark, it is easier to say no than to say yes.

The utilities are especially frustrated. Of the 70 MW approved as potential sites by the central government in 1991, regional planning agencies approved only 55 MW. Subsequently, local agencies cut the approved sites an additional 25 MW. Another 20 MW was rejected by the central government's Environmental Complaints Board, despite the approval of neighboring landowners and area planning agencies.[15] An unexpected source of conflict is Denmark's desire to double its forested land. The Danish Forest Agency has emerged as a strident critic of further expansion.

Denmark is not the only country to convene a national land-use committee to determine where wind turbines will be acceptable. In the Netherlands, even more so than in Denmark, every square meter has an existing "interest," says Energy Connection's L. A. G. Arkestijn. Thus it is absolutely essential that wind energy have high public support, hence high political support. Without such support, there will be few accessible sites in the Netherlands. Wind energy, because it is the new kid in the planning neighborhood, must fight for its place in the planning hierarchy.

Since site planning began in Denmark, excluding many areas from wind energy development, as it has in the Netherlands, the price of remaining acceptable sites has doubled. The increase in Danish land costs has eroded much of the cost reductions made by advances in Danish wind technology.

The difficulty of siting new turbines has led proponents to suggest replacing early machines with contemporary products. This *repowering* would, for example, replace 55-kW turbines, which are often at some of the best sites, with modern designs 10 times as large. By doing so, says Birger Madsen of BTM Consult, it will be possible to meet the nation's wind energy target with only a modest increase in the number of turbines already in place. In 1994, Denmark's parliament proposed paying owners of older turbines 15% of their original investment, up to DKK 200,000 ($33,000), to replace them with modern turbines.

Ringkøbing

"The real development of wind power has taken place in California, [but] it's very difficult to use California" as a model for wind energy development, says Vestas's managing director, Johannes Poulsen. First, the turbines are older, less reliable, and less productive than those now being built. But more important to Danish manufacturers, California's haphazard and aesthetically jarring development paints an ugly picture for the future. Instead, Danes like to point to the northwestern corner of the Jutland peninsula, especially Ringkøbing commune, as a model for integrating wind turbines with the landscape and the concerns of people who live on it.

The commune, or district, includes the city of Ringkøbing, at the northern end of the shallow fjord of the same name. The commune, in cooperation with Vestkraft (the local utility) and Vestas, actively sought development of a wind power plant, first at Tændpibe and subsequently at the adjoining Velling Mærsk site, to reduce the commune's consumption of fossil fuel. The city of Ringkøbing consumes 25 million kilowatt-hours per year, and generation by the wind plant at the 6.3-m/s (14-mph) Tændpibe–Velling Mærsk site amply compensates.

The project is unique for several reasons: until 1992 Tændpibe–Velling Mærsk contained the greatest number of wind turbines at any single site in Europe; it has strong visual appeal; it abuts a cycling route along the scenic east side of the Ringkøbing Fjord; and it is within view of the city of Ringkøbing and popular coastal beaches.[16]

The 35 turbines at Tændpibe, owned cooperatively by 508 local families, adjoin 65 turbines at Vestkraft's Velling Mærsk wind plant. The two projects form a seemingly contiguous whole of 100 similar, though not identical machines. The visual uniformity of Tændpibe–Velling Mærsk masks the presence of wind turbines of four different sizes, representing the progressive development of the technology. Because all were built by the same manufacturer—Vestas, in nearby Lem—and were installed on similar tubular towers, the different models are nearly indistinguishable, even to the trained observer.

Communes such as Ringkøbing can restrict how projects are built; they can, for example, specify that developers use tubular towers or that the turbines should all appear similar. Vestas and Ringkøbing commune are justifiably proud of the result at Tændpibe–Velling Mærsk.

The successful integration of wind energy with communities such as Ringkøbing have produced strong ongoing support for wind energy. In a country where many residents see wind turbines daily, a nationwide survey in 1993 found that more than 90% of respondents rated wind energy an environmentally superior technology to coal, and more than four fifths believed wind energy should expand further. Despite occasional media accounts to the contrary, only 13% found the turbines "very noisy" (32% found them "a little noisy"). Nearly two thirds thought that wind turbines fit "well" or "reasonably well" with the Danish landscape and 77% were willing to pay 10% more for electricity if it was produced by wind turbines. Not surprisingly, in light of research elsewhere, those most familiar with the technology, because they could see the turbines from their homes, were more willing to accept further wind development than the average Dane.[17]

Naturalization

The Danish product cycle of newer and bigger turbines every two years has led to a quandary. As the turbines grow larger they become harder to export, particularly to the U.S. market. Blades are already longer than the standard 40-ft shipping containers. "Local production is absolutely part of our strategy," says Poulsen, Vestas's managing director.[18] This prompts a question dogging the wind industry: Is a Danish wind turbine built in the United States still a Danish wind turbine?

This question was first raised in California when U.S. firms rebuilt and modified early Danish wind turbines and adapted them to the rugged California climate. The success of Danish wind turbines in California is as much due to American perseverance as to Danish engineering and craft tradition. The turbines of many California wind companies may have started life in Denmark, but they have long since become "naturalized."

Yet without Danish wind turbines, wind energy could have died at birth. Danish machines became the foundation of California's wind industry as it struggled to emerge from the late 1980s—wind energy's dark age. The Danes may have ensured wind energy's survival as a technology suitable for a future sustainable society.

Chapter 2 Endnotes

1. Maribo Pedersen, "Wind Energy in Denmark," in *Wind Energy in Denmark: Research and Technological Development,* F. Øster and H. M. Andersen, eds. (Copenhagen: Danish Energy Agency, 1990), pp. 4–6.
2. Bent Rasmussen and Flemming Øster, "Power from the Wind," in *Wind Energy in Denmark: Research and Technological Development,* F. Øster and H. M. Andersen, eds. (Copenhagen: Danish Energy Agency, 1990), pp. 7–11.
3. Ibid.
4. Finn Godtfredsen and Peter Jensen, "Wind Energy in Denmark: Development in Wind Turbine Technology and Economics since 1980," American Wind Energy Association's Annual Conference, Windpower '93, San Francisco, July 1993.
5. Jesse Wheeler, Jr., J. Trenton Kostbade, and Richard S. Thoman, *Regional Geography of the World,* 3rd ed. (New York: Holt, Rinehart and Winston, 1975), pp. 220–221.

6. Finn Godtfredsen, Jorgen Lemming, et al., "Wind Energy Planning in Denmark," paper presented at "The Potential of Wind Farms," The European Wind Energy Association's special topic conference, Herning, Denmark, September 1992.
7. Finn Godtfredsen, Risø National Laboratory, personal communication, March 3, 1993.
8. Birger Madsen, BTM Consult, personal communication, February 17, 1993.
9. If the limit is raised to 20,000 kWh per year, the $2000 they receive in payment is worth the equivalent of $2900 in gross income.
10. Birger Madsen, Evaluation of the Stimulation of Wind Energy in Europe, European Strategy Document on Wind Energy Utilization, Status Report B, The Association of Danish Windmill Manufacturers, Herning, Denmark, September 1990, p. 22.
11. At 7% interest, 0% inflation, O&M of 3% of installed cost, and with a 20-year lifetime. From Godtfredsen and Jensen, "Wind Energy in Denmark."
12. Godtfredsen and Jensen, "Wind Energy in Denmark."
13. In Denmark, roughness class 1 represents a coastal site with a 6.5-m/s (14-mph) average annual wind speed. This is equivalent to 2800 kWh/m^2/per year. Modern Danish wind turbines in roughness class 1 will produce 800 to 1100 kWh/m^2 per year. From Godtfredsen and Jensen, "Wind Energy in Denmark."
14. Godtfredsen et al., "Wind Energy Planning in Denmark."
15. Peggy Friis and Mogens Held, "Commercial and Experimental Windpower in ELSAM Utility Area, Denmark," paper presented at the American Wind Energy Association's annual conference, Windpower '93, San Francisco, July 1993.
16. Nordtank's 18-MW project at Nørrekær Enge superseded in size the 12.6 MW installed at Tændpibe–Velling Mærsk as Europe's most powerful. The 78 turbines are grouped onto two distinct sites overlooking the great Limfjorden west of Ålborg. Like its predecessor, all turbines at Nørrekær Enge appear uniform. Both were surpassed when EcoGen installed 103 Mitsubishi turbines at its 31-MW Penryddlan–Llidiartywaun project in mid-Wales.
17. Gregers Lyster and Ole Persson, AIM Research, Copenhagen, February, 1993. Report of a nationwide telephone survey of 1016 respondents sponsored by Foreningen af Danske Vindmøllefabrikanter, the Association of Danish Windmill Manufacturers.
18. Johannes Poulsen and Martin Sondergaard, Vestas-DWT, personal communication, September 4, 1990.

3

The Research and Development Dilemma

"The wind and waves are always on the side of the ablest navigators." Edward Gibbon, Decline and Fall of the Roman Empire.

To many students of technological development, Denmark's success is even more surprising because it owed so little to government-directed research and development. This blasphemous observation runs counter to the American faith (the word is apt because the frequent repetition of this belief in public discourse sounds much like a mantra) that federal research and development (R&D) will solve the nation's energy ills. "Increase federal R&D spending and all will be saved," chant its followers.

At the 1993 conference of the American Wind Energy Association in San Francisco, a young engineer from the California Energy Commission (CEC) eagerly described a radical new wind turbine to two grizzled veterans. He proudly explained that the CEC was going to spend $400,000 on development of a machine that would revolutionize wind technology. To his surprise the old-timers seemed unimpressed, if not bemused. They asked politely for more details. After a series of queries, their demeanor darkened. "You're serious, aren't you?" One asked incredulously. "How," he continued, "can the CEC fund someone to design a 400-kW wind turbine when the designer never mastered an earlier 80-kW version?"

The veterans argued that the CEC was foolishly throwing good money after bad, noting that some 200 of the designer's turbines littered southern California, most having stood derelict for nearly a decade, silent witnesses to a flawed design and a misplaced belief that government can direct technology development. These turbines, they revealed, were the result of a misguided R&D program begun two decades ago by the U.S. Department of Energy. Shaking their heads in disbelief, the veterans said the CEC could help wind energy much more by removing the designer's derelict turbines from the landscape than by adding to it his new machines that would also probably fail.

The CEC grant was only one of a series of similar decisions that stung the American wind industry during the early 1990s. Previously the U.S. Department of Energy (DOE) through the National Renewable Energy Laboratory (NREL) had awarded grants to three companies for development of "advanced wind turbines." None of the three—Atlantic Orient, Northern Power Systems, and R. Lynette & Associates—were manufacturers of medium-sized wind turbines. Not surprisingly, the designs they proposed were the direct result of the DOE wind energy program during the 1970s. All three original designs had failed commercially. The federal grants infuriated California wind companies, whose R&D needs were going unmet because of funding to the "next generation" wind turbines. European manufacturers breathed a sigh of relief. The federal R&D program in the United States would, once again, pose no threat to their dominance.

Questioning Centrally Directed R&D

The goal of all wind energy proponents is to produce more wind-generated electricity. But opinions diverge as to how best to achieve that end. One camp urges expanded research and development of the technology, believing that this will eventually lead to greater use of wind energy sometime in the future. The other camp urges expanded use today, believing that installing wind turbines will create improved technology as wind-generated electricity is produced.

The success of crash R&D programs, such as the Manhattan project and the *Apollo* landings, have led to an almost mystical faith that massive support for science and technology "would help solve a range of problems and lead to a heaven here on earth."[1] Paradise has yet to appear, and many have since come to question whether "big science" can deliver on the promise.

Those questioning this widely accepted belief include Carl Weinberg, who led Pacific Gas & Electric's research and development program during the early 1990s, and European academics of the post-*Sputnik* era. Both groups contend that the success of wind energy has less to do with government-sponsored R&D, what Weinberg calls *technology push,* than with incentives or *market pull.* Danish political scientist Peter Karnøe goes so far as to suggest that centrally directed R&D not only did little to advance wind turbine technology but may have also retarded the use of wind energy, especially in the United States.

Wind energy is a deceptive field. It is neither the archaic domain of a miller furling the cloth sails on a Dutch windmill that some envision it to be, nor the high-tech realm found in nuclear power plants or aboard supersonic transports. The technology of modern wind turbines falls somewhere in between. There is a paradox, says Karnøe, "in that heavy U.S. R&D funding to the aerospace industry in the 1970s and 1980s" was unable to develop technology that could compete with wind turbines from the Danish farm machine industry.[2] Technical sophistication, he suggests, may have distracted American engineers from sound wind turbine design.

Like Karnøe, Matthias Heymann wanted to know why Danish wind turbines were so successful and why many American designs were not. He examined the style of technological development in Denmark, the United States, and Germany and how this influenced design characteristics. He wanted to know, for example, why the specific weight of Danish wind turbines is typically twice that of German and U.S. machines and why Danish turbines spin at lower speeds.[3] Independently, the two researchers reached similar conclusions: Big science successes, such as atomic bomb development and moon landings, infused German and American engineers with overconfidence that technical challenges could be overcome given enough time and money in the form of R&D, and that such R&D would result in technological breakthroughs.[4] Both agreed that the American and German R&D programs squandered enormous sums on a technological dead end with big wind turbines and clearly failed to produce commercially successful technology. Although these conclusions are widely shared within the wind industry, they anger many in the R&D community on both sides of the Atlantic.

Karnøe's work, especially, strikes a nerve among U.S. researchers. Karnøe, a specialist in the growth and accumulation of technical knowledge, says a bottom-up approach characterizing the Danish industry outperformed the top-down development strategy used in the United States. In a rare display of nationalistic anger, Robert Thresher, NREL's wind program manager, shouted "He's wrong," in response to a question about Karnøe's conclusions at a wind energy conference.

It is true, Thresher says, that there was an emphasis on big machines in the United States, but there was also a small turbine program at Rocky Flats outside Boulder, Colorado. Nearly all U.S.-designed wind turbines installed in California grew out of the small machine program at Rocky Flats, says Thresher. He argues that even U.S. Windpower's model 56-100 owes its existence to DOE's wind program. The design evolved, says Thresher, from DOE funding of research at the University of Massachusetts. "The roots of today's technology were nurtured and grown in the federal program," he says.

Not everyone agrees. No doubt DOE provided some support, says Karnøe, but its influence was weak, indirect, and came at high cost. Grumman's Windstream 25, Alcoa's early Darrieus turbines, Carter's 25-kW model, and U.S. Windpower's 56-50 were all originally developed with internal financing says Vaughn Nelson at West Texas State University's Alternative Energy Institute. Grumman and Alcoa eventually went on to win development grants from DOE, but their original work was independent of DOE's Rocky Flats program.[5]

Forrest (Woody) Stoddard traces U.S. Windpower's early design to work by Canada's Brace Research Institute in the 1960s. He and other graduate students at the University of Massachusetts adapted the Brace design to the university's experimental wind turbine and subsequently to U.S. Windpower's model 56-50.[6]

With the exception of U.S. Windpower's model 56-100, none of the U.S.-designed machines in California can be called a success. After excluding U.S. Windpower's turbines, European and Japanese manufacturers accounted for 70% of California's wind generating capacity in 1992. By the mid-1990s there was no major U.S. manufacturer selling commercially proven wind turbines to independent developers in the United States and there were practically no U.S. wind turbines operating in Europe.[7]

Stoddard, an outspoken critic of the NREL/DOE program, says that "no U.S. government-sponsored designs have survived the aerospace thinking" of the early U.S. wind program. "The most successful [American] company, U.S. Windpower, has succeeded not because of aerospace engineering expertise, but by the sheer volume of failure statistics on many hundreds of early machines."[8] Their success was based on experience gained after commercially deploying the technology.

Thresher is particularly proud of DOE's work on composite-wood blades and teetered hubs. The technology for composite-wood blades was adapted from that of high-performance racing boats by Gougeon Brothers in Bay City, Michigan.[9] Gougeon first built blades for NASA's Mod-0A series and later for U.S. manufacturers Enertech and ESI. Two Enertech designs were the direct product of DOE's small wind turbine program at Rocky Flats. The ESI design resulted indirectly from the DOE program when project engineers working at Rocky Flats left the government to build their own turbine. The ESI became the second commercial wind turbine to use a teetered rotor downwind of the tower. The first, the Carter 25-kW model, was designed outside the federal wind program.

These design elements may not be in use by U.S. companies today, says Thresher, but they will prove essential to development of "next-generation" turbines. He could be right. But Meade Gougeon believes otherwise. Gougeon pulled the family-owned firm out of manufacturing wood wind turbine blades in early 1994. Gougeon had become concerned that the R&D community was on the road to making the same mistakes it did in the 1970s. Referring to Swedish designers, Gougeon lamented that they "want to rewrite the same script" about teetered two-bladed rotors that the United States had acted out in the 1970s and early 1980s. "The fundamental problem is still there," says Gougeon, "and all the engineering in the world won't correct it."[10] The failure of the DOE wind program was not due to want of trying or for a lack of money.

R&D Spending

During the 1970s and early 1980s, U.S. research was mission oriented and technology driven, following the more traditional aerospace and defense industry models. The United States lavished nearly half a billion dollars on the aerospace industry from 1974 to 1992, most for multimegawatt machines (Figure 3.1). During the same period, the Danish government spent the equivalent of $53 million for wind energy R&D (Table 3.1). Like similar programs in Germany, Denmark, Sweden, and Britain, the $330 million spent by NASA/DOE on multimegawatt machines in the United States produced no commercial results that survived into the 1990s.[11]

The United States spent another $120 million on development of small to medium-sized machines through Rockwell's Rocky Flats test center. None of the designs produced by the Rocky Flats program were being manufactured by the late 1980s, although several hundred problem-prone derivatives were operating in California wind plants. In contrast, Denmark spent $19 million on medium-sized wind turbines and dominated worldwide manufacturing from the mid-1980s through the present. Fortunately for the Danish industry, Denmark combined a small amount of R&D funding for medium-sized wind turbines with market incentives. The products commercially available during the mid-1990s were the direct result of the market-driven, bottom-up activities of small entrepreneurial firms.[12]

The lesson from this, say critics, is that wind generation is uncorrelated with R&D expenditures—that research must be market driven, not technology driven.[13] Denmark, the nation with the largest number of manufacturers,

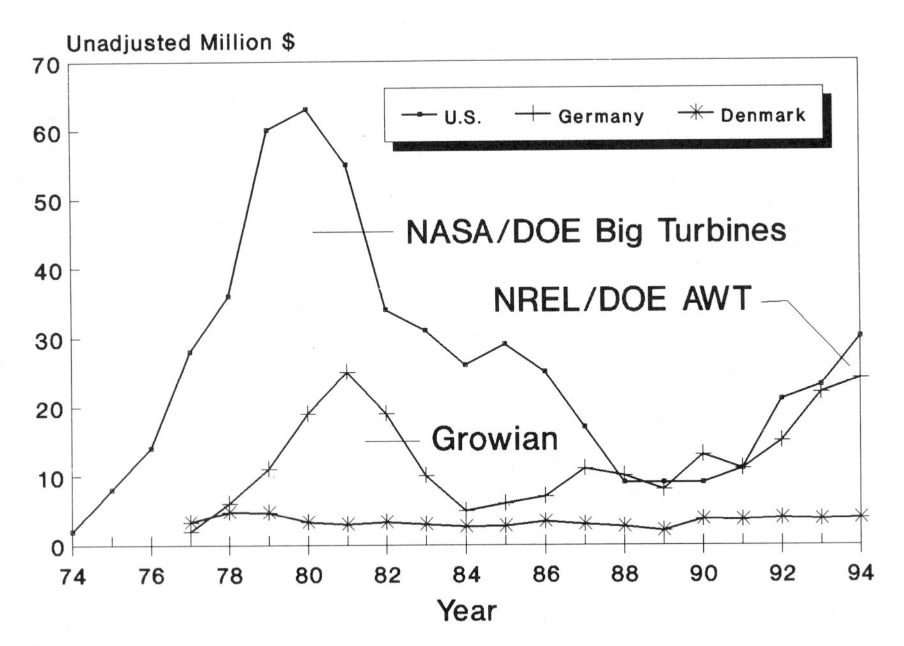

Figure 3.1. Government R&D expenditures. Both the United States and Germany spent far more than Denmark on wind energy R&D. Yet Denmark has produced the largest number of commerical wind turbine manufacturers.

Table 3.1
R&D and Market Incentives (Millions)

	US	D	S	NL	GB	DK
R&D, 1974–1992 ($)	486	178	71	65	60	53
Capital subsidies ($)						
Federal	340	2		48		35
State	560					
Tax exemption, 1980–1992 ($)						115
Green premium		15			16	
	1386	195	71	113	76	203
1992 Generation (kWh)	2800	359	32	99	65	900
kWh/$R&D	5.8	2.0	0.5	1.5	1.1	17.0
1992 kWh/$total	2.0	1.8	0.5	0.9	0.9	4.4
Number of major manufacturers	1	2	0	2	1	4
World market share (%)	30	5	0	5	1	53

US, United States; D, Germany; S, Sweden; NL, The Netherlands; GB, Great Britain; DK, Denmark.

spent the least on R&D. Countries that spent heavily on market incentives generate more wind energy than those that spent heavily on R&D. Those nations that emphasized premium tariffs on generation, that is, paid fair prices for the electricity generated, did better than those that emphasized capital subsidies on the installation of equipment, such as the United States.

After spending nearly 10 times as much on R&D and 20 times as much on capital subsidies as Denmark, the United States generates only three times the energy. Relative to its R&D expenditure, the United States produces only one-third as much wind energy as does Denmark. Overall, for the period 1974–1992 Denmark produced twice as much value in kilowatt-hours of generation for its public investment as did the United States. Moreover, Denmark built a thriving export industry, whereas only one major U.S. manufacturer of medium-sized wind turbines, U.S. Windpower, remained in existence by the mid-1990s.

Sweden, which had no market incentive program until the mid-1990s, produced the poorest results of any nation. Sweden held none of the world's manufacturing market and had no commercial manufacturers of wind turbines, despite its hefty R&D budget. With its combination of market incentives and modest R&D spending, Denmark generated nearly 10 times more electricity per equivalent dollar of public investment than did Sweden.

Despite data confirming that market incentives produce far more of the desired results—that is, more wind-generated electricity and a steady advancement in wind technology—the wind programs conceived in Washington, DC and Brussels continue to emphasize R&D on so-called "advanced wind turbines."

In 1990 NREL began its advanced wind turbine (AWT) program with a goal of halving the cost per kilowatt-hour of today's technology at midwestern sites that are less energetic than those in California. By improving existing U.S. designs from the 1970s and 1980s, NREL expects the first phase or "near-term" turbines to begin emerging in the mid-1990s. The "next-generation" machines, says NREL, will deliver electricity for $0.04 per kilowatt-hour at 5.8-m/s (13-mph) sites on the Great Plains by the late 1990s (see Chapter 7). Altogether the AWT program will cost $75 million over six years and will account for one-third of DOE's wind budget during the 1990s.[14]

After acknowledging that the EC's previous WEGA (Wind Energie Grösse Anlagen) program produced only "experimental"—that is, uneconomical—designs, Brussels will spend 9 million Ecu ($10 million) more on its WEGA II program during the early 1990s. Member states will contribute another 30 million Ecu ($40 million) to the program for constructing seven machines of 1 MW or more distributed diplomatically across Europe. These machines will begin appearing in the mid-1990s.

Many observers speculate that this money, whether in dollars or Ecu, would be better spent on market incentives or on resolving environmental problems in deploying the technology. In the United States, R&D spending through 1992 substantially exceeded the amount the federal energy tax credit cost the Treasury. This R&D expenditure has generated an insignificant amount of electricity. But state and federal market incentives encouraged the installation of nearly 17,000 wind turbines in California from 1981 through 1987, most of which are still in commercial service (Figure 3.2).

If the R&D funds had been used as incentives rather than misdirected to NASA and the aerospace industry, it is conceivable that the United States would not only have developed a more dynamic manufacturing industry but would also have installed more wind turbines. These machines would be producing far more wind-generated electricity, than that produced by the entire R&D program during the past two decades.

At an AWEA conference during the early 1990s, a staffer for the Electric Power Research Institute disparaged Danish manufacturers for using "funny money" to finance exports to the United States. This is a belief endemic among American engineers, a form of mass denial that U.S. products failed in the marketplace not because of inherent flaws but because of sinister Danish government subsidies. Echoing this view, NREL's Thresher says that Danish firms had a significant economic advantage over U.S. companies during the 1980s (Figure 3.3). Danes were successful not solely because of superior design.

There was no funny money, but the exchange rate was favorable to Danish products during the 1980s, giving them a 20 to 30% price advantage over U.S. products. Danish goods were half as expensive in the mid-1980s as they were in the early 1990s, when the exchange rate returned to historical levels.

The absolute values in Table 3.1 are less important than their relative magnitude. Even in the unlikely event that Denmark had invested $50 million

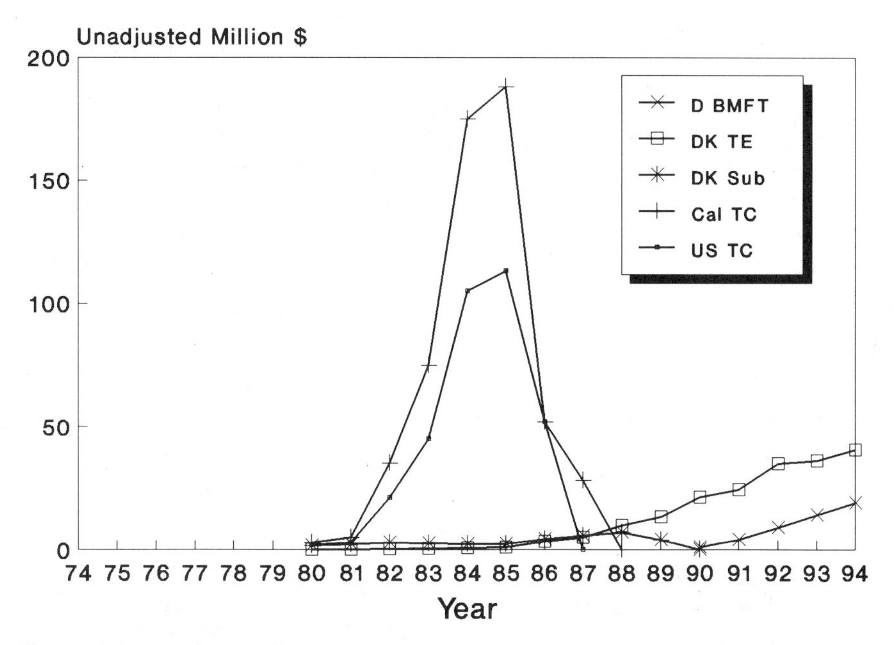

Figure 3.2. Market incentives. In the United States, generous federal (US TC) and state (Cal TC) energy tax credits spurred investment in California wind power plants during the early to mid-1980s. In contrast to U.S. tax incentives and Danish investment subsidies (DK Sub), German (D BMFT) and Danish (DK TE) incentives have been both more long-lasting and directed more specifically toward energy production instead of capital investment. This has produced more sustained and stable growth than the boom–bust cycle in California.

more in R&D than is publicly known, Danish R&D would still be 50% more productive than U.S. R&D, and total Danish public expenditure would still be 75% more productive than U.S. public investment. Seeking scapegoats is a dangerous attitude that prevents acknowledging failure and dooms the American R&D community to repeating its mistakes.

Cost Effectiveness

Thresher also sings a refrain commonly used to justify DOE's path during the mid-1990s. "To get the costs down, the turbines must become lighter and more sophisticated. U.S. technology looks good for the future," he says. Critics such as Stoddard, Karnøe, and Heymann charge that NREL may be making the same mistakes as famed German designer Ulrich Hütter, who erroneously identified technical optimization with economic optimization.[15] Like gas-fired combustion turbines, today's wind machines are already highly efficient.

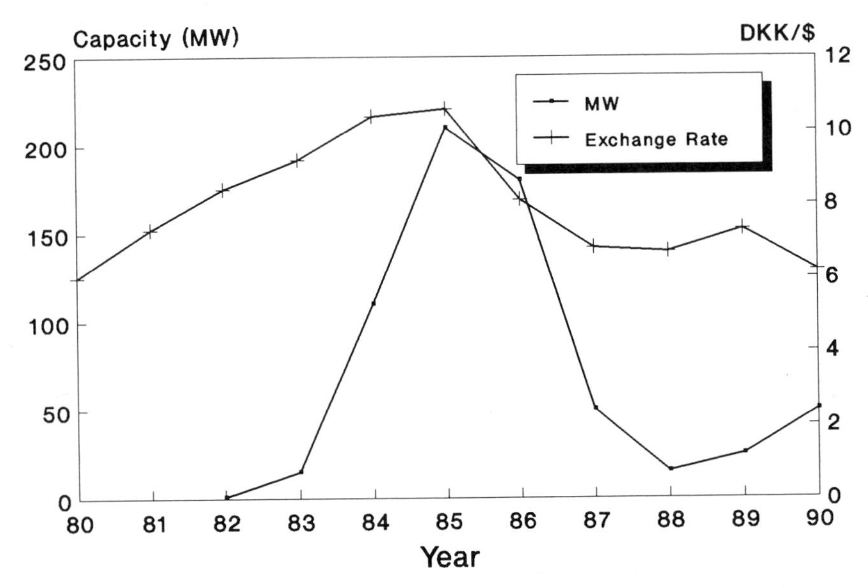

Figure 3.3. Danish wind turbine exports. During the peak of California wind development, the exchange rate was favorable for Danish exports to the United States. In 1985, when more than 200 MW of Danish wind turbines were exported to California, the Krone was trading far outside its historic range of 6 to 7 Kroner per dollar. To Californians, Danish products appeared to cost half as much in 1985 as they did in 1980 or in 1990.

"There is no reason to believe that other configurations, such as vertical-axis machines, will produce major benefits," says, the University of Massachusetts' Jon McGowan.[16]

John Kuhns, founder of one-time Wall Street darling Catalyst Energy and now principal of New World Power, sees as much decrease in the cost of energy from improved financing as from improved technology. Like many others, Kuhns expects only incremental improvements in technology. He is concentrating instead on developing what he calls, in a play on the words of NREL, the "next generation" of financing techniques. Kuhns says that present project financing, for example, is more concerned about risk than about potential reward. From his experience at Catalyst, he believes that the construction risk of wind projects is less than that of hydro, because wind is more modular. But the market does not recognize this yet. Kuhns's goal is to move from allocating risk to emphasizing reward. He suggests issuing new securities that offer investors strong upside potential. These securities will attract the low-cost capital necessary for building less costly wind plants without reliance on exotic new technology.[17]

Styles of Technology Development

After the 1973 oil embargo rekindled interest in wind energy, two groups, who were often at odds, sought to make wind energy work. One fostered a top-down, centrally directed government program acting through aerospace contractors. Initially, this group focused on big, multimegawatt wind turbines that would be operated by electric utilities. Eventually, they expanded their purview to small and medium-sized wind turbines. The names read like a who's who of the Western aerospace industry. In the United States, Hamilton Standard, General Electric, Westinghouse, and Boeing all attempted to develop large wind turbines as part of DOE's research program. The niche was filled by Messerschmidt-Bölkow-Blohm (MBB) and MAN (Maschinenfabrik-Augsburg-Nürnberg) in Germany, the Swedish State Shipyards in Sweden, British Aerospace in England, and Fokker in the Netherlands. Rockwell International, McDonnell Aircraft, Alcoa, Kaman, Grumman, and United Technologies all attempted to design medium-sized machines for DOE's program at Rocky Flats.

The second group, led by an assortment of activists and entrepreneurs, advocated a bottom-up strategy using small wind turbines built by small companies. Both groups found that designing reliable and cost-effective wind turbines was far more difficult than they had first thought. But only the bottom-up strategy succeeded. Despite their technical credentials, the U.S. aerospace industry failed. Danish farm equipment manufacturers, on the other hand, the Danish equivalents of John Deere and Massey-Ferguson, succeeded, and they did so without direct government investment.

Both styles of development can be traced to World War II. On one side of the Atlantic, Palmer Putnam assembled a talented team of engineers and academics to build a giant wind turbine 53 m (175 ft) in diameter for the S. Morgan Smith Co., a manufacturer of hydroelectric turbines. The 1.25-MW Smith–Putnam turbine became a technological guidepost pointing the way to subsequent U.S. designs of large down-wind turbines[18] (Table 3.2).

Across the Atlantic, Johannes Juul was laying the groundwork for what was to become the Danish configuration: an upwind three-bladed rotor operating at slow speeds.[19]Juul had little formal education and trained in the Danish craft tradition by attending La Cour's school for windmill electricians. He based the turbine at Gedser on his experience during World War II with the 60 wind turbines produced by cement company F.L. Smidth. Juul believed in incrementally increasing the size of wartime designs with which he already had experience.[20] He emphasized low cost, simplicity, and the use of readily available materials.

Ulrich Hütter, the "Werner von Braun of wind energy" as one wag described him, was at the opposite end of the academic and institutional spectrum from Juul. At the time Putnam was assembling his team in the United States, Hütter began work for the Ventimotor ("wind motor") company as chief designer.

Table 3.2
Historical Technology Development

Turbine	Diameter (m)	Swept Area (m²)	Power (kW)	Specific Power (kW/m²)	Number of Blades	Rotor Orientation[a]	Control[b]	Tower Height (m)	Date in Service
Poul LaCour, DK	23	408	18	0.04	4	u			1891
Smith–Putnam, US	53	2231	1250	0.56	2	d	p	34	1941
F.L. Smidth, DK	17	237	50	0.21	3	u	s	24	1941
F.L. Smidth, DK	24	456	70	0.15	3	u	s	24	1942
Gedser, DK	24	452	200	0.44	3	u	s	25	1957
Hütter, D	34	908	100	0.11	2	d	p	22	1958

[a] u, upwind; d, downwind.

[b] p, variable pitch; s, stall.

Ventimotor was a creation of the state, one of the first examples of government-directed wind energy R&D. According to Heymann, the Nazi party formed the business in anticipation of Germany's need for sources of energy independent of their soon-to-be-enemies and financed Hütter's early work. Based on his research at Ventimotor, Hütter completed his Ph.D. in 1942 with a thesis on optimal wind turbine design. Hütter concluded that high efficiency and low weight were two design priorities necessary to make wind turbines competitive. This philosophy follows from Hütter's previous work with gliders, where design is driven by attempts at decreasing weight and increasing performance. After the war, Hütter continued his research and in 1958 built a 100-kW prototype embodying his design philosophy. The experimental machine used a rotor 34 m (112 ft) in diameter comprised of two slender fiberglass blades mounted downwind of the tower on a teetering hub.[21]

Hütter's turbine swept twice the area of Juul's Gedser mill but less than half that of Putnam's. It looked more sleek and aerodynamic than the clumsy, cluttered appearance of the Gedser machine and the angular shape of Putnam's design. In contrast to Juul's measured development and Hütter's use of previous wind turbine experimentation, Putnam, with no prior experience, leaped from the small battery-charging machines then in use on the American Great Plains to a multimegawatt machine. Of the three, Putnam was the most unsuccessful. His machine threw a blade in 1945 and was dismantled. Only dusty photos remain of Putnam's bold effort (Figure 3.4).

Hütter fared better; his machine operated experimentally a total of 4200 hours from 1956 to 1968 at a low-wind site near Stuttgart. Although Hütter's turbine suffered severe damage on several occasions, it became renowned for its high efficiency. Hütter succeeded technically because he used more modern materials (fiberglass rather than Putnam's steel), and he understood better than Putnam the aerodynamic forces on a wind turbine because of his previous experimentation. For example, he loaded the rotor only one-fifth as much as Putnam did. At 34 m in diameter, Hütter's 100-kW turbine in 1956 was as large as the largest commerical wind turbines available in the early 1990s, machines that are capable of generating 400 to 450 kW.

Figure 3.4. Photomontage of 53-m (175-ft) Smith–Putnam wind turbine on Lincoln Ridge near Rutland, Vermont in 1941. (Photo courtesy of Carl Wilcox, project engineer.)

The course of wind energy in the United States and Germany would have been far different had Hütter failed as dramatically as Putnam did, for his sophisticated design has captivated the dreams of many German and American engineers ever since the 1960s. Although Hütter's approach has been tried repeatedly by many competent engineers for the past 40 years, it has yet to prevail commercially.

In contrast to Hütter's sleek turbine, Juul's machine at Gedser was an ugly duckling. He used concrete for the tower because it was locally abundant, and a chain drive rather than a gearbox. Instead of high-performance blades, he built simple airfoils and strengthened them with struts and stays. Rather than devise a complex mechanism for changing the pitch of the blades in high winds as had Hütter and Putnam, he opted for pivoting only the tips of the blades. Despite its simplicity, the Gedser machine operated commercially from 1957 to 1967 at a windy coastal site. The Gedser mill was still in good enough condition in 1976 for Risø, with NASA/DOE backing, to restart the turbine for further experimentation. Juul's Gedser mill served as the prototype for nearly all subsequent Danish wind turbines. Today, a modern medium-sized wind turbine operates on Gedser's concrete tower.

To Karnøe, the designs of Putnam and Hütter represented top-down technology development; Juul's, bottom-up. And the outcome during the 1980s with turbines in government-directed wind programs was no different from that during the 1950s. American and German engineers have designed some of the world's most efficient wind turbines, but none of these high-performance machines have ever operated reliably for long periods of time. Wind turbines, after all, are power plants, not race cars. "The Danes built a Model T, while we were intent on building a Maserati," says Meade Gougeon.[22]

Karnøe argues that Putnam's dramatic failure presaged the later failures of both the German and U.S. wind technology development programs.[23] Yet Thresher contends that Danish wind turbines are not the Davids felling the aerospace Goliaths that Karnøe portrays. He argues that Danish designs are as much a result of government-sponsored R&D as are American designs. "Risø's been there all along," he says. "Risø set design standards and they [the manufacturers] built clones."

Karnøe and most other observers disagree. The pioneers of the modern Danish wind industry were divorced from official government research at Risø. They represented a "low-tech" approach dominated by entrepreneurs, machinists, and "hands-on" engineers. For Danish firms there was no alternative but a bottom-up, incremental development of the technology. The Danish design—an upwind fixed-pitch three-bladed rotor of heavy construction—resulted largely from a lack of adequate design tools. Early Danish designers knew how to design a turbine, according to Karnøe, even if they did not always know why the design worked.

Risø did not select the Juul approach as many Americans believe, says Karnøe.[24] By 1979 all the principal Danish manufacturers were already using the "Danish" configuration. It was a design they could manage with in-house skills. These heavy Danish machines were more "forgiving" than American designs and better matched the skills of their proponents.[25] Risø implemented design standards only after a storm struck Denmark in 1981, laying waste to tens of turbines, particularly to S.J. Windpower's flimsy "wind rose." The following day, Risø's Per Lundsager, who acknowledges that Danish wind turbines are "dreadnoughts of iron and steel," resolved to prevent further disasters and required all new turbines to meet minimum design specifications. Prior to the storm, Risø's only specific requirement limited rotor loading to 300 W/m^2—three times the loading of Hütter's turbine but half that of Putnam's design and many contemporary turbines.

U.S. manufacturers steadfastly fought design standards, objecting to the type of approval program popularized by Risø. Americans charged that standards would lock in designs and their attendant defects. Karnøe argues that this is not the case, not even for Danish machines. It was the Danish market, through the actions of the Danish owners' association, and not Risø's certification program that forced standardization of design. Danish manufacturers had to sell "workable" designs, or the Danish owner's association would warn others to avoid them. This led manufacturers to stay with Juul's approach, which they knew worked.

Not Invented Here

Thresher's parting jibe that Danish wind turbines are all clones is telling. Clones are copies; and copying implies a lack of originality. American engineers from the aerospace fraternity deride the commonality of Danish designs. They would never propose such a seemingly simple, robust solution to the wind turbine problem, because the Danes have already done it.

In an aside about various means of orienting wind turbines with the wind, Putnam mentions in *Power from the Wind* that the "Russians copied this [Dutch] idea."[26] This revealing quote may explain why many U.S. engineers during the postwar period despised copying ideas: "The Russians do it, and everyone knows the Russians are automatons who can't think for themselves." Historians may one day link this fear with the "Not Invented Here" syndrome that plagues many U.S. and German industries.

This resistance to commonality forced U.S. R&D toward one-of-a-kind designs in which the research needs were highly specialized. In contrast, the Danes could direct their more limited R&D spending on components common to all manufacturers. In arguing the merits of this "open architecture" one respected American dynamicist notes that "so much time and money has been spent [in the United States] on one-of-a-kind blades" and other technology for NREL's advanced wind turbine program "that we won't get much for our money if they don't make it in the marketplace."

The corollary to "Not Invented Here" is equally blinding. "If we invented it, there must be merit to it." For example, Hütter was the first to propose and build an innovative one-bladed turbine. The one-bladed turbine is, consequently, a "German" concept and best embodies the aerospace principle of maximum weight reduction. Thus it was pursued as part of the German federal R&D program by MBB, the nation's largest aerospace company and the direct corporate descendent of World War II's deadly Messerschmidt.

This corollary could also explain why NREL and DOE are wed to lightweight designs such as the ESI 80: They were seduced by the beauty of their own arguments. Observers note that the NREL/DOE selection and then defense of the AWT designs sound ominously as if they were trying to justify their early work with lightweight machines and prove "we were right after all."

The relationships are sometimes incestuous. Sandy Butterfield, the ESI 80's chief designer, worked for NREL's predecessor at Rocky Flats. Butterfield returned to government service after helping launch ESI. One of the winning proposals in DOE's advanced wind turbine program was a redesign of the ESI 80. The proponent, R. Lynette & Associates, even adopted NREL's designation as its own and dubs its machine the AWT (for "advanced wind turbine"; Figure 3.5).

Stoddard fears that NREL/DOE's design concepts have become institutionalized. "I don't doubt that the UTRC machine would win an award [for design]," he says.[27] But as he had warned a decade earlier, "even a pedestrian European machine can outcompete these products." Stoddard fears that "the federal R&D program has lost its relevance and is now in danger of becoming a dinosaur." He charges that "none [of the designers in the AWT program] have proposed a realistic turbine. . . . We all thought they were ready to do something practical," he says in despair.[28]

Mark Haller echoes Stoddard's lament. Someone with years of hands-on experience fixing flaws in Enertech and ESI turbines in California, Haller characterizes NREL's advanced wind turbine program as a "reinvention of the flat wheel." Many share these views, but few say so openly, for fear of retribution in a business where government R&D spending can buy much-needed equipment and staff.

Karnøe angers the U.S. R&D community because he publicly questions whether NREL's advanced wind turbine program is preparing for the future with failed designs of the past. These "import killers," as NREL/DOE calls them privately, may be nothing of the sort, he warns.

If a design works, particularly if it works well, why not use it? In engineering science there is never one best way to solve a complex problem; there is always a range of solutions. The Danes may not have produced the most sophisticated solution, but it was a solution that worked. There would be more wind turbine manufacturers in the United States in the 1990s had Americans simply bought the manufacturing rights to a successful Danish design. Zond Systems appears to have learned this lesson. But rather than import a Danish design, they imported the designer, Finn Hansen. Former managing director of Vestas, one of Denmark's leading manufacturers, Hansen took control of Zond's design program in 1993 to the chagrin of his Danish compatriots. Zond expects to begin manufacturing a Danish-influenced turbine in the mid-1990s.

Under political pressure to fund "practical" designs, NREL inaugurated a Value Engineered Turbine program in the mid-1990s, ostensibly to improve on existing Danish designs. They awarded the first grant under the program, a modest $1 million, to Zond for its new turbine. Interestingly, the Zond

Figure 3.5. R. Lynette & Associates 26-m (85-ft) wind turbine undergoing tests near Tehachapi, California. The AWT-26 is one of several designs resulting from NREL/DOE's Advanced Wind Turbine program.

turbine was essentially a new design and not a Danish clone despite using three blades upwind of the tower.

Wind Energy and the Aerospace Arts

The approach by Hütter and his devotees grew out of the design principles embodied in aeronautical engineering. But wind turbines are not airplanes.

They were never meant to fly. They are power plants that must operate hours on end with little or no maintenance. Consider that American Airlines spends 5.5 hours of maintenance on a new low-maintenance short-haul jet for every 1 hour in the air. Wind companies can afford only a few hours of maintenance on their wind turbines for every thousand hours of operation. The value of wind turbines is principally in the energy they generate, and energy is the product of power and time. A powerful or highly efficient wind turbine produces little energy if it breaks soon after installation. Efficiency, though important, must take second place to reliability in wind turbine design.

Heymann and Stoddard both argue that sophisticated aeronautical models are no substitute for experience. Neither aircraft wings nor helicopter blades are directly comparable to wind turbine blades. Their only similarity is that all three use airfoils. "Helicopter design codes have been woefully inadequate" for designing wind turbines, says Stoddard. The wind resource is far more demanding than first thought.[29]

Experience and intuition taught early millwrights their craft. This resulted in a body of knowledge sufficient to build wind turbines up to 28 m (90 ft) in diameter. Some lasted for several hundred years. Saying that "a design life of 350 years isn't bad" in their treatise *Wind Turbine Engineering Design,* David Eggleston and Woody Stoddard suggest that millers had a better understanding of the wind than do many modern wind turbine designers.[30]

The failure of the top-down approach, says Karnøe, supports the conclusion "that too much science delayed innovation" in U.S. wind technology.[31] Cumulative knowledge needed to be built up both from R&D and from field experience. In Denmark this required blending the values of Danish craftsmen with the prosaic research needs of fledging manufacturers aided by Risø[32] "In terms of an advanced knowledge base and design tools," he says, "Americans are, without doubt, the most sophisticated" in the world. But "the Danes took the lead by pursuing the bottom-up strategy," and relevant new technological knowledge was acquired by solving problems specific to the open architecture of the "Danish" design.

The Danish industry, says Karnøe, was not handicapped by sophisticated knowledge of aerodynamics. No "ivory tower hierarchy" developed in the Danish wind industry because of its roots in farm machinery manufacturing, where empirical and hands-on knowledge were highly valued. Danish designers relied on simple, pragmatic principles. They used their intuitive or "seat-of-the-pants" knowledge about materials to reach a first-order approximation of how a wind turbine should be built.

That they were right would come as no surprise to civil engineering professor Henry Petroski. In *To Engineer Is Human,* Petroski stresses the importance of intuition in design. Good design, he cautions, is part knowing what questions to ask. "The further we stray from experience," says Petroski, "the less likely we are to think of all the right questions." The sum of empirical knowledge gives the engineer the intuition to create a concept suitable for the task from the start, without reliance on sophisticated technology to rescue a flawed design conceived without that knowledge.[33]

"The more sophisticated engineering tools become, the less we realize that engineering is also an art," explains wind turbine designer Peter Jamieson. "If we had waited for the right analytical tools, we would have never built the first bridge," he says. And many bridges that were conceived without computer-aided design still serve their function well. "Often analytical tools are used to validate decisions reached intuitively. People expect that an absolute analytical evaluation is possible. But there are no absolutes [in engineering]." Even cost is intrinsically "irrational because it is subject to political decisions."[34]

Had Denmark an aircraft industry, it is probable that its wind turbines would have failed in the same manner as those of their German and U.S. competitors from an overreliance on aerospace technology. It is noteworthy that France, whose aircraft industry competes with that of Germany and the United States, sponsored only one entry into the medium-sized wind turbine sweepstakes, Ratier-Figeac. The manufacturer of airplane propellers built a lightweight machine with two flexible blades downwind of the tower, not unlike the UTRC design. The design quickly failed and the French foray into wind energy was equally short-lived.

To Stoddard, an important difference between U.S. and Danish innovation was Danish willingness to proceed incrementally. American designers constantly sought breakthroughs. They wanted to bypass the drudgery of incremental development and bat a home run. Americans leapt from one size to the next with little transition. The Carters jumped from 25 kW to 300 kW, U.S. Windpower from 50 kW to 100 kW, then 300 kW. Danish turbines grew in modest steps: from 35 kW to 55 kW, then to 75 kW, 100 kW, 120 kW, 150 kW, 200 kW, 250 kW, 400 kW, 450 kW, and 500 kW by 1994. The DOE program leapt from 200 kW to the 1000-kW Mod-1, a mere fivefold increase during the first round of scaling. Ignoring the failure of the Mod-1, DOE continued with the program by scaling-up to 2.5 MW in the Mod-2, a tenfold increase over the 200 kW Mod-0A. In contrast, U.S. Windpower reached for a threefold increase only after refining its 56-100 design over nearly a decade.

Prior to NREL/DOE's advanced wind turbine program, no part of federal R&D came in for more criticism from California's wind industry than that of vertical-axis wind turbines. (Few of California's wind companies have ever given much credence to multimegawatt turbines.) From 1974 to 1985, DOE invested $28 million in Sandia National Laboratory for development of Darrieus technology. The result, says program manager Henry Dodd, is nearly 100 MW of commercial wind turbines that have returned the public's investment from a decade of electricity generation.

Dodd has a point. The capacity of conventional wind turbines derived from DOE designs operating in California—that is, non-Darrieus turbines—had fallen to less than 60 MW by the mid-1990s. Only a token amount of capacity originally installed by Windtech (16 MW) and Dynergy (17 MW), offspring of the UTRC design, remained. Two-thirds of the ESI capacity originally installed (60 MW) was still operating, while four-fifths of Enertechs installed (22 MW) were still in service. Altogether, there is a capacity of about 40 MW being generated by conventional wind turbines stemming from DOE's wind program still operating in California, less than half the capacity of the Darrieus design.

Dodd attributes Sandia's modest success (California had nearly 1700 MW of wind capacity in 1992) to a "different approach" from that of DOE's main program. Sandia sought an "evolutionary process instead of trying to make a quantum leap" by going to megawatt-sized machines. Sandia began humbly with a small machine 5 m (15 ft) in diameter and gradually built up to the 17-m (56-ft) commercial prototype that eventually was adopted by FloWind for use in California (Figure 3.6).

The U.S. vertical-axis program has also been far more successful than those of Canada, Germany, and Britain. In Canada, two manufacturers, DAF-Indal and Adecon, have attempted to commercialize the technology with limited success. Only Adecon remains, and it has stumbled along with only a small project in Alberta to show for its effort. At one time, Germany's Dornier planned to commercialize Canada's giant Éole, only to abandon the project soon after it began. Heidelberg Motor has installed a single prototype at the Kaiser-Wilheim-Koog test center but has done little to commercialize its innovative direct-drive machine.

Britain has produced two prototypes of Peter Musgrove's novel H-rotor. This is more a tribute to the staying power of its principal proponent, Ian Mays, a former student of Musgrove, than to the commercial merits of the design. With British and German government R&D waning in the early 1990s, European development of the H-rotor design appears to have reached a dead end.

Another unsung success of the U.S. research program was the result of practical, hands-on work by the U.S. Department of Agriculture's Bushland experiment station near Amarillo, Texas, and analysis of the nation's vast wind resource by Battelle's Pacific Northwest Laboratory. Although the work at Bushland and that by Battelle was never as glamorous as that by NASA, they contributed measurably to the use of wind energy. Bushland's focus on rural applications and water pumping led directly to development of modern wind-electric pumping systems that improves the lives of Third World villagers, such as the women of Ain Tolba, Morocco. And Battelle's Wind Resource Atlases proved to doubters that the United States has enough wind resources to make a major contribution not only to the nation's electricity supply, but enough to meet a major portion of its total energy needs.

Although learning through failure was the most effective way to gather technological knowledge in the 1980s, it will not necessarily be as successful in the 1990s, says Karnøe and others, because the machines have grown much larger and the costs of mistakes have risen proportionally. Danish manufacturers have acquired sophisticated technology and have become more analytical than they were during the formative years. By the mid-1990s,

were sketchy, often only comparing the measured power curve to the manufacturer's power curve, a measure of little utility.

Prospective wind turbine buyers in Germany fared little better than their counterparts in the United States. At the first German test center on the island of Pellworm, domestic manufacturers met their "Waterloo," according to press accounts. Only the one Danish machine at the test center worked reliably; the rest failed.[37] The situation at the test center was similar to that at Rocky Flats, although Rocky Flats never had the foresight to test Danish machines.

Rocky Flats' small turbine program ultimately failed, though not as dismally as NASA/DOE's big machine program. (Enertech's 44/40 model, a product of the program, did achieve limited technical success before succumbing commercially.)

NREL dumbfounded the U.S. wind industry in 1993 when it announced that it would reopen Rocky Flats as the National Wind Technology Center, in a politically motivated decision aimed at defense conversion. (Rocky Flats is best known for its contribution to nuclear weapons.)

Nor could the American Wind Energy Association (AWEA) play the same independent role in information exchange as the Danish owners' association. During its existence AWEA has been dominated by either manufacturers or developers, but never users. AWEA has also had close ties to DOE.

AWEA–DOE Symbiosis

The American Wind Energy Association (AWEA) is the largest, most sophisticated, and most professional wind industry association in the world. No other trade group, not even Denmark's FDV, comes close. This is as it should be. By the early 1990s the United States was still operating more wind turbines than were operated by any other country in the world. After all, California's wind plants were generating nearly $300 million annually in gross revenues by 1993, providing a sufficient pool from which to draw membership fees. This is not the sole reason for AWEA's success. AWEA has had two skilled managers: Tom Gray led AWEA through the nearly disastrous post–tax credit era, and subsequently, Randy Swisher, a savvy Washingtonian adept at winning friends and outmaneuvering opponents. Another reason for AWEA's success is a symbiotic relationship between DOE and AWEA that has evolved over two decades. AWEA and similar trade associations are part of what political pundits call a "Washington political class" that has arisen since power steadily flowed from the states to the federal government following World War II.

Not surprisingly, the fortunes of the U.S. wind industry's trade association and those of DOE are inextricably linked. Both parties benefit by the public's and Congress's increased interest in wind energy. AWEA adroitly lobbies Congress for increased R&D funding, and DOE in turn hires AWEA for contract work that it is especially suited to perform. The result: AWEA's budget has risen step for step with that of DOE. In 1993 50% of AWEA's operating budget came from consulting contracts with DOE and the U.S. Agency for International Development. When Gray left Washington, DOE's spending on wind was down to $11 million per year, and AWEA's budget stood at less than $500,000 per year. By 1994 AWEA had succeeded in boosting DOE's wind budget nearly fivefold to $49 million, and its own budget to $2 million.

The challenge for AWEA since its formation in the basement of a Detroit church by a motley group of hippies was to walk the tightrope between cajoling DOE into what AWEA wanted and responding to DOE's need for political support. If AWEA pushed too hard or was too openly critical of DOE's direction, political support for DOE might waver, wind energy's momentum in Congress slacken, and DOE falter, taking with it the U.S. wind program.

This situation is not unique to AWEA. Washington interests have been logrolling ("You help me and I'll help you") since the republic was founded. The Solar Energy Industries Association and the photovoltaics industry have collectively earned hundreds of millions of dollars in R&D contracts during the past two decades. Nor is this unique to the United States. Wind energy associations in other countries often contract with their governments. But none have been as successful as AWEA.

By the early 1990s DOE and AWEA had become so intertwined that DOE would regularly attend AWEA's board of directors' meetings to issue status reports on federal programs. DOE also cosponsors AWEA's annual conference, subtly dictating the number of professional presentations by DOE scientists and contractors. This cozy relationship is fraught with the risk that success at boosting DOE's wind budget becomes *the* measure of success, an end in itself rather than a means to an end—wind energy development. The potential for confusing who was serving whose interests so alarmed Ken Karas, AWEA's 1993 president, that he proposed a dramatic dues increase to make AWEA more independent of DOE largess. Yet in early 1994 AWEA was giving front-page treatment in its newsletter to a proposed 70% increase in DOE's budget, praising "DOE's commitment" to wind energy.[38]

Diverting Attention

Whenever renewables seem stymied, environmentalists, regulators, and politicians respond that more R&D is needed. This cry arises from an outmoded belief that technological and social innovations spring from the womb of large centralized organizations. This model of innovation no longer produces results either in government or commerce. The call for more R&D diverts attention from what is needed most, structural change in the market. For Congress, R&D is often politically preferable to market incentives because it gives the semblance of action without creating programs that could threaten the status quo and stir heated debate (Figure 3.7).

Figure 3.7. Lift translator. The photograph and accompanying caption stating that this "machine is being developed under a grant from the Department of Energy" lent credibility to this absurd invention and drew public attention from legitimate research needs. This technically flawed concept eventually reached the cover of Popular Science magazine as a wind energy panacea. (Courtesy of U.S. Department of Energy.)

Congress knows that U.S. public opinion typically favors more renewable energy R&D. But pollsters may be reaching the wrong conclusions from their surveys. The public has long demanded increased development of renewable energy through government support and encouragement. This endorsement of federal support is often uncritically translated by pollsters as a demand for more R&D funding. What the public may be saying is "We want renewables," not necessarily, "We want more R&D on renewables." Many would prefer market incentives to R&D if the distinction between the two were clear: market incentives result in more renewables, whereas more R&D often results in just more R&D.

The anticipated results from R&D gives prevaricators reason to postpone decision making. For example, today's technology requires high-wind sites to compete effectively against fossil fuels. These locations are unique. Many, such as California's mountain ridgetops or Denmark's coastline, are also scenic. Why sacrifice these sites with today's technology when better technology is on the horizon? The reason is simple: When that technology finally arrives, even better technology is certainly just around the corner. Waiting until the "best" technology is available is like the dog chasing its tail; there is no end.

During the early 1990s federal emphasis on R&D and the "next generation" of wind turbines hurt wind energy in the United States by dangling "a better technology on the horizon" in front of decision makers. Entrepreneur Dick Farrell recounts how NREL representatives, speaking to groups in the Pacific Northwest, damaged near-term prospects for his firm by constantly referring to the supermachines on NREL's drawing boards. His was not an isolated experience. Zond's Ken Karas tells a similar story about a hearing in Hawaii at which public utility commissioners questioned why they should approve the rate Zond requested when NREL/DOE was advertising wind energy for $0.04 per kilowatt-hour by the end of the decade. To Karas's dismay, the PUC chose to wait. Sara Miller, counsel to the Coalition for Energy Efficiency and Renewable Technology, bitterly recalls how NREL undercut her testimony before a Nevada legislative panel with the same message.

But the government-R&D enterprise has yet to deliver a reliable wind turbine. When NREL/DOE seeks to build specific machines such as those of the advanced wind turbine program, it is too slow for industry to benefit. It took NREL two to three years to develop advanced wind turbine blades. This is about the same amount of time as the entire turbine development cycle in Denmark, where through the 1980s and early 1990s a new model was introduced every two years. "When we needed blades," says Zond's Kevin Cousineau, "we built our own" rather than waiting for NREL. By the time the NREL blades were ready, Zond had already installed hundreds of their own sets. "When NREL decided to build a strain gauge test," continues Cousineau, "we'd had our own in operation for several years."

Technology progresses in response to a stimulus. This can be a top-down R&D program resulting from a congressional appropriation or a bottom-up R&D program led by a market for the technology. As seen, creating a market for wind energy is more powerful at driving the technology forward than is government R&D. A market will never develop if everyone waits for "next generation" technology. The Sierra Club's energy chair, Rich Ferguson, says

NREL's posturing that wind energy will work best sometime in the future after further R&D serves only to keep renewables in a "green ghetto."

Redirecting R&D

Government-sponsored R&D is a chimera. R&D will not produce wind energy. Roger Little of Spire, a manufacturer of solar cells, sums up the attitude of business: "R&D won't make an industry by itself." Public investment should go to wind turbine owners in compensation for the pollution-free electricity produced by wind energy rather than into a never-ending search for technological panaceas that benefit only the R&D community.

In the United States, big turbines resulted not only from a questionable belief in economies of scale common among engineers of the day, but also because the "breakthroughs" they represented were marketable to Congress. Today, the selling of "breakthroughs" promised by the AWT program in the United States mirrors that of the big machine programs of nearly two decades ago and is eerily reminiscent of the "too cheap to meter" days of nuclear power—a promise that was never kept.[39]

There are indications that NREL/DOE is slowly responding to legitimate research needs in the United States. But to regain the confidence of wind energy proponents, they must move swiftly to cancel the advanced wind turbine program and redirect the savings into prosaic, admittedly unsexy subjects such as geartrains, generators, yaw drives, airfoils, and wind resources. These and other research needs go unmet because of the AWT program. We should study the Danish cooperative model of wind development. It is a success story that could be replicated in other countries, especially on the American Great Plains. Most important for the United States, a significant portion of the R&D budget should be directed toward demonstrations of environmentally compatible development practices suitable for the U.S. market. In rural areas, where people still live with their machines, we must ensure that wind turbines will be good neighbors.

The wind industry's technological goal should be the design of productive, quiet, and aesthetically pleasing wind turbines that are as environmentally benign as it is humanly possible to make them. Yet nearly all government-sponsored R&D has focused on productivity, and often, sadly, to the exclusion of other concerns. The future of wind energy could well be determined by how quickly research priorities can be redirected to meet the needs of the 1990s.

It is too late to salvage the waste of the past two decades, but there is time to reorient R&D programs. Public opinion can change the direction of publicly funded R&D. If R&D fails to meet the real-world needs of the wind industry, it becomes irrelevant, nothing more than "welfare for the educated," as San Gorgonio Farms's Bill Adams caustically calls it.

Chapter 3 Endnotes

1. George E. Brown, Jr., "Can Scientists Make Change Their Friend," *Scientific American,* June 1993, p. 152. George Brown was the chairman of the committee on Science, Space and Technology in the U.S. House of Representatives.
2. Peter Karnøe, "Entrepreneurial Organization and the Accumulation of Knowledge in Modern Wind Technology," paper presented at the Roesnaes Workshop on the Process of Knowledge Accumulation and the Formation of Technology Strategy, Institute of Political Studies, University of Copenhagen, May 1990.
3. Matthias Heymann, "Why Were the Danes Best? Wind Turbines in Denmark, West Germany and the USA, 1945–1985," paper presented at the Society for the History of Technology annual meeting, Cleveland, OH, October 18–21 1990, p. 3.
4. Matthias Heymann, "Why Were the Danes Best? Social Determinants of Wind Turbine Development in Denmark, West Germany, and the USA," unpublished manuscript, Universität Stuttgart, p. 33. Peter Karnøe, "Approaches to Innovation in Modern Wind Energy Technology: Technology Policies, Science, Engineers, and Craft Traditions in the United States and Denmark, 1974–1990," unpublished paper, Institute for Organization and Industrial Sociology, Copenhagen Business School, December 1992, p. 58.
5. Vaughn Nelson, "SWECS Industry in the U.S.," unpublished paper, Alternative Energy Institute, West Texas State University, Canyon, TX, January 1984.
6. Forrest (Woody) Stoddard, "Wind Turbine Blade Technology: A Decade of Lessons Learned," World Renewable Energy Congress, Reading, England, September 23–28, 1990.
7. There were 10 Carter 300-kW turbines operating in Great Britain in 1993 out of 6000 turbines in northern Europe. U.S. Windpower sells wind power plants or develops projects using its own turbines and does not sell turbines to other developers.
8. Forrest Stoddard, "California Windfarms: An Update for the 90's," British Wind Energy Association's annual conference, Swansea, Wales, April 10–12, 1991.
9. Stoddard, "Wind Turbine Blade Technology."
10. Meade Gougeon, Gougeon Bros., personal communication, June 8, 1993.
11. Although much of American wind energy R&D may have been squandered on questionable programs, it pales in comparison to spending on other energy technologies, DOE spent $2 billion through 1991 on its "Clean Coal" program and expects to fund another round during the mid-1990s. Blair Swezey, "The Regulatory Outlook for Renewable Electric Generation in the U.S.," National Renewable Energy Laboratory, Golden, CO., paper presented at the 6th Annual Western Conference, Advanced Workshop in Regulation and Public Utility Economics, July 7–9, 1993, p. 5.
12. Karnøe, "Approaches to Innovation in Modern Wind Energy Technology," p. 6.
13. Andrew Garrad, "Wind Energy in Europe: Time for Action!" European Wind Energy Association Strategy Document, August 31, 1990, p. 25.
14. "Wind Energy Multi-year Program Plan," U.S. Department of Energy, Washington, DC, February 5, 1992.
15. Heymann, "Why Were the Danes Best? Social Determinants," p. 17.
16. Jon McGowan, "America Reaps the Wind Harvest," *New Scientist,* August 21, 1993, pp. 30–33.
17. John Kuhns, New World Power Co., oral presentation, American Wind Energy Association's annual conference, Windpower '93, San Francisco, July 1993.
18. Karnøe, "Approaches to Innovation in Modern Wind Energy Technology," p. 26.

19. Heymann, "Why Were the Danes Best? Wind Turbines in Denmark," p. 4.
20. Karnøe, "Approaches to Innovation in Modern Wind Energy Technology," p. 27.
21. Heymann, "Why Were the Danes Best? Wind Turbines in Denmark," p. 7.
22. Gougeon, personal communication.
23. Karnøe, "Approaches to Innovation in Modern Wind Energy Technology," p. 29.
24. Ibid, p. 55.
25. Ibid, p. 75.
26. Palmer Putnam, *Power from the Wind* (New York: Van Nostrand Reinhold, 1948), p. 113.
27. United Technologies Research Center, a division of Hamilton Standard and one-time Rocky Flats contractor.
28. Forrest (Woody) Stoddard, memorandum to American Wind Energy Association R&D committee, February 26, 1992.
29. Stoddard, "Wind Turbine Blade Technology."
30. David Eggleston and Forrest Stoddard, *Wind Turbine Engineering Design* (New York: (Van Nostrand Reinhold, 1987), pp. 2–8.
31. Karnøe, "Entrepreneurial Organization."
32. Karnøe, "Approaches to Innovation in Modern Wind Energy Technology," p. 8.
33. Henry Petroski, *To Engineer Is Human* (New York: Vintage Books, 1992); see Chapter 15, "From Slide Rule to Computer: Forgetting How It Used to Be Done," pp. 189–203.
34. Peter Jamieson, personal communication, February 5, 1993.
35. Karnøe, "Approaches to Innovation in Modern Wind Energy Technology," p. 10.
36. Heymann, "Why Were the Danes Best? Social Determinants," p. 25.
37. Ibid, p. 29.
38. "Clinton Budget Request Includes 70 Percent Increase for Wind," *Wind Energy Weekly,* 13:584, February 14, 1994, pp. 1–2.
39. Lewis Strauss uttered the phrase "Our children will enjoy in their homes electrical energy too cheap to meter" in a speech on atomic energy, September 16, 1954.

4

Death Knell for the Giants

"Le vent a tourné." French expression for "Times have changed"; literally, "the wind has changed direction."

Centrally directed R&D's most spectacular failure was in the ultimately unsuccessful attempt to build the giants of the wind turbine world: the multimegawatt machines. These huge wind turbines, machines with rotors more than 60 m (200 ft) in diameter, have a long history of offering false hope (Figure 4.1). The behemoths have never delivered as much energy or worked as reliably as their proponents promised, nor have they performed nearly as well as the medium-sized wind turbines employed commercially around the world. Worse still, they have devoured enormous sums of public funds, money that could have been used more productively elsewhere, and they have delayed the use of wind energy by seriously damaging the public's faith in the technology. Big-machine R&D served the interests of the R&D community and affiliated aerospace contractors, not wind energy. But the beasts, and the programs that support them, die hard.

In the mid-1970s governments in North America and Europe launched ambitious programs to build a series of multimegawatt turbines. These machines began appearing on the landscape in the late 1970s and early 1980s. By the late 1980s most had already become history, and their program directors artfully designated them "technical successes"; that is, they never operated long enough to destroy themselves.[1] Yet in terms that a wind plant operator or banker understands, they were all failures. They failed in the three areas that matter: hours of operation, energy generated, and a contribution toward making wind technology more competitive with conventional fuels.

Another round of big turbine R&D began in the early 1980s, with the turbines appearing in the late 1980s and early 1990s. These machines performed somewhat better than their forebears, renewing hope that big machines had a future after all—if only there were just one more round of government-sponsored development to bring the technology to fruition.

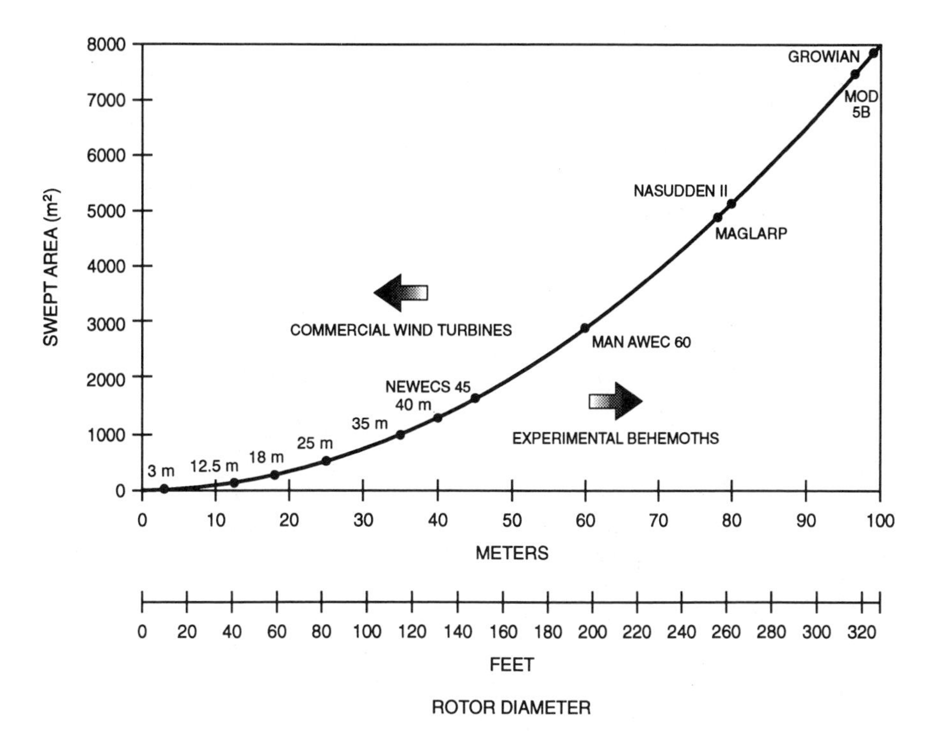

Figure 4.1. ***Rotor diameter and swept area. The area swept by a conventional wind turbine increases exponentially with the rotor diameter. By the mid-1990s, commercially available wind turbines ranged in size from 1 to 45 m in diameter. Larger wind turbines are experimental.***

May Day in Madrid

At the European Community's 1990 wind energy conference in Madrid the EC's director of wind energy research and development held a private seminar on the future of multimegawatt turbines. The by-invitation-only meeting was intended to chart the way for another round of funding for large wind turbine development in Europe. But the packed session did not go the way that Wolfgang Palz expected. He sought consensus on a new program to pursue the "lightweight" design philosophy espoused by Erich Hau and his team of EC contractors. After a lengthy defense of multimegawatt turbines and the desirability of a "next generation of large wind turbines," Hau instead sparked a spirited debate on big wind turbines and government research and development that spilled out into the hallways.

European governments, including the EC in Brussels, had promoted megawatt-sized wind turbines for more than a decade with little success. Brussels' WEGA (Wind Energie Grösse Anlagen) program wound down in the early 1990s with the completion of MAN's AWEC 60 in Spain and Italy's Gamma 60 on Sardinia. In the face of the overwhelming success of medium-sized turbines in northern Europe, Hau, Palz, and some EC contractors were looking for a new round of funding to build large, experimental machines in the mid-1990s.

Preaching an old sermon that wind energy would make its mark in densely populated Europe only when large wind turbines were used, Hau's team analyzed existing multimegawatt technology and "showed clearly the advantages of advanced lightweight concepts" that use two-bladed rotors, teetered hubs, and variable-speed operation.[2] "The challenge," said Hau of the German Institute for Solar Energy (ISET), "is to select the technology now that will lead to a cost-effective design" rather than leave it up to the whims of the private sector. It was this "selection" of a single design approach that concerned most observers. It also worried representatives of small EC nations, such as Denmark, that lack the aerospace industries that typically win such contracts.

Fearing that the private sector might simply try to stretch the then current megawatt designs to multimegawatt size, Hau stressed that redesigning existing big turbines would never succeed. They needed to leap ahead, he said, citing the failure of ELSAM's 2-MW Tjæreborg turbine as an example of how scaling-up Danish medium-sized designs could go awry.

Tjæreborg failed, says Birger Madsen of BTM Consult, not because the fundamental Danish concept was flawed, but because it grew too big, too fast—just the remedy that Hau was prescribing for ailing big turbines. Commercial Danish turbines were already growing larger in steady increments, said Madsen. Every two years Danish industry issued a new iteration. A more sensible way to reach the megawatt range, according to Madsen, was to extend the then-existing 400-kW turbines to 750 kW and proceed from there.

Hau was impatient with the Danish growth rate. "We could wait until they get big enough," said Hau, "but that could be too long." Hau's critics charge that it may indeed take several years for today's commercial turbines to grow to the megawatt size, but it may not take any longer than the six-year program of leapfrogging growth envisioned by Hau. Potential vendors of megawatt turbines also expressed concern that Hau's lengthy program could retard commercial wind development rather than foster it.

Despite being a dauntless defender of megawatt machines, Hau admits that the current crop of megawatt turbines, although working somewhat better than first-generation designs, are still too costly. The second-generation machines such as MAN's 1.2 MW AWEC are so costly, he warned, that "some could argue that the costs will never come down."

The audience remained skeptical. Then Hau played his ace. He pleaded that the poor showing of the EC's megawatt program could lead to less funding for a host of other wind projects as the EC became disenchanted with the technology. Both Hau and the EC's Palz would hold more essential wind energy research hostage unless the industry backed their program. They then urged a "new partnership with industry" for development of the next-generation large turbine. Still, few in Madrid were quick to accept their offer.

The goal, as spelled out by Hau's team, was to design a megawatt turbine that could compete with Vestas' then-current top-of-the-line machine. "If we can't meet this target," said Hau, "then the future of megawatt machines is uncertain." Yet to meet even this modest goal, "a fairly big step must be taken," he said.

Hau's study for the EC found that although existing megawatt turbines have proven technically successful, their specific costs in dollars per square meter of swept area are four times those of existing commercial turbines. The results of the EC's WEGA program showed that "the economics of large WECs [wind energy conversion systems] are markedly inferior to smaller WECs."[3]

The Danes in the audience remained sanguine about this unexpected acknowledgment of their commercial success. Some argued privately later that they hoped the EC proceeded with the new program, thus ensuring that the likely contractors, Europe's large aerospace companies, would be kept out of the commercial market for several more years.

Is Bigger Better?

If the then-current commercial designs were as cost-effective as future "next-generation" megawatt turbines, why wait for the big machines? Theoretically, because big machines, like other power plants, would gain economies of scale, offering further cost savings, and because the big machines would use less land per kilowatt-hour generated than would small wind turbines.

Presumed economies of scale dictated that "bigger is better" be the dominant technical ideology among utilities and government R&D programs in North America and Europe during the 1970s. Since the turn of the century, utilities had seen remarkable improvements in thermal efficiency and the resulting cost of energy as conventional power plants increased in size. Government wind turbine designers hoped to short-circuit the time-consuming process of stepping from one size to the next by circumventing the intermediary stages and vaulting to the biggest machines conceivable. They ignored ample evidence in the 1970s that each succeeding gain in conventional power plant efficiency was more hard won than the last and that the benefits of size had reached a plateau. In a more serious error they also overlooked a fundamental difference between wind turbines and other power plants.

A thermal power plant resembles a giant tea kettle, and it takes proportionately less material to cover a larger and larger kettle than to cover a smaller one. Thus big power plants use less material for a given output than do a series of smaller plants. Not so with technologies like wind, where the forces increase at a rate nearly equal to that of increased energy production. Loads on the rotor increase faster than the energy intercepted as rotor diameter increases. What worked well for steam-driven turbines was inapplicable to wind machines.

There was another misperception. Like that of earlier proponents in Europe and the United States, Hau's justification for big machines was founded primarily on the use of taller towers. Wind speed generally, though not always, increases with height. Big turbines require taller towers than do small turbines to keep their rotors clear of the ground. Thus bigger wind turbines will intercept higher-speed winds for a given site.

Don Smith challenged this assumption in a landmark paper on large wind turbines in the early 1980s.[4] Experience with Pacific Gas & Electric's Boeing Mod-2 confirmed his thesis that basing increased generation on increases in tower height may be misleading. PG&E, for example, found that the Mod-2 was located in a zone of negative wind shear because of unusual meteorological conditions in Solano County. However, Smith noted that big turbines would gain some aerodynamic efficiency relative to small wind turbines. For example, the aerodynamic efficiency of a turbine 10 m (30 ft) in diameter is about 4% less than that of the Mod-2 rotor, said Smith.[5] Yet rotor efficiency, though important, is not the sole determinant of cost-effectiveness.

By 1990 the specific costs (dollars per square meter of rotor area) for large machines were still four times that of medium-sized turbines. The higher energy capture attributed to big turbines because of their taller towers and greater aerodynamic efficiency was still insufficient to compensate. But hope springs eternal and Hau believed there were opportunities for substantially reducing the cost of big wind turbines by cutting their weight in half.[6]

In Madrid, Hau argued further that the specific yield (kilowatt-hours of generation per square meter of swept area) will increase at least 50% by doubling the size of the turbine. That is, today's best-performing turbines would increase their yields from 1000 kWh/m^2 to 1500 kWh/m^2 if they doubled their rotor diameter. In a startling conclusion, Hau claimed that doubling tower height on large-diameter turbines would increase specific yield 2.7 times! But to deter any would-be developer from simply doubling the tower height for a medium-sized turbine, Hau warned that although this would indeed improve energy yield, it would not be enough to offset the increase in tower costs.[7]

There is scant evidence to support Hau's claims. Energy yield from multimegawatt turbines consistently falls below that of commercial turbines, primarily because of poor reliability. Even the best performing big turbines are seldom available for operation more than 70% of the time over a multiyear period, whereas medium-sized machines regularly reach 97 to 99% availability.[8]

The eventual success of multimegawatt turbines through improvements in reliability and decreases in maintenance costs was an article of faith among Hau's team, and they went to some lengths to prove their case. Poul Nielsen of DEFU (*Danske Elværkers Forenings Udredningsafdeling;* the Research Association of Danish Utilities) noted that operation and maintenance of big wind turbines was far more costly than the figure of 4 to 8 ore ($0.007 to 0.013) per kilowatt-hour for medium-sized machines. Surprisingly, Nielsen added: "But among utility people it seems to be agreed that the target O&M costs for large machines should not exceed 6 ore ($0.01) per kilowatt-hour," although this has never been achieved.[9]

It was no different in the United States during the 1970s. Wind program managers of the day "understood how to play with numbers," says wind consultant Bill Vachon, suggesting that economic justifications for some designs were produced after the size and configuration had already been determined.

Like reliability, the cost of O&M becomes a chicken and egg question. If big machines are too expensive because they break too often, no one will build them. But if no one builds them, the manufacturing and O&M costs will never come down. Thus big machine proponents can always point to the incompleteness of the research record. T. van Holten told the audience at the EC's seminar in Madrid that despite all the previous data, "there's absolutely no practical cost experience with these machines because they were all experimental."

One ot the arguments for large machines is their better land utilization than that of smaller turbines. Yet in a study for Danish utility ELKRAFT, Nielsen found that the specific costs (Ecu/m^2) of installation were greater for multimegawatt turbines than for the medium-sized machines then available. Acquistion of the site and installation of the turbines at a theoretical wind plant comprised of multimegawatt turbines would cost 16% more than would an equivalent plant of hypothetical 750-kW machines, and three times that of actual wind plants using 200-kW turbines the utility had already built.[10]

Regardless of Hau's assertion that multimegawatt machines were the wave of the future, few in Madrid were convinced. They had heard the same tiring arguments for more than a decade. Even Palz, frustrated by the course of the debate, admitted at one point that the optimum turbine may not be greater than 1 MW after all, but instead, may be found in the humble range of 500 kW to 1 MW.

Better land use has always been the last great hope of big machine proponents. To Hau and his team "the best use of available land" will prove decisive in the development of wind energy in northern Europe because of environmental constraints on development. With almost religious certitude, Hau concluded that "it is a matter of fact that larger turbines offer considerably more power and energy for a given area."[11]

Contrary to the implication of early government-sponsored reports, the amount of land required for geometric arrays, paradoxically, is independent of wind turbine size. As turbines become larger, they occupy more land per unit because of the greater spacing required between them. However, the relative spacing between turbines, governed by their rotor diameter, remains the same. Don Smith, who at one time worked on Pacific Gas & Electric's big turbine program, demonstrated that increasing rotor diameter does not inherently improve land use in geometric arrays of turbines.

Only in linear arrays will increased size increase land utilization rates. As a result, installing multimegawatt turbines in linear arrays has become the new battle cry for big turbine proponents. The developable European wind resource could be doubled, said Hau, if turbines 60 to 80 m (200 to 260 ft) in diameter instead of the 30- to 40-m (100- to 130-ft) machines common during the mid-1990s are used in linear arrays.[12] Hau said, "We expect that an 80-meter turbine could become competitive."[13]

Although Hau and others still argue that big machines make sense, Peter Jamieson, a former wind turbine designer at James Howden, remains skeptical. Jamieson warns that often big machine proponents extrapolate from the improved performance of today's medium-sized turbines. It is dangerous, he says, simply to extrapolate from experience with small and medium-sized machines, because there is a bias at the lower end of the size scale. The cost of peripherals, what engineers call "the balance of system," remains fairly fixed and is independent of size. Scaling-up spreads more kilowatt-hours over these constant fixed costs, thus making big machines look more attractive.

Wind energy increases with the square of the rotor diameter, while weight and cost typically increase with the cube, says English engineer David Milborrow. But in smaller machines, Jamieson adds, the cost and weight of the rotor can be less than a cubic function of diameter. This misleads designers of bigger machines, who try to extrapolate from the better weight-to-energy performance of small turbines to that of much larger rotors, where gravity becomes a more influential force.[14]

Jamieson suggests a simple comparison between one wind turbine 100 m (328 ft) in diameter, and 100 wind turbines 10 m (33 ft) in diameter. They both intercept the same amount of wind, and if they are installed on the same height towers should capture approximately the same amount of energy. However, the 100-m turbine, a machine the size of MAN's Growian, will require 10 times the mass of all the smaller machines combined.[15]

Wind turbines are like heat exchangers, Jamieson contends. They require lots of surface area to perform their task productively. Thus, he says, it is better to have lots of small machines than to have one big one, just as it is better to have lots of small fins than to have one big one on a radiator. There is "less mass in a multiplicity of structures" intercepting the same area. There is also an economic advantage from the improved reliability because of the redundancy of multiple machines, regardless of their greater infrastructure cost for more wiring and transformers.[16]

European planners and utility engineers in the mid-1990s commonly believe that machines in the range 300 to 600 kW will produce energy at the lowest cost. But concern about the number of sites available is driving development toward 1-MW machines.[17] Work by Jens-Peter Molly, director of the German Wind Energy Institute, suggests that the size of wind turbine most likely to

35. Karnøe, "Approaches to Innovation in Modern Wind Energy Technology," p. 65.
36. Heymann, "Why Were the Danes Best?" p. 14.
37. Poul Nielsen, "Development of Wind Energy in Denmark," paper presented at the American Wind Energy Associations's annual conference, Windpower '93, San Francisco, July 1993.
38. Karnøe, "Approaches to Innovation in Modern Wind Energy Technology," p. 39.
39. "Double Study," *Windpower Monthly,* 10:3, March 1994, pp. 11–12.

5

Historical Background

"Ech kier die Nuet/On schaff och Bruet (Ich wende die Not/und schaffe euch Brot). Inscription on the beard of a windmill in northern Germany: "I turn away need and furnish your bread."

Inventive minds have long sought to harness the wind. Weary Egyptians may have been the first, when they sailed up the Nile against the current. Crude vertical-axis panemones have ground grain in the Afghan highlands since the seventh century.[1] (Figure 5.1). By the seventeenth century, windmills were such a commonplace technology that Cervantes's fictional Don Quixote was tilting at them near Consuegra on the plains of La Manchà. According to French historians, as many as 500,000 windmills were being used in China by the nineteenth century and possibly an equal number were scattered across Europe.[2]

A European Tradition

Traditionally, the history of Western wind technology begins with the first documented appearance of the European or "Dutch" windmill in Normandy in the year 1180.[3] Presumably, the vertical-axis windmills of Persia spread from the Middle East across the Mediterranean to Europe, and in so doing evolved into horizontal-axis windmills. From France, the technology spread across the channel to southern England (1191), into nearby Flanders (1190), then on into Germany (1222), and subsequently north to Denmark (1259). Finally, "Dutch" windmills reached Poland in the fourteenth century.[4] For a more detailed account, see Dennis Shepherd's chapter "Historical Development of the Windmill" in *Wind Turbine Technology.*[5]

In *Harvesting the Air: Windmill Pioneers in Twelfth-Century England,* historian Edward Kealey threw the preceding scenario on its head with his

Figure 5.1. Persian panemone. One of the earliest known wind machines used for mechanical power, the panemone is a simple drag device. (Courtesy of Sandia National Laboratories.)

controversial thesis that the technology is indigenous to Europe, originating in southern England. According to Kealey, the English post mill predates the mills in Normandy, possibly by as much as a century, and originated in England, not in the Orient. He cites records documenting the operation of post mills during the mid-twelfth century, long before the first mention of windmills on the continent. Probably most heretically of all, Kealey argues that the European windmill arose independently of the vertical-axis mills used in Persia, suggesting that the European post mill is technologically unique and not a derivative of those in the East (Figure 5.2).

Kealey ventures farther out on the academic limb to propose that the direction of technology transfer was the reverse of that commonly believed, and that crusaders took windmills to the Holy Land in 1191. Historians commonly cite a reference to German knights building the first windmill in Syria during the Third Crusade.[6] Kealey suggests that the Germans could originally have been Saxons, who as Norsemen had previously "settled" in England. Although this academic debate is of interest solely to historians and students of technological dissemination, a lesser known theme of Kealey's is pertinent to modern wind turbines and their technology.

Figure 5.2. English post mill. Believed to be the forerunner of the better known tower mills common in northern Europe. By the nineteenth century, blades on European windmills incorporated twist and leading-edge camber, characteristics of modern airfoils. (Courtesy of Sandia National Laboratories.)

Windmill Liberation

For many of the alternative energy and environmental activists of the 1970s, wind turbines were more than mere machines; they were vehicles of social change. Wind turbines offered these idealists a means for building a more sustainable society, a mechanism for living within natural bounds instead of outside them. Few then knew that there was a strong historical case for such a vision.

Kealey, as well as French historians, have found that windmills have at various times fueled great social upheavals. According to Kealey's research, many early English windmill pioneers were progressive citizens with reputations for what we would today call social activism. The development of the English post mill occurred during a period of social and agricultural modernization and was a product, as well as a cause, of these changes.[7]

Just as the sail liberated slaves from Mediterranean galleys, the proliferation of post mills across the English countryside in the twelfth century put power into the hands of those who were previously powerless and liberated women from grinding grain by hand. "The windmill was an instrument of social progress," says Kealey. "It enlarged the community of skilled mechanics and lightened the daily work load of countless women." Kealey lovingly describes the English post mill as "appealing, productive, and even mysterious," but the "windmill was above all a triumph of ingenuity over toil."[8]

The challenge to the growth of wind energy in the twelfth century was, as it is today, not due to the technology's limitations but to resistance from those with the most to lose. The land, the forests, the water: all were part of the feudal estate. The wind, however, was not. Church and feudal lords feared losing their lucrative milling rights to commoners who harnessed the free power of the wind.[9]

According to these historians, the windmill provided a technology for the "liberation theology" of the era, and devotees went forth spreading the word. Early wind advocates, including an obscure cleric, "broadcast an invention," says Kealey, "that challenged the foundations of medieval society."[10] Their proselytizing spread windmills across central England.

The technology took root at a time of rebellion against the tyranny of feudal monopolies. Often, windmills were built in direct opposition to the feudal lord who controlled the nearby water mills. Water power was never free of conflicting claims, even at the time of the Magna Carta, which gave the rights of passage and other uses of stream courses to the nobility but limited the nobility's right to build structures in waterways.

To an entrepreneur, wind was advantageous on several counts. There were more sites available for windmills than there were for water mills, and users were not tied to the river courses, where most prime sites had already been developed. The lower cost of windmills to water mills encouraged their proliferation at a time when a growing urban population needed a new energy source. Wind filled the void. The advent of the windmill and the growth of cities simultaneously breached the hold of feudal lords.[11]

Feudalism eventually adapted to the threat. During the period 1162–1180, for example, the archbishop of southern France regulated windmills by demanding 5% of the grain ground there.[12] In England, lords sometimes destroyed windmills that were a commercial threat to their water mills, or seized a windmill on a pretext. Yet despite the setbacks to their owners, the modest English post mill flowered, jumped the Channel, and grew into the towering windmills of the lowlands.

According to Jaap de Blecourt, the millwright who reassembled the De Zwaan windmill in Holland, Michigan, wind's heyday in the Netherlands contributed to the country's golden age. Windmills "fit" the rural Dutch landscape because they were *the* available power source during the seventeenth century. Blecourt's comment reflects a common perception among the Dutch. "Windmills belong to the Dutch landscape, to such an extent, that we cannot imagine this landscape without them," said Frederick Stokhuyzen in his book *The Dutch Windmill.*[13]

Only by tapping the wind could Jan Leeghwater (literally, "empty water") and the engineers who followed him drain the polders and make the Netherlands what it is today. As late as 1850, 90% of the power used in Dutch industry came from the wind. Steam supplied the rest. The 700 windmills in the Zaan district north of Amsterdam formed the core of what would become the center of Dutch manufacturing.[14]

Only in the late nineteenth century did the use of wind wane. Yet in 1904, wind still provided 11% of Dutch industrial energy. The switch from wind to steam was based on more than cost; reasons included changes in social conditions, agricultural practices, and the mood of the rural populace.[15] This could be a harbinger of why wind's star may be in the ascendancy now that it has become economical once again. Wind offers other attributes now considered important, such as its ability to generate electricity renewably without combustion or the creation of radioactive waste.

In 1896, at the height of the industrial revolution, wind still pumped 41% of the polders in the Netherlands. Only after cheap coal became available from the nearby Ruhr did steam pumping erode wind's dominance. Even then, steam was not clearly superior. Steam required larger polders to perform optimally, and individual polder mills were cheaper than equivalent steam pumps to operate through the turn of the century.

Just as is the case today, the capital costs of wind were higher than those of coal. Although the wind was free, capital was not, and much as it is today, the coal was cheap once publicly constructed canals linked the Netherlands with the Rhine and the Ruhr. Steam also required less than one-third the labor used by wind to drain a large polder such as the Haarlemmermeer, where Amsterdam's Schipohl airport now sits. Steam was also available upon demand; wind was not. A long lull could delay spring planting until the windmill pumped the polder dry. The steam engine put control over drainage into the hands of the community. For the first time, farmers could manage the water level to maximize crop yields, something not possible with the polder mill. (Fortunately, today's large interconnected electrical networks have allayed concern about wind's intermittency.) In addition, steam engines could be placed wherever they were needed, whereas polder mills needed well-exposed sites.[16] For these reasons, the European windmill began a long decline which was only arrested in the 1970s by preservation societies and industrial archeologists (Table 5.1).

Windmill performance increased greatly between the twelfth and nineteenth centuries with the introduction of metal parts. In the seventeenth century, these parts were some of the first examples of the standardization that eventually led to mass production. By the time the "Dutch" or European windmill began to fall out of favor at the turn of the century, the typical machine used a rotor spanning 25 m (80 ft). The stocks on some reached 30 m (100 ft) in length, the length of the tallest tree that could be shipped. Curiously, one of

Table 5.1
Bloom and Decay of European or "Dutch" Windmills

	The Netherlands		*Germany*[a]		*Britain*		*Denmark*		*France*	
Year	Units	MW	Units	MW	Units	MW	Units	MW	Units	MW
1700									16,000	400
1750										
1800									20,000	500
1850	9,000	225								
1900	4,000	100	18,000	450	8,000	200	3,000	75		
1950	1,500	38	5,000	125						
1990	1,000	25	1,000	25	100	3	100	3		

[a] Northern provinces.

the largest windmills ever built was erected in San Francisco's Golden Gate Park early in the twentieth century. With a diameter of 114 ft (35 m), this giant cold pump 40,000 gallons (150 m^3) per hour.[17] Most mills were capable of producing the equivalent of 25 to 30 kW in a mechanical form suitable for grinding grain, shredding tobacco, sawing timber, milling flax, pressing oil, or pumping water for polder drainage.

Later innovations included automatic fan tails for pointing the rotor into the wind, automatic movable louvers instead of sails, airbrakes, and airfoil-shaped leading edges that pointed the way toward modern wind turbines. At their height, there were some 1500 MW of European windmills in documented use, a level not seen again until 1988. It was only in the late 1980s that wind turbines of equivalent size were once again plentiful. Modern airfoils and materials enable today's machines to extract 10 times more power from the wind than had the European windmill 100 years earlier.

The Windmill That Won the West

Three technological innovations made settlement on the Great Plains possible, wrote historian Walter Prescott Webb: the Colt 45, barbed wire, and the farm windmill. The water-pumping windmill was so essential for life in what was then known as the Great American Desert that settlers warned newcomers that "no women should live in this country who can't climb a windmill or shoot a gun." Promoters extolled the virtues of a land where "the wind pumps the water and the cows chop the wood."[18]

In the semiarid lands west of the Missouri River the wind did indeed pump the water. Unlike the eastern United States, few streams coursed across the surface of the prairie, and seldom were aquifers within reach of simple hand-dug wells. Water was there, but at depths that required pumping by machines—wind machines. (Homesteaders, who seldom could find wood for their hearth on the treeless landscape, instead burned cow chips from their bovine lumberjacks.)

T. Lindsay Baker traces the fascinating development of the farm windmill in his exhaustive *Field Guide to American Windmills.* Like Webb, science writer Volta Torrey places the farm windmill in its social context in his entertaining account, *Wind-Catchers.* All three note that in 1854, Daniel Halladay invented the first fully self-regulating windmill. Until then, turning the spinning rotor out of the wind or reefing (rolling up) the sails during storms had to be done manually. Halladay changed all that by constructing a multiblade rotor (similar to that of today's farm windmill) made up of several movable segments. Rather than attaching these segments to the hub directly, he pivoted them about a ring. In high winds, the segments would swing open into a hollow cylinder, allowing the tempest to pass unimpeded. Early models of Halladay's "rosette" or "umbrella" mill, as they were called, used a tail vane to point the rotor, or "wheel," windward. Later "vaneless" versions did away with the tail vane by orienting the wheel downwind of the tower.

Halladay's patent mills were immediately popular with farmers and ranchers for watering livestock. Because his mills could be left unattended, they were ideal for remote pastures where water was scarce. But the fledgling industry began to grow only after the boom in railroad construction that followed the Civil War.

Water was as essential as coal to running a steam locomotive. As the transcontinental railroads pushed westward across the plains, the water-pumping windmill came into its own. Huge windmills (even by today's standards), with rotors up to 60 ft (18 m) in diameter, pumped a steady stream into storage tanks at desolate way stations. Through skillful marketing, the Eclipse emerged as the "railroad" mill.

Invented in 1867 by Leonard Wheeler, the Eclipse used fewer moving parts and was both cheaper to produce and easier to maintain than Halladay's windmill. Wheeler built a solid wheel instead of the sectional wheels that Halladay popularized. Wheeler protected his windmill in high winds by furling the rotor toward the tail. The Eclipse did this automatically by using a pilot vane. This small vane extended just beyond but parallel to the rotor disk. During storms, wind striking the pilot vane would push the vane and rotor toward the tail, effectively taking the wheel out of the wind. The idea was so successful that even Halladay's U.S. Wind Engine and Pump Co. began producing similar versions under the "Standard" trade name.

The stage was set. The technology existed and an industry was in place. The nation's westward migration both caused and was aided by the growth of a great midwestern industry building windmills. By the late nineteenth century, 77 firms were assembling them in one form or another.[19] Farm catalogs of the day bristled with choices. During the height of the farm windmill's glory in 1909, manufacturers employed 2300 to service the mass market for wind pumping on the Great Plains.[20]

Demand grew rapidly as the railroads poured settlers onto the plains. The industry blossomed like the homesteader's gardens that cheap windmills soon made possible. The farm windmill was on its way to becoming an American institution. It quenched thirsts and provided on occasional bath, and baptisms in its tank even offered salvation. The creaking windmill sang a prairie lullaby to many a homesteader. The gentry found it useful as well. The windmill brought running water and the convenience of Sir Thomas Crapper's flushable toilet to their country villas and suburban estates.

This was an age of invention and empire building. Steel and the factory system were driving the industrial revolution to new heights in the United States. Into this milieu entered LaVerne Noyes, Chicago industrialist. In 1883 Noyes hired an engineer, Thomas Perry, to develop a thresher. While with his previous employer, the U.S. Wind Engine and Pump Co., Perry had conducted over 5000 tests on different windmill designs. His was the first scientific attempt to improve wind machines. By using a steam-driven test stand, Perry designed a rotor nearly twice as efficient as those then in use. When the U.S. Wind Engine and Pump Co. showed no interest, Noyes stepped in and encouraged Perry's work. Five years later, when Perry's trade secrets were released, Noyes introduced the Aermotor. Derisively tagged the "mathematical" windmill by competitors, the Aermotor incorporated both Perry's design and Noyes' manufacturing sense.

Although it was not the first to use metal blades (Mast, Foos and Co.'s Iron Turbine used them in 1872), Aermotor's stamped sheet-metal "sails" revolutionized the farm windmill. Another innovation was Aermotor's method of furling the rotor in high winds. Rather than use a pilot vane as on the Eclipse, Perry offset the Aermotor's wheel from the center of the tower. High winds striking the rotor disk force it to fold toward the tail, eliminating the need for the pilot vane. The furling force was counterbalanced by a spring that held the rotor into the wind. This arrangement worked so reliably that it is still widely used.

Noyes' ability to mass-produce windmills at Aermotor's Chicago plant reduced their cost to one-sixth that of their competition and led Aermotor to dominate the industry.[21] Perry's design was so successful that it was widely copied. More than 1 million examples are still in use worldwide. Eight times out of ten, the abandoned windmill that motorists see while driving down backroads in America's heartland is likely to be an Aermotor (Figure 5.3).

The industry peaked in the early part of the twentieth century, but collapsed quickly after the introduction of electricity by the Rural Electrification Administration during the 1930s (Table 5.2). Durability played no small part in the industry's demise. On many homesteads, the family windmill has been in continuous use for generations. Heller-Aller Co., makers of Baker windmills, reported in 1989 that one of its machines had run for 61 years.

After a century of refinement, the farm windmill continues to perform its job well. It remains a signature of the Pennsylvania Dutch and Amish settlements in Ohio, Indiana, and Iowa. They are also to be seen in Mexico at

Figure 5.3. Multiblade farm windmill. The multiple curved sheet-metal blades take advantage of the air rushing between the blades to create a "slot" effect such as that found on jib-rigged sailboats.

Merida, the tourist mecca on the Yucatan peninsula. The Amish, who use them for domestic water, and ranchers of the southwest, who use them for stock watering, account for the few thousand sold in the United States each year.

"They still make sense in a lot of places," says Vaughn Nelson of West Texas State University's Alternative Energy Institute. To replace the 60,000 operating farm windmills on the Southern High Plains alone, he says, would cost nearly $1 billion (Table 5.3).

Table 5.2
U.S. Installed Wind Capacity Through 1970

Year	Windmills hp (thousands)	Windmills MW	Estimated Number of Units[a] (thousands)	Sailing Vessels hp (thousands)	Sailing Vessels MW	Total MW
1850	14	11	42	400	300	311
1860	20	15	60	597	448	463
1870	30	23	90	314	236	258
1880	40	30	120	314	236	266
1890	80	60	240	280	210	270
1900	120	90	360	251	188	278
1910	180	135	540	220	165	300
1920	200	150	600	169	127	277
1930	200	150	600	100	75	225
1940	130	98	390	26	20	117
1950	59	44	177	11	8	53
1960	44	33	132	2	2	35
1965	30	23	90	2	2	24
1970	24	18	72	1	1	19

[a] At 0.25 kW/unit.

Source: Historical Statistics of the United States: Colonial Times to 1970, Part 2 (Washington, D.C.: U.S. Bureau of the Census, 1976), p. 818.

Not unlike North American's Great Plains, Argentina's sweeping pampas are home to the world's largest remaining concentration of wind pumps. While most are used by gauchos for stock watering, a large number are used to water garlic farms in Medanos province. Many Argentinean windmills were built domestically, using the classic multiblade form which many outside the United States call the "Chicago" style, in reference to the multitude of American windmill manufacturers once located in an arc around southern Lake Michigan. The farm windmill has also long been essential to life in Australia's outback. There may be as many as a quarter million wind pumps still operating in South Africa.[22]

For much of the world, water remains a commodity difficult to obtain. Finding it, and finding the energy to get it to the surface, are part of the daily struggle for life. Exports kept the U.S. windmill business alive after rural electrification. The few U.S. manufacturers remaining agree that an important market is abroad. Spreading drought, and never-ending population pressure could create a Renaissance for water-pumping windmills, an arcane and often overlooked technology. Long-time wind pump designer Peter Fraenkel estimates that the worldwide market could reach 15,000 machines per year.[23] Now U.S. manufacturers compete head to head with other producers of multiblade windmills, such as Argentina's FIASA and Australia's Southern Cross, and with newly designed mechanical wind pumps and dual-duty wind generators.

Table 5.3
Water-Pumping Windmills (Wind Pumps) Worldwide

	Units	MW
Argentina	600,000	150
Australia	250,000	63
Brazil	2,000	1
China	1,700	0
Columbia	8,000	2
Curaçao	3,000	1
Nicaragua	1,000	0
South Africa	100,000	25
U.S. Southern Plains	60,000	15
	1,025,700	256

Sources: Wind Energy for Rural Areas, ECN, the Netherlands, 1991; *Windpower Monthly,* November, 1993; Vaughn Nelson, West Texas State University, personal communication.

Note: Most are 2.5-m (8-ft) traditional multiblade farm windmills.

Early farm windmills coupled the rotor directly to a piston pump in the well by means of a crank and pump rod. The windmill would lift the piston pump with every revolution of the rotor. In light winds, the weight of the water in the well would often stall the rotor, bringing it to a halt. One of the great innovations introduced by Perry and others was "back-gearing." Today, most, if not all, American farm windmills are back-geared and use a transmission to increase the rotor's mechanical advantage in light winds. This increases the farm windmill's complexity but enables it to pump water more reliably in light winds. Most back-geared windmills lift the piston once every three revolutions. Back-gearing works for rotors up to 16 ft (5 m) in diameter. Larger windmills are again crank-driven or "direct acting" up to about 30 ft (9 m) in diameter, where they become geared up to compensate for the slow rotor speeds and long stroke.[24]

U.S. farm windmills are available in sizes ranging from 6 to 16 ft (about 2 to 5 m) in diameter. Australia's Southern Cross can be ordered in sizes up 25 ft (8 m) in diameter. The most common size in the United States is the 8-ft mill, which is capable of pumping less than 10 gallons per minute (2300 liters per second) from a depth of 100 ft (30 m).

Experimenters have developed two devices for improving the operation of the traditional farm windmill. The simplest is a counterbalance to the weight of piston and rod. Farm windmills tend to speed up when the pump rod begins its downward journey. On the upstroke, the rotor slows down as it lifts both the weight of the rod and the weight of the water in the well. To steady the speed of the rotor and maintain a more optimum relationship between the rotor and wind speed, some designers have added weights or springs to counterbalance the weight of the pump rod.

The other approach tackles a more fundamental problem with multiblade wind pumps. For a given pump size and depth, if the stroke of the windmill is adjusted for optimum production in high winds, the windmill will perform poorly, if at all, in low winds. Ideally, the stroke should vary with wind speed to more closely match pumping capacity with the wind available. In tests, a variable-stroke design by Don Avery of Hawaii has doubled the output from the traditional farm windmill.

Although these innovations sound appealing, neither mechanism is widely used. Manufacturers continue building farm windmills in the same way that they have for the past 100 years. Dutch researchers, on the other hand, have successfully designed modern versions of the farm windmill. The modern wind pumps developed by CWD in the Netherlands use only 6 to 8 blades of true airfoils, in contrast to the 15 to 18 curved steel plates found on the American farm windmill. These modern mechanical wind pumps are simpler and less costly to build than the American farm windmill. Both aspects are important for Third World countries such as India. Some 1500 of these modern water pumpers have been built by semiskilled labor in Africa, and another 2000 in India. Although nearly twice as efficient as the U.S. farm windmill, modern wind pumps are less rugged and may be best suited for regions with light winds.

More promising than advancements in mechanical wind pumps has been the development of new wind-electric pumping systems. For the same rotor diameter, today's wind-electric pumping systems can deliver about twice as much water as the traditional farm windmill, depending on how well each pump is matched to its load.

The windmill that "won the West" is enshrined in the American psyche. At the National Museum of American History in Washington, DC, the first image that visitors see beyond the popular pendulum exhibit is a collage containing a nineteenth-century farm windmill, a portrayal physically linking American history and wind energy. In the American context, the farm windmill also links farm windmill technology with that eventually used to generate electricity and ultimately with that of modern wind turbines.

While Poul la Cour was experimenting with European windmills for generating electricity in Denmark, the American Charles Brush was adapting the technology he knew best, the multiblade farm windmill, to perform the same task. His 56-ft (17-m)-diameter wind turbine generated 12 kW for charging batteries at Brush's Cleveland, Ohio estate. It produced enough electricity for his domestic needs and rivaled that generated by fossil fuel for banking magnate J. P. Morgan's New York mansion.[25] For comparison, modern wind turbines of the same size are capable of generating 75 to 100 kW, six to eight times more than that of the multiblade rotor of Brush's era (Table 5.4).

Work on the farm windmill both predates and is contemporaneous with advances in aerodynamics. Flint and Walling introduced their Star Zephyr model in 1937 only after extensive tests in a wind tunnel, says historian T. Lindsay Baker. As a result of their tests, Flint and Walling changed the

Table 5.4
Technological Development

	Diameter (m)	Capacity (kW)
Dutch windmill	25–30	25–30
Modern wind turbine	25–30	250–300
Brush multiblade	17	12
Modern wind turbine	17	75–100
Aermotor	5	0.5
Modern wind turbine	5	5–6

shape of the familiar farm windmill rotor. The sheet-metal blades began to look more like those of modern wind turbines. The Star Zephyr's blades subtly tapered from the hub to the tip, in contrast to other farm windmill blades, in which the blades tapered from tip to hub. They also rounded the tips, abandoning the square corners that are still common on farm windmills today. The blades were not only curved, as on other farm windmills developed since Perry's Aermotor, but were slightly twisted from hub to tip, as on modern wind turbines. The tail vane also took on a "modern" appearance. In sharp contrast to the Aermotor's horizontal tail vane, the Star Zephyr's tail vane became more vertical, with a swept-back, streamlined shape popular during the golden age of aeronautics. Flint and Walling continued building this successful design into the 1950s, pointing the way toward more modern machines.[26]

Farm windmills and their place in rural life continue to intrigue Americans. Much later than the Dutch, who first called for preservation of their windmill heritage in 1923, Americans are slowly beginning to turn their attention toward preservation. In the mid-1990s, four museums were under development to display rare farm windmill collections: the National Windmill Project in Lubbock, Texas; Kregel Windmill Museum in Nebraska City, Nebraska; Canadian National Windpower Centre in Etzikom, Alberta; and Kendallville Windmill Museum and Historical Society, Kendallville, Indiana.

In explaining his fascination for farm windmills, Bryce Black, a self-described "hippy homesteader," cites the fact that they are "aeliotropic"; in other words, they follow the wind like the sunflower follows the sun. Black, who has carved out a niche for himself as a farm windmill repairman in western Wisconsin, where some of the machines are still in use, says "there's something timeless" about farm windmills. To Black, the farm windmill may have been forged from fire but provides an almost living link between sky, earth, and water.

Small Wind Turbines

Next to their ability to pump water mechanically, small wind turbines are best known for their ability to generate power at remote homesteads. They have distinguished themselves in this role for decades. During the 1930s, when only 10% of U.S. farms were served by central-station power, literally hundreds of thousands of small wind turbines were in use on the Great Plains. These "home light plants" provided the only source of electricity to homesteaders in the days before the Rural Electrification Administration brought "high-line" electricity to all. Until then, wind generators were glorified farm windmills adapted to driving direct-current generators. But the advent of the aircraft propeller altered how people thought of wind and mechanical energy. By 1931, says Mick Sagrillo, patents were being issued for wind generator airfoils, which would forever change the appearance of windmills.[27]

During the 1930s, the market for small turbines flourished as crude "crystal" radio sets were rapidly replaced with more powerful—and power-consuming—radios using vacuum tubes. Batteries initially met the need, but batteries needed frequent charging. The solution was the "windcharger." Through skillful promotion by Zenith Radio and wind turbine manufacturer Wincharger, small radio chargers began sprouting from rooftops across the Midwest.

The radio soon whetted the homesteader's appetite for electrical appliances and power tools. In response, windchargers grew steadily in size. By the end of the era, Wincharger was building turbines 14 ft (4 m) in diameter (Figure 5.4). Another grand old name of pre-REA windchargers, Jacobs Windelectric, at one time employed 260 people building what Marcellus Jacobs, in a bit of marketing genius, called the "Cadillac of windchargers"[28] (Figure 5.5).

In response to the resurgence of interest in small wind turbines following the oil embargoes of the 1970s, many starry-eyed idealists roamed the plains buying used windchargers. Some of the budding windsmiths who once reconditioned junk windmills were tending wind power plants in the mid-1990s. Several new manufacturers were also born, some of which have continued to build modern windchargers into the mid-1990s. Despite the collapse of the market for small wind turbines in the 1980s, at least 4500 of them were operating interconnected with local utilities, and another 1000 turbines were charging batteries in the United States in the mid-1990s (Table 5.5).

Mainland China imported some early American designs from the 1970s but soon began building its own versions. By the early 1990s, the Chinese Wind Development Center was reporting that 110,000 turbines had been installed and that Chinese factories were building 10,000 microturbines per year. These turbines are so small (producing only 50 to 200 W) that they can be carried on horseback. Most are used in Inner Mongolia, where they supply about 20% of the nomadic herdsman with portable power. Another 20,000 are used in Gansu, Qinghai, Xinjiang, and Tibert provinces.[29] Across the border, the Re-

Figure 5.4. Wincharger. The so-called "Chevrolet" of 1930s' era battery-charging wind turbines. This later model Wincharger used weights to vary the pitch on two of the four extruded aluminum blades.

public of Mongolia was beginning domestic production of the successful English microturbine, Marlec Engineering's Rutland, for its herdsmen. Marlec has built 20,000 of its Rutland 50-W model, most for export.[30]

Ed Wulf was building his dream home in southern California's Tehachapi Mountains when he learned that the local utility would charge him $50,000 to bring in power. "I can do better than that," he said, and he did. Wulf only had to look out his picture window for the solution. His window opens onto one of the world's largest wind power plants. The 5000 wind turbines across from Ed Wulf churn out enough electricity to serve the residential needs of 500,000 energy-hungry Californians. Surely they could spare some for him. For regulatory and logistical reasons, the area's wind power plants were unable to help Wulf. He had to go it alone. But the very existence of those plants

Figure 5.5. 1980s Jacobs. Not to be confused with the 1930s version by the same name. This model uses a blade-activated governor to regulate power, and self-furling to limit thrust in high winds. More than 600 of this version were installed in California and more than 300 on the big island of Hawaii, including nearly 200 at Kahua Ranch, some of which are shown here. This model Jacobs powers an alternator mounted vertically in the tower by a right-angle drive.

proved that wind energy would work for him. Wulf set out to install his own stand-alone power system, a hybrid that uses the area's abundant wind and solar energy. Now Wulf's single Bergey wind turbine attracts nearly as much attention from tourists as the big machines on the nearby hillsides.

The growing demand for hybrid power systems such as the one that Wulf built have led to a booming off-the-grid market in the United States. In 1993, modern homesteaders spent nearly $1 million for wind energy equipment, according to a survey by *Home Power* magazine.[31]

Table 5.5
Small Wind Turbines

	Units	MW
United States		
Interconnected	4,500	10
Battery-charging	1,000	1
	5,500	11
China		
Inner Mongolia	90,000	9
Other provinces	20,000	2
	110,000	11
Great Britain and export	20,000	1
Ex-Soviet Union	2,500	1
	138,000	24

The growing popularity of small wind turbines is due largely to their greatly improved reliability. In contrast to the designs of a decade ago, which used complicated drive trains and mechanical governors more suitable for medium-sized wind turbines, modern small turbines are the height of simplicity. Today's small wind turbines, such as those listed in Appendix B (see the table "Characteristics of Selected Small Wind Turbines"), typically employ an upwind rotor and are passively directed into the wind by a tail vane (Figure 5.6). Most use direct-drive permanent-magnet alternators and passively furl the rotor in high winds. No simpler means for controlling wind turbines has ever been devised.

Some small wind turbines, such as those built by Bergey Windpower, furl the rotor toward the tail in much the same way as Perry's Aermotor did at the turn of the century (Figure 5.7). For example, the rotor on the Bergey 1500 operates at variable speed from startup through furling: the rotor rpm increasing with increasing wind speed. Similarly, voltage and frequency increase with wind speed. Above the furling speed, the rotor begins to swing toward the tail vane. This occurs when the thrust on the rotor, which is slightly offset from the yaw axis, overcomes the restraining force of the hinged tail boom. Power drops dramatically as furling proceeds. When high winds subside, the turbine returns to its operating position automatically. Others, such as Northern Power Systems' HR series, World Power Technologies' Whisper models, and Southwest Windpower's Windseeker furl the rotor vertically (Figure 5.8).

Blade stall and furling are the only means for limiting the rotor's speed and power during both normal and emergency conditions. There are no brakes either to stop the rotor in high winds or to park the rotor for servicing. Nor are there yaw brakes to prevent the wind machine from yawing about the top of the tower in response to changes in wind direction as there are in medium-sized wind turbines. The designers of these machines stress simplicity over complexity, and the turbines are designed for little or no maintenance.

Self-furling turbines are well suited for the stand-alone, battery-charging applications found beyond the utility line. Through use of a synchronous inverter, a limited number of small turbines, such as the Bergey Excel, can also be used by homes, farms, and businesses with existing utility service. Bergey Windpower and Marlec Engineering carry the theme of simplicity one step further and intergrate the hub and rotor housing into one assembly. Bergey goes so far as to combine the mainframe and stator assembly into one unit (Figure 5.9).

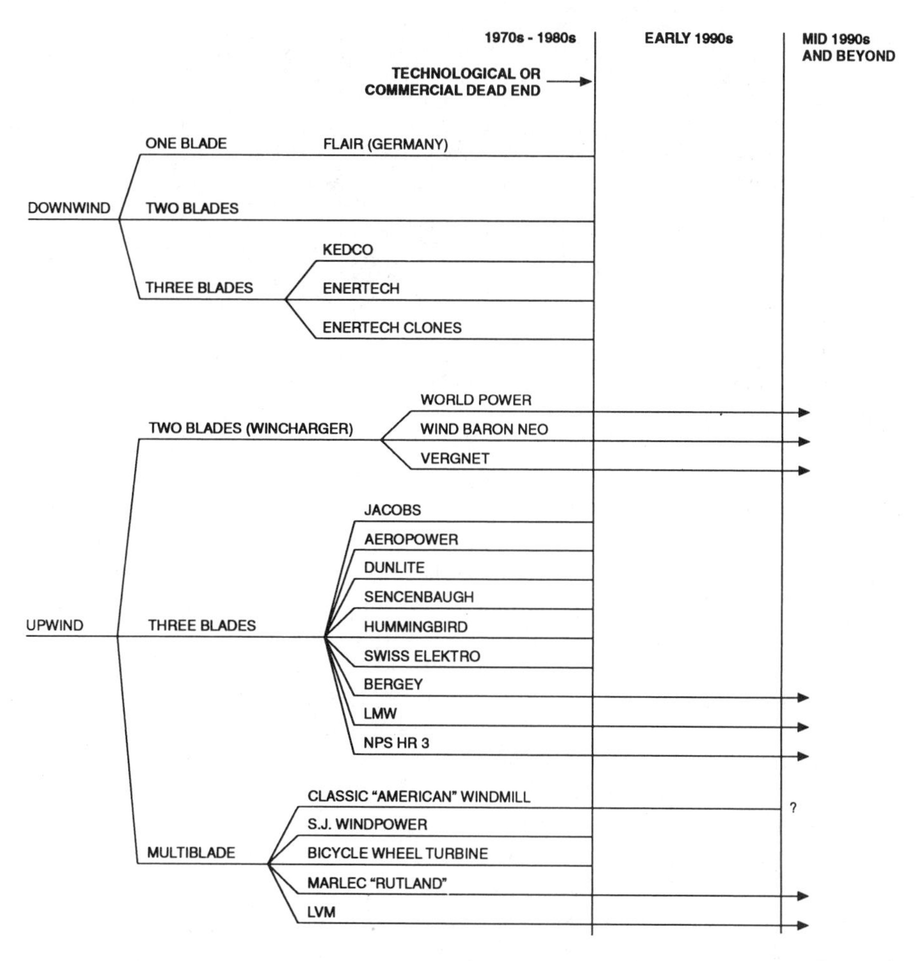

Figure 5.6. Technology pathways for small wind turbines. Most small wind turbines today use upwind rotors with two or three blades.

Figure 5.7. Furling. Passively yawing the rotor out of the wind is a common and effective means of limiting the thrust on a small wind turbine in high winds.

Heavier, more solidy built small wind turbines have proven more rugged and dependable than lightweight machines. The direct-drive Jacobs windcharger was frequently described as "massive" in comparison to models of its competitors, such as the Wincharger. To this day, the Jacobs windcharger still commands a higher price than the Wincharger because of its well-earned reputation for durability. Mick Sagrillo of Lake Michigan Wind & Sun follows what he calls the "heavy metal school" of small wind turbine design. Heavier, more massive turbines, he says, typically run longer without repairs. According to Sagrillo, who rebuilds small wind turbines for a living, "you won't need an attorney" to keep the heavier turbines in operation (Figure 5.10).

Once bedeviled by poor performance and unreliability, thousands of small wind turbines now work dependably day in, day out, worldwide. Some have even proven more dependable in remote power systems than the traditional engine generators they were originally designed to supplement. In a comprehensive study during the 1980s, Wisconsin Power & Light found that modern, integrated wind turbines, represented by the Bergey Excel in Table 5.6, can

Figure 5.8. Microturbine. Southwest Windpower's Windseeker is carrying its share of the load in a hybrid power system for a remote homesite near Tehachapi, California. Although the tower shown here is simple and inexpensive, it should be taller to elevate the wind turbine above the turbulence around the building. The Windseeker furls vertically in high winds.

be available for operation more than 95% of the time. More complex turbines performed less satisfactory at the rural sites where the turbines were installed. The utility encountered no safety problems from any of the machines during the entire monitoring period. "There were also no power quality problems at the site," says the author Scott Pigg.[32] Power factor and voltage flicker, a concern to many utilities, were well within the utility's desired range.

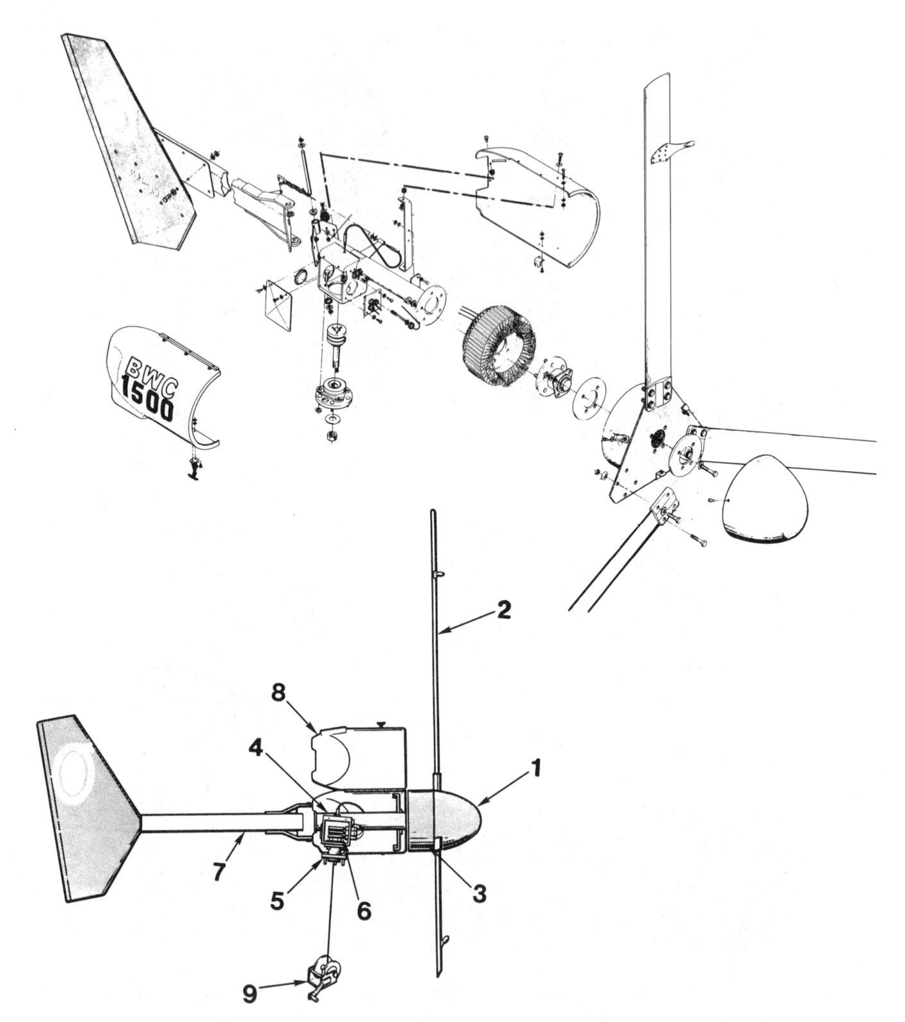

Figure 5.9. Bergey 1500, an integrated direct-drive small wind turbine. Permanent magnets (not visible) are mounted inside the rotor housing. Power is drawn from the stator windings. This design sets the blades at a high angle of attack for starting the turbine in low winds. As rotor speed increases, the weight twists the torsionally flexible blade, changing blade pitch progressively toward the optimum running position. Nomenclature: 1, spinner or nose cone; 2, pulltruded fiberglass blades; 3, permanent-magnet alternator; 4, mainframe; 5, yaw bearing; 6, slip rings and brushes; 7, tail vane; 8, nacelle cover; 9, winch for furling the rotor out of the wind. (Courtesy of Bergey Windpower Co., Norman, OK.)

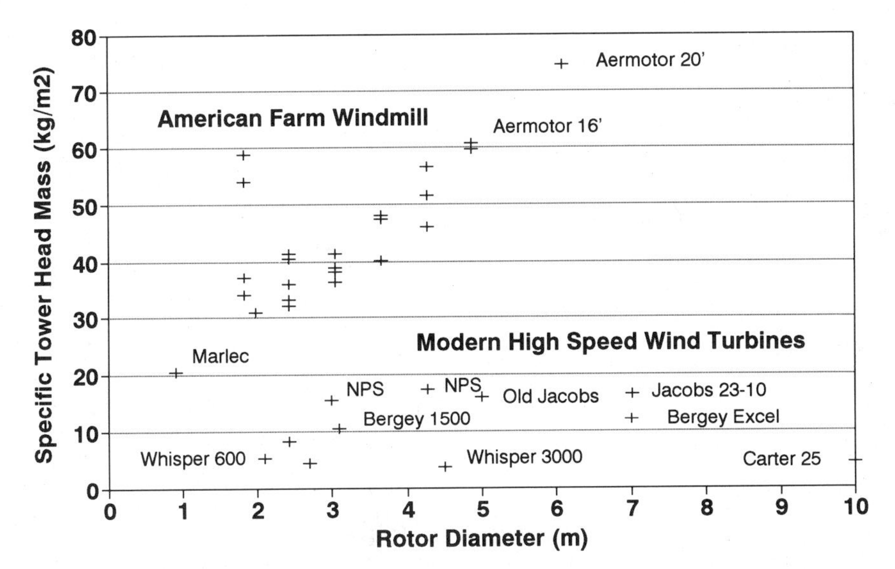

Figure 5.10. Specific rotor and nacelle mass for small wind turbines. Farm windmills are too material intensive for generating electricity cost-effectively, but are ideal for low-speed, high-torque applications such as pumping water. Modern high-speed turbines use materials more efficiently than farm windmills. Marlec's Rutland is a multiblade microturbine.

U.S. manufacturers of small turbines, who were shipping about 400 machines per year during the late 1980s, expect further refinements of these designs in the years ahead as incremental improvements continue to enhance reliability, boost performance, and lower costs.[33] Mike Bergey believes that he can reduce manufacturing costs about 25% by increasing the volume of production to 4000 units per year, 10 times the total U.S. production of small wind turbines (Figure 5.11).

Bergey estimates that technological advancements can boost productivity 25% from specific yields of 400 kWh/m^2 to 500 kWh/m^2. Bergey believes he

Table 5.6
Operating Histories of Four Small Wind Turbines by Wisconsin Power & Light Co.

Turbine	Site	kW	Capacity Factor (%)	Specific Yield (kWh/m^2/yr)	Average Wind Speed (mph)	Availability (%)	Operating Hours	Days Down	Repair Costs
Jacobs	Fond Du Lac	10	12	270	12	75	41,000	491	$5830
Enertech	Arlington	20	21	270	12	76	21,600	279	5356
Carter	Berlin	25	12	270	12	82	39,000	363	6810
Bergey	Delavan	10	13	310	13	96	26,700	89	100

Source: Scott Pigg, Wisconsin Power & Light, 1988.

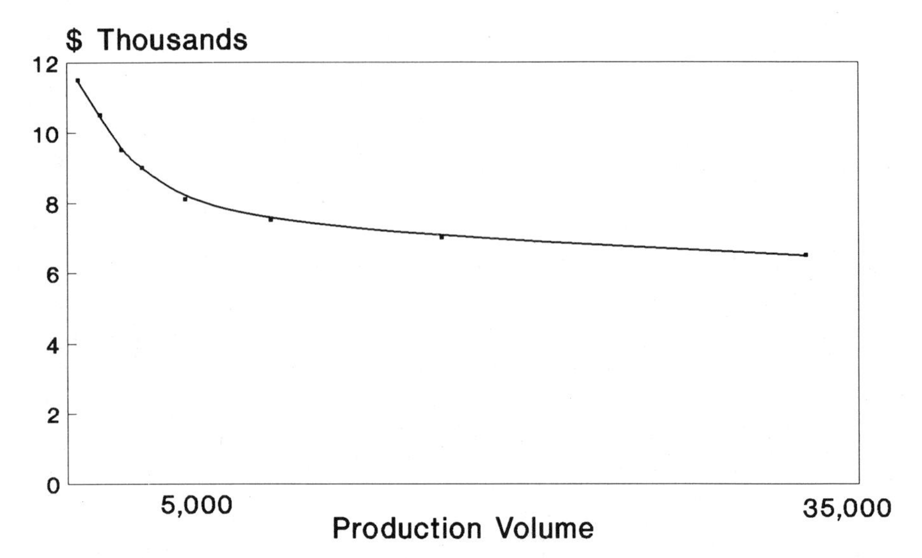

Figure 5.11. Production economies. Estimated factory net price for Bergey Excel (turbine only) in 1992 dollars relative to production volume. Bergey Windpower estimates that increasing volume to 4000 units per year will cut the price about 25%. (Courtesy of Bergey Windpower Co., Norman, OK.)

can do so by using larger neodymium–iron–boron (NeFeB) permanent-magnet generators and advanced power electronics, and by increasing hub height from the typical 24-m (80 ft) tower to 40 m (130 ft). He envisions significantly lowering the noise level of small turbines and simplifying installation by eliminating on-site concrete work. New inverters, says Bergey, will provide an adjustable power factor for simpler interconnection with local utilities and, with a small amount of customer-owned storage, will be adaptable for use as uninterruptible power supplies.[34] These quasi-independent power systems will provide greater benefits to utilities than does current technology.[35]

The accomplishments of small wind turbines in achieving improved reliability and providing valuable services to rural peoples of the globe, although remarkable, are overwhelmed by those of medium-sized wind turbines, the machines that slew the megawatt monsters.

Chapter 5 Endnotes

1. Edward J. Kealey, *Harvesting the Air: Windmill Pioneers in Twelfth-Century England* (Berkeley, CA: University of California Press, 1987), p. 9.
2. Jean-Claude Debeir, Jean-Paul Deléage, and Daniel Hémery, *In the Servitude of Power: Energy and Civilization Through the Ages,* trans. John Barzman (London: Zed Books, 1991), p. 64.
3. Ibid., pp. 78–79.

4. Wolfgang Frode, *Windmühlen* (Hamburg: Ellert & Richter, 1987), p. 12.
5. Dennis Shepherd, "Historical Development of the Windmill," in *Wind Turbine Technology* (New York: ASME Press, 1994), pp. 1–46.
6. Dennis G. Shepherd, "Historical Development of the Windmill," National Aeronautics and Space Administration, Lewis Research Center, Ohio, 1990, p. 11.
7. Kealey, *Harvesting the Air,* Chapter 3, "Sighting the Target," pp. 59–93.
8. Ibid., p. 1. Under good conditions even crude Egyptian sailboats could harness the equivalent of the muscle power from several dozen slaves, some several kilowatts; from Debeir, Deléage, and Hémery, *In the Servitude of Power,* p. 27.
9. Kealey, *Harvesting the Air,* Chapter 6, "Free Benefit of the Wind," pp. 132–153.
10. Kealey, *Harvesting the Air,* "Pushing Round the Tail Pole," pp. 197–209.
11. Debeir, Deléage, Hémery, *In the Servitude of Power,* pp. 78–79.
12. Jean-Marie Homet, *Provence des moulins a vent* (Aix-en-Provence, France: Edisud, 1984), p. 6.
13. Frederick Stokhuyzen, *The Dutch Windmill,* trans. Carry Dikshoorn (New York: Universe Books, 1963), p. 11.
14. Ibid., p. 14.
15. Harry W. Lintsen, "From the Windmill to Steam Engine Waterpumping: Innovation in the Netherlands During the 19th Century," in *Wind Energy: Technology and Implementation,* Proceedings of the European Wind Energy Conference (Amsterdam: Elsevier, October 1991), pp. 7–12.
16. Ibid.
17. Volta Torrey, *Wind-Catchers: American Windmills of Yesterday and Tomorrow* (Brattleboro, VT: Stephen Greene Press, 1976), p. 150.
18. Walter Prescott Webb, *The Great Plains* (New York: Grosset & Dunlap, 1931), p. 320.
19. Torrey, *Wind-Catchers;* p. 88.
20. Webb, *The Great Plains,* p. 337.
21. "A Chronological History of Aermotor Windmill Corp.," Aermotor Windmill Corp., San Angelo, TX, undated.
22. Bruce Griffiths, "Alleviating Drought in Africa," *Windpower Monthly,* 9:11, November 1993, 29–32. Griffiths reports 100,000. But Van Der Linde suggests there may be as many as 280,000 of the 2.5-meter conventional multiblade wind pumps. Andries Van Der Linde, personal communication, P.E. Technikon, Unit for Development Research Into Energy, Port Elizabeth, Republic of South Africa, September 12, 1992. Data on U.S. Southern Plains from Vaughn Nelson, West Texas State University, personal communication. Data on Australia from *Australian Encyclopedia,* Vol. 9, 1958, Angus and Robertson. The Australian Bureau of Statistics estimates about 100,000 in 1993. Data on Argentina, Brazil, China, Columbia, Curaçao, and Nicaragua from *Wind Energy For Rural Areas: Water Pumping and Power Generation,* proceedings of international workshop, ECN, Petten, the Netherlands, October 10–14, 1991.
23. Peter Fraenkel, IT Power, Eversley, United Kingdom, "Why Windpower: A Consideration of Factors That May Determine the Success or Failure of Small Wind Turbines in the Market Place," *Wind Energy for Rural Areas: Water Pumping and Power Generation,* Proceedings of International Workshop, ECN, Petten, The Netherlands, October 10–14, 1991, p. 225.
24. Bruce Griffiths, "Alleviating Drought in Africa," *Windpower Monthly,* 9:11, November, 1993, pp. 29–32.
25. Robert Righter, "A Few Words About This Picture: A Wind-Fueled Electric Power Plant in the Backyard, 1888," *Invention & Technology,* Spring/Summer 1991, p. 30.

26. T. Lindsay Baker, *A Field Guide to American Windmills* (Norman, OK: University of Oklahoma Press, 1985), p. 248.
27. Mick Sagrillo, "How It All Begin," *Home Power,* 27, February/March 1992, pp. 14–17.
28. Vaughn Nelson, "SWECS Industry in the U.S.," unpublished paper, Alternative Energy Institute, West Texas University, Canyon, TX, January 1984. See also Sagrillo, "How It All Began."
29. Shi Pengfei, Chinese Wind Energy Development Center, "International Cooperation for Rural Areas in China." in *Wind Energy for Rural Areas: Water Pumping and Power Generation,* Proceedings of the International Workshop, ECN, Petten, The Netherlands, October 10–14, 1991, pp. 23, 25, 26.
30. J. F. Fawkes, "Wind Energy: Mongolian Experience," *Proceedings of the 2nd World Renewable Energy Congress,* Reading University, Reading, England, September 1992, p. 1689.
31. Richard Perez, "Who Owns the Sun?" *Home Power,* 37, October/November 1993, p. 89.
32. Scott Pigg, "A Performance Analysis of Four Small Wind Energy Systems" Madison, WI: Wisconsin Power & Light, November 1988, p. 7.
33. Mike Bergey, "Comments on the Maturation and Future Prospects of Small Wind Turbine Technology," American Solar Energy Society's conference, Solar '90, Austin, TX, April 21, 1990.
34. Mike Bergey, "The Application of Advanced Small Wind Turbines to Distributed Generation," Windpower '93, annual conference of the American Wind Energy Association, San Francisco, July 1993.
35. For more on small wind turbines see Paul Gipe, *Wind Power for Home and Business* (White River Junction, VT: Chelsea Green Publishing, 1993).

6

The Giant Killers

"There is something in the wind." Shakespeare, Comedy of Errors.

Since its 1970s rebirth, modern wind technology has matured faster than any other energy technology. It is significant that wind turbines were generating commercial quantities of electricity within a decade of their reintroduction, in contrast to the development of nuclear energy. It was a decade after President Eisenhower's 1953 speech before the United Nations extolling the virtues of the "peaceful atom" that nuclear power began to prosper, and it was only in the mid-1960s, nearly 20 years later, that nuclear plants were built on a truly commercial scale.[1] The wind turbines that contributed most to the growth of wind technology were not giant multimegawatt behemoths but humble machines of more modest dimensions.

Power in the Wind

For all the advancements in wind technology since the development of European and American windmills, there remain fundamental principles that govern the operation and performance of modern wind turbines. Countless inventors—and unfortunately, not a few wind farm promoters—would have been well advised to understand these principles before they set off on ill-conceived journeys to make their fortunes.

As E. W. Golding said so succinctly in his classic *The Generation of Electricity by Wind Power,* "wind is merely air in motion." From the formula for kinetic energy and the mass of air moving through a unit of area, the power in the wind in watts is therefore

$$P = \tfrac{1}{2}\rho A V^3$$

where ρ is the air density, A the area intercepting the wind, and V the wind's velocity, or speed in common parlance.[2]

Air density varies with temperature and elevation. Warm air is less dense than cold air and packs less energy. What pilots call "density altitude" affects the power in the wind in much the same way that it affects the carrying capacity of an airplane wing. This was of little concern as long as wind turbines were installed along the shores of the North Sea. But changes in air density do make a difference in many areas where wind turbines are now deployed. For example, on a blistering summer day atop 5000-ft (1500-m) Cameron Ridge in the Tehachapi Pass when the temperature reaches 95°F (35°C), the air is 80% as dense as at sea level during a comfortable spring day. This leads to a proportionate drop in the power available. Conversely, Minnesotans seeking wind development to offset an aging nuclear plant proudly boast that the upper Midwest's frigid winter winds hold more power than the winds of California's warm climes.

Unless otherwise specified, wind turbine manufacturers assume that air density is 1.225 kg/m^3, representing air pressure at sea level and a temperature of 15°C (59°F).[3] (For changes in air density at higher or lower temperatures, see Figure A.1 in Appendix A, and for changes in air density due to changes in elevation, see Figure A.2.) However, the effect of changes in temperature or elevation on wind power is dwarfed by changes in wind speed.

Power in the wind is a cubic function of speed. Doubling the wind speed, say from 5 m/s to 10 m/s, increases eightfold the power available. Even a small increase in wind speed can substantially boost the power in the wind. For example, consider the power available at one site with a wind speed of 6 m/s and another with 7 m/s. Although the wind speed is only 17% greater, there is 60% more power in the wind at the windier site. This is why wind companies expend so much effort in finding the windiest sites.

$$\frac{P_2}{P_1} = \left(\frac{V_2}{V_1}\right)^3$$

$$P_2 = \left(\tfrac{7}{6}\right)^3 P_1 = 1.6P_1$$

Meteorologists find it more accurate to discuss wind resources in terms of power density—the power per unit of area intercepting the wind—than simply in terms of wind speed.

$$\frac{P}{A} = \tfrac{1}{2}\rho V^3$$

If the value for air density is substituted for ρ in the equation, the power density in watts per square meter is

$$\frac{P}{A} = 0.6125V^3$$

where speed is in m/s. (In the English system: $P/A = 0.05472V^3$, where wind speed is in mph, and $P/A = 0.08355V^3$ where wind speed is in knots.)

Wind-Speed Frequency Distributions

Wind speeds vary with time in a constant ebb and flow: calm one day, howling the next. The question then arises: Does the wind speed represent one instant or the average for some period? The use of an average wind speed introduces a complicating element because the average of the cubes is greater than the cube of the average. The average of the cube of different wind speeds over time is greater than the cube of the average speed. For example, the cube of an average speed of 10 m/s is 1000, whereas if the wind blows half the time at 15 m/s and half the time at 5 m/s, the average of the cubes is

$$\frac{15^3 + 5^3}{2} = \frac{3375 + 125}{2} = 1750$$

although the average speed is still 10 m/s. Thus it is necessary to define the distribution of wind speeds over time, that is, the amount of time the wind blows at 5 m/s, 6 m/s, and so on. Typically, annual average distributions are used to assess wind energy's economic potential.

Annual speed distributions vary widely from one site to another, reflecting climatic and geographic conditions. Power density may even vary significantly among sites with the same average annual wind speed, due to differences in their speed distributions (Table 6.1).

The relationship between the power density derived from the average speed and the power density from a speed distribution is the energy pattern factor (EPF) or cube factor. This is a rough measure of the distribution's shape. Sites in the Caribbean often have high average speeds, but the area's famed trade winds blow steadily for much of the year, with few periods of extremely high winds. For example, the EPF at Culebra (1.4) is closer to unity than that of the San Gorgonio Pass (2.4), where periods of calm alternate with periods of extremely strong, power-producing winds. Because of the cubic relationship between speed and power, brief periods of high winds contribute more to the total power of the wind in Palm Springs than they do at Culebra.

Table 6.1
Effect of Speed Distribution on Wind Power Density for Sites with Same Average Speed

Site	*Annual Average Wind Speed*		Actual Wind Power Density (W/m^2)	EPF or Cube Factor
	m/s	mph		
Culebra, Puerto Rico	6.3	14	220	1.4
Tiana Beach, New York	6.3	14	285	1.9
San Gorgonio, California	6.3	14	365	2.4

Source: Battelle PNL, *Wind Energy Resource Atlas,* 1986.

Meteorologists have found that the Weibull probability function best approximates the distribution of wind speeds over time at sites around the world where actual distributions of wind speeds are unavailable. The Rayleigh distribution is a special case of the Weibull function, requiring only the average speed to define the shape of the distribution.[4] (Figure 6.1). The EPF for the Rayleigh distribution is 1.91.

When the EPF is added to the formula for power density, the annual average $P/A = 0.6125V^3$ EPF, in W/m^2 per year. Because energy is a measure of power over time, and there are 8760 hours in a year, the annual average power density is often equated with the wind resource's annual energy content (Table 6.2).

Wind Speed and Height

Wind speed also varies with height above the ground. Trees, shrubs, buildings, and other obstructions retard the flow of the wind. These effects typically decrease with increasing height, permitting the wind to flow more freely. Data on wind speeds often include the height at which the wind was measured. When the height is unspecified, the wind speed refers to either a standard international height of 10 m (33 ft) above ground level, or the height at which the wind turbine operates. Manufacturers, for example, usually refer the performance of their turbines to wind conditions at hub height.

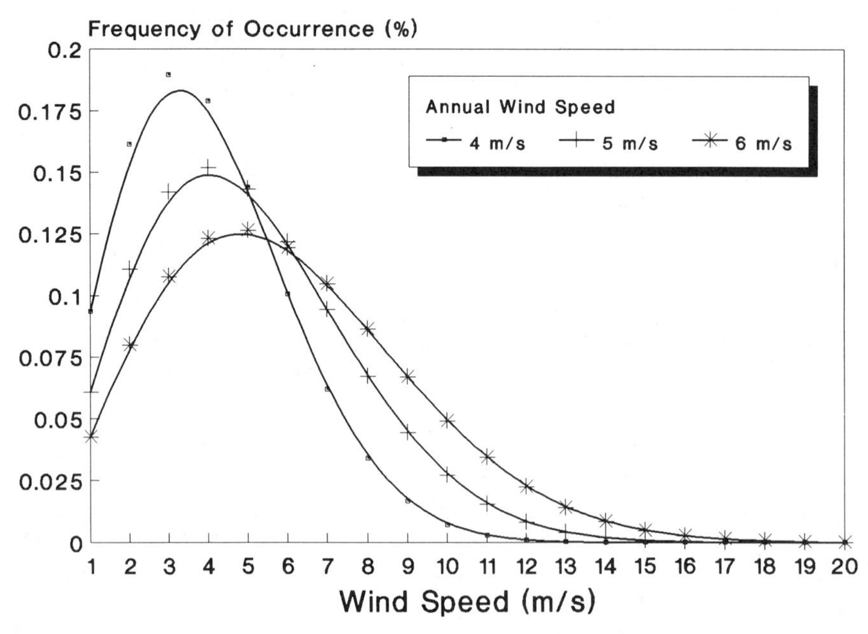

Figure 6.1. Rayleigh wind-speed distribution.

Table 6.2
Annual Wind Power Density for Rayleigh Distribution

Annual Average Wind Speed		Annual Power	Annual Energy
m/s	mph	Density (W/m²)	Density (kWh/m²)
4	9.0	75	656
5	11.2	146	1281
6	13.4	253	2214
7	15.7	401	3515
8	17.9	599	5247
9	20.2	853	7471

A problem arises when it is necessary to estimate the wind speed at a height different from that at which the wind was measured. The rate at which wind speeds increase with height varies with vegetation, terrain, and climate. The increase in wind speed with height is greatest over rough terrain or numerous obstacles, such as in the suburbs, and smallest over smooth terrain, such as the surface of a lake.

There are two approaches to estimating the increase in wind speed with height: the power law method common in North America, and the logarithmic method common in Europe. Logarithmic extrapolation is derived mathematically from a theoretical understanding of how the wind moves across the surface of the earth and was used by Risø when preparing its European Wind Atlas. In contrast, the power law equation is derived empirically from actual measurements.

Using the power law, wind speed increases with height

$$\frac{V}{V_o} = \left(\frac{H}{H_o}\right)^{\alpha}$$

$$V = \left(\frac{H}{H_o}\right)^{\alpha} V_o$$

where V_o is the wind speed at the original height, V the wind speed at the new height, H_o the original height, H the new height, and α the surface roughness exponent (Table 6.3).

Both systems require the user to estimate surface roughness. Where the rate of increase in wind speed with height is unspecified, it is commonly assumed in North America that the "$\frac{1}{7}$ power law" applies; that is, the surface roughness exponent is 0.014, representing open plains. Empirical results indicate that the $\frac{1}{7}$ power law fits many, though certainly not all, North American sites (Table 6.4).

The European approach has worked well along the coastlines of the North German Plain, particularly in Denmark. In open areas with few windbreaks, such as coastal sites (roughness class 1 in the European system) the logarithmic

Table 6.3
Typical Surface Roughness Exponents for Power Law Method of Estimating Changes in Wind Speed with Height

Terrain	Surface Roughness Exponent, α
Water or ice	0.1
Low grass or steppe	0.14
Rural with obstacles	0.2
Suburb and woodlands	0.25

model produces a result similar to that of the $\frac{1}{7}$ power law.[5] Further inland, results from the two methods diverge. For inland sites, where roughness class 2 applies, the logarithmic model finds 23% more energy in the wind than that of the $\frac{1}{7}$ power law.[6]

Although Europeans rightly note that the $\frac{1}{7}$ power law often underestimates wind speed at hub height, there is no overriding reason for North Americans to abandon the power law. The overwhelming experience in the United States during the past 20 years has shown a tendency by proponents of wind energy to overestimate the increase in speed with height, leading to grossly optimistic projections of annual energy production. The effect of height on wind speed is so great that it behooves meteorologists to measure actual wind speeds at hub height rather than rely on estimates produced by either system. Yet in the absence of actual measurements, the $\frac{1}{7}$ power law is a reasonable, if sometimes conservative, approximation.

Table 6.4
Changes in Wind Speed with Height for Selected DOE Candidate Wind Turbine Sites

Site	Speed (m/s) at Height:		Speed Increase Ratio	Surface Roughness Exponent, α
	9.1 m 30 ft	45.7 m 150 ft		
Finley, North Dakota	6.1	9.1	1.49	0.25
Block Island, Rhode Island	5	7.4	1.48	0.24
Boardman, Oregon	3.8	5.5	1.45	0.23
Huron, South Dakota	4.7	6.8	1.45	0.23
Russel, Kansas	5.3	7.3	1.38	0.20
Clayton, New Mexico	5.4	7.3	1.35	0.19
Minot, North Dakota	6.5	8.4	1.29	0.16
Amarillo, Texas	6.3	8.1	1.29	0.16
San Gorgonio Pass, California	6.2	7.7	1.24	0.13
Livingston, Montana	6.8	8.4	1.24	0.13
Kingsley Dam, Nebraska	5.3	6.5	1.23	0.13
Bridger Butte, Wyoming	7	8.4	1.20	0.11

Source: Battelle PNL, *Wind Energy Resource Atlas,* 1986.

Using the $\frac{1}{7}$ power law, a doubling of height increases wind speed 1.1 times, or 10% (Figure 6.2). Increasing height fivefold, for example, from 10 m (33 ft) to 50 m (160 ft) increases wind speed 1.26 times, or 26%. Power increases with height as the cube of wind speed:

$$P = \left(\frac{H}{H_o}\right)^{3\alpha} P_o$$

The distribution of wind speeds may change slightly with changes in height. Consequently, the simple extrapolation of a wind speed distribution at a reference level may lead to errors when estimating the power in the wind at a new height (Table 6.5). For example, in the San Gorgonio Pass the roughness exponent is 0.13 for changes in wind speed from a fivefold increase in height. Power, on the other hand, increases faster than a simple cubic function would indicate, increasing the power produced at the original height by 2.03 times that at the original height. In this case there is nearly 15% more power in the wind at the new height than that derived by simply scaling up the speed distribution from the original height (1.87 times original power).

Again, it is preferable to measure the actual speed distribution at the height at which the turbine will operate. Where the distribution is unavailable, the power at the new height can be estimated using the power law, given its

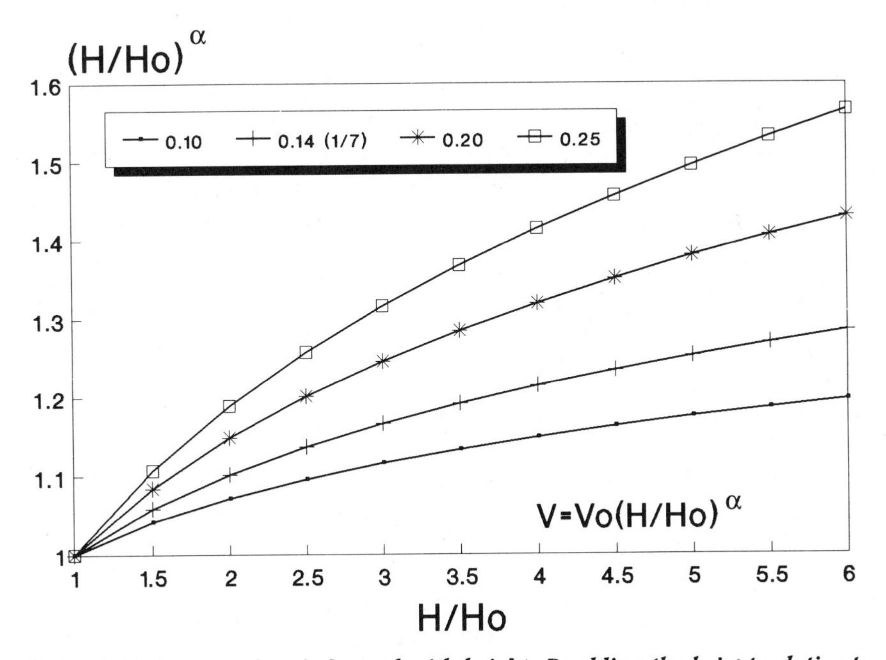

Figure 6.2. Increase in wind speed with height. Doubling the height relative to a reference height increases wind speed 1.1 times (10%) for an α value of 0.14, the $\frac{1}{7}$ power law.

Table 6.5
Changes in Power Density with Height for Selected DOE Candidate Wind Turbine Sites

Site	*Power Density (W/m^2) at Height:* 9.1 m 30 ft	45.7 m 150 ft	Power Increase Ratio	Surface Roughness Exponent, α
Finley, North Dakota	234	737	3.15	0.24
Block Island, Rhode Island	133	407	3.06	0.23
Boardman, Oregon	102	278	2.73	0.21
Huron, South Dakota	131	332	2.53	0.19
Russel, Kansas	173	373	2.16	0.16
Clayton, New Mexico	162	334	2.06	0.15
Minot, North Dakota	271	533	1.97	0.14
Amarillo, Texas	228	464	2.04	0.15
San Gorgonio Pass, California	351	712	2.03	0.15
Livingston, Montana	457	794	1.74	0.11
Kingsley Dam, Nebraska	160	286	1.79	0.12
Bridger Butte, Wyoming	371	589	1.59	0.10

Source: Battelle PNL, *Wind Energy Resource Atlas,* 1986.

limitations (Figure 6.3). Increasing the height fivefold, for example, from 10 m to 50 m, increases the power available nearly two times when using the $\frac{1}{7}$ power law.

Wind Resources

Wind turbines are most commonly found today in the United States in areas with power densities greater than 300 W/m^2 or average wind speeds above 6.5 m/s (15 mph) at 10 m (33 ft) above ground level. This corresponds to the wind resource designated class 5 by Battelle Pacific Northwest Laboratory and that found on the west coast of Denmark's Jutland peninsula, roughness class 1 in Risø's classification system. This resource is equivalent to a power density of 400 to 500 W/m^2 or an average wind speed of 7.5 m/s (17 mph) at 30 m (100 ft) above the ground, the typical tower height of the early 1990s. At tower heights of 50 m (164 ft), at which most new wind turbines will be operating during the mid-1990s, the same resource is equivalent to a power density of 500 to 600 W/m^2, or an average annual wind speed of about 8 m/s (18 mph).

Wind speeds on the crests of ridges in California's Tehachapi Pass average 18 to 19 mph (8 to 8.5 m/s). The wind shear, the rate of increase in wind speed with height, is slightly greater than that calculated using the $\frac{1}{7}$ power law. Sites atop Altamont Pass's rolling hills are less windy, 13 to 18 mph (6 to 8 m/s), and the wind shear is less marked, only 0.10. Because of local meteorological conditions, the wind shear in the Altamont becomes negative

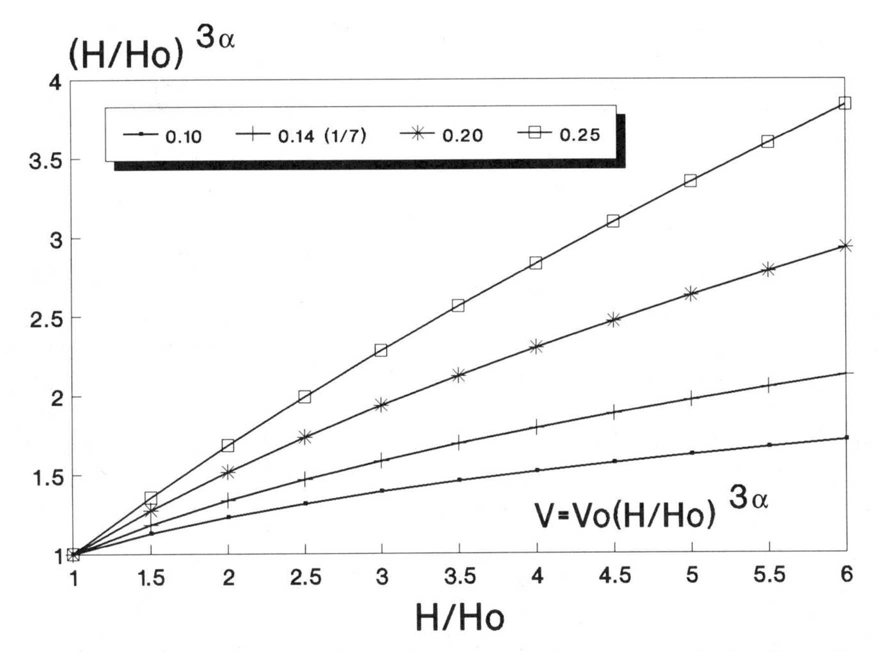

Figure 6.3. Increase in wind power with height. Increasing the height to five times that of a reference height, for example from 10 m to 50 m, increases available wind power nearly twofold, for an α value of 0.14, the $\frac{1}{7}$ power law.

above a certain height. Sustained winds greater than 60 mph (27 m/s) are frequent in Tehachapi but rare in the Altamont.

California and Denmark are not unique; there are ample winds in many locations around the world to generate cost-effective electricity. At hub height along the coast of the Netherlands, for example, the average speed is 7.5 m/s. At less well exposed sites nearby in Germany's Lower Saxony, wind speeds at hub height average 6.8 m/s, while average speeds can exceed 9 m/s in the mountains of Wales. But in complex terrain, such as on Welsh mountaintops and in California's mountain passes, it is much more difficult to define the wind resource than it is in the lowland countries of northern Europe.

One of the lessons learned from development in California is the necessity of measuring the wind resource in complex terrain. During the early 1980s, it was common to monitor the wind with one anemometer per 150 to 350 proposed turbines of 50 kW each or 10 to 20 MW of proposed capacity.[7] As a result, many operators greatly overestimated the amount of wind their turbines would intercept. By the late 1990s, California developers had seen the light and were monitoring the winds with one anemometer for every 0.5 MW of new capacity (one for about every two or three 200-kW turbines).

Rotor Swept Area

Once the power density of the wind resource is known, it is simple to approximate the amount of power a wind turbine can capture:

$$P = \frac{P}{A} \times \mathrm{A} \times \text{conversion efficiency}$$

where A is the area swept by the wind turbine's rotor. In 1920, Albert Betz found that the aerodynamic efficiency of a wind turbine rotor was limited to 16/27, or 59.3% of the power in the wind. No wind turbine rotor has ever reached this level, although many have exceeded 40%.

The rotor is only one of three stages in converting the power in the wind into electricity. The other two are the transmission for matching the rotor speed to that of the generator, and the generator itself. The overall annual conversion efficiency of modern medium-sized wind turbines is approximately

rotor ×	transmission ×	generator ×	yawing and gusts	
40% ×	95% ×	95% ×	95%	= 35%

Under optimal conditions, then, a well-designed wind turbine at a site with an average annual wind resource of 7 m/s mimicking a Rayleigh wind-speed distribution (see Table 6.2) will capture

$$35\% \times 3500 \text{ kWh/m}^2 \text{ per year} = 1225 \text{ kWh/m}^2 \text{ per year}$$

If the wind turbine intercepts 1000 m^2 of the wind stream, it will generate about 1.225 million kilowatt-hours per year. Wind turbines seldom perform this well—for a variety of reasons—but the derivation demonstrates why Tacke Windtechnik declares in its advertisements that Der Rotor ist der Motor, and why nothing says more about a wind turbine's ability to capture the energy in the wind than its frontal area: the amount of area intercepting the wind stream.

Conventional wind turbines are comprised of three essential components: rotor, nacelle, and tower (Figure 6.4). As Tacke says in its play on words, the rotor is the heart of any wind turbine. The rotor is the prime mover; it powers the wind turbine's drive train, which (on a conventional wind turbine) sits atop the tower inside the nacelle. The tower simply raises the rotor above the ground, where it can better catch the wind.

For conventional wind turbines, the frontal area is simply the area of a circle described by the spinning rotor.

$$A = \pi R^2$$

where R is the rotor's radius in meters. Because the area swept by the rotor determines how much work a wind turbine can perform, the diameter of the rotor on a conventional turbine becomes a convenient shorthand for the size of the machine.

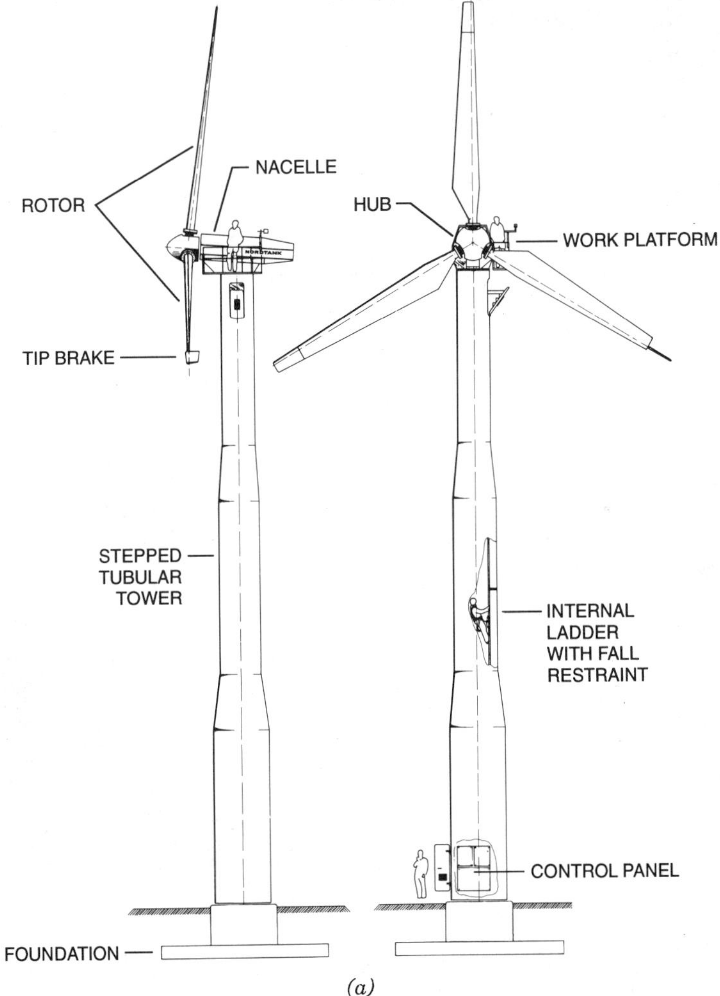

Figure 6.4. (a) Typical first-generation Danish wind turbine. Note tip brakes (pitchable blade tips), internal ladder, work platform, and "rocket" or stepped tubular tower. (Courtesy of Nordtank Energy Group, Balle. Denmark.) (b) Open array of first-generation Danish turbines near Tehachapi, California. These Nordtank turbines use a fixed-pitch rotor 15 m (50 ft) in diameter upwind of the tower.

Giant Killers: Medium-Sized Wind Turbines

It is medium-sized wind machines, turbines 10 to 50 m (30 to 150 ft) in diameter that slew the giant multimegawatt turbines of the German, British, and U.S. wind programs and now dominate global wind technology. These turbines have gradually increased in size from about 10 to 12 m (30 to 40 ft) in diameter in the early 1980s to 35 to 40 m (110 to 130 ft) in diameter in the mid-1990s. These later wind turbines sweep 10 times the area of the earlier machines. The area intercepting the wind has increased from 125 m^2 to 1250 m^2 (Figure 6.5).

Where wind turbines are capable of extracting 35 to 40% of the energy in the wind as in the example, they will yield about 500 kWh/m^2 of rotor swept area per year in a 5-m/s wind regime, 900 kWh/m^2 per year in a 6-m/s wind regime, and 1200 kWh/m^2 per year in a 7-m/s wind regime. The annual generation of wind turbines at windy sites in California or on the coast of Denmark have increased from 150,000 kilowatt-hours per unit per year to 1.5 million kilowatt-hours per year as the turbines have increased in size (Table 6.6).

Increasing size, although no panacea for lowering the cost of energy produced by wind turbines, does provide an important economic benefit. Up to a point, operators gain savings from bigger wind turbines by spreading maintenance costs over greater generation. It takes the same number of windsmiths to service a 25-m 200-kW turbine as it does an 18-m 100-kW machine. But the bigger turbine intercepts twice the area and captures twice

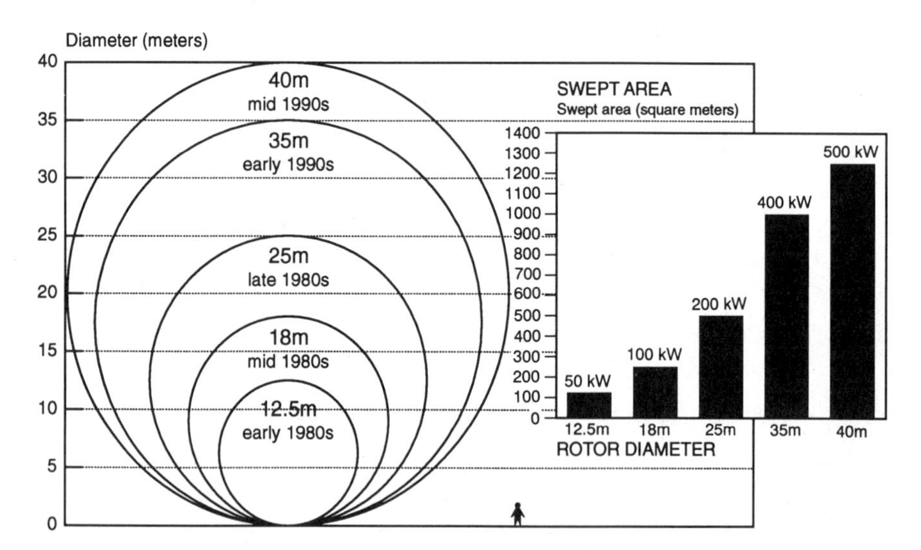

Figure 6.5. Increasing size. By the mid-1990s, medium-sized wind turbines were 10 times the size of those in the early 1980s, increasing from 125 m^2 to 1250 m^2 of rotor swept area.

Table 6.6
Technology Development of Medium-Sized Wind Turbines

Period	Rotor Diameter (m)	Swept Area (m^2)	Capacity (kW)	*Nominal Annual Energy Output for Hub Height, Wind Speed, and Maximum Specific Yield:* 5 m/s 11 mph 500 kWh/m^2	6 m/s 13 mph 900 kWh/m^2	7 m/s 16 mph 1200 kWh/m^2
Early 1980s	12.5	125	50	62,500	112,500	150,000
Mid-1980s	18	250	100	125,000	225,000	300,000
Late 1980s	25	500	200	250,000	450,000	600,000
Early 1990s	35	1000	400	500,000	900,000	1,200,000
Mid-1990s	40	1250	500	625,000	1,125,000	1,500,000

the energy. The 25-m turbine cuts the labor cost per kilowatt-hour in half over that of the smaller machine. For wind turbines greater than 40 to 50 m (100 to 160 ft) in diameter, the costs of the large cranes, which are needed for all but the simplest of repairs, consume any further economies of scale.

By the mid-1990s, there were 18,000 wind turbines 10 to 24 meters in diameter operating throughout the world, mostly in Europe and California. Another 3000 wind turbines 25 to 33 m in diameter were installed during the early 1990s. Analysts consider wind turbines of this size a proven technology. By mid-1994, some 1000 machines 33 to 40 m in diameter had also been installed, representing the state of the art for medium-sized wind turbines. These are big machines, but they are considerably smaller than the behemoths once thought most economical (Figure 6.6).

Turbine Rating

In this discussion, the measure of size has been rotor diameter, not generator capacity. There is a good reason for this. Wind turbine ratings, in kilowatts, give only a crude indication of how much electricity a wind turbine can produce. Worse yet, ratings in kilowatts can deceive the unwary. Rotor diameter is always a much more reliable indicator.

Power ratings grew out of the role that electric utilities played in the industry's formative years. Much of the early work was done by or for the utility industry. As a result, when utility engineers talked among themselves about wind turbines, they described turbines in the terms they knew best: generator size. They blithely applied the same terminology to wind turbines as they did to reactors at Three Mile Island or diesel generators at Nome, Alaska. Wind turbines, though, are different.

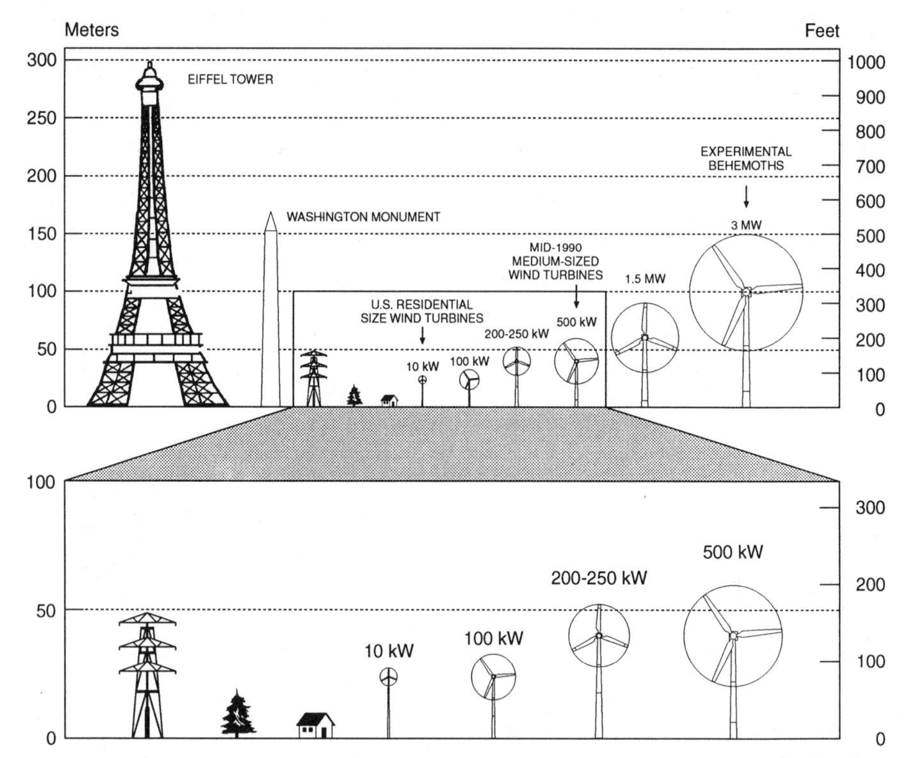

Figure 6.6. Relative size. Comparison of contemporary wind turbines with other structures on the landscape.

In conventional power plants, the operator controls the fuel to produce the power desired. Utilities try to run large-base-load plants as close to full output as possible, leaving a little room as "spinning reserve." Even when they bring small plants on line, they prefer to run them as close to their full capacity as possible for best performance. It makes sense, then, to call a nuclear reactor 1000 MW if the utility operates it at nearly 1000 MW most of the time.

Wind plants are dependent on the wind—the operator has no control over the fuel. All the plant operator can do is ensure that the turbines are in service when the wind is present. Sometimes the wind may not be blowing, and even when it does the wind may often be insufficient to drive the turbine at its rated power. Unlike conventional plants, wind turbines seldom operate for long periods at their "rated" power.

Utility engineers had to derive a different yardstick for comparing wind turbines to conventional power plants. Early wind turbine designers created a hybrid rating system that sufficed: the power output at some arbitrary wind speed. This method would work reasonably well if all agreed on the speed at which wind turbines would be "rated." But there is no international consensus on what this speed should be. Rated wind speeds vary from 10 to 16 m/s (22 to 36 mph).

This results partly from tailoring wind turbines to different wind regimes, and partly from different approaches to maximizing total generation. This rating approach also results from the early concept that turbines would reach their rated capacity, then limit output to the rated amount for wind speeds up to cutout when they would turn themselves off (Figure 6.7). This is represented as a straight line on the power curve, a chart of the power produced at various wind speeds. Few wind turbines operate this neatly in the real world.

This is most apparent in wind turbines using aerodynamic stall to regulate power in high winds. Typically, power in these turbines peaks at higher than rated wind speeds, then declines until the cutout wind speed is reached (often 25 m/s for Danish wind turbines). The "rated" power only approximates the power these turbines will produce at wind speeds above "rated." And among pitch-regulated turbines, the power will fluctuate above and below the rated value as the blade pitch mechanism adjusts to changing wind conditions. Power curves are only approximations of what the wind turbine will produce at any given instant, depending on whether the wind speed is increasing or decreasing.

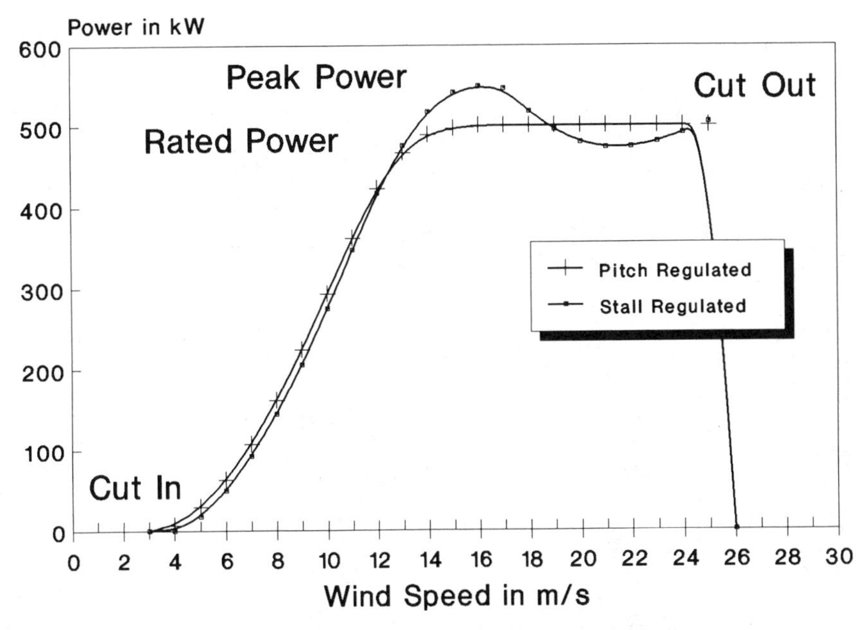

Figure 6.7. Sample power curves for 500 kW wind turbines. These turbines are quite different, although both have the same rating. One is 37 m (120 ft) in diameter and regulates power by aerodynamic stall. The other is 39 m (130 ft) in diameter and regulates power by varying blade pitch.

This rating designation ultimately leads to confusion. The Carter model 25, for example, uses a rotor 10 m in diameter to drive a 25-kW generator, loading the rotor to 0.33 kW/m^2 of swept area. The Fayette 95 IIS (Figure 6.8) uses a rotor about the same size to drive a 95-kW generator for a specific capacity rating of 0.95 kW/m^2, nearly three times that of the Carter turbine. An unsuspecting investor could easily be led to believe that the Fayette turbine is three times more productive than its competitor, the Carter 25, because of its higher rating. The peak outputs from the two turbines do differ markedly, because of differing rotor design. Generation differs as well. But the Fayette turbine, when it is operating, generates much less than three times the energy of the Carter turbine, despite what would be indicated by its higher rating.

Specific ratings today range from 0.3 to 0.5 kW/m^2 (Table 6.7). Some turbines, such as the Carter 300 and WindMaster 300 kW, have specific ratings of up to 0.7 kW/m^2, but these are outside the norm. Other manufacturers of the same-size turbines would rate them at 200 kW or 0.4 kW/m^2. Historically, specific ratings were even lower.

In the early 1970s, NASA designed its 100-kW Mod-0 prototype with a rotor 38 m in diameter, giving it a specific loading of only 0.1 kW/m^2. Its successor, the Mod-0A, used a generator twice the size of the Mod-0, pushing the specific rating up to 0.2 kW/m^2.

Several years later, the private sector began introducing its own turbines. The early Enertech E44 (13 m), for example, was rated at 20 kW for about 0.14 kW/m^2. This was similar to Bergey Windpower Co.'s small 1-kW

Figure 6.8. Fayette 95-IIS. First-generation American turbine notorious for its unreliability and excessively high power rating.

Table 6.7
Typical Medium-Sized Wind Turbine Ratings

Rotor Diameter (m)	Swept Area (m^2)	Range of Rated Capacity (kW)	Specific Rated Capacity (kW/m^2)
12.5	125	40–60	0.3–0.5
18	250	80–100	0.3–0.4
25	500	200–300	0.4–0.6
35	1000	300–400	0.3–0.4
40	1250	500–750	0.4–0.6

(2.8-m) model at 0.16 kW/m^2. Later models of both companies pushed ratings higher. Bergey rated its 7-m (23-ft)-diameter Excel at 10 kW or 0.26 kW/m^2. Enertech introduced its 44/40 as a 40-kW commercial version of its earlier machine, with a specific rating of 0.28 kW/m^2. Subsequently, Enertech increased the rating further to 60 kW for a specific capacity of 0.43 kW/m^2.

During the 1970s and early 1980s, there was a clear trend among U.S. manufacturers toward higher rotor loading in successive models. Turbines introduced from 1983 to 1985, for example, had higher loadings than those introduced during the 1970s. During the years of peak development in California, manufacturers were tempted to raise the kilowatt rating. A higher rating lowered a turbine's relative cost, in dollars per kilowatt, in the eyes of unsophisticated buyers, and was used to competitive advantage by some manufacturers. Since the tax credits expired, there has been less incentive to boost a turbine's kilowatt rating artificially.

Some engineers argue that ratings are important, as they tell project designers how to size the wind plant's transformers, collectors, and substation. Few use a turbine's rating as such a guide. To size the power collection system properly, the peak capacity given on the generator's nameplate must be used, and generators are rated in a different manner than are wind turbines. Those who develop and finance wind projects today are more interested in cost and annual generation than they are meaningless "ratings," as well they should be. Wind turbine manufacturers continue to "rate" their turbines out of a necessity to identify their product as much as for any other reason.

Tailoring a wind turbine's performance to a specific wind regime does require a skillful matching of rotor and generator performance. If the rated speed is too low, at a very energetic site too much energy will be lost in high winds. Conversely, if the rating is too high, performance at a less energetic site will suffer. Picking a rated speed is as much an art as a science, because gearboxes and generators are not manufactured in continuous increments. Only discrete sizes are available. The rated capacity and the size of the gearbox and generator are partly determined by the manufacturers of the generator and gearbox. Small variations from the optimal rating are not critical for most temperate latitude sites where wind turbines are currently deployed.

The influence of rating on overall performance is more important where wind turbines will operate in wind regimes with characteristics far different from those of northern Europe or the continental United States, such as the Caribbean. The optimal rated power at tradewind sites, such as on the island of Curaçao off the coast of Venezuela, should be somewhat lower than that of machines now in use elsewhere because there are few periods with high winds. Generators oversized for the occasional high winds of temperate climates would operate nearly all the time at a fraction of their capacity on Curaçao, resulting in lower efficiency.[8]

To avoid the rating dilemma, most European manufacturers refer to their products by rotor diameter. Thus Vestas' V39 is a wind turbine 39 m in diameter. American manufacturers eventually adopted a hybrid designation: diameter followed by generator size. For example, Enertech's 44/40 or U.S. Windpower's 56-100 designate wind turbines respectively, 44 and 56 ft in diameter, powering 40- and 100-kW generators.

Eric Miller of U.S. Windpower notes that the rating of conventional power plants is also subject to interpretation. The ratings of many cogeneration (combined heat and power) plants, for instance, vary seasonally with changes in air density because they use aeroderivative turbines. U.S. Windpower has abandoned the hybrid designation for its newest model, once known as the 33-300 because of its 33-m (110-ft) rotor and 300-kW generator. The new turbine is dubbed simply the 33M-VS, which refers to its rotor diameter in meters and its variable-speed operation. The rating in kilowatts of the new turbine, says Miller, becomes a function of wind regime. The turbines will have a higher rating (360 kW) for windy sites along the Columbia River Gorge than for those in Minnesota (340 kW) or in California's Solano County (300 kW). Like their competitors, U.S. Windpower will tailor the turbine's power curve to the site as required.

Danish manufacturers take a slightly different approach. They will supply different rotors for different wind regimes. For example, Vestas supplies a 29-m (95-ft) rotor with its 225-kW model to the Midwest (0.34 kW/m^2) while it ships the same turbine with a rotor 27 m (90 ft) in diameter (0.39 kW/m^2) for a windy ridge in the Tehachapi Pass.

This illustrates the weakness of the rated power designation; the rotors differ in size, the power curves differ, and the rated speed differs. But both turbines reach the same rated power. The turbine with the 29-m rotor reaches its rated capacity at a slower wind speed (14 m/s) than that of the 27-m rotor (15 m/s). The larger rotor shifts the power curve slightly to the left, so the turbine will generate more power at low wind speeds. The turbine will endure greater loads at higher winds than the smaller rotor, but winds at these speeds occur less frequently at Midwestern sites than at Tehachapi, justifying the trade-off between greater energy generation and greater wear and tear.

Engineers use the power curve to estimate the turbine's annual energy output (AEO). By matching the power curve to the wind speed distribution for a specific site, they can more accurately project production than by using the turbine's swept area and guessing at its overall efficiency.

For example, consider the power curve for the 500-kW variable-pitch turbine shown in Figure 6.7. This turbine uses a rotor 39 m (130 ft) in diameter (1195 m^2 of swept area) to drive a 500-kW generator that reaches its rated capacity at 16 m/s (36 mph) (Figure 6.9). At a 7-m/s (16-mph) site with a Rayleigh distribution, winds at this speed occur about 75 hours per year. Winds at the rated speed contribute some 37,000 kWh per year to the turbine's total generation. Winds above rated produce about 90,000 kWh per year, only 7% of total generation at this site. Most of the energy generated by this wind turbine at this site is produced at wind speeds lower than rated, once again illustrating that rated capacity says little about how much energy the turbine will generate. See Table 6.8 for the same example presented in tabular form.

At a 7-m/s site with a Rayleigh distribution, the energy density is 3515 kWh/m^2 per year (see Table 6.2). The 500-kW wind turbine in this example captures 1080 kWh/m^2 of swept area per year, or 31% of the energy in the wind. The performance projection for this 500-kW machine is typical for medium-sized turbines. Small wind turbines are considerably less productive at such windy sites (Table 6.9 and Figure 6.10).

Productivity

As with wind turbine rating, there are several measures of wind turbine productivity in common use, some more meaningful than others: generation per turbine (kWh/unit), generation per unit of capacity (kWh/kW), capacity factor (%), and specific yield or generation per unit of area swept by the turbine's rotor (kWh/m^2).

Annual generation per turbine or annual energy output (AEO) is used by developers, investors, and homeowners to gauge performance because it is easily understood and directly comparable to performance projections. If a homeowner is buying a single turbine, the projected generation per unit will clearly state how much energy can be expected. In the same way, the homeowner can also easily monitor performance by comparing what the turbine did deliver with what was expected.

Most manufacturers provide AEO estimates with their product literature in either tabular or chart form. For example, the manufacturer of the 500-kW turbine in Table 6.8 projects an annual output of 1.24 million kilowatt-hours in roughness class 1, using the Danish system of wind resource classification.

Annual generation per unit of capacity in kilowatt-hours per kilowatt of rated capacity is more useful to project planners where a broad measure of productivity is more important than the number of specific machines. This

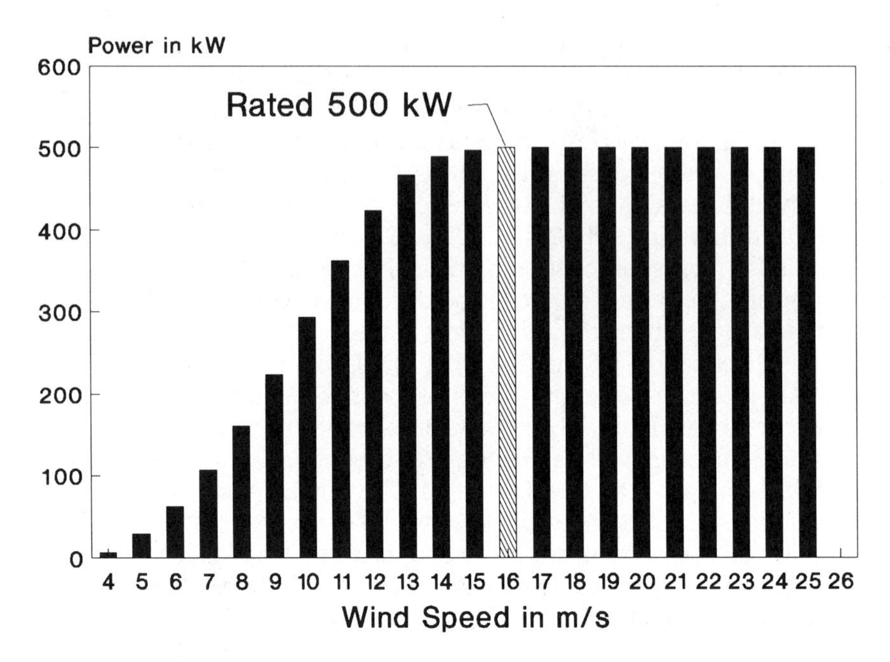

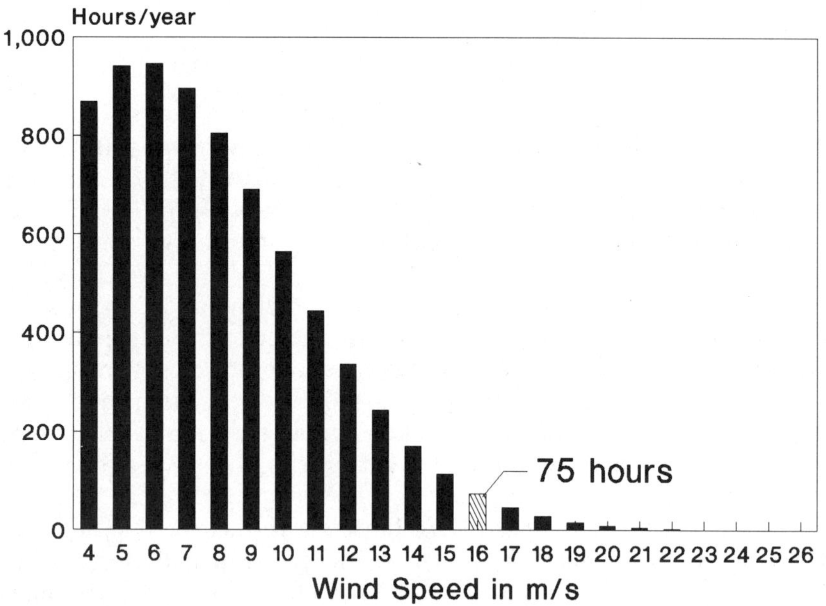

Figure 6.9. Power curve and annual energy output. Matching the power curve to the wind-speed distribution enables estimating the annual energy output.

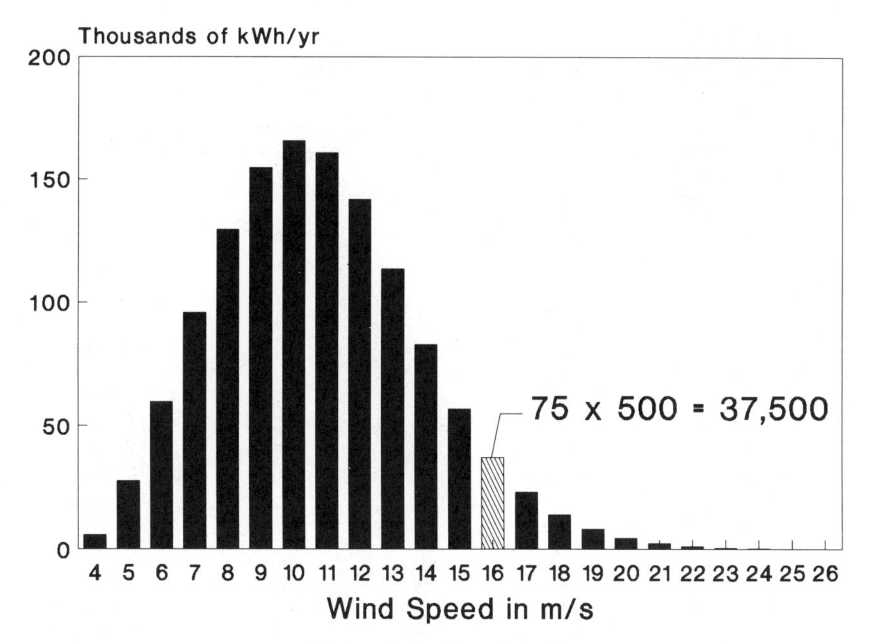

Figure 6.9. (Continued)

measure is easily convertible to total expected generation once the total project capacity in megawatts is known. The 500-kW turbine in the previous example produces 2600 kWh/kW of capacity at a 7-m/s site. This figure of merit, like the capacity factor, is influenced by the rated capacity. If this turbine was rated at 750 kW, as are some turbines this size, it would only produce 1700 kWh/kW of capacity.

Annual capacity factor, a related parameter in common use within the electric utility industry, is the quotient of actual generation to potential generation if the wind turbine operated at rated power for the entire year. It too, is dependent on the rated capacity. The 500-kW turbine in the example delivers a capacity factor of 30%. But if the machine were rated at 750 kW, the capacity factor would be only 20%, and appear less productive, even though the turbine had still generated the same amount of electricity. Capacity factors are useful only when the specific capacity of the turbine in kW/m^2 is known. The capacity factor and specific generation per rated kilowatt are useful when the swept area is unknown or uncertain in statistical summaries (for example, see Table 6.10).

Specific yield or annual generation per area swept by the rotor in kWh/m^2 per year is the ideal measure of reliability, efficiency, and a site's wind resource. Specific yield is solely a function of wind regime and wind turbine performance and is independent of the turbine's rating in kilowatts.

Table 6.8
Annual Energy Output for 39-m 500-kW Wind Turbine Using Power Curve[a]

Wind Speed Bin (m/s)	Frequency of Occurrence %	Frequency of Occurrence hr/yr	Capacity kW	Energy kWh
4	0.0992	869	7	6,000
5	0.1074	941	29	27,700
6	0.1080	946	63	59,500
7	0.1023	896	107	95,900
8	0.0919	805	161	129,700
9	0.0788	690	224	154,600
10	0.0645	565	293	165,600
11	0.0507	444	362	160,800
12	0.0383	335	423	141,800
13	0.0278	243	467	113,600
14	0.0194	170	489	83,100
15	0.0131	114	497	56,800
16	0.0085	74	500	37,100
17	0.0053	46	500	23,200
18	0.0032	28	500	14,000
19	0.0019	16	500	8,200
20	0.0011	9	500	4,600
21	0.0006	5	500	2,500
22	0.0003	3	500	1,300
23	0.0002	1	500	700
24	0.0001	1	500	300
25	0.0000	0	500	200
26	0.0000	0	0	0
			Annual energy output	1,287,200

[a] 7 m/s annual average wind speed at hub height with Rayleigh distribution; equivalent to Danish roughness class 1.

Reliability, available wind energy, and turbine rating affect each measure in varying degrees. Turbine rating has a direct effect on generation per unit and an inverse effect on generation per kilowatt and on capacity factor. Increasing a turbine's rated capacity may increase generation slightly by enabling the turbine to capture energy in higher winds while lowering overall capacity factor. As discussed previously, specific rated capacity is a function of wind turbine design. Once turbine design is known, capacity factors can be related to specific yields (Figure 6.11).

The energy in the wind produces a sizable effect on all measures of productivity because of the cubic relationship between speed and power. As noted elsewhere, an increase of only 1 m/s, from a 6-m/s site to a 7-m/s site, will boost productivity 60%. Such variations in wind speed are common from one site to another in California's mountain passes.

Improved reliability has probably played the most important role in improving productivity since the late 1970s, when most wind turbine models were

Table 6.9
Typical Specific Yields

Wind Speed at Hub Height (m/s)	Nominal Wind Speed (mph)	Nominal Energy in the Wind (kWh/m²/yr)	*System Efficiency (%)*		*Nominal Annual Yield (kWh/m²)*	
			Low	High	Low	High
Medium-Sized Wind Turbines						
4	9	700	25	30	160	200
5	11	1300	30	40	400	500
6	13	2200	30	40	700	900
7	16	3500	25	35	900	1200
8	18	5200	25	30	1300	1600
9	20	7500	20	25	1500	1900
Small Wind Turbines						
4	9	700	20	30	130	200
5	11	1300	20	30	300	400
6	13	2200	15	25	300	600
7	16	3500	10	20	400	700
8	18	5200	10	15	500	800
9	20	7500	10	15	700	1100

introduced. Designers and operators have simply made wind turbines work better and more often. During the early years in Denmark, major failures occurred in half of those turbines installed at any one time. The failure rate, though, declined rapidly. Similarly, many California projects in the early 1980s were available for operation only 60% of the time. Availability, the wind industry's measure of reliability, improved rapidly. All projects installed in California after 1987 were available for operation more than 95% of the time, and many are consistently available for operation 97 to 99% of the time (Figure 6.12). Danish utilities have seen the same improvement. From 1987 through 1990, ELSAM averaged 98% availability from the 43 MW it was then operating.[9]

The improvement in availability is a major technological achievement. According to Claire Lees of Field Service and Maintenance, "no other prime mover must run as many hours without major repairs" as modern wind turbines. High availability became such an expected part of wind turbine operations by the mid-1990s that trade publications found newsworthy any hint that a company's availability had fallen to less than 95%.

Improved reliability, improved airfoils, and adoption of taller towers have doubled specific yields in California and Denmark during the past decade. According to data from the California Energy Commission's Performance Reporting System, BTM Consult, and Denmark's Risø National Laboratory,

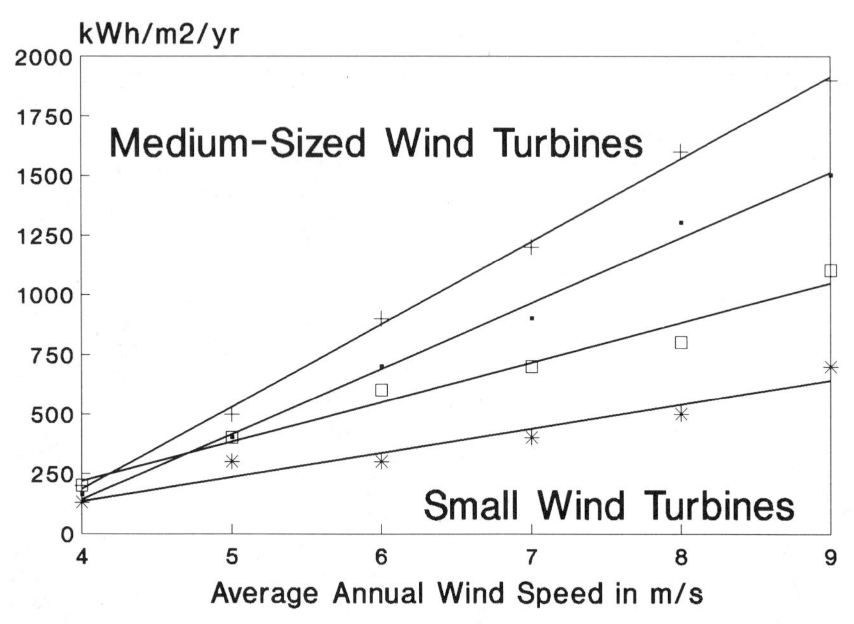

Figure 6.10. Typical specific yields. Energy yield from small and medium-sized wind turbines for hub-height wind speeds derived from manufacturer estimates and field experience. At a 7-m/s site, for example, medium-sized wind turbines produce about 1000 to 1250 kWh/m² per year. Thus a 35-m turbine, which intercepts 1000 m² of the wind stream, will capture 1 to 1.25 million kilowatt-hours per year.

Table 6.10
Danish 1992 Performance Summary

Nominal Size Class					
kW	m	Units	Average 1992 kWh/unit	Average kWh/kW	Average Capacity Factor
55	14–16	144	109,188	1985	0.23
75	17	76	165,280	2204	0.25
95	18	70	214,765	2261	0.26
99	19	82	217,822	2200	0.25
150	21	185	356,609	2377	0.27
200	23–25	74	400,309	2002	0.23
225	27	43	564,434	2509	0.29
250	25	57	396,768	1587	0.18
		731			

Source: Wind Stats, Winter 1993, Vol. 6, No. 1, p. 6.

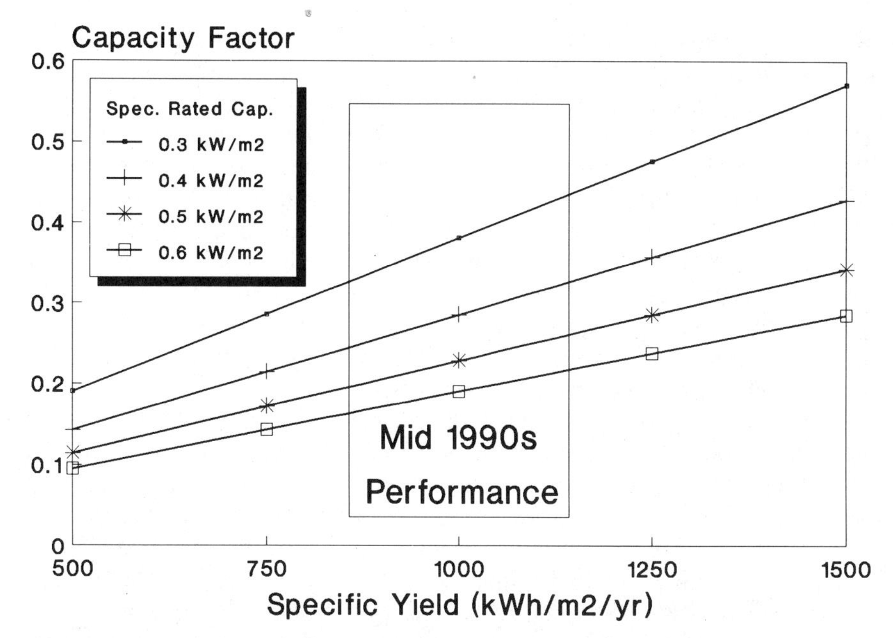

Figure 6.11. Equivalent capacity factor. The capacity factor for various specific energy yields relative to specific rated capacity. Specific yield is a function of wind regime and wind turbine performance. Specific rated capacity is a function of wind turbine design.

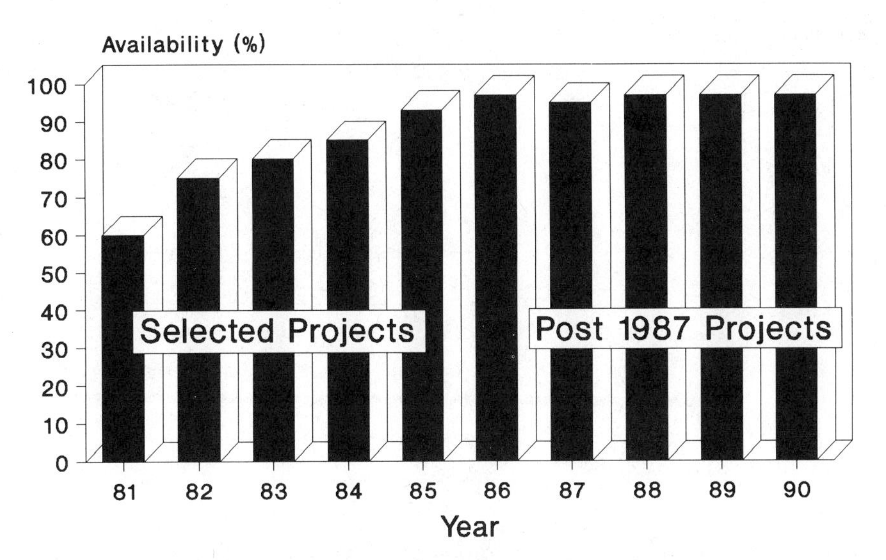

Figure 6.12. Percentage of time California wind power plants are available for operation.

the specific yield of individual wind turbines and wind power plants in California and Denmark has steadily increased since the early 1980s (Figure 6.13).

Although there has been a significant increase in specific yield of each succeeding design iteration, there has also been a steady improvement over time within each size class. For example, the average performance of early 55-kW Danish turbines increased 25%, from about 400 kWh/m^2 per year to about 500 kWh/m^2 per year. Later models of this class, machines 14 to 16 m (45 to 50 ft) in diameter, incorporated lessons learned from field experience with the earlier turbines, pushing productivity to nearly 500 kWh/m^2 per year. But productivity improved most dramatically as Danish manufacturers introduced newer, larger turbines and reached an average specific yield of 850 to 900 kWh/m^2 per year for the 450-kW model.[10]

The same pattern can be seen in California where the fleet average increased 40%, from 500 kWh/m^2 per year in 1985 to 700 kWh/m^2 per year during the early 1990s, as wind companies mastered the art of maintaining wind turbines. As in Denmark, newer turbines were more productive than earlier designs. The specific yield of wind turbines installed since 1985 reached 850 kWh/m^2 per year during the late 1980s. Improvement in the statewide capacity factor matched that of specific yield, increasing from 13% in 1985 to about 20% during the early 1990s. The capacity factor of turbines installed since 1985 averaged 23 to 25%.[11] Productivity will continue rising in California, although less dramatically than during the 1980s, as inoperative turbines installed during the industry's formative years are returned to service or are replaced

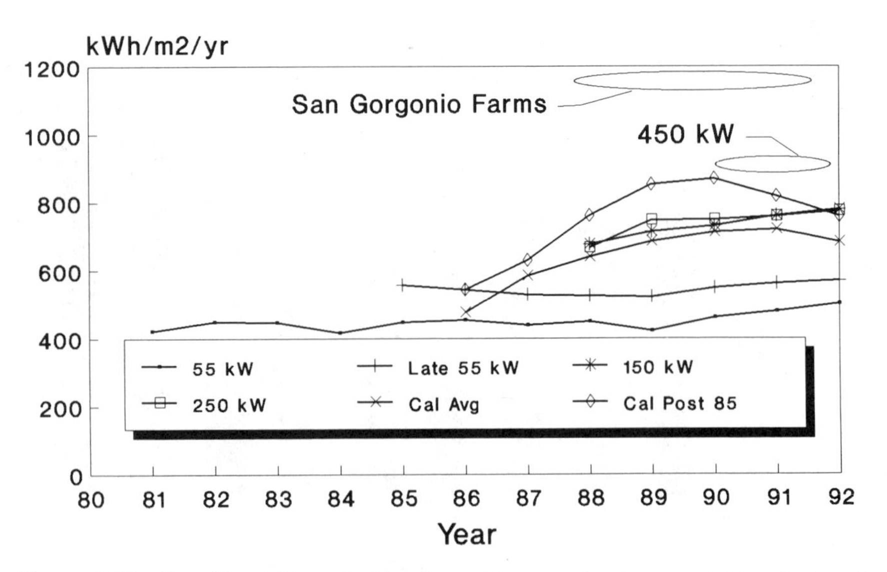

Figure 6.13. Specific yield in California and Denmark. Energy yield in kWh/m^2 per year has increased over time and with the introduction of successive wind turbine designs.

Figure 6.20. Typical second-generation European wind turbine. WEG MS-2 in California's Altamont Pass. The MS-2 uses a variable pitch rotor 25 m (80 ft) in diameter upwind of the tower.

design philosophies. Not only did European turbines use rotors upwind of the tower, they operated them at much lower speeds than those of American wind turbines. Moreover, European wind turbines simply appeared much more massive than American machines: their blades were thicker and their nacelles bulkier. They projected a sense of solidity and durability missing from the frail, frantically flailing arrays of American turbines.

Their appearance betrays an underlying difference in design. First-generation European wind turbines (the nacelle and rotor) weigh up to three times as much as American designs relative to the area swept by their rotors. The specific tower head mass of a Vestas V17, at 27 kg/m^2 of rotor area, contrasts sharply with that of U.S. Windpower's model 56-100 at 10 kg/m^2,

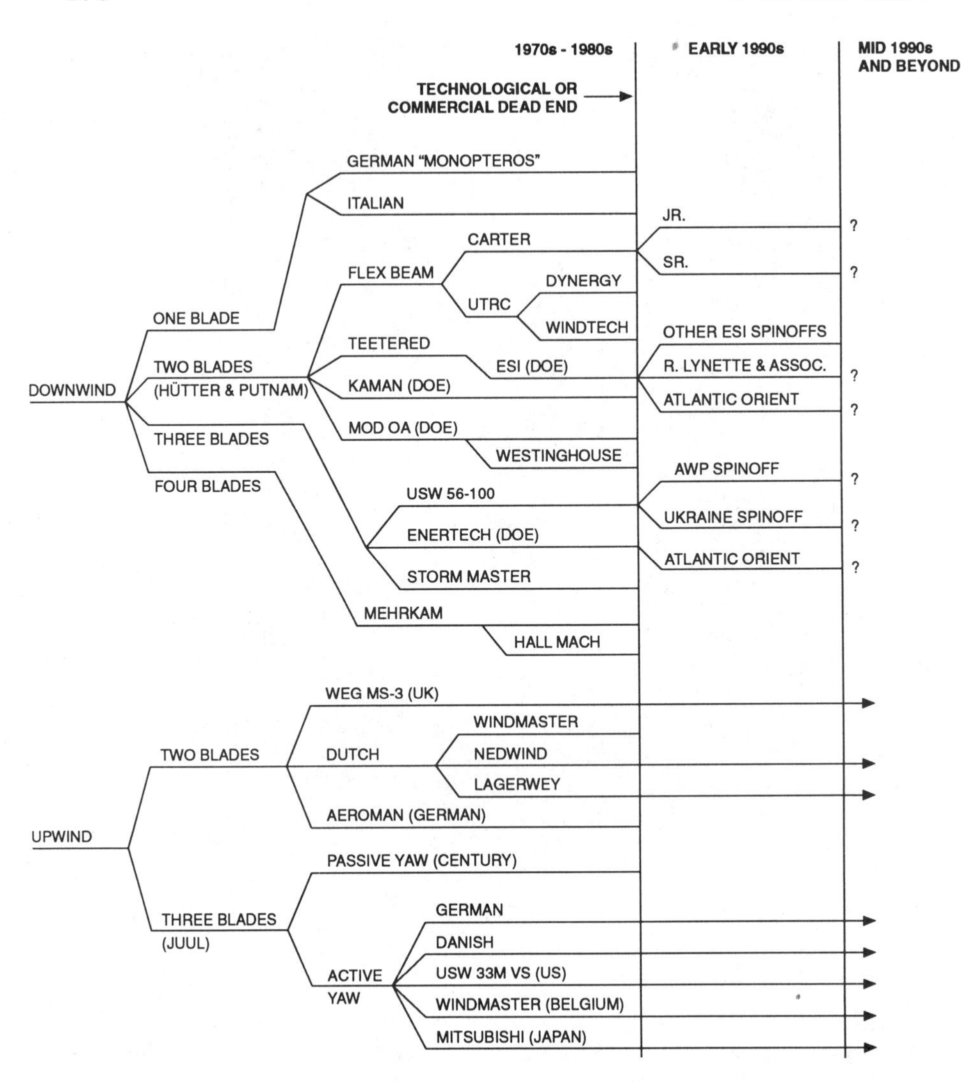

Figure 6.21. Technological pathways for HAWTs. Derivation and possible fate of medium-sized horizontal-axis wind turbines. Most successful medium-sized wind turbines use rotors with three blades upwind of the tower. The future of downwind technology is uncertain.

even though both turbines intercept about the same area. Carter and Storm Master turbines weigh even less, with specific mass below 10 kg/m^2 (Figure 6.22).

Critics of early lightweight American designs label them "flimsy," while their defenders argue that these turbines use materials more efficiently and "do more with less," as Buckminster Fuller admonished engineers. Designers of lightweight turbines charge that Danish designs are "bulky" and that the Danish approach to engineering problems is to "throw more steel at it."

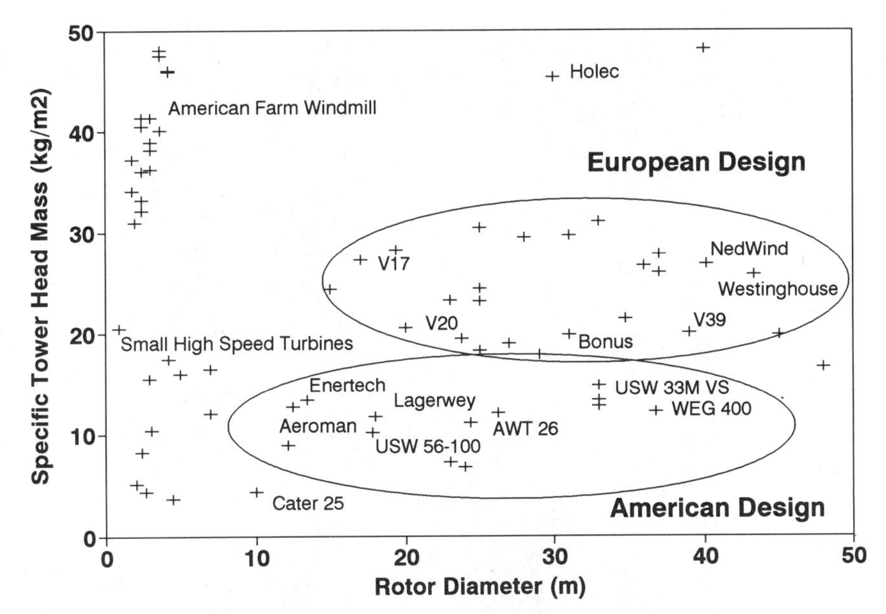

Figure 6.22. Specific rotor and nacelle mass. Typical European wind turbines are more massive relative to their swept area than are American designs.

Danish and Dutch engineers retort that steel is cheap in comparison to wind turbines that fail in the field. The verbal battle thus seesaws endlessly back and forth.

There are important exceptions to this American–European division over design. The specific mass of the Dutch Lagerwey (12 kg/m^2), the German Aeroman (13 kg/m^2), and the WEG MS-3 (13 kg/m^2), which use only two blades, is less than that of some wind turbines manufactured in the United States. The multiblade American farm windmill is also extremely massive for its small size. Multimegawatt turbines, because of the increasing influence of gravity as size increases, weighed in at up to 70 kg/m^2.

The trend in Europe is toward lighter machines, such as the Bonus Combi and Vestas V39, while that in the United States is toward heavier machines, such as U.S. Windpower's 33M-VS, a hybrid that falls midway between the two camps at about 15 kg/m^2. The relative weight of the 33M-VS is lower than that of Danish designs but is 50% heavier than its predecessor, USW's model 56-100.

Few wind turbines epitomize U.S. lightweight design philosophy better than those designed by the Carter family (Figure 6.23). Jay Carter, Jr.'s unique rotor design passively pitches the blades toward stall to regulate power in high winds. There is no cutout speed, and Carter operates his turbines in hurricane-force winds. The stall-regulated rotor uses highly flexible blades that sweep progressively downwind, like the fronds of a palm tree, as wind speed increases. Carter rates his lightweight, high-performance turbine at

Figure 6.23. Carter 300.

300 kW, operating the 24-m (80-ft) rotor under an unusually high load. Most wind turbines of comparable size would drive only 200-kW generators. The Carter designs, including Jay Carter, Sr.'s Wind Eagle, have been perennial favorites of U.S. engineers, offering "promise" since the late 1970s (Figure 6.24). To the Carter family's credit, their pioneering designs evolved independently of the DOE wind program and without federal subsidies. Through sheer force of will and persistence, the Carters have continually improved the performance of their designs, although their future is uncertain.

Other turbines characterizing lightweight designs of the early 1980s include Storm Master, Windtech, and ESI. Like the Carters, Ed Salter is an independent American proponent of lightweight, flexible rotor systems. Salter, who has experimented with novel wind turbine designs for nearly two decades, created a three-blade downwind design he called Storm Master. This 12-m 40-kW turbine has only slightly more specific mass (9 kg/m^2) than that of the Carter

Figure 6.24. Carter, Sr.'s Wind Eagle. The ultimate "American" design: lightweight, downwind, two-bladed rotor, passive yaw control, passive pitch control, passive azimuth control. Two Carter 25s in the background.

turbines (4 to 7 kg/m^2) (Figure 6.25). Some 400 of the machines with their whiplike pulltruded fiberglass blades were eventually installed in California. Only a portion were still in service by the mid-1990s.

One of the most spectacularly unsuccessful designs to evolve from the federal wind program in the United States was that of United Technologies Research Center. This first-generation American design featured a downwind rotor incorporating passive yaw and passive pitch control of slender pulltruded fiberglass blades. As the Carters had before, UTRC used a torsionally flexible spar that allowed the blade to change pitch or twist toward stall via an ungainly system of pitch weights and metal straps. The flexibility of the spar eliminated

Figure 6.25. Storm Master. The turbine shown here uses rigid wood-composite blades as a replacement for the original slender fiberglass blades.

the need for bearings between the blades and the hub. Although technically elegant, the flexibility of the blades induced complex and ultimately uncontrollable dynamics. Windtech commercialized a 15.8-m (52-ft) version, and more than 200 of the problem-prone machines were installed in California. Dozens of them stand derelict in the Tehachapi Pass today, nearly a decade after they ceased working (Figure 6.26).

ESI was another commercial spin-off from the federal wind program. The ESI turbines are distinctive for their fairly large diameter (they were an unreliable U.S. equivalent of first-generation Danish designs), the high speed of their rotors, and the tip brake at the end of each blade. The tip brakes were intended to protect the turbine during high wind emergencies. The design uses two fixed-pitch wood-epoxy blades downwind of its hinged lattice tower.

Figure 6.26. Windtech. Commercial derivative of DOE small wind turbine program.

The turbines were serviced either standing upright or laid on the ground. Because of their high rotor speed and their tip brakes, the ESI machines are among the noisiest wind turbines in commercial use. Their characteristic "whop-whop" particularly annoyed neighbors near Palm Springs. Two versions were introduced: nearly 700 of the ESI-54 (16.4-m) and 50 of the ESI-80 (24-m) were installed in California (Figure 6.27).

These turbines have consistently and substantially underperformed European designs and U.S. Windpower's successful model 56-100. Carter and Enertech produced, on average, 400 to 600 kWh/m^2 per year while ESI, Fayette, Storm Master, and Windtech generate 200 to 400 kWh/m^2 per year at best (Table 6.12). For comparison, the average specific yield from all turbines in California during the same period was about 700 kWh/m^2 per year, and nearly

Figure 6.27. ESI-54. Cluster of early ESI-54s at Whisky Run on the coast of Oregon. The turbines have since been removed.

800 kWh/m^2 per year when these turbines were excluded (refer to Figure 6.13).

Drivetrains

The difference between "American" and "Danish" design can be seen in their distinctly different drivetrains: the arrangement of shafts, gearboxes, and generators that convert the motion of the spinning rotor into electricity. Designers of lightweight turbines frequently used integrated drivetrains to conserve materials and simplify assembly (Figure 6.28).

Integrated drivetrains were first developed by Hütter in Germany, so it is not surprising that MAN would have opted for this configuration. As on Carter, Enertech, ESI, and Windtech turbines, the Aeroman drivetrain is integrated with the frame of the wind turbine. The gearbox housing acts as the frame supporting the main (slow-speed) shaft of the rotor as well as the generator. Unlike most U.S.-built machines of the period, the 12.4-m Aeroman uses an upwind rotor with variable-pitch blades. MAN engineers also placed the brake on the main shaft near the rotor, whereas American designers universally placed the brake on the high-speed side of the transmission.

Rather than integrate the drivetrain, Danish manufacturers fabricated a metal frame or bed plate, to which they mounted the main shaft, transmission,

Table 6.12
Performance of Selected American-Designed Turbines in California

	Carter		Enertech		ESI		Fayette		Storm Master		Windtech	
Year	kWh/m²	Units	kWh/m²	Units	kWh/m²	Units	kWh/m²	Units	kWh/m²	Units	kWh/m²	Units
1985		305		488		716		1,468		310		212
1986	669	398	244	533	198	722	421	1,455	118	294	112	207
1987	424	368	434	485	244	587	502	1,400	226	286	51	128
1988	257	298	614	485	299	354	423	1,370	152	167	16	130
1989	504	363	483	396	333	353	365	1,362	412	100	128	15
1990	584	168	294	475	261	433	194	1,351	290	10	108	5
1991	531	165	455	475	261	343	243	1,351	355	105	125	10
1992	495	140	412	451	263	343	167	1,351	404	20	—	5

Source: Results from the Wind Project Performance Reporting System, Annual Reports (Sacramento, CA: California Energy Commission, 1985–1992).

generator, and other components (Figure 6.29). The separate main shaft and its support bearings on Danish machines allowed the transmission to be replaced readily without requiring removal of the rotor. Danish manufacturers Bonus and Vestas placed the brake on the main shaft, while Nordtank, Micon, and others placed the brake on the output side of the transmission.

There has been a tendency among proponents of integrated drivetrains to abandon the nacelle cover, considering it an unnecessary amenity for protecting drivetrain components from the elements. Neither of the prototypes built by Atlantic Orient and Northern Power Systems in NREL/DOE's AWT program incorporate nacelle covers. Early U.S.-designed turbines that used integrated drivetrains also lacked work platforms for servicing the turbines.[15]

The design of U.S. Windpower's model 56-100 diverges from other American turbines of the period even though it passively orients the rotor downwind of the tower. Like that of Danish turbines, the model 56-100's drivetrain uses modular construction with a separate main shaft assembly, gearbox, and generator. U.S. Windpower also encloses the nacelle and provides work platforms. Unlike Danish manufacturers, which frequently introduced new designs, USW has refined their variable-pitch rotor through several iterations (Figure 6.30). No Danish turbine commercially employed variable pitch until the late 1980s when Vestas introduced its V25 series (Figure 6.31). By then, U.S. Windpower had nearly a decade of experience with variable-pitch control.

At about the time that Vestas began marketing its variable-pitch series, British manufacturer WEG introduced its MS-2 (refer to Figure 6.20), which differs from the Vestas design by hanging the rotor directly on the transmission's input shaft, an arrangement popularized by Belgium's WindMaster in the early 1980s. WEG followed this pattern with its subsequent MS-3. Several manufacturers have since adapted this approach to their turbines. Examples are U.S. Windpower's 33M-VS and Zond's Z-40.

Danish manufacturer Bonus has modified the traditional modular drivetrain by placing the gearbox at the end of an unusually long main shaft (Figure 6.32). This allows torsional flexure in the drivetrain to absorb some of the

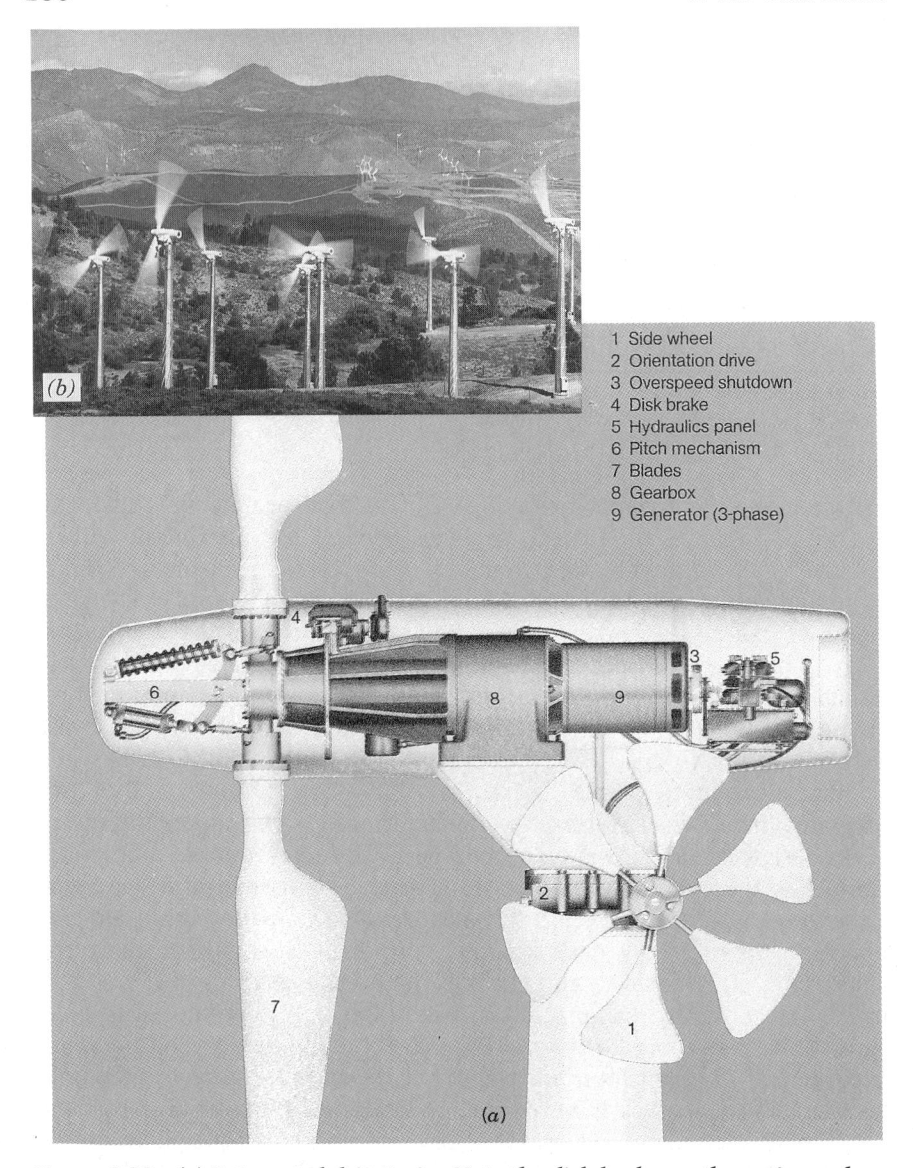

Figure 6.28. (a) Integrated drivetrain. Note the disk brake on the main or slow-speed shaft. The fan tail orientation has been superseded by an active yaw drive (wind vane and yaw motor). (Courtesy of Jacobs Energie, Heide, Germany.) (b) Aeromans atop Cameron Ridge near Tehachapi, California.

torque loads in gusty winds. The transmission also counterbalances the weight of the rotor. Bonus also uses internal yaw gears to orient the nacelle. This provides more contact between the yaw pinion on the nacelle and the bull gear on the tower than do external yaw rings.

As turbines have grown larger, engineers have sought novel ways to handle the wind's thrust on the rotor, the rotor's weight, and the torque the rotor

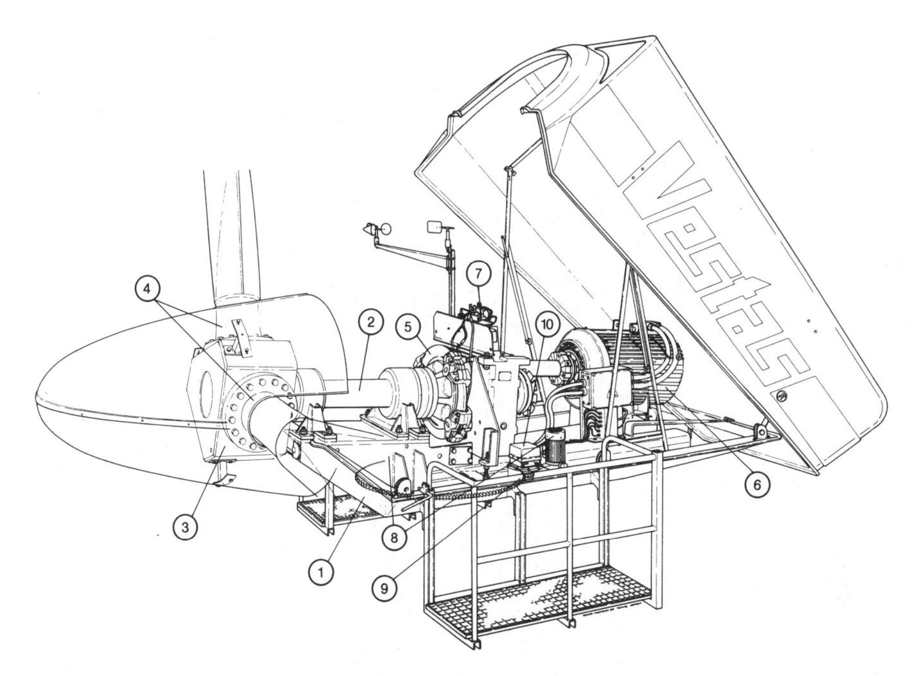

Figure 6.29. First Generation Danish Drive-Train. Vestas V20 is a 20-meter derivative of earlier Vestas designs using fixed-pitch blades. Nomenclature: 1. Bed plate or frame, 2. Main shaft, 3. Rotor hub, 4. Blades, 5. Disk brake on main or slow speed shaft, 6. Generator with dual windings, 7. Gear box or transmission, 8. Yaw drives, 9. Yaw ring, 10. Yaw gear and yaw brake. Note the work platform, large nose cone, and nacelle cover. (Courtesy of Vestas-Danish Wind Technology, Lem, Denmark.)

transmits to the transmission. The Bonus 450-500 uses an unusual cast frame rather than the more common welded assembly that supports the main bearings within the hub. Unlike conventional Danish designs, the main shaft does not carry the weight or thrust of the rotor. It only transmits torque (Figure 6.33). When placed within the hub the main bearings carry no overhanging load.

Mitsubishi has attacked the problem in a similar manner. It uses an extension of the nacelle's frame to support the rotor. The Mitsubishi design is also novel in other ways because it uses bulkheads and a tubular frame to provide structural rigidity. This construction is common in ship building (Figure 6.34). Both Mitsubishi Heavy Industries and Germany's Husum Schiffswerke, which uses a similar technique, build ships. Mitsubishi uses a form of blade attachment that has since fallen out of favor. The blade spindle, or kingpin, is seldom found among commercial variable-pitch wind turbines in this class. Most variable-pitch turbines use a large-diameter turntable bearing instead. The version of this turbine found in California has been plagued by numerous hydraulic leaks that stain the blades and towers.

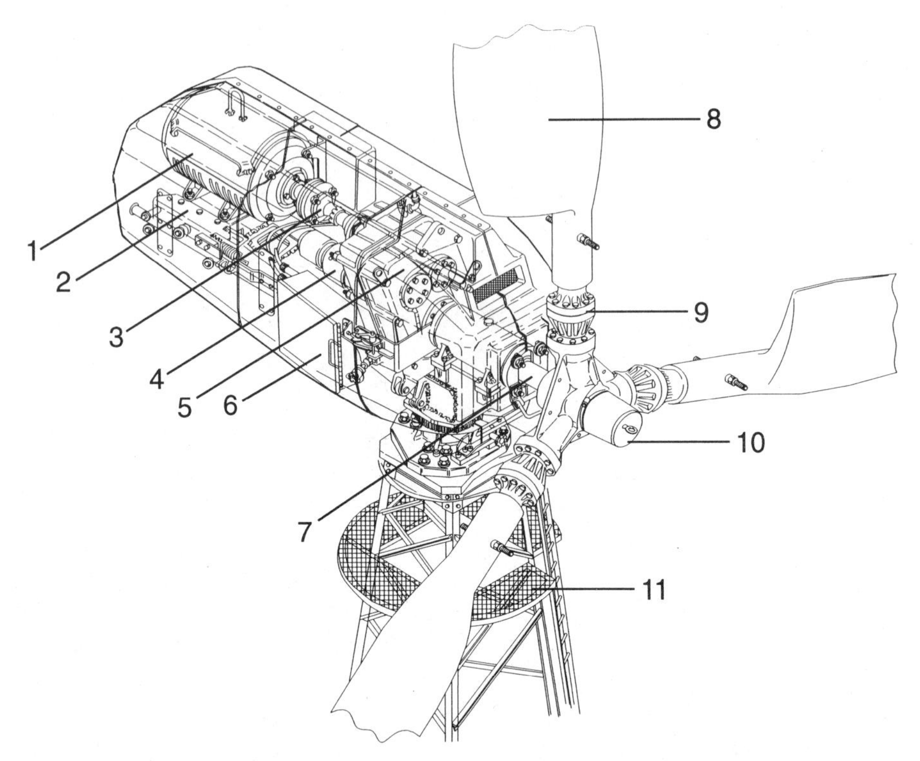

Figure 6.30. U.S. Windpower 56-100. A downwind variable-pitch turbine with passive yaw. Nomenclature: 1, generator; 2, bed plate or frame; 3, coupling; 4, pitch mechanism; 5, gear box; 6, top control unit; 7, main shaft; 8, blade; 9, blade root; 10, hub; 11, work platform; (Courtesy of KENETECH Windpower, San Francisco, CA. Copyright KENETECH Windpower, Inc., 1988.)

Bonus' Combi and 450 are serviced from inside the nacelle, as are WindMaster, WEG, Vestas (V25 and larger), Zond Z-40, and Mitsubishi. But the Mitsubishi nacelle is cramped and difficult for large-framed Americans to service.

Transmissions

Designers specify transmissions that use either parallel shafts, the conventional choice for medium-sized wind turbines, or planetary transmissions. Parallel shaft gearboxes occupy more space and weigh more than planetary transmissions, but are quieter. ESI, Windtech, Carter, and Mitsubishi all use planetary transmissions, and the resulting mechanical noise is a characteristic of the turbines.

Transmissions or gearboxes typically can increase the speed of the main shaft up to a maximum of 6:1 per stage. For example, a two-stage transmission would increase rotor speed a total of 36:1.[16] The high rotor speed of some

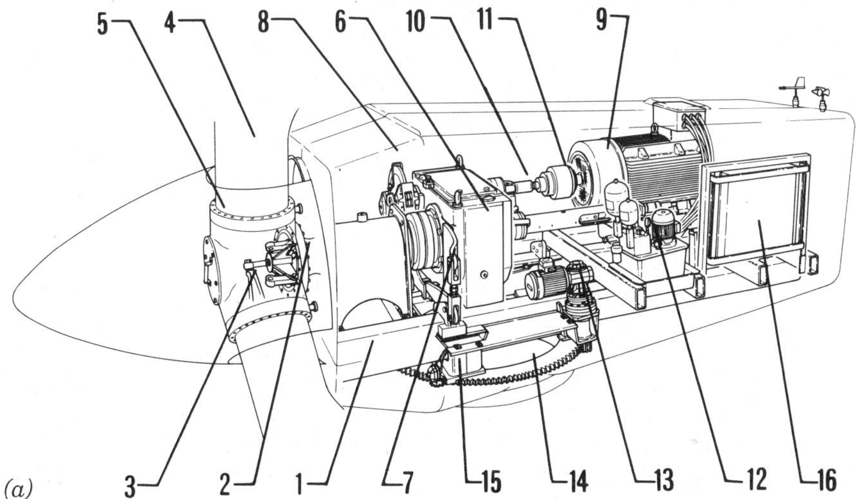

Figure 6.31. Vestas V27. (a) *An upwind variable-pitch Danish turbine. Nomenclature: 1, bed plate or frame; 2, main shaft; 3, pitch mechanism; 4, blade; 5, blade root; 6, gear box or transmission; 7, gear box torque relief; 8, disk brake; 9, generator; 10, coupling; 11, clutch; 12, hydraulic unit; 13, yaw drive; 14, yaw ring; 15, yaw brake; 16, top control unit. (Courtesy of Vestas-Danish Wind Technology, Lem, Denmark.)* (b) *Coastal cluster of Vestas V27s sandwiched between the Cumbrian Mountains and the Irish Sea at the former Haverigg air base, Great Britain.*

early American designs, coupled with their 1800-rpm generators, enabled their manufacturers to cut costs by reducing the size of the gearbox needed. Most medium-sized wind turbines use two-stage transmission, although larger turbines will use three stages (Table 6.13).

The number of stages influences cost and efficiency. At rated power, transmission consumes 1 to 2% of the rotor's power per stage. Because the transmission consumes a certain minimum amount of power, losses in the gearbox

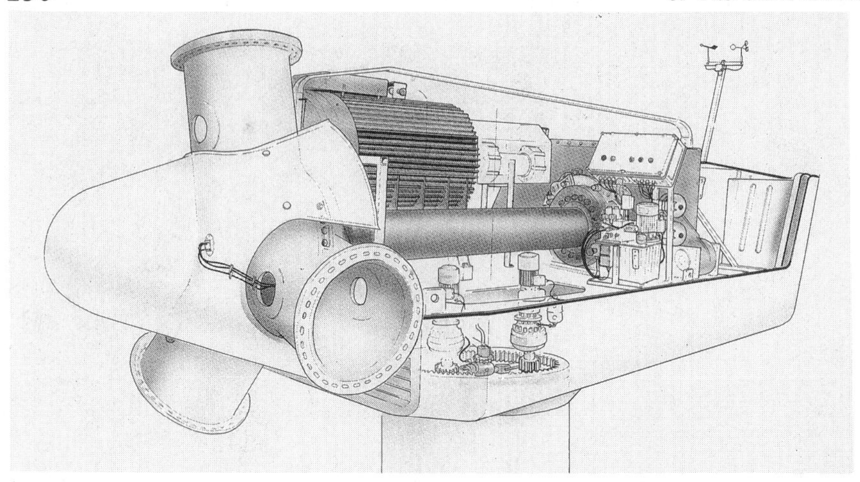

Figure 6.32. Bonus Combi. The 31-m Bonus Combi uses a conventional Danish configuration: rigid three-bladed upwind rotor. The fiberglass canopy completely encloses the nacelle to reduce noise. (Courtesy of Bonus Energy, Brande, Denmark.)

can become considerable at low power.[17] Designers should be able to boost low-power efficiency through dual output shafts, as in USW's 33M-VS or NedWind's 40-m model. For low-speed operation only one generator would need to be energized, and multiple shafts would permit using the minimum number of gear stages possible for a given size.

Many parallel shaft transmissions on larger turbines are mounted or hung on the main shaft rather than mounted directly on the frame of the turbine. Spring dampening these shaft-mounted gearboxes allows the transmission to absorb fluctuations in torque caused by gusty winds, introducing needed compliance into the drivetrain.

Gearboxes are not intrinsic to wind turbines. Designers use them only because they need to increase the speed of the slow-running main shaft to the speed required by mass-produced induction generators. Designers can opt for special-purpose, slow-speed generators and drive them directly without using a transmission. For this reason, specially designed permanent-magnet alternators have revolutionized the reliability and serviceability of small wind turbines. By mid-1994 one German manufacturer of medium-sized turbines, Enercon, had installed 100 units of its novel direct-drive design, in which the three-blade, 40-meter rotor drives a large 500-kW ring generator with specially designed electromagnets.

Aerodynamics

How to win the most energy from the wind striking a wind turbine rotor is exceedingly complex. Aerodynamicists Stoddard and Eggleston devote an en-

Figure 6.33. (a) Bonus 450 drive train. (Courtesy of Bonus Energy, Brande, Denmark.) (b) Bonus 450 in operation at Bryn Titli, Wales.

tire book, *Wind Turbine Engineering Design,* to the subject. Suffice it to say that modern wind turbines use very little material to capture the energy in the wind stream. They employ low-solidity rotors; that is, the two or three blades used occupy a small portion of the area swept by the rotor: from 3% for two-bladed U.S. turbines to 8% for three-bladed Danish designs.[18] The

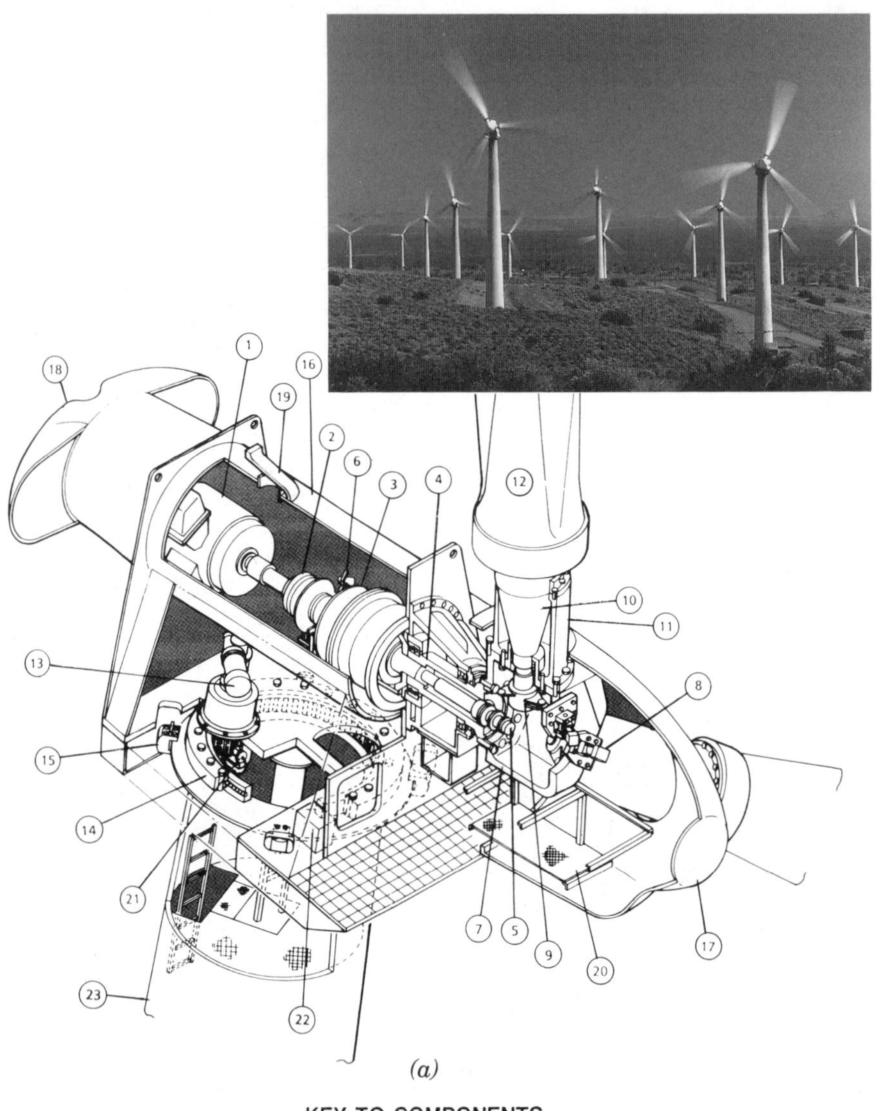

KEY TO COMPONENTS

① Generator	⑨ Link	⑰ Front capsule
② Coupling	⑩ Blade spindle	⑱ Rear capsule
③ Main gear	⑪ Blade guide	⑲ Manhole
④ Main shaft	⑫ Blade	⑳ Maintenance floor
⑤ Rotary joint	⑬ Yaw gear device	㉑ Yaw gear stopper
⑥ Main shaft brake (hand brake)	⑭ Yaw bearing	㉒ Oil unit
⑦ Rotor head	⑮ Yaw brake	㉓ Tower
⑧ Hydraulic cylinder	⑯ Nacelle	

Figure 6.34. Ship or Bulkhead Frame. (a) Mitsubishi 31-m wind turbine. (Courtesy of Mitsubishi Heavy Industries, Nagasaki, Japan.)

Table 6.13
Transmission Gear Ratio for Selected Turbines

	kW	Rotor Diameter (m)	Rotor rpm	Generator rpm[a]	Gear Ratio	Number of Stages
USW 56-100	100	17.6	71	1500	21	2
V20	100	20	46	1010	22	2
Carter	300	24	65	1522	23	2
V27	225	27	35	1010	29	2
USW 33M VS	330	33	11–53	500–2400	45	2
DWT	400	34.6	33	1010	31	3
V39	500	39	30	1522	51	3
E40	500	40	18–41	18–41	1	0

[a] For 50 Hz in Europe.

blades on these turbines typically taper and twist from hub to tip (Figure 6.35). Why they do so is more difficult to explain.

Apparent Wind and the Angle of Attack

The sum of lift and drag acting on a wind turbine blade generates a force that pulls the blade on its journey through the air, much as it pulls a sailboat through the water (Figure 6.36). Lift is determined by the blade's angle of attack, the angle with respect to the apparent or relative wind. Lift increases with increasing angles of attack until the airflow over the blade becomes turbulent. Lift then deteriorates rapidly, drag increases, and the airfoil is said to stall. The angle of attack at which this occurs varies from one airfoil to another.

The angle of attack is a function of the blade's angle to the plane of rotation—its pitch—and the apparent wind. The apparent wind is the wind encountered by the blade as it moves through the air: the resultant of the airflow due to the blade's own motion and the airflow from the wind. The apparent wind is dependent on the relative magnitude and position of each.

If wind speed increases while the blade's speed through the air remains constant, the position of the apparent wind rotates toward the wind direction. As the apparent wind changes position, it changes the angle of attack. Designers must decide how best to deal with this relationship, because there is an angle of attack for every airfoil where the lift/drag ratio is optimum and performance reaches a maximum.

To maintain an optimum angle of attack as wind speed increases, a fixed-pitch blade must increase its speed proportionally. Thus to maximize aerodynamic performance, the rotor must spin faster as the wind speed increases so that the blade speed will increase and the ratio of blade speed to wind speed (the tip-speed ratio) remains constant. Most small wind turbines operate this way. They do so because it is not only more efficient, but also simpler.

Figure 6.35. Taper and twist. Wind turbine blades taper and twist from hub to tip. Micon 108 near Tehachapi, California.

Before continuing it is important to note that the pitch of blades on nearly all medium-sized wind turbines is fixed through their operating range. Blades on variable-pitch turbines are no exception. Contrary to popular belief, these turbines vary the pitch of their blades only after they have reached their rated capacity.

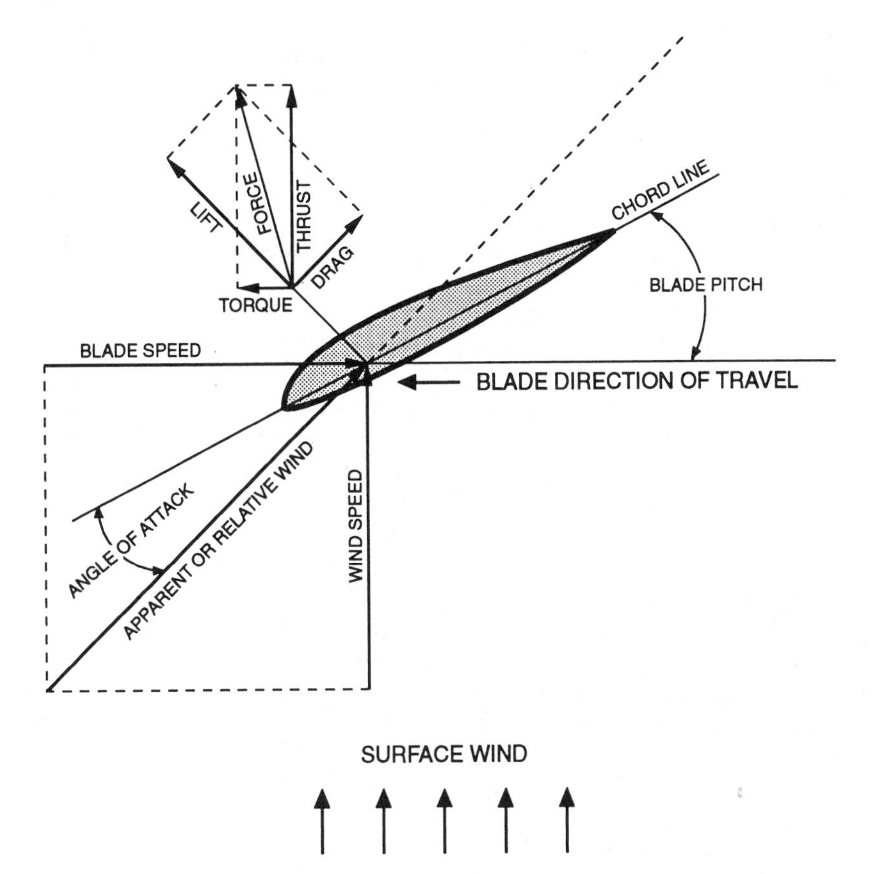

Figure 6.36. Airfoil. The relative wind and the pitch of the blade determines the angle of attack and the airfoil's resulting performance. The wind speed shown is not the true free-stream velocity, because wind speed decreases as the rotor extracts energy from the wind.

In contrast to small wind turbines, most medium-sized wind machines operate at constant speed. As wind speed changes, these turbines continue to spin at the same speed because of the induction generators they use. Their rotors operate at varying tip-speed ratios, both above and below the optimum. Designers are willing to sacrifice some performance for the simplicity of fixed-pitch blades driving constant-speed generators. On stall-regulated turbines, the airfoil begins to stall and performance declines as wind speed increases. This reduces the rotor's power in high winds, making it easier for designers to build protective controls to keep the rotor from destroying itself. On pitch-regulated turbines, the blades change pitch in high winds, dumping excess power.

The amount of force driving the blade forward is not only a function of the airfoil's lift and the blade's angle of attack, but also of the area of the blade and its speed through the air.

Taper and Twist

The speed at which a point on the blade moves through the air varies with the radius it describes. The tip travels faster through the air than a point near the hub. The direction and magnitude of the apparent wind will also vary with radius. The apparent wind increases in strength, and its position shifts toward the plane of rotation, as radius increases. To maintain the angle of attack (to optimize performance) as blade speed increases with increasing radius, the pitch of the blade must decrease toward the tip. As a result, wind turbine blades are twisted from root to tip. The pitch is greatest at the root and smallest at the tip, where the blade is parallel to its direction of travel.

Wind machine designers long ago learned that blade area (the product of the number of blades times their length and width) governs the amount of torque a rotor can produce. The more blade area the rotor has for the wind to act on, the more torque it will produce. Greater solidity generates greater torque. The American farm windmill uses multiple "sails" that nearly cover the entire rotor disk (80% solidity). It was designed to deliver high torque for pumping water in the low winds of late summer, and it does its job remarkably well.

The multiblade farm windmill looks like it would capture more wind than a modern machine with two or three slender blades. Intuitively, we feel that the rotor should have more blades to capture more wind. If this were true, however, the optimum rotor would cover the entire swept area with blades, in effect producing a solid disk. The wind would pile up in front of the rotor and flow around rather than through it. Some wind must move through the disk and it must keep moving to make way for the air behind. A spinning rotor can capture the energy in the wind, despite the gap between blades, either by using many wide blades moving through the air slowly or by using two or three slender blades moving swiftly.[19]

The wide vanes of the farm windmill deflect the wind substantially as it moves through the rotor. The combined effect of the multiple blades on the wind as the wind moves across them causes the wind leaving the rotor to spiral like a corkscrew. Albert Betz demonstrated mathematically that there was an optimum balance between the amount of wind striking the rotor and the amount flowing through it. To maximize the amount of work the wind can perform, designers minimize this deflection and spiraling of the wind stream in the turbine's wake.

This spiraling effect is greatest, as is the amount of power lost, in rotors producing high torque. High-solidity rotors such as the farm windmill produce plenty of torque, but they also lose more energy in the wake than do lower-solidity rotors delivering less torque.

Power is a product of torque and rotor speed. To deliver the same amount of power from a rotor with reduced torque, the rotor's speed must be increased. Stoddard and Eggleston summarize this relationship by noting that a turbine producing high torque at low rotational speed (the farm windmill) is less efficient than a turbine rotating at high speed and producing little torque.[20] Conversely, decreasing torque while increasing rotor speed will improve the rotor's efficiency.

Lift provides the force needed to produce torque and is a function of blade area and speed. As radius increases, blade speed and hence lift both increase. Thus torque increases with increasing radius. To minimize torque and rotation of the wake as radius increases, blade area must decrease. For constant lift over its length as blade speed increases toward the tip, the chord (width) of the blades must decrease. The blade, as a result, tapers from the root to the tip (Figure 6.37). A Vestas V27 blade, for example, tapers from 1.3 m (4.3 ft) in width at the hub to 0.5 m (1.6 ft) in width at the tip. These blades are also twisted and have a pitch of 13° at the hub and 0° at the tip. The twist is identical to that of the Gedser mill, the forerunner of modern Danish turbines.

Tip-Speed Ratio

Although efficiency improves with increasing rotor speed, there are practical limits. Noise is directly proportional to the speed of the blade tip. The tip speed of European machines has increased from 40 m/s on first-generation designs to about 60 m/s during the mid-1990s. Higher tip speeds are unlikely

Figure 6.37. Unloading Vestas V27 blade near Tehachapi, California.

because of concern about noise.[21] Danish turbines characteristically operate at lower tip speeds and lower tip-speed ratios at their rated output than do high-performance turbines (Table 6.14).

Blade speed increases with rotor speed and radius. To maintain an optimum tip-speed ratio for best aerodynamic performance, rotor speed will vary inversely with rotor diameter. As turbines increase in size, the speed of their rotors decreases.

Operating the turbine at variable speed permits the designer to maintain the relationship between tip speed and wind speed, the tip-speed ratio, for the optimum angle of attack. Until recently this was impractical in medium-sized wind turbines. Most turbines were limited to operating at constant speed by their induction generators. Consequently, the rotor's performance declined when wind speed increased above its target wind speed, altering the relative wind and the angle of attack. Designers turned this limitation to advantage by judiciously using aerodynamic stall to degrade performance in high winds on constant-speed machines.

Self-Starting

Low-solidity rotors do have a drawback: they may not be self-starting. One solution for rotors using fixed-pitch blades is to spin the rotor up to a speed where it can drive itself. This is a common practice for Darrieus turbines. There are also conventional wind turbines, such as the ESI and Enertech, which motor the rotor up to speed. But most designers are willing to sacrifice a slight amount of performance to gain a self-starting capability. Stoddard and Eggleston note that one advantage of twist and taper is that it can provide sufficient torque to start the rotor in light winds.[22]

Darrieus turbines are typically not self-starting, although it is now known that Darrieus turbines can self-start under the right wind conditions. These

Table 6.14
Tip-Speed Ratios at Rated Power for Selected Turbines

	Diameter (m)	rpm	Rated Wind Speed (m/s)	Tip Speed (m/s)	Constant or Variable Speed	Tip-Speed Ratio
Farm windmill	3–7				v	1
Dutch windmill	25	25	12	33	v	2.7
Bonus Combi	31	31	15	50	c	3.4
Vestas V39	39	30	16	61	c	3.8
Vestas V27	27	44	15	62	c	4.1
USW 56-100	17.6	72	13	66	c	5.2
Mitsubishi	28	48	13	70	c	5.5
Enercon E40	40	36	13	75	v	5.8
WEG 400	36.7	40	12	77	c	6.4
Monopteros 50	56	43	11	125	v	11

conditions, though infrequent, do occur. Normally, the rotor must be motored up to speed.

An H-rotor with articulating blades is self-starting. The pitch of each blade is set according to a predetermined schedule and the position of the blade relative to the wind. The blade's angle of attack is optimized at each position of its orbit around the rotor's axis. Controlling blade pitch with respect to the wind gives the rotor a reliable self-starting capability not found in the ϕ-configuration Darrieus. But like so many other designs, giromills have never lived up to expectations. They are also more material-intensive than are conventional turbines.

Two or Three Blades

A great debate rages among engineers as to whether wind turbines should use one, two, or three blades. (There is general agreement that no more than three are ever needed except for applications requiring high starting torque.) The dispute revolves around aerodynamic efficiency, complexity, cost, noise, and aesthetics.

Only one blade is needed to capture the energy in the wind. To sweep the rotor disk effectively, a one-bladed turbine must operate at higher speeds than a two-bladed turbine, thereby reducing the gear ratio required for the transmission, and its associated mass. Proponents argue that one slender blade delivers optimum engineering economy because of the resulting rotor's extremely low solidity: the blade itself covers only a small portion of the entire frontal area.

But rotors using two and three blades are more efficient. British engineer John Armstrong argues that one blade captures 10% less energy than two blades.[23] Glid Doman, who designed Hamilton Standard's WTS 3 and 4, further suggests that one-bladed turbines capture 14% less energy than two-bladed machines. And although one blade may be cheaper, they say, the rotor it is attached to is just as heavy as one with two blades. First, the blade on a one-bladed turbine must be stronger than a comparable blade on a two-bladed turbine, because it must capture twice as much energy in the wind for an equivalent output. Second, a one-bladed rotor must compensate for the weight of the missing blade by using a massive counterweight. Because of its higher speed and greater aerodynamic loading, one blade will also emit more noise than two. Ultimately, one-bladed rotors may provide no cost savings. Some manufacturers claim they can build three simple blades for the cost of a single high-performance blade and the sophisticated hub required for a one-bladed rotor.

The advantages of rotors with two blades over those with three are similar to those of rotors with one blade over those with two. The rotor is cheaper, lighter, and operates at higher speeds, leading to lower-cost transmissions. Two blades are easier to install than three because they can be bolted to the hub on the ground in the position they will assume on the assembled turbine. The disadvantages of two blades are similar as well. Because of their higher speeds and greater rotor loading, they are typically noisier.[24]

Conventional wisdom holds that three-bladed machines will deliver more energy and operate more smoothly than either one- or two-bladed turbines. They will also incur higher blade and transmission costs as a result.[25] Armstrong estimates that rotors with three blades can capture 5% more energy than two-bladed turbines while encountering less cyclical loads than one- and two-bladed turbines when reorienting the nacelle to changes in wind direction.[26]

The dynamic or gyroscopic imbalance of two-bladed turbines with changes in wind direction is a classic engineering challenge. The phenomenon can best be seen in the jerky or wobbly motion of small two-bladed turbines as they yaw with the wind. When the rotor is vertical, there is little resistance to yaw, but as the rotor nears the horizontal, the inertia retarding the rotor from reorientating itself reaches a maximum. This occurs twice every revolution. Three blades effectively eliminate this imbalance, as does teetering the two-bladed rotor at the hub. According to California aerodynamicist Kevin Jackson, a two-bladed rigid rotor encounters 10 times the dynamic loads that a three-bladed rotor does. "This is why two-bladed rotors are always teetered" on medium-sized wind turbines, explains Jackson. With teetering, the two-bladed rotor experiences even fewer cyclic loads than does the common three-bladed design.

Putnam found that in his case the addition of a third blade would add 2% to total generation, not enough to warrant the extra blade. He opted for two hinged blades downwind of the tower with dampened coning to minimize flapping.[27] Golding cites research based on Putnam's work, published by the U.S. War Production Board just after World War II, concluding that two blades were optimum.[28] Many American and British engineers have since followed this prescription. Less prone to technical resolution is the perception that one- and two-bladed turbines are less aesthetically appealing than those with three blades (see "Visual Design" in Chapter 8).

Blade Materials

Blades can be made out of almost any material. Wood has been a popular choice. Early farm windmills used crude wood slats, and windchargers of the 1930s used wood almost exclusively. Wood is still the material of choice for many small wind machines. Solid wood planks work well for microturbines, but laminated wood is preferable in slightly larger turbines, for better control of the blade's strength and stiffness. Manufacturers of medium-sized turbines have successfully used wood composites made up of layers of wood veneer sandwiched together with a resin and molded into an airfoil shape by a process widely used to build the hulls of high-performance boats. Wood-composite blades fabricated by Michigan's Gougeon Brothers have earned a reputation for strength and reliability in wind turbines up to 43 m (142 ft) in diameter.

Steel replaced wooden blades on the farm windmill at the turn of the century and has been used ever since. Steel is strong, inexpensive, and well understood. For these reasons, Boeing used steel to build the blades for the rotor on its massive 300-ft (91-m)-diameter Mod-2 and the 320-ft (98-m)-diameter Mod-5B. The Gedser mill used sheet-metal blades and steel struts and stays, while the Nibe B turbine used a steel spar. Steel is limited by its great weight, and no commercial medium-sized wind turbines use the material.

Aluminum is lighter, and for its weight, stronger. Unfortunately, it has two limiting weaknesses: it is expensive, and it is subject to metal fatigue. Aluminum has consistently failed in wind turbine applications. Most of the problems encountered by Darrieus turbines in California have been due to the fatigue of their aluminum blades. No manufacturer of medium-sized wind turbines uses aluminum blades.

Another drawback of metal blades, whether of steel or aluminum, is television and radio interference. Metal reflects television signals, and this can cause "ghost" images on nearby TV sets. This has proven to be far less of a problem than first thought, even among the existing wind turbines using metal blades, including the 500 Darrieus turbines still operating in California.

Most medium-sized wind turbines use blades made of fiberglass [glass-fiber-reinforced polyester (GFRP) to Europeans]. Like wood, fiberglass is strong, relatively inexpensive, and has good fatigue characteristics. It also lends itself to a variety of designs and manufacturing processes. Fiberglass can be pulltruded, for example. Instead of extruding the material through a die, fiberglass cloth (like the cloth used in fiberglass auto body repair kits) is pulled through a vat of resin and then through a die. Pulltrusion produces the siderails for fiberglass ladders and other consumer products. Pulltruded blades can easily be identified by their constant width and thickness. Several U.S. manufacturers, including Storm Master and Windtech, have built wind turbines using pulltruded blades. The technology has been employed most successfully by Bergey Windpower in its series of small wind turbines.

As with wood composites, the techniques used to build fiberglass boats have been adapted successfully by Danish, Dutch, and American companies to build wind turbine blades. In this process, manufacturers place layer upon layer of fiberglass cloth in half-shell molds of the blades. As they add each additional layer, they coat the cloth with a polyester or epoxy resin. When the shells are complete, they add structural stiffeners and a longitudinal spar running the length of the blade, then literally glue the two halves together to form the complete blade. Nearly all medium-sized European wind turbine blades are made using variations of this technique.

Manufacturers have used steel blade spars with limited success. Spars are most often fabricated from fiberglass in the shape of U-channels, box beams, or I-beams. Carter and others have used filament winding successfully to build strong spars. In this process, fiberglass strands are pulled through a vat of resin and wound around a mandrel. The mandrel can be a simple shape like a tube, or a more complex shape like that of an airfoil. Originally developed for spinning missile cases, filament winding delivers high strength and flexibility. Although some blades have been made entirely from filament winding, the process is most often used to produce only the spar. Pulltrusion is also suitable for fabricating inexpensive spars.

Blade technology made significant strides during the 1980s. The specific weight of a wind turbine blade (its mass relative to the area swept by the rotor) tends to increase with increasing rotor diameter, because of an exponential increase in the loads the blade must endure. This increase in specific weight has been surmounted by more sophisticated design and assembly. According to Risø's Finn Godtfredsen, specific blade weight has decreased 80% relative to simply scaling early blades to the larger turbines availabe during the mid-1990s.[29] Contemporary fiberglass blades are one-third to one-half the specific weight of blades built during the early 1980s (Table 6.15). High-performance technology can cut the specific weight of contemporary designs again by half.

In the United States, wood-composite blades cost slightly more than standard fiberglass blades but weigh one-third less, says aerodynamicist Kevin Jackson. Carbon-fiber-reinforced epoxy blades weigh even less than wood composites for comparable strength. However, only the French manufacturer Atout Vent was building carbon-fiber wind turbine blades by the mid-1990s.

Rotor weight is a function of specific weight and number of blades. Most medium-sized wind turbines use three blades. Because of the increased load on each blade, two-bladed turbines have typically been constructed of higher-performance materials. Most of the turbines using wood composites, such as R. Lynette's AWT 26 and the WEG 400, and high-performance technology using epoxy resins, such as the Carter 300, employ only two blades. The exception, Howden, used wood composites on a three-bladed upwind rotor (Figure 6.38).

Table 6.15
Reduction of Mass per Blade

	Single Blade Mass/Swept Area (kg/m^2)	Material[a]
Early 1980s fiberglass technology	2–3	GFRP
Contemporary fiberglass technology	1.5–2	GFRP
Early 1990s fiberglass technology	1–1.5	GFRP
Wood-composite technology	0.7–0.9	Wood
High-performance technology	0.5–0.7	GFRE-CFRE

[a] GFRP, glass-fiber-reinforced polyester; GFRE, glass-fiber-reinforced epoxy; CFRE, carbon-fiber-reinforced epoxy.

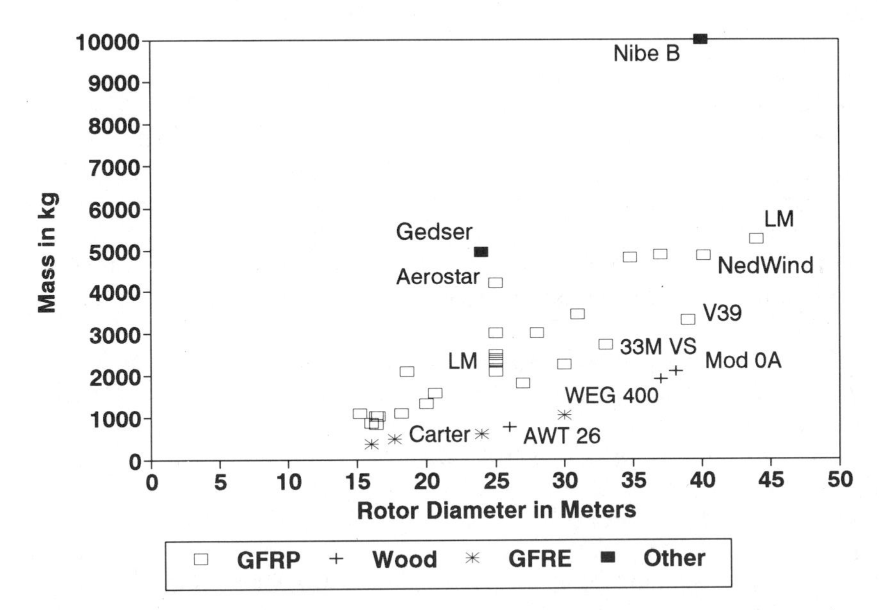

***Figure 6.38.** Rotor blade mass relative to rotor diameter and technology: glass-reinforced polyester, wood composites, and glass-reinforced epoxy. The Gedser mill used steel blades, and Nibe B used a steel spar. Aerostar supplied blades for early Danish wind turbines.*

Most wind turbine blades are adaptations of airfoils developed for aircraft. They were not optimized for wind turbines operating at moderate wind speeds, were susceptible to soiling, and were never intended to operate beyond stall. (Aircraft designers try to avoid stall at all costs, a deadly effect in aircraft.) The power generated by stall-regulated turbines using aircraft airfoils often reached excessive peaks before the airfoils stalled. These peaks dramatically shortened the lives of generators and transmissions.[30] Another unexpected weakness of aircraft airfoils was their sensitivity to surface roughness. As long as wind turbines were deployed in humid regions, frequent rains washed the blades clean. But once wind turbines were installed in areas with scant rainfall, dust and dead insects adhering to the blades robbed them of power. Dirty blades cut performance of some wind turbines in arid California by up to 40%. The problem was so severe that one firm, Fluidyne, created a niche for itself by washing dirty wind turbine blades with a high-pressure spray.

The new generation of special-purpose airfoils designed specifically for wind turbine applications, such as NREL's thin airfoil family, enhance performance 10 to 30% at low speeds, are less subject to soiling, and more effectively control power above rated capacity. Limiting power in high winds is an important means of extending turbine life. All wind turbines incorporate a means of controlling power and limiting rotor speed in high winds.

Overspeed Control

The rotor is the single most critical element of any wind turbine. The rotor confronts the elements and harnesses the wind. A principal wind turbine design feature is its ability "to survive" high winds, according to Andrew Garrad of Garrad Hassan. "The wind can be very brutal and will wreak revenge on any designer that does not give it adequate respect," said Garrad in a lecture at London's Imperial College.[31] How well a wind turbine controls the forces acting on the rotor, particularly in high winds, often determines its reliability.

The simplest method for controlling the rotor in high winds is to decrease the area of the rotor intercepting the wind. As frontal area decreases, less wind acts on the rotor. This not only reduces the power of the rotor but also reduces thrust on the blades and the tower. Halladay's umbrella mill exemplifies the concept. This nineteenth-century water-pumping wind machine automatically opened its segmented rotor into a hollow cylinder in high winds, allowing the wind to pass through unimpeded. Later, Leonard Wheeler used the same concept (changing the area of the rotor intercepting the wind), choosing rather to furl the entire multiblade rotor out of the wind toward its tail vane. As the rotor furled toward the tail, the rotor disk took the shape of a narrower and narrower ellipse, gradually decreasing the area exposed to the wind. Millions of machines using Wheeler's approach to overspeed control have been put into use on farm windmills around the world. Today, nearly all small wind turbines use vertical or horizontal furling to control rotor speed.

Peter Musgrove's contribution to wind technology was designing a way to reduce the rotor's intercept area on straight-bladed, vertical-axis wind turbines. The weakness of the H-rotor configuration is the tremendous forces trying to bend the blades at the cross-arm. These bending forces can be reduced and the speed of the rotor controlled by hinging the blades. In the Musgrove turbine, the blades are hinged to the cross-arm in such a manner that in high winds the blades tip away from the vertical, varying the geometry of the rotor as the blades approach the horizontal.

When most people first consider the problem of controlling a rotor in high winds, they immediately think of changing blade pitch. This probably results from our exposure to propeller-driven airplanes. Like changing intercept area, changing blade pitch affects the power available to the rotor. Increasing or decreasing blade pitch will affect the amount of lift that the blade produces.

Blades can be pitched toward stall or toward feather (Figure 6.39). Twisting the blade until it is nearly parallel to its direction of travel (perpendicular to the wind) causes the airfoil to stall. Because blade pitch is usually set a few degrees into the wind, the blade need change pitch only a few degrees to effect stall. Stall destroys the blade's lift, limiting the rotor's power, but does not reduce thrust on the rotor or the tower. On upwind machines, thrust on the blades bends them toward the tower. Designs dependent on blade stall as

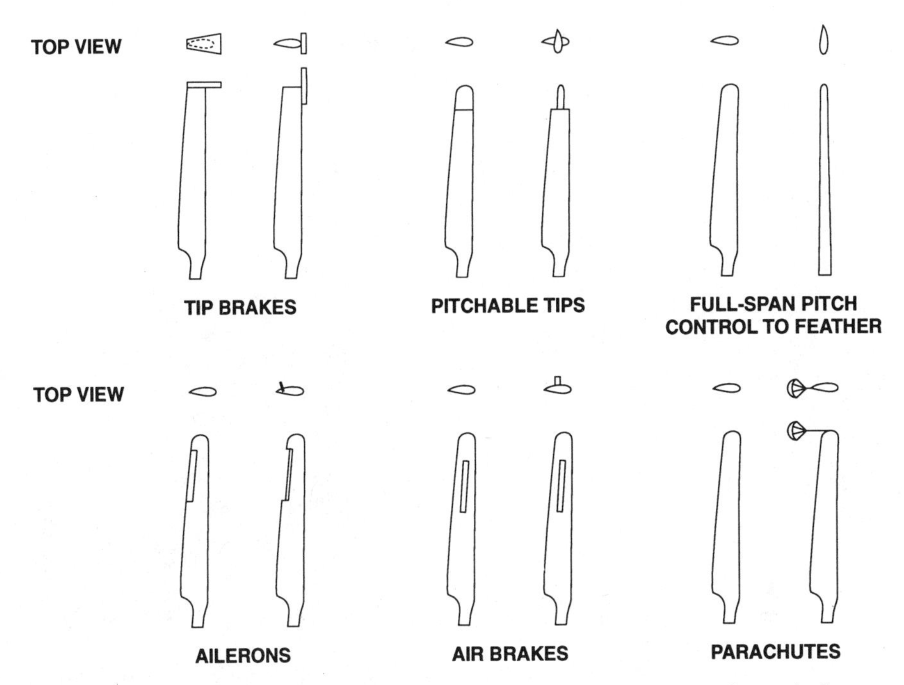

Figure 6.39. Overspeed control. Means to limit power and control speed of rotor under emergency conditions.

the sole means of overspeed protection have a poor survival record. Downwind turbines using stall regulation have had fewer problems, as the blades are forced to cone farther away from the tower. Still, they too have had an overall poor reliability record.

A blade can also be pitched toward feather by turning the blade until it is at right angles to its direction of travel (a pitch of 90°), or parallel to the wind. The pitch mechanism must act through a greater distance to pitch a blade to feather than to stall. One contemporary design, NedWind's 500-kW model, pitches the blade to stall for controlling power in winds above the turbine's rated capacity. When winds exceed the cut-out speed, they pitch the blades to feather. Pitching the blade to full feather cuts thrust on the blade in winds above the cut-out speed to one-fifth that of a blade perpendicular to the wind.

Because controlling the pitch of the blade along its entire span requires a sophisticated pitch-change mechanism and large bearings where the blades meet the hub, variable pitch control has been limited to larger turbines (Figure 6.40). Two early U.S. designs attempted to eliminate the bearings and mechanical governor. Windtech attempted to commercialize the bearingless rotor concept developed at United Technologies by attaching the blades to the hub with a torsionally flexible spar. At high wind speeds, the blades twist

Figure 6.40. WEG MS-3. WEG turbines operating on the Mynydd-y-Cemais plateau in the Cambrian Mountains of Wales. This two-bladed upwind turbine uses a teetered rotor with full-span variable-pitch blades.

toward zero pitch, stalling the rotor. The design uses no moving parts: no bearings, knuckles, or sliding shafts.

Several medium-sized wind turbines currently on the market use pitch control successfully to limit power in high winds. Most of these are machines greater than 25 m (80 ft) in diameter, where the higher costs and complexity of pitch control can be justified.

Almost all wind machines without pitch control use aerodynamic stall to some extent for limiting power from the rotor. This is particularly true of medium-sized wind turbines, which use fixed-pitch rotors to drive induction generators. In winds above the rated speed, the tip-speed ratio for these turbines declines because the speed of the rotor remains constant. The angle of attack increases with increasing wind speed for wind turbines operating at constant speed, causing the blades increasingly to stall.

Designers seldom rely on blade stall as the sole means of overspeed protection. Stall is most effective on induction wind machines with fixed-pitch rotors. Induction generators, however, are dependent on the utility for controlling the load. During a power outage, the generator immediately loses this load. The rotor, no longer restrained to run at a constant speed, accelerates immediately. Stall now becomes ineffectual for regulating power until a new equilibrium is reached. Unfortunately, this occurs at extremely high rotor speeds.

On a small upwind machine with a tail vane, the rotor can be prevented from destroying itself if it is furled out of the wind. Since tail vanes are limited to small turbines, medium-sized upwind machines and all fixed-pitch downwind machines must use a different strategy. Brakes are the most popular.

Once brakes have been selected as the means of limiting rotor speed during a loss-of-load emergency, they are also frequently used during normal operation. In a typical fixed-pitch wind machine, the brake is applied at the cut-out speed to stop the rotor.

Brakes can be placed on either the slow-speed shaft or on the high-speed shaft of the transmission. Brakes on the high-speed shaft are common because the brakes can be smaller and less expensive for the needed braking torque than those on the main shaft. When on the high-speed shaft, the brakes can be found between the transmission and the generator or on the tail end of the generator. In either arrangement, braking torque places heavy loads on the transmission and couplings between the transmission and generator. Moreover, should the transmission or high-speed shaft fail, the brake can no longer stop the rotor.

In general, brakes on fixed-pitch machines should be located on the main shaft, where they provide direct control over the rotor. But the lower shaft speeds require more braking pressure and greater braking area. As a result, the brakes are larger and more costly than those on the high-speed shaft. Danish manufacturers Bonus and Vestas use brakes on the main shaft of their stall-regulated models. Nordtank, Micon, and others apply the brakes on the high-speed shaft.

Brakes can be applied mechanically, electrically, or hydraulically. Most operate in a fail-safe manner that requires power to release the brake. The brake engages automatically when the wind machine loses power. The problem with brakes of any kind is that they can fail.

Experience has taught wind turbine designers that wherever a brake is used to control the rotor, there must be an aerodynamic means of limiting rotor speed should the brake fail. There are three common choices for aerodynamic overspeed protection on stall-regulated medium-sized wind machines: tip brakes, pitchable blade tips, and spoilers. Analysts believe that one of the reasons for early Danish success was the insistence by the Danish Owners Association that wind turbines sold in Denmark must have an aerodynamic means of protecting the rotor from self-destruction.

Tip brakes are plates attached to the end of the blade, which are activated by centrifugal force should the rotor reach excessive speed. When deployed, they increase the drag of the blade. They are simple and effective, and they have saved many a fixed-pitch rotor from destruction. Tip brakes, however, have been likened to keeping one foot on the accelerator and the other on the brake. Tip breaks keep the rotor from reaching destructive speeds but do nothing to reduce the lift of the blade or the thrust on the wind turbine and tower. Tip brakes are also noisy and reduce the performance of the rotor under operating conditions by increasing drag at the tip, where blade speed is greatest. Enertech adopted tip brakes on its turbines after discovering that stall alone was insufficient to protect the machine in high winds. ESI also used tip brakes.

Most of the power captured from the wind is captured by the outer third of the blade. Consequently, it is unnecessary to change the pitch of the entire blade to limit the rotor's power. The performance of the blade in the tip region can be reduced by using spoilers or pitchable blade tips. In one Danish design, centrifugal force activates spoilers along the outer portion of each blade. The spoilers pop out of the blade, change the shape of the airfoil, and destroy the blade's lift, reducing its power. Spoilers or air brakes are nothing new. They appeared on English windmills in 1860.[32]

Many stall-regulated Danish wind turbines use pitchable blade tips. Medium-sized turbines 15 to 25 m (50 to 80 ft) in diameter use passive controls to activate the blade tips. At higher-than-normal rotor speeds, the tips are thrown away from the rotor by centrifugal force, which causes them to slide along a grooved shaft. As they move along the shaft, the tips pitch toward feather. This action decreases lift where it is greatest while dramatically increasing drag. Both spoilers and pitchable blade tips have proven highly successful, although refinements have been necessary. (In early designs, frequently only one or two of the blade tips would activate at the same time.)

Boeing's Mod-2 and Mod-5B used a similar approach. Instead of using passive controls, Boeing hydraulically drove the tips toward the feathered position, actively controlling power in high winds. Today, most fixed-pitch turbines greater than 25 m (80 ft) in diameter actively regulate the pitch of the blade tips to control power, much like Boeing (Figure 6.41). Just as a turbine with full-span pitch control will feather the blades before applying the brake to bring the rotor to a halt, stall-regulated turbines in this size range pitch the blade tips toward feather before applying the brake.

Some medium-sized turbines have also used parachutes for emergency overspeed control. Parachutes are unsatisfactory because they are difficult to reset. NASA's Lewis Research Center at Sandusky, Ohio, briefly experimented with ailerons, or more properly flaps, on the Mod-0A. During the early 1990s, Northern Power Systems's Clint (Hito) Coleman dusted off the concept and applied it to the NW 250, one of the machines in NREL/DOE's advanced wind turbine program (Figure 6.42). During the mid-1990s, Zond also began experimenting with aileron control on its prototype 40-m (130 ft) turbine.

The generator itself can be used to bring the rotor to a halt. This dynamic braking was first used successfully by Enertech on its E-40. R. Lynette's AWT 86 and Atlantic Orient's 50/15, two turbines in NREL/DOE's advanced wind turbine program, use tip brakes in conjunction with dynamic braking.

Generators

Generators convert the torque of the spinning rotor, in this case the prime mover, into electricity. In its simplest form, a generator is nothing more than a coil of wire spinning in a magnetic field.

***Figure 6.41.** Variable-pitch blade tips. Most medium-sized fixed-pitch turbines actively control pitchable blade tips to limit power and aid in braking. Group of 300-kW Howden turbines, each 33 m in diameter, in the Altamont Pass.*

The cost of generators increases proportionally with size.[33] For a given voltage, it costs less to increase generator speed than to increase the size of the generator. But there are trade-offs. Generators operating at higher speeds require either higher gear ratios or rotors that operate at higher speeds, or a combination of both.

Many early U.S. wind turbines operated at high rotor speeds. These machines were noisy and trouble-prone. Danish designs operating at more modest rotor speeds were more reliable. When used in California, Danish turbines typically drove six-pole generators at 1200 rpm, while their U.S. competitors used four-

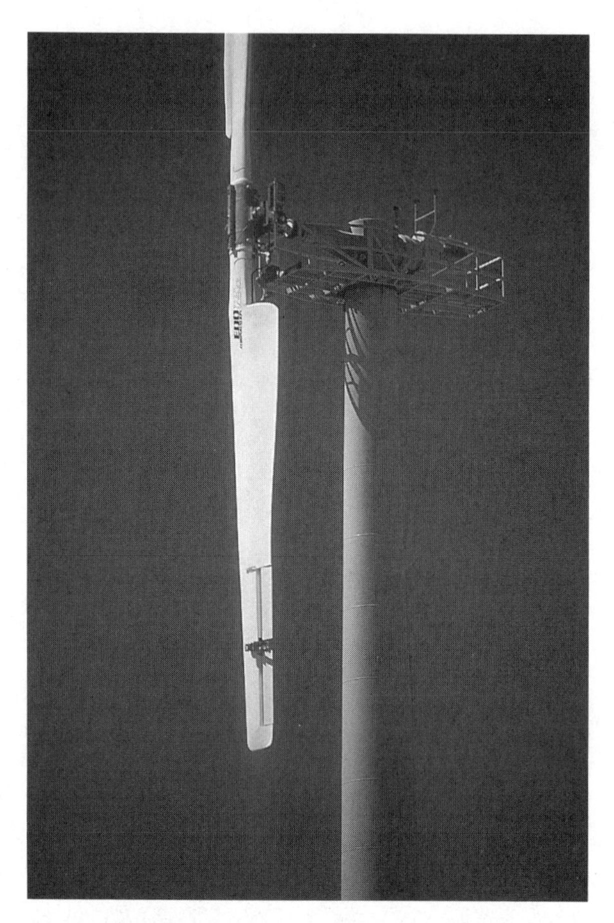

Figure 6.42. Ailerons or flaps. Northern Power Systems developed ailerons for the rotor of its DOE/NREL funded design.

pole generators running at 1800 rpm (Table 6.16). Today none of the early U.S. designs is still being built.

Manufacturers of small wind turbines may opt for slow-speed generators tailored to their wind turbine. Bergey Windpower, Northern Power Systems, and others have done just that by designing direct-drive slow-speed alternators. In today's alternator it is the field, rather than the armature, which revolves. Power is drawn off the stator.

In a conventional alternator, the field revolves inside the stator. But small turbine manufacturers Bergey Windpower and Marlec Engineering, for example, spin the field outside the stator. For those that use permanent magnets in the field, such as the Bergey wind turbine, there is no need for slip rings, or other moving contacts, insider the alternator.

Table 6.16
Generator Speed (rpm)

	Europe (50 Hz)	North America (60 Hz)
Four-pole	1500	1800
Six-pole	1000	1200

Alternators generate alternating current, the frequency of which varies directly with the speed of the rotor and indirectly with the number of poles in the generator. For a given number of poles, frequency increases with increasing generator speed.

Wind machines driving electrical generators operate at either variable or constant speed. In variable-speed operation, rotor speed varies with wind speed. In constant-speed machines, rotor speed remains relatively constant, despite changes in wind speed.

Small wind turbines typically operate at variable speed. This simplifies the turbine's controls while improving aerodynamic performance. When these small wind machines drive an alternator, both the voltage and frequency vary with wind speed. The electricity they produce is incompatible with the constant-voltage, constant-frequency alternating current (ac) produced by the utility, but can be used as is for resistive heating or pumping water at variable rates, or it can be rectified to direct current (dc) for charging batteries.

To generate utility-compatible electricity, the output from a variable-speed alternator must be conditioned. Although it is possible to use rotary inverters for this task, variable-speed turbines typically use a form of synchronous inverter to produce constant-voltage 60-hertz (Hz) (50-Hz in Europe) (ac) like that of the utility (Figure 6.43). Most of these inverters use the utility's alternating current as a signal to trigger electronic switches that transfer the variable-frequency electricity at just the right moment to deliver 60-Hz ac at the proper voltage.

Although some manufacturers of medium-sized wind machines build variable-speed turbines, most operate the rotor at or near constant speed. These machines produce utility-compatible power directly via induction (asynchronous) generators.

Induction generators have two advantages over alternators: they are inexpensive, and they can supply utility-compatible electricity without sophisticated controls. An induction generator begins producing electricity when it is driven above its synchronous speed: 1200 or 1800 rpm in North America. Induction generators are not true constant-speed machines. As torque increases, generator speed increases 2 to 5%, or 36 to 90 rpm on an 1800-rpm generator. This increase of 1 to 3 rpm in rotor speed is imperceptible in a wind turbine operating at a nominal speed of 50 rpm. As torque increases, the magnetic field in the induction generator also increases. This continues until the generator reaches its limit, which is about 5% greater than its

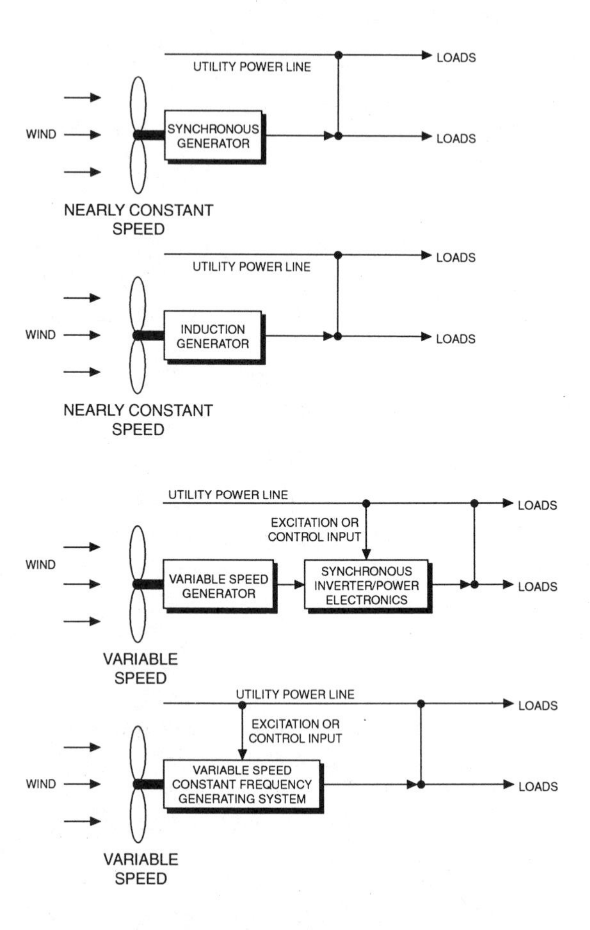

Figure 6.43. Utility-compatible wind machines. Means for interconnecting wind turbines with electric utilities. Most interconnected wind turbines use induction generators. (Reprinted from Wind Power for Home & Business, *courtesy of Chelsea Green Publishing Company.)*

synchronous speed. Induction generators are readily available in a range of sizes and are easily interconnected with the utility. Medium-sized wind turbines use induction generators almost exclusively.

Dual Generators

Generators operate inefficiently at partial loads. In a 500-kW wind turbine, for example, where the generator is designed to reach its rated capacity at a wind speed of 16 m/s (35 mph), the generator operates at partial load much of the time (Table 6.17). At a site with an average wind speed of 7 m/s, the

Table 6.17
Typical Generating Hours for Rayleigh Distribution

Average Wind Speed m/s	Average Wind Speed mph	Generating Hours/Year	Percent of Total Hours	Hours Below Rated Capacity	Percent of Generating Hours	Hours on Small Generator	Percent of Generating Hours
5	11	5970	68	5960	100	4480	75
6	13	6710	77	6670	99	4150	62
7	16	7200	82	7020	98	3650	51
8	18	7540	86	7090	94	3150	42

Note: Cut-in, 4 m/s; cut-out, 28 m/s; small generator, 4–7 m/s, rated, 16 m/s.

generator will operate 97% of the time at less than rated capacity and about half the time at less than 100 kW.

Efficiency drops off rapidly when the generator is operated at less than one-third of its rated value.[34] For example, the efficiency falls nearly 15% (from 95% at rated output) when a 500-kW generator operates at 100 kW. To avoid this, designers of constant-speed wind turbines often use dual generators or dual windings: one main generator and a smaller generator one-fifth to one-third the capacity of the main generator. The small generator operates at nearly full load in low to moderate winds from cut-in at 4 m/s (9 mph) to about 7 m/s (15 mph) (Figure 6.44). When the wind speed reaches 7 m/s, the small generator switches off and the main generator switches on. Thus both generators operate more efficiently than either one alone. At many sites the small generator will operate more than 50% of the total generating time, although it delivers less than half the total generation because of the much greater energy produced by higher-speed winds.

The two generators may be in tandem and driven by the same shaft, or they can be side by side, with the small generator driven by belts from the main generator. During the mid-1990s, most new constant-speed turbines used one generator with dual windings. The generator operates on six poles during light winds and uses four poles in higher winds. Not all turbines under development in the mid-1990s will use dual generators or dual windings. Turbines in NREL/DOE's advanced wind turbine program will use only one generator.

The use of dual generators permits the turbine to operate at two speeds, enabling designers to drive the rotor at a higher aerodynamic efficiency over a broader range of wind speeds than with only one generator. In low winds, generators with dual windings operate at two-thirds the speed that they would operate at normally (Table 6.16). A turbine that operates at 50 rpm at its rated wind speed will operate at 33 rpm when all six poles of a dual-wound generator are energized. Dual-speed turbines, while incapable of taking full advantage of the optimum tip-speed ratio over the entire operating range, can capture most of the efficiency advantages of variable-speed turbines, at only a small increase in cost for the extra windings.[35]

Figure 6.44. Dual generators. Many constant-speed wind turbines operate at two speeds by using dual generators, as shown here, or by using dual windings within one generator.

The advantage of one single generator with dual windings becomes problematical as turbines grow ever more powerful. Because a generator's power is proportional to its volume, while losses are proportional to its surface area, larger generators are also more efficient than smaller ones. This could add perceptibly to the improved performance of larger turbines over that of their smaller predecessors. The tenfold increase in generator size from the early 1980s to the mid-1990s has added as much as a 3% improvement in generator efficiency, providing a 1% improvement in a turbine's overall conversion of wind energy to electricity.[36]

Medium-sized wind turbines may have reached a critical size threshold. Generators and their controls are mass-produced in limited sizes. There are fewer generators to choose from for machines 300 to 500 kW in size than for those 100 to 300 kW in size. There may be an optimum, argue some designers, among generators that are mass-produced.

The concept has been explored in Europe. When contracted to rebuild a damaged 750-kW turbine at Masnedø, Danish Wind Technology chose to change the traditional Danish design by opting for a transmission with dual output shafts for driving twin 400-kW generators instead of one 800-kW generator. Similarly, Dutch manufacturer NedWind uses twin 250-kW generators on its 40-m 500-kW model, and U.S. Windpower uses twin generators on its 33M-VS (Figure 6.45).

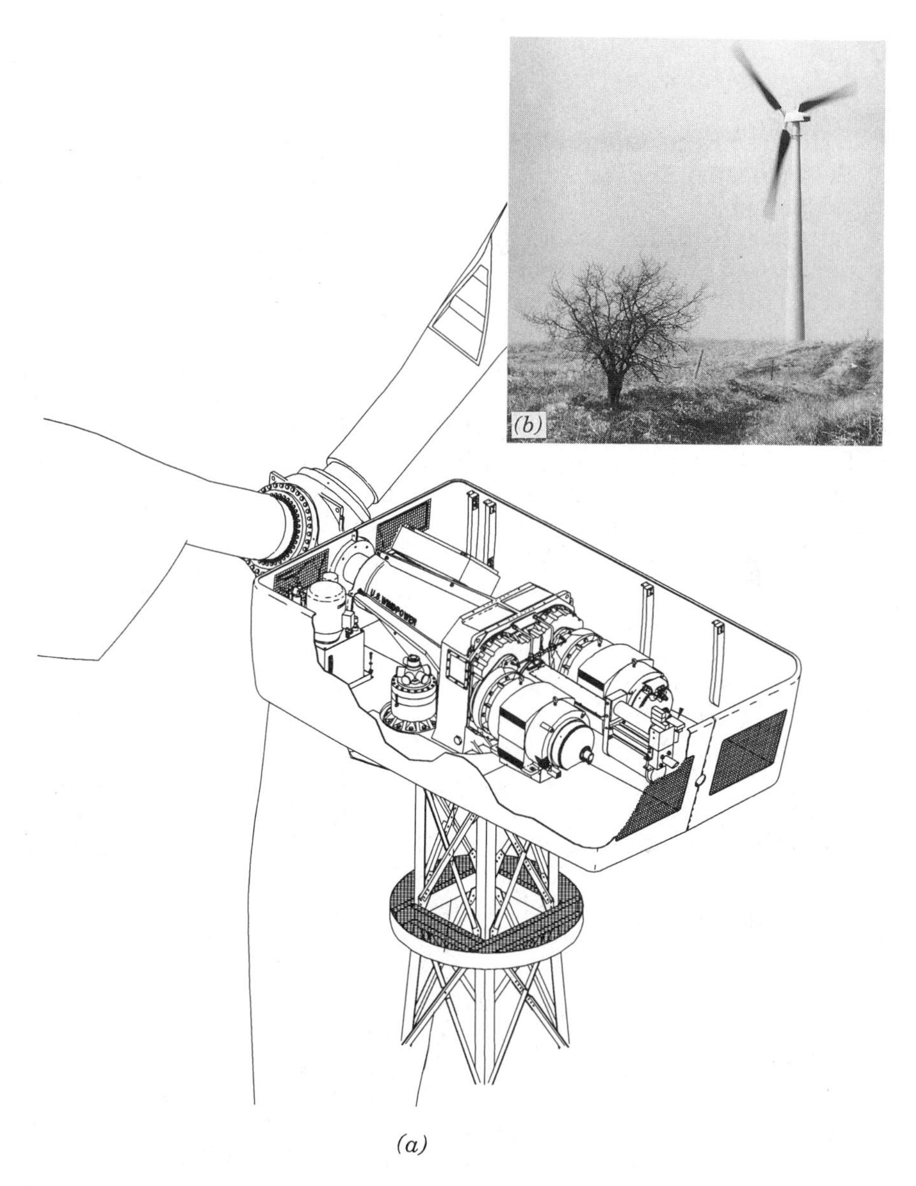

Figure 6.45. (a) *U.S. Windpower 33M VS. Note integral drive train, dual generators, work platform, and characteristic "shark fin" blades. (Courtesy of KENETECH Windpower, Inc., San Francisco, CA. Copyright KENETECH Windpower, Inc., 1993)* (b) *One of 73 turbines in USW's 25-MW wind plant on Buffalo Ridge, Minnesota.*

Variable-Speed Versus Dual-Speed Operation

Probably more has been written about the advantages of variable-speed operation than about any other aspect of wind energy as a result of a sophisticated marketing campaign begun by U.S. Windpower in 1990. One journalist writing in the influential *EPRI Journal* overzealously described the concept as a "breakthrough."[37]

The variable-speed idea is not new. Bergey Windpower has been building grid-connected small wind turbines operating at variable speeds for more than a decade. The Dutch first employed the concept on medium-sized and larger machines during the mid-1980s.

Variable speed is attractive because it enables designers to gain greater rotor efficiencies by allowing rotor speed to vary with wind speed. There may be additional benefits as well. Slower rotor speeds in light winds lower noise emissions just when the aerodynamic noise of the blades is most noticeable. Variable-speed operation may also reduce dynamic loads on the turbine's drivetrain, thus extending turbine life. When operating at variable speed, the rotor stores the energy of gusty winds as inertia as its speed increases, rather than forcing the drivetrain to absorb the increased torque instantaneously.

Variable speed offers greatest improvement in low-wind-speed regimes or with turbines having high rated wind speeds, says ABB's Gerald Hehenberger, because the turbine will operate more of the time below its rated output. This is the area in which variable-speed operation boosts efficiency most, by maintaining a constant tip-speed ratio. At sites with higher wind speeds, or with turbines using lower-rated speeds, the turbine operates more often at rated output. Variable speed then contributes principally to reducing loads on the turbine, not to enhancing performance. Lower loads only show up as a plus over the long term. "No one knows the optimum solution for all wind regimes in the world," says Hehenberger, who explains that variable speed would be a benefit in the strong gusty winds of Tehachapi but not in the steady tradewinds of the Caribbean.

Few manufacturers are convinced that variable speed is sufficiently attractive to justify its greater cost. Nevertheless, it has long attracted the interest of designers.

During the mid-1980s a consortium of two Dutch aerospace companies designed a series of variable-speed wind turbines at the behest of the Dutch government. The consortium first built a 26.5-m 300-kW turbine, the NEWECS 25. Three of the two-blade upwind machines were built, two in the Netherlands and one on Curaçao, an island in the Netherlands Antilles. Only the turbine on Curaçao is still operating.

The NEWECS 25 was a forerunner of the 1-MW turbine now operating at Medemblik on the shore of the Ijsselmeer, the large inland sea formed by closure of the Zuiderzee. Both versions use a variable-pitch rotor to drive a variable-speed generator by means of a planetary transmission. The turbines generate utility-compatible electricity by conditioning the generator's output with power electronics.

The NEWECS 25 on Curaçao has logged more than 43,000 hours of grid-connected operation since its installation in September 1985. The well-maintained turbine has been in nearly continuous service, and the power electronics have continued to perform to the local utility's satisfaction. The turbine, located at Boca San Pedro on Curaçao's northeastern shoreline, is one of the few medium-sized wind machines operating in the Caribbean, according to Margo Guda of Curaçao's Fundashon Antiyano Pa Energia. Guda characterizes the NEWECS 25 as the unsung success story of Dutch wind turbine development.

Building on this experience, Dutch manufacturer Lagerwey commercialized the variable-speed concept for the Dutch and German markets. Lagerwey had installed nearly 250 of its 15- to 18-m variable-speed turbines by the early 1990s.

In a part of the world where nearly all wind turbines use three-bladed rotors to drive conventional induction generators, Lagerwey's design stands apart. The northern Dutch provinces of Friesland and Groningen are now dotted with the gaunt two-bladed wind turbines that have come to symbolize Dutch private-sector development of wind energy. Lagerwey uses passive pitch control to regulate the speed of the flexible two-bladed rotor, a design feature that U.S. manufacturers Carter and Windtech both attempted to master during the early 1980s. Lagerwey carries the design one step further and marries this flexible rotor system to a variable-speed generator, a feat not yet attempted in North America.

Variable-speed operation incurs a one-time penalty of a 10% increase in installed costs for the necessary power electronics but will increase energy capture at continental sites about 10% per year over that of similarly sized machines operating at constant speed.[38] This was confirmed in a study by the Wind Energy Group for Britain's Department of Trade and Industry.

The DTI-WEG tests also confirmed a less widely known belief that the same results can be obtained by operating so-called constant-speed machines at two speeds: one speed for low winds, one for high. Most "constant-speed" wind turbines, such as those built by WEG, operate at two speeds. WEG's Ervin Bossanyi concludes that two-speed operation is superior to variable speed unless the maximum rotor speed and power of the variable-speed turbine are markedly increased. There is little or no performance advantage to variable-speed over dual-speed operation until power rating and rotor speed are increased 10 to 20% over the base case, he says (Table 6.18). According to Bossanyi's estimates, a variable-speed wind turbine 33 m in diameter will need to operate up to a power rating of 400 kW to capture as much energy as an equivalent dual-speed design.[39]

Those skeptical of the power electronics necessary for using variable speed have often wondered how much energy is lost in the electronic conversion.

Table 6.18
Percent Increase in Generation of Two-Speed and Variable-Speed Wind Turbines over 300-kW Single-Speed Wind Turbines 33 Meters in Diameter

	Percent Increase for Average Annual Wind Speed of:		
	6.5 m/s	7.5 m/s	8.5 m/s
300 kW, two-speed	10	6	4
300 kW, variable speed	10	6	4
400 kW, two-speed (WEG 400)	18	19	20
400 kW, variable speed	17	17	19
330 kW, variable speed, 52 rpm	13	10	9
360 kW, variable speed, 58 rpm	15	14	14

Source: E. A. Bossanyi, WEG, 1993.

Bossanyi sums up his findings by noting that "two-speed operation may actually be more efficient overall, with higher electrical efficiency compensating for slightly lower aerodynamic efficiency." He does acknowledge that variable-speed operation permits lower rotor speeds in low winds, significantly reducing noise, which is an important concern in Europe.

The WEG study also examined possible cost savings by using variable speed to lower peak mechanical loads. Bossanyi says that the cost to WEG for a transmission to a variable-speed turbine represents only a 10% savings over that of a transmission for their constant-speed machine. Material savings in the tower and elsewhere were even less significant.

As of the mid-1990s, the performance improvement from turbines using variable-speed operation in the field was hard to ascertain. One method is to examine published estimates of projected production, although it is difficult to compare performance estimates directly from one manufacturer to another. Some manufacturers are more conservative than others when estimating production.

One manufacturer of variable-speed machines that does publish production estimates is Germany's Enercon, which began building medium-sized wind turbines using variable speed in 1984. By the mid-1990s, Enercon had installed nearly 200 of its subsequent design, a 33-m 300-kW version. Enercon believes that variable-speed machines capture 8 to 11% more energy from the wind than do constant-speed machines.

According to product literature, Enercon's E33 variable-speed wind turbine produces about 5% more kWh/m^2 of swept area than the WEG 400 or the Bonus 450, but about 3% less than the Vestas V39 and about 10% less than the Carter 300 at sites with a 7-m/s average wind speed and a standard Rayleigh speed distribution. At 6-m/s sites, Enercon's E33 is evenly matched with the V39 and the Carter 300 and produces about 10% more than the WEG 400 and the Bonus 450. Actual production data from Germany indicate that there

29. Godtfredsen, Jensen, and Morthorst, "Wind Energy in Denmark." The authors contend that the expected weight of a 19.5-m blade is 4.7 kg/m^2 when scaled up from early 1980s technology, where a 7.5-m blade weighed 1.8 kg/m^2. The actual specific weight of the 19.5-m blades is 0.9 kg/m^2.
30. James Tangler et al., "Atmospheric Performance of the Special-Purpose SERI Thin-Airfoil Family: Final Results," Solar Energy Research Institute (NREL), Golden, CO, undated.
31. Garrad, "Forces and Dynamics."
32. Frederick Stokhuyzen, *The Dutch Windmill,* trans. Carry Dikshoorn (New York: Universe Books, 1963), p. 62.
33. Eggleston and Stoddard, *Wind Turbine Engineering Design,* p. 319.
34. Johnson, *Wind Energy Systems,* p. 140.
35. Cavallo, Hock, and Smith, *Renewable Energy Systems.*
36. Johnson, *Wind Energy Systems,* p. 140.
37. Leslie Lamarre, "A Growth Market in Wind Power," EPRI Journal, December 1992, pp. 4–15. The article failed to acknowledge that EPRI partly funded development of the turbine and has a royalty agreement with the manufacturer on sales of the machine.
38. Cavallo, Hock, and Smith, *Renewable Energy Systems,* pp. 143–144. Variable-speed operation offers the potential for a 5 to 15% improvement in performance, according to Bob Lynette of R. Lynette & Associates. Robert Lynette, "Status and Potential of Wind Energy Technology," American Wind Energy Association's annual conference, Windpower '90, Washington, DC, September 1990, pp. 1–3.
39. E. A. Bossanyi, "Viability of Variable Speed," paper presented at the British Wind Energy Association's annual conference, York, England, October 6–8, 1993.
40. Mark Haller, oral comments at Windpower '94, annual conference of the American Wind Energy Association, Minneapolis, MN, May 11, 1994.
41. Cavallo, Hock, and Smith, *Renewable Energy Systems,* p. 144.

7

Wind Energy's Declining Costs

It started in 1989 when John Schaefer published a short article in the Electric Power Research Institute's technical journal assessing the remarkable progress of California's wind industry. Schaefer concluded that at good sites wind energy cost about $0.08 per kilowatt-hour. To put that in perspective, he startled the utility industry by suggesting that the cost was "just about the same as that from more conventional sources."[1]

Then, in a widely circulated *Scientific American* article in the fall of 1990, Carl Weinberg described how the cost of wind energy would fall to $0.04 per kilowatt-hour by the year 2000, one-tenth of its cost in the early 1980s.[2] As Weinberg was head of Pacific Gas & Electric's research and development department, his projection carried clout.

Not to be outdone by one of its member companies, EPRI followed in 1991 with a feature article in its journal on the prospects for renewables. EPRI moved Weinberg's time line forward, estimating that wind energy would cost $0.05 per kilowatt-hour by the mid-1990s. EPRI projected that the cost would drop to only $0.035 per kilowatt-hour by the year 2000.[3]

The floodgates were open and soon a torrent of articles in the popular and trade press were discussing wind energy's newfound cost-effectiveness. A surprisingly well-timed marketing campaign by U.S. Windpower skillfully ushered the deluge along by emphasizing that its new technology would push wind below the magic $0.05 per kilowatt-hour mark. It was as if former vice president Thomas Riley Marshall had been reborn as a wind energy promoter and proclaimed that what the country needed was no longer a good 5-cent cigar but a good 5-cent (per kilowatt-hour) windmill.

In their eagerness to tell wind energy's story, researchers and U.S. government agencies quickly climbed aboard the bandwagon, each bent on bettering the other's projections of how cheap wind energy would become. Some reached absurd lengths, estimating that wind energy would cost as little as $0.02 to $0.026 per kilowatt-hour.[4]

Europeans shook their heads in disbelief. Had Americans gone berserk, they mused, or was wind energy really so inexpensive? They feared that Americans, in their zeal to sell wind energy to reluctant utilities, were nearing the line that the nuclear industry had crossed decades before. If the trend of ever-lower projections continued, wind energy would soon become "too cheap to meter."

They were not alone. The prospect disturbed Ken Karas, then president of the American Wind Energy Association, to such an extent that he publicly addressed the haphazard handling of cost-of-energy calculations at AWEA's 1992 conference. Karas told attendees that the industry was in danger of misleading policymakers with these projections. The advancement of wind energy is a success story so convincing that it need not be oversold, he argued. "After installing 260 MW of wind capacity," Karas warned, "I don't think I could ever build a project for $0.02 per kilowatt-hour."[5]

The cost of wind energy, said Karas, depends on the assumptions used. Widely differing results can all be equally correct. For example, the California Energy Commission's 1992 Energy Technology Status Report estimates that electricity from future wind plants will cost from $0.041 to $0.077 per kilowatt-hour, depending on who owns the plant and whether the cost does or does not include inflation.[6]

Karas particularly criticized the use of simple cost of energy (COE) calculations common in engineering economics. This model produces a useful index for comparing one technology to another, said Karas, but it is too inflexible to reflect real-world conditions. The model fails to determine what it would actually cost a utility or wind company to build and operate a wind power plant. This is acceptable as long as everyone understands the model's limitations. Problems arise, though, when decision makers, particularly those at regulatory commissions, begin to blur the distinction between these indexes and the actual cost of building a power plant. As a former banker, Karas prefers to estimate his expenses and revenues year by year, from which he can derive a more realistic levelized cost of energy. This is the amount he will need to pay for his project.

Potentially even more misleading is how the various estimates deal with inflation. Because anticipating the rate of inflation over the next 20 years is similar to gazing into a crystal ball, engineers frequently ignore inflation altogether and give costs of energy in *constant* or *real* dollars. This is valid only when comparing technologies whose capital and operating characteristics are similar, Karas stressed. But inflation is a fact of life that increases the revenues required to pay for a power plant's generation, especially to pay for future fuel costs in fossil-fueled plants. In contrast, renewables such as wind energy have minimal future costs subject to inflation. Economists account for inflation by presenting the levelized cost of energy in *nominal* dollars. Thus the cost of energy in constant dollars is significantly less than that in nominal, or inflation-adjusted, dollars and distorts the relationship between fossil fuels and renewables over the life of the plants. The CEC estimates, for example, that an investor-owned utility could produce wind-generated electricity in 1997 for $0.041 to $0.047 per kilowatt-hour in constant 1989 dollars. Yet the same electricity will cost $0.06 to $0.068, or nearly 50% more, when inflation is included.[7]

To make the best case for wind energy, proponents often cite costs in constant dollars, although they could never build a power plant for that amount. They do so because the political process compares the cost of energy from new wind power plants to the cost of energy from conventional plants that were installed 10 to 20 years ago. These fossil-fired plants are nearly amortized and their cost of energy primarily reflects currently low fuel costs. Proponents thus confront decision makers with two sets of numbers: constant dollars for persuading them that wind energy makes economic sense, and nominal dollars for determining how much they actually need to be paid. Naturally, this leads to confusion, and regulators seize on the lower value, thwarting development.

To standardize reporting, Karas urges that COE calculations be expressed in nominal dollars. The resulting estimates should clearly state the assumptions on which they are based. Neither the EPRI article nor the article in *Scientific American* met Karas' minimum requirement that the reader be alerted as to whether the estimates used constant or nominal dollars. (As will be explained, they used constant dollars.)

The Cost of Wind Energy

What does wind energy cost? To find the answer, it is first necessary to understand the assumptions behind the calculations. Regardless of its limitations, the COE methodology is widely used. Its simplicity permits quick comparisons between different assumptions.

The levelized cost of energy is derived from installed cost, annual generation, operations and maintenance cost, fuel cost, and the fixed charge rate, a factor accounting for the cost of taxes, insurance, interest on debt, and the rate of return on equity. The cost of energy is

$$\text{COE} = \frac{\text{installed cost}}{\text{annual generation}} \times \text{fixed charge rate} + \text{annual O\&M/kWh} + \text{fuel costs/kWh}$$

The fixed charge rate varies with the type of ownership and whether constant or nominal dollars are used. In the United States, publicly owned utilities have lower fixed charge rates than investor-owned utilities because of lower financing costs and exemption from federal and state taxes (Table 7.1). By comparison, the Union of Concerned Scientists used a fixed charge rate of 10.5% (constant dollars, private utility) in their pioneering work on the potential of renewables in the midwest[8] (Table 7.2).

Outside the fixed charge rate, the cost of energy from a wind plant is determined primarily by the installed cost and the amount of electricity generated. (There are no fuel costs in a wind plant.) To monitor the economic development of a technology, analysts often use specific costs in dollars per

Table 7.1
Example of Fixed Charge Rates (%)

	Karas	California Energy Commission
Investor-owned utility		
Nominal	14.3	12.0
Constant	9.6	8.0
Publicly-owned		
Nominal	10.0	8.4
Constant	6.6	4.5

Source: Karas: Windpower '92; CEC: *1992 ETSR.*

kilowatt of capacity or dollars per square meter of swept area, and projected generation derived from capacity factors or specific yields (kWh/m^2 per year) as substitutes for the cost of individual projects. EPRI and PG&E made their estimates after noting the rapid decline in the specific cost of California wind plants at the same time that specific yields were increasing.

There is general agreement on the performance of modern wind turbines. But determining the installed cost is another matter. Installed cost is often confused with installed price. Although related, the two are distinctly different. Unless the analyst has access to proprietary data, the only values publicly available are the prices that buyers pay for installed wind plants. It is the price paid for wind projects that determines the cost of energy to utilities or their ratepayers.

The price of wind turbines sold in California during the early 1980s was a function of their cost and the wind resource at the site. Turbines at energetic

Table 7.2
Fixed-Charge-Rate (%) Assumptions for Investor-Owned Utilities

	Union of Concerned Scientists	Karas	California Energy Commission
Inflation	5	4	3.8
Return on equity (nominal)	10.8	11	12.65
Interest on debt (nominal)	8.7	8	9.25
Debt ratio	50	50	47.5
Life	30	30	30
Combined tax rate	38	40	40.1
Property tax	2	1.2	0.5
Fixed charge rate (nominal)		14.3	12
Fixed charge rate (constant)	10.5	9.6	8

Source: UCS: *Powering the Midwest,* 1993; Karas: Windpower '92; CEC: *1992 Energy Technology Status Report.*

sites commanded a premium because they would generate more revenues than turbines elsewhere. Thus a 65-kW Danish turbine could sell for $180,000 at one location and $100,000 at another.

Wind Power Plant Price

By tracing the published transaction price, not the cost of building the projects, analysts have found that the price paid for wind power plants in California and Denmark fell dramatically during the 1980s (Figure 7.1). The published price includes the cost plus a profit. Developers typically take out some of their profits up front, inflating the price.

In California the average installed price in 1992 dollars tumbled from nearly $4000 per kilowatt in 1981 to a low of $1200 per kilowatt in 1987. The price paid by Danish utility ELSAM in 1991 currency dropped from nearly 1500 Ecu ($2000) per kilowatt in 1987 to about 900 Ecu ($1200) per kilowatt in 1990.[9] For reasons unique to Britain, the price of projects in England and Wales averaged $2000 per kilowatt during the early 1990s.

The line in Figure 7.1 obscures the price variability among wind plants in California. For example, an early project that installed American-built ESI turbines in 1982 cost nearly $6000 per kilowatt (Table 7.3). Installed price dropped quickly with the introduction of Danish machines and increased

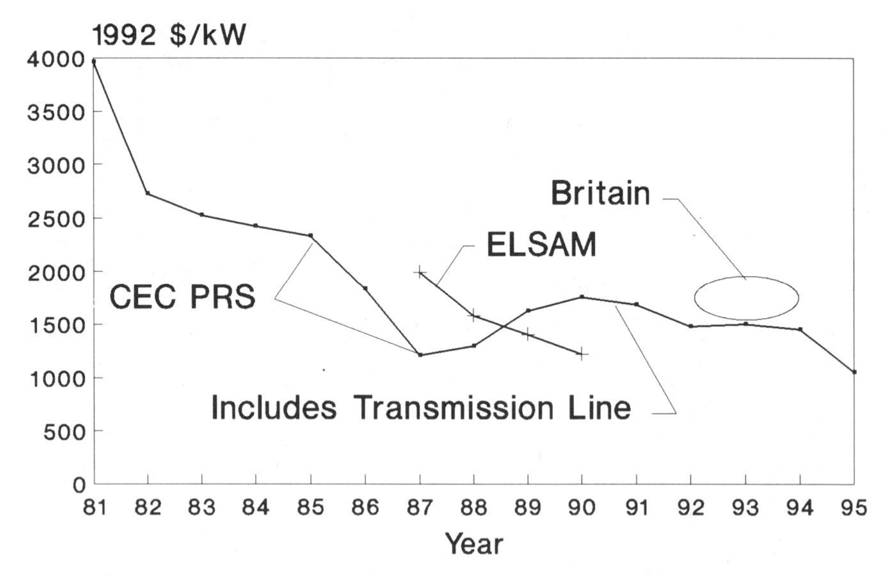

Figure 7.1. Wind power plant price. Actual price paid for wind power plants in California, Denmark, and Great Britain.

Table 7.3
Sample of Early 1980s Price of Wind Turbines in California

Model	Year	Cost ($)	Capacity (kW)	Diameter (m)	Swept Area (m^2)	Approx. Cost in 1992 ($/kW)	Specific Cost ($/$m^2$)
ESI 54	1982	225,000	50	16	213	5580	1310
Micon	1983	96,154	65	16	201	1780	580
Enertech	1983	62,500	40	13	141	1880	530
Vestas	1984	160,000	65	15	177	2910	1070
Micon	1984	180,000	65	16	201	3270	1060
Aeroman	1985	125,000	40	13	123	3620	1180
Micon	1985	190,000	110	20	299	2000	740

competition for investors. It was common at the time for a 65-kW Danish wind turbine 15 to 16 m in diameter to sell for anywhere from $100,000 to $180,000, depending on the site and the revenues it might be expected to generate.

Prior to 1985, the California Energy Commission informally estimated the average installed price from limited telephone surveys. After 1985 developers were required by state law to submit reports to the CEC of the amount of capacity installed, the performance, and the price paid for new projects. The CEC's performance reporting system subsequently published the price of wind projects from 1985 to 1987, when it discontinued the practice because the paucity of new installations threatened to reveal proprietary data of those still installing turbines.

Again the average price in Figure 7.1 fails to show the variability in the California market (Table 7.4). Some turbines were installed for as little as $400 per kilowatt according to the CEC reports. However, there is no information on whether those turbines remain in service.

During the late 1980s and early 1990s there were far fewer projects than

Table 7.4
Wind Power Plant Price Reported by California Energy Commission (1992 $/kW of Installed Capacity)

Year	Low	High
1985	770	2670
1986	1150	2710
1987	400	1600

Source: Results from the Wind Project Performance Reporting System, Annual Reports (Sacramento, CA: California Energy Commission, August 1986, p. 17; January 1988, p. 24; and August 1988, p. 23).

during the mid-1980s, but the projects were bigger, 20 to 80 MW, versus the 10 to 20 MW of the tax-credit era. Because there were fewer projects, unique site conditions easily distorted the installed price. This is especially the case with projects installed during 1989, 1990, and 1991, which included a costly 45-mile high-voltage transmission line from Tehachapi to near Los Angeles.

Data for 1992 and 1993 is from two small projects in Tehachapi and that for 1994 is from the initial 5-MW phase of the Sacramento Municipal Utility District's wind plant in Solano County. The same developer, U.S. Windpower, installed a 25-MW project in Minnesota for Northern States Power in 1994 for $1100 per kilowatt. The full 50-MW project for SMUD will be completed at a price of $1050 per kilowatt in 1996.

During the early 1990s Danish utilities were installing their first 100 MW of wind capacity. The published price for ELSAM's Danish wind plants is more representative of the true costs to utility owners than the much larger projects installed by nonutility generators in California.

One of California's largest developers reported privately in 1992 that if up-front profit and extraordinary expenses (such as those for a lengthy transmission line) were excluded, wind plants could be built in the United States using either foreign or domestic wind turbines for about $1000 per kilowatt, although no project had been built for this price by 1994.

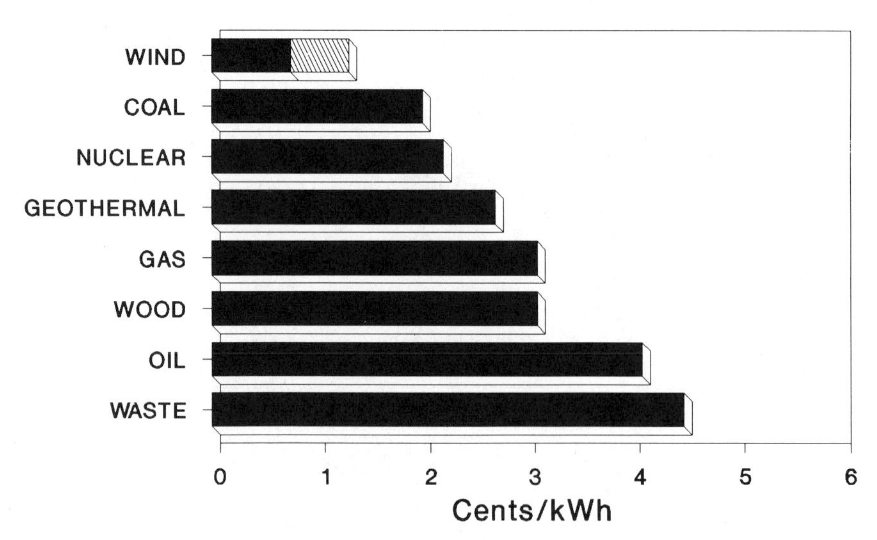

Figure 7.2. Operation, maintenance, and fuel costs in U.S. power generation. Wind power plants are less expensive to operate and maintain because the fuel is free.

Maintenance Costs

Another factor influencing the cost of energy is the cost of operating and maintaining the wind turbines. Proponents have often argued that unlike the fuel for a conventional power plant, the fuel for a wind plant, the wind, is free. Now after more than a decade of experience, it is possible to compare the actual cost of operating, maintaining, and fueling conventional plants to that of wind power plants.

Based on a sampling of projects in the middle to late 1980s, EPRI concluded that it cost $0.005 to $0.017 per kilowatt-hour to operate and maintain a fleet of existing turbines in California. The average cost hovered around $0.01 per kilowatt-hour.[10] This represents about half the cost of operations, maintenance, and fuel for coal and nuclear plants, and about one-third the cost of that for gas-fired plants (Figure 7.2).

Jens Vesterdal calculates that ELSAM pays from 0.0034 Ecu ($0.004) per kilowatt-hour to 0.0144 Ecu ($0.019) per kilowatt-hour to operate and maintain their 43 MW of wind plants in Jutland with a weighted average of 0.0089 Ecu ($0.012) per kilowatt-hour in 1991 currency.[11] U.S. Windpower told the California Energy Commission during the early 1990s that it could operate and maintain wind plants for $0.013 per kilowatt-hour in 1989 dollars. Zond reported it could do so for $0.01 per kilowatt-hour.[12] In its study of renewable energy in the midwest, the Union of Concerned Scientists assumed that operations and maintenance would cost $0.011 per kilowatt-hour in 1992 dollars at energetic sites.

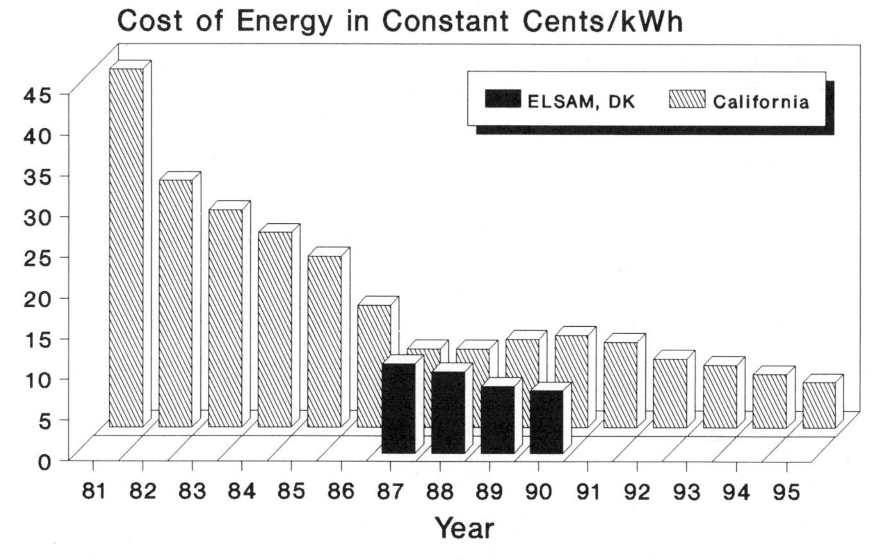

Figure 7.3. Relative cost of wind energy in California and Denmark.

Relative Cost of Wind Energy in California

Incorporating the price of installed wind plants in California and Denmark from Figure 7.1, their performance as discussed in Chapter 6, and the cost of operating and maintaining them ($0.02 per kilowatt-hour in the early 1980s, $0.013 per kilowatt-hour in the late 1980s, and $0.011 per kilowatt-hour in the early to mid-1990s) enables calculating the cost of energy from wind plants in California and Denmark (Figure 7.3).

The cost of energy in Figure 7.3 is in constant dollars derived from a fixed charge rate comparable to that used by the Union of Concerned Scientists. The data used in Figure 7.3 are similar to those used by the authors of the articles in the *EPRI Journal* and *Scientific American.* The cost of energy from California wind plants fell from nearly $0.45 per kilowatt-hour in the early 1980s to less than $0.10 per kilowatt-hour in the early 1990s, and is expected to reach $0.05 per kilowatt-hour in 1996. This is the origin of the "5 cent per kilowatt-hour" wind energy that has been bandied about in the United States since the early 1990s.

Specific Cost and Size

The foregoing applies to the cost of energy from medium-sized wind turbines at windy sites. Typically, installed price decreases with increasing size. Medium-sized wind turbines cost considerably less per unit of swept area than

Table 7.5
Typical Wind System Installed Cost

Rotor Diameter (m)	(ft)	Swept Area (m^2)	(kW)	Approx. Installed Cost ($)	Specific Cost ($/m^2$)
Microturbines					
1	3	0.75	0.25	2,500	3,300
2	7	3	0.75	5,000	1,700
Small Turbines					
3	10	7	1.5	10,000	1,400
7	23	40	10	30,000	750
Medium-Sized Turbines					
18	59	250	100	125,000	500
25	82	500	200	250,000	500
35	115	1000	400	500,000	500
40	131	1250	500	600,000	480

do small wind turbines. (Table 7.5). Above 20 m (65 ft) in diameter, price is less subject to economies of scale.

The cost of installing single turbines is typically greater than that of installing a multiunit array. However, NOVEM's Ruud De Bruijne has seen just the opposite result in the Netherlands, where single turbines and small clusters of machines are less costly than wind power plants. By using the existing infrastructure of roads and power lines, Dutch farmers installing one or more turbines avoid the expense that a larger wind plant would entail. Projects requiring special foundations, such as those on dikes or breakwaters, cost considerably more than those on level terrain. Turbines installed offshore cost more than at any other site (Figure 7.4).

Most manufacturers were moving toward 500-kW turbines for use in wind power plants by the mid-1990s, but not, as commonly believed, because their specific costs are less than those of smaller 300 to 400-kW machines. The price of 500-kW turbines on the German market per square meter of rotor swept area is about the same as that for 100-kW turbines. Surprisingly, the specific price for some of the smaller machines is less than that for the 500-kW turbines (Table 7.6). German developers opt for the larger machines

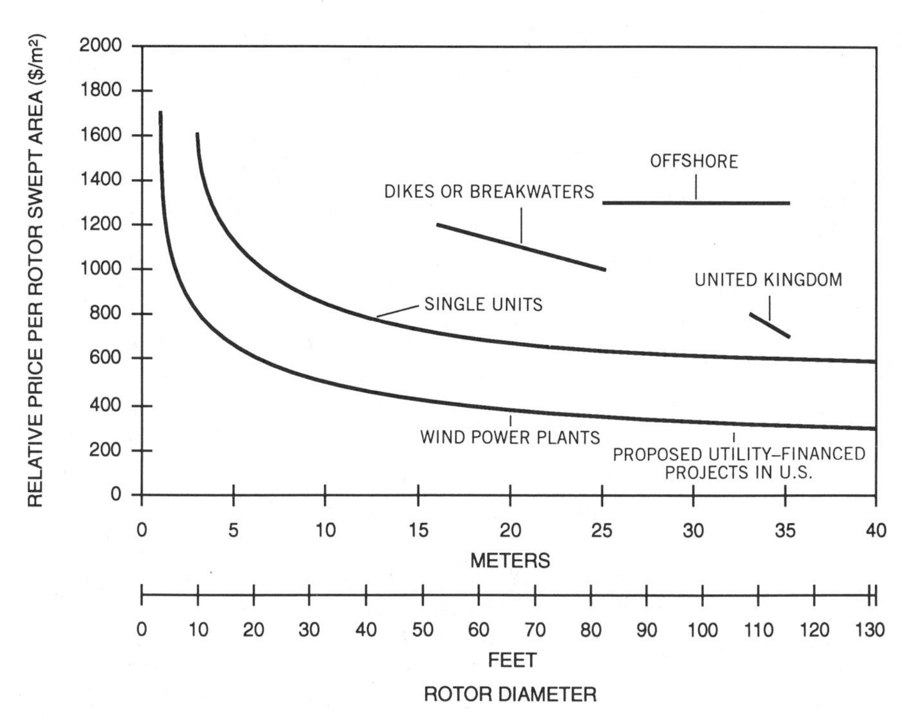

Figure 7.4. Relative installed price by rotor swept area. Medium-sized wind turbines are considerably less costly per square meter of swept area than are small wind turbines.

Table 7.6
Wind Turbine Prices on German Market

	Diameter (m)	Capacity (kW)	Swept Area (m^2)	Price ($)	Specific Price ($/$m^2$)
Small Turbines					
Enercon[a]	19.4	80	296	140,000	470
Tacke	21	80	346	160,000	460
Lagerway	18	80	254	120,000	470
Vestas	20	100	314	130,000	410
Medium-Sized Turbines					
Enercon[a]	40	500	1,257	620,000	490
Bonus	37	450	1,075	540,000	500
Nordtank	37	450	1,075	560,000	520
Tacke	37	500	1,075	550,000	510
Vestas	39	500	1,195	560,000	470

Source: Wind Energie Aktuell, September 1993.
Enercon includes transformer.

for two reasons: they can generate more energy from a single site, and they can spread their operations and maintenance cost over more kilowatt-hours.

Effect of Installed Cost on the Cost of Energy

The hyperbole surrounding the 5-cent (per kilowatt-hour) windmill in the early 1990s was driven primarily by decreases in installed costs expected in the middle of the decade from wind turbines then under development. The turbine and its tower contribute two-thirds to three-fourths of the cost of a wind plant (Table 7.7).[13] Reducing the cost of the turbine can reduce total installed costs substantially. When the fixed charge rate and the cost of operations and maintenance are held constant, the cost of energy is most sensitive to the energy in the wind (the performance among modern medium-sized wind turbines is comparable) and the installed cost (Figure 7.5).

The cost of energy to a private utility is presented in Figure 7.5 as a function of specific yield and installed cost. The units for installed cost (dollars per square meter of rotor swept area) in Figure 10-5 are uncommon in the utility industry, but are more explicit than the familiar units of dollars per kilowatt because of variations in the generator rating of wind turbines. A wind turbine that costs $400 per square meter of rotor swept area is equivalent to a turbine costing $1000 per kilowatt if it has a specific rated capacity of 0.4 kW/m^2—a 500-kW model using a 40-m rotor, for example. A turbine that costs $300 per square meter is equivalent to a turbine costing $750 per kilowatt (Table 7.8).

Table 7.7
Installation Expenses for Danish and California Wind Plants (% of Total Costs)

	Zond	ELSAM
Wind Turbine	73	68
Foundation	4	9
Installation	4	
Roads	3	2
Electrical	10	8
Grid connection	4	6
Land		3
Control building		1
Miscellaneous	2	3
	100	100

Source: Zond Systems: Ken Karas, Wall Street seminar, 1993; ELSAM: Jens Vesterdal, "The Potential of Wind Farms," 1992.

Most wind plants installed in Europe and California during the early 1990s cost \$500 to \$600 per square meter and were yielding 900 to 1100 kWh/m^2 per year at energetic sites on the North Sea coast and near Tehachapi, as well as near Palm Springs. Both NREL and U.S. Windpower were projecting that they could build wind plants using their new technology for only \$300 per square meter in the mid-1990s. U.S. Windpower guaranteed that their 50-MW project for SMUD would deliver 950 kWh/m^2 at SMUD's Solano County

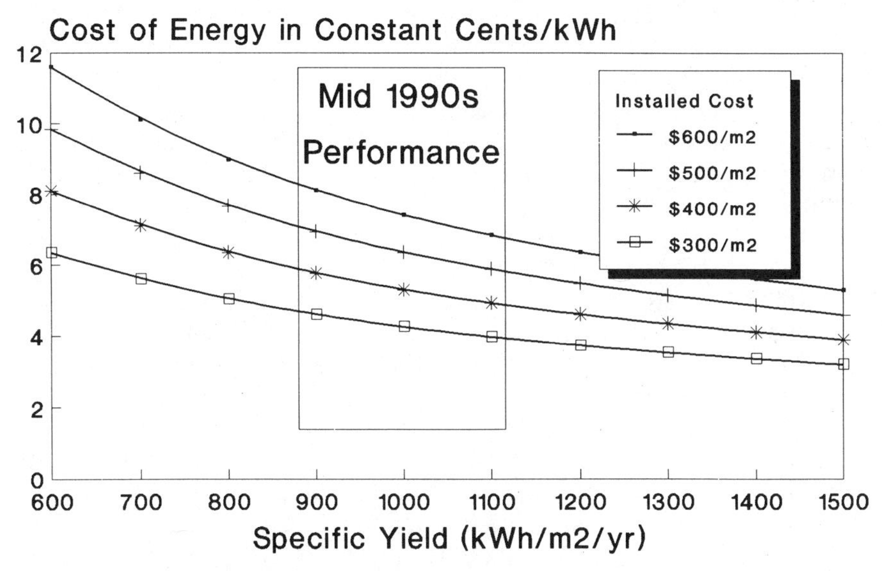

Figure 7.5. Effect of installed cost on the relative cost of energy. Because wind energy is capital intensive, lower installed costs substantially reduce the cost of energy. Wind regime and the performance of the wind turbine determine specific yield.

Table 7.8
Specific Installed Cost in Terms of Rated Capacity

Installed Cost ($/m²)	Equivalent Installed Cost ($/kW) for Specific Rated Capacity:		
	0.4	0.5	0.6
800	2000	1600	1333
700	1750	1400	1167
600	1500	1200	1000
500	1250	1000	833
400	1000	800	667
300	750	600	500

site. The SMUD transaction is the first major sale of an entire wind plant to a utility in the United States. As such, the SMUD project avoids the financing costs and developer profits that burden other wind plants in California.

Future Costs

Every two years the California Energy Commission compares the cost of energy from more than 230 electrical generating and end-use technologies. The resulting Energy Technology Status Report reviews the commercial and economic feasibility of each technology. The study provides critical input into the CEC's energy policy recommendations and its evaluation of any new power plant of more than 50 MW proposed for the state. For these reasons, the Status Report becomes an important reference for the CEC and other state agencies. Because of the state's pivotal role in energy policy, the report also carries weight with federal agencies and researchers around the world.

During the early 1990s one U.S. manufacturer told the California Energy Commission that it could install wind plants for about $700 per kilowatt in 1989 dollars. Another reported that it could install the same size project for $1000 per kilowatt.[14]

In its 1992 report, the CEC assumed that a wind plant could be built in 1997 for as little as $780 per kilowatt in 1989 dollars based on manufacturers' data. They also assumed that such a wind plant would deliver a capacity factor of 25% and cost $0.013 per kilowatt-hour to operate and maintain. They estimated that an investor-owned utility, under such conditions, could generate electricity for $0.06 to $0.068 per kilowatt-hour in nominal currency (Figure 7.6).[15]

By 1994, no wind plant had been built at this price, but one manufacturer had publicly announced that they would install a wind plant for $1050 per

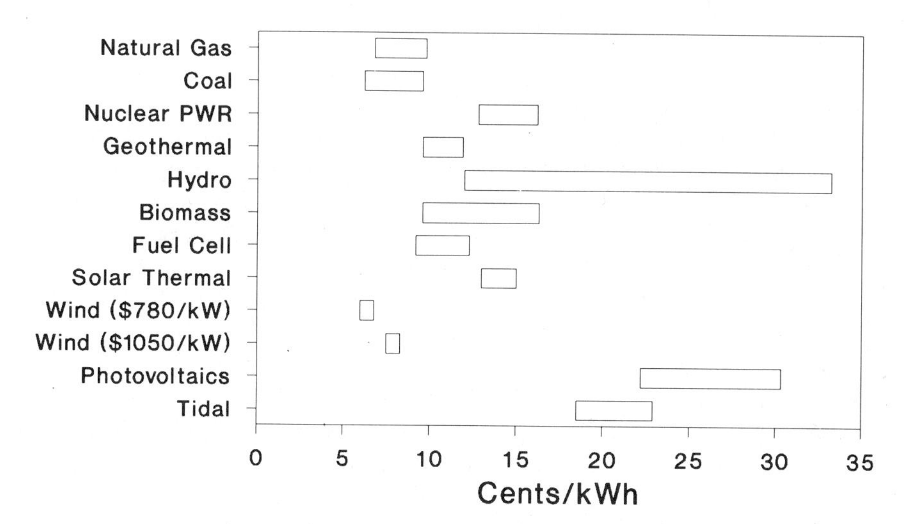

Figure 7.6. ***Levelized cost of electricity. An estimate, in nominal dollars (1989), of the relative cost among various technologies in California.***

kilowatt in 1996. At the author's request, the CEC recalculated the cost of energy using an installed cost of $1050 per kilowatt and obtained $0.075 to $0.083 per kilowatt-hour.

In late 1994 Britain announced the third tranche of the Non Fossil Fuel Obligation for England and Wales and the first contracts of the Scottish Renewable Order. The bids are revealing because they include new 15-year contracts. The longer contracts, nearly twice the length of previous NFFO tranches, enabled developers to find less costly financing than before. This in turn halved the price for new wind projects compared to previous NFFO tranches. Wind companies bid an average of 4.3 pence ($0.069) per kilowatt-hour for projects in England and Wales larger than 4 MW. The lowest winning bid was 3.98 pence ($0.064) per kilowatt-hour. The results were similar in Scotland, but stronger winds permitted developers to bid as low as 3.8 pence ($0.061) with an average price for wind bids of 3.99 pence ($0.064) per kilowatt-hour.

Thus the cost of wind energy in nominal dollars is less than $0.05 per kilowatt-hour only under the most optimistic conditions, that is, $780 per kilowatt, and only in the United States, where the federal production tax credit of $0.015 per kilowatt-hour applies.[16] Nevertheless, wind energy remains one of the least costly sources of new electrical generation and is competitive with new fossil fuel–fired plants. According to the CEC's estimates, wind-generated electricity is about half the cost of new nuclear capacity, even using the less optimistic estimates of wind's installed cost.

Other Factors

Frequently, the cost of energy comparison between technologies overlooks factors that are difficult to quantify in terms of price. These include the cost of decommissioning plants once they have reached the end of their useful lives, the value of modularity, and social or environmental costs. Another is the cost of capital. The ability to finance debt influences not only the cost of energy but also determines who is most likely to own the project.

Decommissioning

The difference in cost between decommissioning nuclear power plants and wind power plants is substantial, but often ignored. For example, the two units at Pacific Gas & Electric's Diablo Canyon nuclear plant will cost $580 million to decommission. Dismantling Southern California Edison's 2400-MW three-unit complex at San Onofre will cost $870 million in 1988 dollars. Decommissioning the damaged Unit 2 reactor at Three Mile Island will cost more than $1 billion when completed, nearly as much as it cost to build the plant. Overall, decommissioning nuclear plants will cost from $0.3 billion to as much as $3 billion per 1000-MW reactor in 1985 dollars, or $300 to $3000 per kilowatt.[17]

Although individual turbines at wind plants may need replacement after 20 to 30 years, there is no compelling reason for abandoning the site. New, more cost-effective turbines can be added, the infrastructure upgraded if necessary, and the plant returned to use. This is one of wind energy's most significant attributes: it is sustainable. Because there is no accumulation of wastes (as in a nuclear plant) or exhaustion of the fuel source (as in a fossil fuel–fired plant), production of wind-generated electricity is sustainable indefinitely.

It is still useful to look at the costs of dismantling because removal of individual turbines may be necessary from time to time. The public also wants assurance that removal costs are within the financial reach of salvage companies should the wind plant become inoperative.

Decommissioning a wind plant, in its simplest form, requires removal of the turbines. The U.S. Bureau of Land Management (BLM) estimated that removal of a turbine and tower would cost $1500 to $2500. Removal of the foundation and wiring could add another $1500 to $2500; revegetation of disturbed soil, another $500 to $700. The BLM estimated that total removal would cost from $3000 to about $5500 per turbine for the 100-kW turbines typical of the mid-1980s.[18]

The BLM's estimate, about $50 per kilowatt, represents the high end of expected removal costs. To date, dismantling has cost much less. Several early 40-kW turbines were dismantled in Tehachapi for $1500 each. In the Altamont

Pass, U.S. Windpower dismantled 200 of its 50-kW turbines, removing the towers, transformers, and the top 2 ft of the foundation for $925 per site, about $20 per kilowatt.

A turbine's salvage value often offsets much of the removal cost. In 1989 a contractor to Riverside County began removing 98 turbines from a site near Palm Springs for $23,000, only $230 per turbine. U.S. Windpower received salvage bids of $1025 per turbine for their 200 machines in the Altamont Pass, resulting in a net credit of $100 per unit removed.[19]

The lower end of the range of cost estimates for decommissioning nuclear power plants is six times greater than the highest estimate of the cost of dismantling a wind power plant.

Modularity

The modularity of wind technology has contributed significantly to the industry's rapid growth. Modularity is also a powerful lure to further industry expansion because it offers substantial cost savings over traditional, large power stations.

Wind turbines individually cost a fraction of the cost of a conventional power plant, and nonutility generators or the utility itself can install as many wind turbines as it needs, when it needs them. This modularity has driven the technology at a faster pace than that of conventional energy development because wind technology has a much shorter lead time. Modularity and short lead time result in quick adoption of technological refinements.

In an era of uncertain growth in electricity demand, modularity offers utility planners not only flexibility but cost savings as well. When growth is uncertain, the addition of a large central station may commit too much of the utility's resources for too long. The utility runs the risk that some of the capacity may stand idle once completed. It must absorb the financing costs until the plant enters the rate base as well as any plant costs determined by regulators as not contributing to useful capacity. Instead of spending a decade and committing large amounts of capital to constructing one plant, developers can bring wind plants on line within one year. An illustration of the speed with which wind plants can be built occurred in 1985, when Zond Systems installed more than 1000 wind turbines (representing about 80 MW) at four sites in six months.

One study by Los Alamos Laboratory found that "utilities could afford to pay as much as four times more in overnight construction costs for 5 year lead-time plants than for 15 year lead-time plants."[20] For example, if a conventional plant cost $1000 per kilowatt, a utility could afford up to $4000 per kilowatt for a plant with only a five-year lead time. Thus even where new wind capacity is more expensive than conventional base-load plants, the benefits of modularity may make wind energy attractive.

More specifically, modularity could save in excess of $0.01 per kilowatt-hour. According to Oak Ridge National Laboratory, building a series of short-lead-time power plants would cut the price of electricity 18% over the 15-year period considered, or $0.013 per kilowatt-hour, relative to building one 600-MW coal-fired power plant over four years.[21]

Social or Environmental Costs

Each technology also incurs social or environmental costs not reflected in COE calculations. In 1989, Michael DeAngelis and Sam Rashkin of the CEC's research and development office attempted to account for the hidden costs of conventional sources, such as the air pollution from coal-fired power plants, or, conversely, the benefits from nonpolluting sources, such as wind, solar, and conservation. Rather than undertake the difficult and controversial task of placing a monetary value on these effects, they sought simply to rank technologies by their social and environmental impacts.[22]

DeAngelis and Rashkin, both with long experience monitoring California's renewable industry, adapted a weighting technique, first developed for land-use planners, to the effort. They gave each impact an arbitrary value of 1 to avoid the conflict inherent in determining the degree or severity of each impact. The actual numbers are not as important, say the authors, as the realization that social and environmental costs are real—and significant.

Decisions on what kind of power plant to build depend primarily on price, that is, what the utility and its customers will have to pay. Social values or costs are not normally included. To gauge the effect of "price" on their ranking, the CEC boosted the value of price to five times that of the "social" costs. Thus in the "price" case, DeAngelis and Rashkin awarded the price advantage of fossil fuels a 5 and environmental benefits a 1. They did this to compensate for any bias against fuels with heavy social and environmental costs, such as coal and oil (Table 7.9).

By weighting the costs of indirect subsidies, pollution, energy security, and waste disposal, as well as the price of the resulting electricity, the authors found that wind energy costs a fraction of that from oil, nuclear, coal, and natural gas plants. Even when the importance of price was emphasized (the "price" case), the CEC found that wind energy, as well as energy efficiency, emerged ahead of the others. "Renewable and conservation technologies have the lowest external costs, while fossil fuels have significant to very high external costs," they reported. Wind energy costs less overall than all other technologies, except solar thermal electric generation and conservation.

DeAngelis and Rashkin concluded "that there are certain technologies which remain at the ends of the list in both cases. Oil combustion, nuclear fission, fuel cells, and coal are among the highest cost." Similarly, efficiency, wind, biomass, hydroelectric, and solar thermal electric "are among the lowest cost

Table 7.9
Ranking of Economic, Social, and Environmental Costs (Highest to Lowest Cost)

Test Case	"Price" Case
Oil	Fuel cells
Nuclear	Oil
Coal	Nuclear
Natural gas	Coal
Cogeneration	Ocean thermal
Fuel cells	Photovoltaics
Municipal solid waste	Municipal solid waste
Geothermal	Cogeneration
Ocean energy	Geothermal
Hydroelectric	Solar thermal electric
Photovoltaics	Natural gas
End use (cons./solar)	Hydroelectric
Biomass	Biomass
Wind	Wind
Solar thermal electric	End use (cons./solar)

Source: M. DeAngelis and S. Rashkin, California Energy Commission, 1989.

technologies in both cases." Advocates of renewables prefer that a monetary value ultimately be placed on social costs to give the market accurate price signals.

Cost of Capital

Small wind turbines may be purchased outright, but few people buy medium-sized wind turbines with cash. Most COE calculations, such as those done by Karas and the CEC, include the cost of financing a portion of the project with bank loans. Karas assumed that an investor-owned utility could finance 50% of the wind plant's cost at 8% interest, whereas the CEC financed 48% at 9.10 to 9.25% for the life of the plant. The terms are less favorable to a nonutility generator. Banks perceive that independent producers encounter greater risk than utilities and demand higher interest rates over a shorter period (15 years, or half the 30-year life of a project) than that for utilities. This raises the revenues required in a project's early years, increasing the levelized cost of energy.

EPRI self-servingly argues that "utility ownership is cheaper because utilities, which have access to much more capital, can get lower financing rates than the smaller independent developers and entrepreneurs." The bottom line, says EPRI, is that "the cost of energy from a wind power plant owned by an independent power producer could be 30 to 40% greater than that of energy from the same plant owned by a utility."[23] But the regulatory changes wracking U.S. utilities in the mid-1990s increase their risk relative to independents, raising the cost of capital to utilities above that once envisioned by EPRI.

There is also a cost associated with each financial transaction. For small projects, the transaction cost of debt financing is spread over fewer kilowatt-hours than for larger plants. This increases the cost of energy for small projects and is one reason why projects in the United States are typically so much larger than those in Europe: developers attempt to spread their transaction costs over as many kilowatt-hours as possible.

Catherine Mitchell at the University of Sussex notes that the inflexibility of Britain's banking system makes it much harder to find debt financing for small projects than in Denmark, Germany, and the Netherlands. During the early to mid-1990s Mitchell found that financing was far less costly on the continent than in England and Wales.[24] (Table 7.10). The banking system discourages small projects and dispersed applications in Britain, while encouraging their proliferation on the continent using the same technology.

Bidding and Price

Rising productivity from advancements in technology, coupled with the falling costs of building and operating wind turbines, have led to dramatically lower costs for wind-generated electricity. As wind technology has matured, the cost of wind-generated electricity has plummeted, enabling wind energy to compete with traditional sources for the first time since European windmills graced the landscape of the Old World.

Yet the market still presents numerous barriers to wind energy. As North America, Britain, and other countries increasingly turn to market mechanisms, such as bidding, to determine the type of power plants that will be built, wind energy's nonprice attributes are being ignored in the "dash to [natural] gas," the mid-1990's fuel of choice.

A survey by NREL found that competitive bidding in most states has a "systematic bias against renewables" because of the emphasis on using

Table 7.10
European Cost of Capital (%) for Wind Projects, Early to Mid-1990s (Constant)

	Utility Owned	Privately Owned
Denmark	5–7	5.5–8.4
The Netherlands	2.6–3.1	6.2
Germany		3.3
Great Britain		8.3–10.1

Source: Catherine Mitchell, University of Sussex, 1993.

"proven" (that is, traditional) technologies, the undervaluation of environmental costs, and the exclusion of fuel-price uncertainty for fossil fuel plants or, conversely, the inclusion of wind energy (with its low operation, maintenance, and fuel costs) as a hedge against fuel-price volatility.[25]

The "tyranny of economic screening," says Harvey Sachs, seldom includes those attributes that make wind energy and other renewables so attractive. Estimating the future cost of energy is not an exact science, says Sachs, director of the University of Maryland's Center for Global Change, yet utilities and regulators act as if it were. Sachs argues that there is no reason for using hard-edged values of avoided cost to screen renewables. Yet the regulatory process discourages using a range of expected values to screen what is and is not cost-effective. The cost of all new sources are based on assumptions: assumptions about the future cost of a fossil fuel, such as natural gas, for example. And gas may not be as inelastic to demand as many hope. In 1992 the spot price shot up 60%. In contrast, there is less risk to estimating wind energy's future cost because there are no fuel costs.

And "how you count the beans does affect the answer you get," says economist Shimon Awerbuch. Intangibles, such as diversity, risk, and air quality "are not counted, because we don't know how to count them." Average levelized prices, says Awerbuch, conceal much of what happens to prices over time. "Fossil fuel saddles future generations with rising fuel and environmental costs," not reflected in a levelized price.[26]

"The market is underpricing [natural] gas," says Ken Oberg of ESI Energy (a corporate sibling of Florida Power & Light), "because it is not accounting for all of gas's risk, particularly contract risk," the risk that suppliers can and will break their contracts if prices escalate rapidly. The market should instead discount future cost projections of fossil fuel–fired technologies because of their potential price volatility. The risk increases with the length of the planning horizon. (The price forecast for next year is more likely to be on target than the forecast for 20 years hence.) By deemphasizing the price volatility of fossil fuels, the cost of energy appears lower in early years than that for technologies with high capital costs but low fuel costs, like those of renewables.

Don Bain of Oregon's Department of Energy echoes Oberg's concerns and warns that the market is also underpricing the environmental risk of gas because of the potential for future controls on emissions of global warming gases. Since these risks have been widely publicized, says Bain, utilities and their regulators can never argue that they were ignorant of the consequences. There is the potential that just as some investments in nuclear power in the United States during the 1970s and 1980s were deemed "imprudent," dependence on gas generation may force regulators to similar conclusions a decade from now.

Without an open system like that of northern Europe, where contracts with attractive tariffs are available to all those wishing to install wind turbines, or without contracts specifically set aside for renewables, wind energy is forced to bid against traditional fuels. To compete against the low initial cost of gas-fired plants, wind companies eager to win new utility contracts will make every effort to underbid fossil fuels as well as their industry competitors. There is a risk, like that of any other business, that wind companies will underprice their competitors by scrimping on environmental measures, safety, and other "amenities." Bidding can lead to a vicious competitive spiral where, as the old adage advises, "you get what you pay for."

S. David Freeman, who once managed the Tennessee Valley Authority and diversified the Sacramento Municipal Utility District's generating portfolio to include renewables, warns of "a fundamental conflict emerging between the headlong rush toward competition for price and price only and the needs of society to develop a sustainable energy economy."[27]

Wind turbines produce electricity from a clean, renewable resource in a relatively benign manner. Wind-generated electricity is inherently more valuable than electricity produced by fossil or nuclear sources without these attributes. The benefits of wind generation go well beyond its cost of energy. Although the cost of wind generation is now competitive with that of traditional sources using traditional accounting, for wind energy to fulfill its role, the tariff paid must ultimately reflect its value. Wind energy should never become "too cheap to meter."

Chapter 7 Endnotes

1. John Schaefer, "Wind Systems," *EPRI Journal,* July/August 1989, pp. 49–52.
2. Carl Weinberg and Robert Williams, "Energy from the Sun," *Scientific American,* 263:3, September 1990, pp. 147–155. PG&E engineer Don Smith had earlier published the data on which the chart was based. Smith's analysis showed the cost of wind energy falling from $0.52 per kilowatt-hour in 1981 to $0.05 per kilowatt-hour in 1986 in constant dollars for projects installed in the Altamont Pass. Smith, who used a CRF of 6.5% in his calculations, estimated that the cost of wind energy would drop to $0.033 per kilowatt-hour when installed costs fell to $700 per kilowatt-hour in the future." See Don Smith, "The Wind Farms of the Altamont Pass," Annual Review of Energy, 1987, p. 175.
3. John Douglas, "Renewables on the Rise," *EPRI Journal,* June 1991, pp. 17–25.
4. Ken Karas, "Wind Energy: What Does It Really Cost?" paper presented at Windpower '92, annual conference of the American Wind Energy Association, Seattle, WA, September 1992. Karas referred to projections by DOE/NREL.
5. Ibid.
6. "1992 Energy Technology Status Report," Report Summary Sacramento, CA: California Energy Commission, November 1992, pp. 65–70.
7. Ibid. Note that the capital cost of the wind plant used in this calculation is hypothetical. No wind plant has ever been built for the $780 per kilowatt assumed by the CEC.
8. Michael Brower, Michael Tennis, Eric Denzler, and Mark Kaplan, *Powering the Midwest* (Cambridge, MA: Union of Concerned Scientists, 1993), p. 25. The assumptions used by UCS are similar to those used by Karas and the CEC.

9. Jens Vesterdal, "Experience with Wind Farms in Denmark," paper presented at the European Wind Energy Association's special topic conference, "The potential of Wind Farms," Herning, Denmark, September 8–11, 1992.
10. Schaefer, "Wind Systems." The range was based on work by R. Lynette & Associates during the mid-1980s. See Robert Lynette, "Wind Power Stations: 1984 Experience Assessment," Research Project 1996-2, Interim Report for the Electric Power Research Institute Bellevue, WA: R. Lynette and Associates, January 1985.
11. Vesterdal, "Experience with Wind Farms in Denmark."
12. Alec Jenkins et al., "Technology Characterization Report," for ER-92 Sacramento, CA: California Energy Commission, November 22, 1991, pp. E6–2, E6–3.
13. The cost of the land on which the turbines are located is accounted for as an operating expense by companies that lease sites or as part of the installation costs when the site is purchased outright. ELSAM paid a one-time fee for 30-year use of the land on which its turbines rest. Other differences between Zond's construction of wind plants in southern California and ELSAM's installations in Denmark are Zond's use of truss towers set on piers. These are less costly than the tubular towers and the attendant foundations used by ELSAM.
14. Jenkins et al., Technology Characterization Report.
15. California Energy Commission, 1992 Energy Technology Status Report.
16. The National Energy Policy Act of 1992 grants a tax credit of $0.015 per kilowatt-hour of wind-generated electricity sold to a third party in the United States for a 10-year period after installation. The credit can be claimed only for new generation commencing after January 1, 1994 and before June 30, 1999. Under U.S. law the credit can be carried forward or back and includes adjustments for inflation. The credit against tax liability is reduced $0.001 per kilowatt-hour for every $0.0002 per kilowatt-hour increase in the national average payment for wind-generated electricity above $0.08 per kilowatt-hour, adjusted for inflation. (Only wind turbines installed after 1989 determine the average payment.) The credit is also reduced proportionately by any other government grant or government-provided, low-interest financing used in the installation. Public utilities, cooperatives, and government entities qualify for $0.015 per kilowatt-hour payment (adjusted for inflation) during a 10-year period for the production of new wind-generated electricity commencing after October 1, 1993 and before September 30, 2003 only if funds are appropriated by Congress. As of 1994 no funds had been appropriated.
17. Cynthia Pollock, *Decommissioning: Nuclear Power's Missing Link,* (Washington, DC: Worldwatch Institute, April 1986).
18. WIMP, Phase III, "Wind Implementation Monitoring Program," Draft Report, Riverside County, Riverside, CA, October 1987, pp. E18–E19.
19. Joanie Stewart, letter to Solano County Department of Environmental Management, Final Environmental Impact Report for the Montezuma Hills Wind Park, January 1989.
20. R. J. Sutherland and R. H. Drake, "The Future Market for Electric Generating Capacity: A Summary of Findings," Los Alamos, NM: Los Alamos National Laboratory, December 1984.
21. Eric Hirst, "Benefits and Costs of Small, Short-Lead-Time Power Plants and Demand-Side Programs in an Era of Load-Growth Uncertainty", Oak Ridge, TN: Oak Ridge National Laboratory, March 1989, pp. 15–17.
22. Michael DeAngelis and Sam Rashkin, "Social Benefits and Costs of Electricity Generation and End-Use Technologies," Staff Report Sacramento, CA: California Energy Commission, 1989.

23. Leslie Lamarre, "A Growth Market in Wind Power," *EPRI Journal,* December 1992, pp. 4–15. Investor-owned utilities fund EPRI's work.
24. Catherine Mitchell, "The Financing of Wind Energy in the UK Compared to Other European Countries and Its Implications," paper presented at the British Wind Energy Association's annual conference, York, October 6–8, 1993.
25. Blair Swezey, testimony before the Nevada Legislative subcommitte to Review Present Efforts to Conserve and Develop Energy Resources, National Renewable Energy Laboratory, Golden, CO, January 31, 1994, p. 4.
26. Shimon Awerbuch, "New Economic Cost Perspectives for Solar '94, annual conference of the American Solar Energy Society, San Jose, CA, June 29, 1994.
27. S. David Freeman, excerpts of remarks to the International Symposium on the Electric Power Industry, in Coalition Energy News, Center for Energy Efficiency and Renewable Technologies, Spring 1994, Sacramento, CA, pp. 6–7.

II

ENVIRONMENTAL COSTS AND BENEFITS: THERE'S NO FREE LUNCH

Moi, je travaille avec le vent qui est la respiration du bon Dieu. "Me, I work with the wind, the breath of the good Lord," says the miller in Lettres de mon moulin by Alphonse Daudet.[1]

We pay a price for energy. This price includes not only economic costs but social and environmental costs as well. No form of energy, no technology for generating electricity—including wind energy—comes without some environmental and social cost. In Part II we examine the costs and benefits of wind energy. We begin with the most serious issue facing the technology: community acceptance. And nothing in the pantheon of wind energy's real or perceived ills rivals that of wind's aesthetic impact on the landscape.

8

Machines in the Garden: Aesthetics, Opinion, Design, and Acceptance

N'allez pas là-bas. Ces monstres utilisent la vapeur, C'est une invention du diable. "Don't go there. The monsters use steam. It's an invention of the devil," the miller advises in Alphonse Daudet's Lettres de mon moulin.[2]

Any photograph of California wind turbines reveals one of the biggest obstacles to the development of wind energy. They are very visible and very ugly.—Thomas Lippman, writing for the Washington Post.[3]

The aesthetic pollution is obvious when a piece of land is transformed into a whirling wind factory.—Steve Ginsberg, writing in the Santa Monica Bay Audubon Imprint.[4]

Lavatory brushes in the sky.—Comment by conservative MP Sir Bernard Ingham.[5]

Much more aesthetically pleasing and certainly quieter than Sir Bernard Ingham.—Lynton Atkinson, writing in guest book at Delabole wind farm.

The wind turbines are "tall productive spires surrounded by green hills and grazing cows."—Bonnie Allen, in a posting on EcoNet after seeing the Altamont Pass for the first time.[6]

A "landscape of trilliums."—Sandy Lloyd, after driving through the Tehachapi Pass, referring to a flower that she remembered fondly from springtime walks in Michigan's north woods.

Diamonds in the sky.—Comment by a member of the Tehachapi Heritage League during a discussion of wind energy in 1993.

Are wind turbines ugly, or are they beautiful? Who is right? Is it the cynical big city reporter, the Audubon activist, the English parliamentarian, the environmentalist, the passing tourist, or the long-term resident? Whose aesthetic tastes best represent society at large? Do any of them? Is it important? Of these questions, only the last has a definite answer: yes, it is important. How people view wind turbines on the landscape and their acceptance of them will be *the* determining factor in whether wind energy ultimately fulfills its potential.

Aesthetics

The most frequently mentioned objection to the use of wind energy is the perceived aesthetic impact that wind turbines have on the rural vista. Unfortunately the anecdotes introducing this section highlight the sometimes polarized responses that wind turbines elicit. Few can look at wind turbines without forming an opinion. Yet these opinions are often fluid and subject to change depending on the time of day, the viewer's mood, and numerous other factors. It should come as no surprise that opinions of wind turbines on the landscape will differ from one person to the next in a pluralistic society with a host of differing cultural backgrounds, beliefs, and expectations.

In this chapter we examine how people view structures on the landscape, the symbols they attach to them, and the two great philosophical currents that influence our response. Subsequently, we review opinion surveys to discern what they tell us about the public's reaction to wind turbines and the measures that wind turbine designers and the developers of wind power plants may take to maximize acceptance. Contrary to popular belief, there is no universally consistent and invariable view of what is or is not pleasing to the eye. One of the best examples of this is public reaction to the Eiffel Tower.

Eiffel's Tower

The criticism of Gustave Eiffel's plan to erect a great tower in the heart of Paris is instructive, for there are many similarities between the initial objections to the Eiffel Tower and aesthetic objections to today's wind turbines. Not insignificantly, several thousand wind turbines in California are mounted atop towers incorporating the famous compound taper that provides the sweeping curves characteristic of the Eiffel Tower.

Eiffel won a commission from the French government to build what would become the centerpiece of the 1889 *Exposition Universelle:* an iron tower 300 m (1000 ft) tall. One of Eiffel's grandiose goals for the tower was to display the marvels of science in the service of humanity. He hoped the tower and accompanying iron-framed buildings of the exposition would set a pattern for mass-produced housing. Eiffel never intended for the tower to serve a practical function, although he later used the tower himself for experiments on airfoils and meteorology. It has also been used as a radio and television transmitter once those technologies became available.

Some of France's most illustrious figures, including the architect of Paris's renowned opera house, Charles Garnier, denounced the tower as a "grotesque monster," calling it a "symbol of decadence and a menace to art and culture." Paris literati, including writers Alexandre Dumas, Jr. and Guy de Maupassant and poet Sully Prudhomme, vociferously opposed the tower on aesthetic grounds.

In 1887 these and other artists issued a highly publicized petition calling for the project's abandonment. Although they waited until only four months before construction began to launch their vitriolic attack, they spared no literary device in stating their case.[7]

> We writers, painters, sculptors, and architects, lovers of the hitherto unspoiled beauty of Paris protest with all our might and all our indignation, in the name of unrecognized French taste and in the name of menaced French art and history, against erecting in the very heart of our capital the useless and monstrous Eiffel Tower, which the people's malice, frequently imbued with common sense and a spirit of justice, has already named the Tower of Babel.[8]

Parisian artists tried to outdo each other in their descriptions of the monstrosity in the florid style of late-nineteenth-century literature. Huysmans called it a "hollow candlestick"; Maupassant, an "ungraceful giant skeleton"; François Coppée, an "insane pyramid."[9] Would Paris "allow itself to be deformed by [such] monstrosities, by the mercantile dreams of a maker of machinery" critics asked rhetorically before warning that the "gigantic, black factory chimney" would stretch across the Champ de Mars "like a black blot, the odious shadow of the odious column built up of riveted iron plates."[10]

Just as happens today, opponents marshaled their forces and arrayed their arguments against the tower. And as all too often happens in such situations, facts are buried under a blizzard of misinformation. In Eiffel's case, a learned mathematician predicted that the tower would collapse after it reached the third floor. (It did not.) Thus the tower was not only a threat to artistic sensibilities, but also a threat to public health, safety, and well-being. Nearby residents of the Champs de Mars became the fiercest opponents, fearing that the tower would fall on their rooftops. They relented only after Eiffel assumed personal responsibility for any disaster.[11]

The viciousness of Eiffel's critics and their personal attacks mirrors those seen in many controversies surrounding large public works today, including the construction of wind power plants. Like proposals to plant wind farms on pastoral landscapes, Eiffel's tower elicited strong emotions. Supporters heaped adulation upon the tower, and opponents decried it with equal vigor.

Elitism certainly played a part in the controversy. Class distinctions linger to this day in France, despite the thousands of aristocratic heads that fell during the French Revolution. Parisian literati vilified Eiffel for deigning to deface their Paris. Composer Charles Gounod, in an attempt to associate Eiffel with crassness, added sarcastically, "Even commercial America wouldn't want the Eiffel Tower."[12] Léon Bloy called the tower a "tragic street lamp" produced by an "ironmonger's haughtiness."[13] Eiffel was, after all, the descendent of German immigrants, a mere engineer, who stooped further to work in commerce. This democratic (as the tower was built for the millions of people who would attend the exposition) and commercial (as Eiffel financed the tower himself) venture represented the antithesis of Parisian high society's aspirations. This has not been lost on later observers. Conservative commentator George Will has described the tower as a work of democratic art, designed to be viewed by the multitudes, not just a museum-going minority.[14] Despite whatever the majority of Parisians truly thought of its appearance, it was popular. In its first year alone, the Eiffel Tower nearly paid for itself from the 2 million people who visited it.

In a modern-day replay of Parisian attacks on the Eiffel Tower, more than 60 members of British intelligentsia charged in the Times Literary Supplement that wind turbines visible on moorland made famous by the Bronte sisters were an "assault on our literary and artistic heritage . . . and wholesale despoilation of a landscape with uniquely important literary associations." Sounding much like their counterparts across the Channel a century earlier, they called for a moratorium on further wind development within a 20-mile radius of the Bronte home at Haworth in the South Pennines—an area of 1300 square miles (3300 square kilometers)—to keep the machines out of view.[15]

In time, Eiffel won over his critics. Even Gounod relented, eventually performing a "concert in the clouds" atop the tower. A magazine article appearing after the tower's completion in July 1889 noted a spectacular change of opinion among those who had formerly opposed the tower. "When the barriers opened, when the crowd was able to touch the monster, to stare at it from all sides, wander among the legs, climb its flanks, the last resistance weakened among the most recalcitrant." Sully Prudhomme retracted his earlier salvo against Eiffel: "I had, fortunately, judged and condemned only by default, and in front of the finished and victorious work I feel today more comfortable than others in appealing against the sentence that I meted out."[16] To Paul Gauguin, the exposition was "the triumph of iron, not only as far as machines are concerned, but also in the realm of architecture."[17] Engineers, led by Eiffel, were developing a new form of decoration, a "Gothic tracery in iron," Gauguin said.[18]

The success of the tower took on symbolic importance. For Eiffel, the structure remained through the rest of his life a "symbol of strength and of difficulties overcome."[19] Today, French children's books recount the battle as an example of public-spirited perseverance surmounting narrowmindedness, fear, and intolerance. The tower now symbolizes Parisian architecture and, ironically, has became the theme of many works of art. Nearly all "accept" it as a natural part of the Paris skyline. Few could imagine Paris without it, and few would openly advocate its removal. Most would be appalled by the idea. It is as much a celebrated part of Paris as the Louvre's Mona Lisa. The tower, made a national landmark in 1964, is to Amélie Chazelles "the indispensable ornament on the Parisian landscape." To Chazelles "The Eiffel Tower is Paris."[20]

Machines in the Garden

French intellectuals of the nineteenth century were not alone in their strident attacks on the machines that symbolized the ascendancy of the industrial revolution. American and British authors of the era placed the machine in opposition to tranquility and made them an "emblem of the artificial," says Leo Marx in his landmark book about the role of technology on the landscape, *The Machine in the Garden.* In literature of the era the machine came to represent, says Marx, "the unfeeling utilitarian spirit" of the age.[21] Wordsworth, for example, fought extension of the railroad to Lake Windermere, and Thoreau decried the railroad that passed by Walden Pond. Yet Marx finds a degree of ambivalence toward technology even in Thoreau, who used the railroad's direct path through Walden's woods as a shortcut to town, pausing to ponder the flowers growing on its linear embankments.

Romantics associated the machine with crude masculine aggressiveness that violated the tranquility of the pastoral ideal. To Hawthorne, it was the shriek of a locomotive's whistle. (To a modern American, it may be the angry buzz of a neighbor's leaf-blower.) The Romantics portrayed the male features of the machine as thrusting their presence onto the feminine repose of the landscape.[22] This sexual metaphor is ill applied to the dilemma posed by wind turbines in Marx's pastoral garden, but serves to illustrate the intensity with which some view artifacts and machines on the landscape. Their views are not without foundation. Machines still do violate the earth in the coal fields of Appalachia, Wyoming, and Indiana, where giant draglines paw at the ground, gouging miles of deep furrows from which pour rivers of acidic waters.

Landscapes themselves are neutral. It is our response to them that is not. To understand our response to differing landscapes and objects on the landscape, we need not only to examine their cultural and historical context, but also to explore the long-running conflict between conservation and preservation.

Britain's National Trust, which has done more to preserve rural landscapes in England and Wales than most other groups, issues a frank warning to readers of its guidebook before attempting to explain why it selects one site over another. The Trust advises that "to analyze these feelings" of what is and is not beauty crosses "the boundary between reason and mysticism."[23]

Indeed, observers see objects on the landscape through more than their aesthetic eye. We judge landscapes both aesthetically and morally, says historical geographer Richard Francaviglia.[24] Several researchers have noted that social values play a role in how the public views landscapes; "technological development," for example, often has negative connotations in industrialized countries.[25] Thus our judgment of what we see is based on our values and the symbols that have come to represent those values.

Aesthetics are also tied to both a time and a place. Our view of the landscape and the value we place on it has changed dramatically since the turn of the century. At one time the forest embodied savagery, darkness, and mystery. To many, a cleared forest symbolized our dominion over nature and the success of our efforts to push back the darkness. Prior to the Romantic era, "beautiful scenery" was associated with agricultural and industrial productivity. For many in the nineteenth century, mining, with its large-scale alteration of the landscape, symbolized progress.[26] Even as late as the early twentieth century some writers were describing industrial landscapes as approaching a form of aesthetic perfection.

After the Romantics revolted against the "taming of the lowlands" by the headlong rush of the industrial revolution then sweeping Britain, the emphasis shifted from rural productivity to the value of "wildness," that is, the untilled, the untamed.[27] This is the genesis of the classic dispute between wildness (or wilderness) and the pastoralism of the Virgilian landscape, where shepherds tend their contented flocks.

The realization that a growing population could exhaust the world's remaining undeveloped or unaltered landscapes leads many to place a higher value on preserving the landscape status quo than on altering it with new developments of any kind.[28] This primitivism, as Leo Marx calls it, can be found in the writings of John Muir, who extolled the wondrous grace of nature and exhorted humanity to find salvation in nature's solemnity.[29] Muir's writings launched a movement to preserve wild lands from conversion to pastoral uses and from the encroachments of the urban world. Cattle were merely "hoofed locusts" to Muir in his campaign to preserve rapidly vanishing wilderness in California's Sierra Nevada mountains. His efforts ultimately led to the formation of the Sierra Club, one of the oldest and most powerful environmental groups in the United States.

Unlike Muir, Gifford Pinchot, Muir's one-time ally, believed that natural resources could be used responsibly. Timber could be harvested from forests, cattle could graze on pastures, farmers could till arable land—all sustainably. Pinchot eventually became the first U.S. Secretary of the Interior and created the Forest Service. During his career, he set aside huge tracts of land as forest preserves in his native Pennsylvania and throughout the United States. Environmental historian Roderick Nash credits Pinchot and his creed of conservationism with protecting more wild lands from unregulated abuse than Muir's and the Sierra Club's campaign for preservation.

Both philosophies were in marked contrast to the ruling belief of the late nineteenth and early twentieth centuries that natural resources were infinitely abundant and there for the taking. How these resources were used was immaterial. Both Muir and Pinchot challenged this concept and applied moral teachings to treatment of the land. The conflict between followers of Muir and those of Pinchot often overshadows their shared beliefs and mutual opposition to wanton abuse of natural resources. Unfortunately, many of the terms most useful in describing Pinchot's philosophy have been usurped during the 1990s by the so-called "Wise Use" movement in the United States. All too often, mention of the words "wise use" or "responsible use" triggers a defensive reaction in environmentalists who have spent decades trying to preserve remnants of wild lands.

This discussion is no mere academic exercise. It cuts to the heart of the debate about the placement of wind turbines on the landscape and explains why one "environmentalist" sees wind energy as part of a solution to a problem and another "environmentalist" sees wind energy as just another problem. To confound engineers, developers, and politicians even further, environmentalists carry within them both philosophies in varying degrees, and the relative importance of preservation to conservation ebbs and flows in a sea of varying emotional and rational responses to the issues of the day. Just as the shepherd straddles worlds of both nature and man, environmentalists must continuously resolve the conflict between preservation and conservation.

This conflict partly explains why many environmentalists support the concept of wind energy in the abstract (conservation) but may object to specific projects (preservation) in what has been called the NIMBY (Not In My Backyard) syndrome. Martin Edge at Robert Gordon Institute found this true in Scotland when observers were confronted with the difference between wind energy as a principle and wind energy as an actual structure on the landscape.[30]

The dichotomy between wild lands and the pastoral garden is unresolvable. The two philosophies will always coexist uneasily side by side within each of us. Roderick Nash articulated the issue for the wilderness wing of the environmental movement, of which he is one of the contemporary shining lights, in his 1979 article "Problems in Paradise."[31] Nash visited the utopian garden ideal of many alternative energy advocates and found it wanting. He feared that it would consume too much of the treasured wild lands by using diffuse energy sources such as the wind and sun.

Nash, a professor at the University of California's Santa Barbara campus, has written extensively on the philosophy of environmentalism and wilderness. He goes further than most environmentalists in his concern that renewable technologies could demand too high an aesthetic price. In one controversial article, he went so far as to suggest that nuclear power may be preferable to renewable energy, a proposal that many in the environmental camp would consider a sacrilege. "From the standpoint of scenic pollution and the destruction of wildness," he said, "there are distinct advantages to the hard energy option . . . a nuclear plant modifies a relatively small area compared to a large-scale solar installation, a hydropower dam, or a windmill complex."

Sounding an alarm about the giant multi megawatt wind turbines then fostered by the U.S. Department of Energy, Nash warned that they would not appear on the landscape as "picturesque structures surrounded by tulip beds."

Instead, "forests of the machines would have to be deployed," and they would probably be placed "in exposed, highly visible locations such as on ridge lines and the shores of oceans and lakes." Ominously, Nash drew a parallel between wind energy and the nemesis of California (and influential Santa Barbara) environmentalists: "If offshore oil rigs offend, can a much greater number of windmills be any better?" Unlike later devotees who believed that renewables were land-intensive and are therefore undesirable, Nash is less quick to state an unequivocal answer. To Nash, "For those who covet unmodified vistas and the solitude of the high windy hills, the choice between a single nuclear plant and thousands of enormous windmills is, once again, not easy."[32]

Nash's provocative piece touches the heart of the dilemma: renewables will migrate toward so-called marginal lands that constitute "de facto wild lands," because the land has not been deemed "useful" for other purposes.[33] These are not lands officially designated as parks, preserves, or areas of outstanding natural beauty, but they are undeveloped lands that offer solitude. Some are pastoral in the traditional sense, where sheep and cattle graze peacefully. Some, such as the mountain crests of the southern Sierra Nevada bordering Zond's Sky River project, qualify as truly wild lands. To Nash, the middle or pastoral landscape offers few of the benefits offered by true wild lands: solitude and the beauty of unmodified nature. To him, the optimum is an alternation between the truly wild (parks and preserves), and the truly civilized (cities and nuclear power) rather than the "constant ruralism" fostered by the decentralization of population and energy sources envisioned in Amory Lovins's "soft energy" path.[34]

For wind energy to gain its place in the sun with the least anguish for its proponents, it must steadily move toward the garden: the middle landscape where the tilled or grazed land is no longer truly wild. As long as public policy steers wind energy toward undeveloped "wild lands" on windy ridge crests, whether in the western United States or in areas of outstanding natural beauty in the United Kingdom, there will be bitter conflicts between soft-energy proponents and preservationists, although what they seek to preserve is physically far different in each country.

But is this middle landscape clearly discernible? No, it too is subject to interpretation. To some, wind turbines represent "an unwarranted intrusion into 'natural' landscapes" of the English countryside, says Dave Elliott of the Open University. However, he cautions, "The current UK landscape is far from natural: it has been almost continuously redefined over the centuries."[35] The much-loved hedgerows, for example, which give the English landscape its tidy, well-kept appearance, are simply fences cultivated to separate fields before the advent of barbed wire.

Elliott uses the term *appropriate landscapes* to describe his perceptions of the existing English countryside. In contrast to the inappropriate landscapes of monocultures, including cornfields in East Anglia and tree plantations in Wales, his appropriate landscapes preserve the biodiversity inherent in mixed agriculture.[36] Elliott's view parallels that of Britain's National Trust. The Trust finds that farming adds greatly to the scenery as long as it is farming that the Trust feels is not industrial or "factory farming" in scale. The Trust has even reintroduced sheep at some sites, to "preserve a pastoral scene" rather than allow the land to revert to scrub, its wild state.[37] Yet the Trust fights other rural land uses, such as the "slab-sided" straight lines of tree farms, for their artificialness.[38]

To complicate matters further, modern wind turbines are unconventional structures to many. Wind energy creates a landscape not seen for at least several hundred years, a landscape that is neither "industrial" nor truly "pastoral," says Angus Duncan, one of Oregon's representatives on the Northwest Power Planning Council and formerly a FloWind executive. Certainly, considering the scale of California development, wind plants are *new* landscapes, says Robert Thayer, a professor of environmental design at the University of California at Davis.[39] Yet unlike mining or urban development, says the Open University's Dave Elliott, wind plants can be removed if ultimately found unacceptable, and if no major roads were required during their construction, there should be no long-term impact.[40]

Effectively, wind energy is a new land use, one that is seen competing with recreation, vacation homes, wildlife, and other public uses. Wind energy may be and often is compatible with many of these uses, but the public has insufficient experience with wind turbines to know that.

The British use an apt expression to describe aesthetic impact: the loss of "visual amenity" caused when wind turbines are seen as intruding on a rural vista. Single turbines, small clusters, and even large arrays can be assimilated into a pastoral idyll, depending on the landscape and the type, size, and number of wind turbines. English parliamentarians are fond of describing their ideal as "a landscape with wind turbines, not a wind turbine landscape," implying some acceptable number of turbines in the first case.

Europeans want to avoid the wind turbine landscape characteristic of California's windy passes. The Open University's Alexi Clarke calls the dense arrays of small to medium-sized machines found on the first floor of the Whitewater Wash near Palm Springs an impenetrable "thicket."[41] Clarke acknowledges that this effect partly results from the use of wind turbines whose towers represent a greater proportion of the structure than do their rotors. As wind turbines become larger, he expects the rotor to become more dominant and the absolute spacing between turbines to increase. This lessens the thicket effect.

Even in the Tehachapi Pass, with its intense concentration of turbines, there are vantage points, such as atop Cameron Ridge, from which the wind turbines, appearing as white pinwheels against the dark backdrop of the Tehachapi Mountains, are absorbed by the landscape. From vantage points in the Tehachapi valley, members of the Tehachapi Heritage League, whose goal is preserving vestiges of the area's history, have described the wind turbines as "twinkles on the hillsides."

The View from Danish Forestry

American environmentalists may find it curious that the Danish agency for forestry would concern itself with the aesthetic impact of wind turbines on the Danish landscape. This is less surprising in its European context, where forestry, especially reforestation, often clashes with open-space preservation. Wind turbines in Denmark will ultimately compete with reforestation for the country's remaining open space. The pace of Danish wind development threatens reforestion, prompting the agency to call for restrictions of wind energy on aesthetic grounds. They specifically want to limit the number of single turbines installed, preferring to concentrate the turbines in wind plants.

In a video that stirred the ire of the Danish wind industry and raises as many questions as it attempts to answer, the agency on one hand described modern wind turbines as creating technical landscapes comparable to those created by utility transmission towers, while on the other it described traditional windmills as a harmonious and "picturesque" element of the Danish landscape. The agency said the proportions and design of traditional windmills are in harmony with surrounding buildings and the landscape because of the materials used, their color, and their height relative to other structures. In contrast, modern wind turbines were labeled as "large technical facilities" and "new versions of a common phenomenon: a technical landscape."[42]

"Wind turbines divert attention from the landscape," intones the video, "and change its proportions." When wind turbines stand atop a kame (hillock), for example, they direct attention away from the kame and its unique topographic shape. Wind turbines on coastal cliffs "reduce and level the natural contrast between the steep cliffs and the sea surface." Although this is possibly true, even the educated, sensitive observer will find it hard to evaluate what this means while watching the video. In one scene of wind turbines near the coast, the narrator says the turbines "disrupt the aesthetic experiences of this unique coastal area." A non-Danish viewer may find it impossible to tell why the view has been "disrupted." It is far less apparent than the authors portray.[43]

Despite the video's criticisms of what many Americans, at least, would find to be aesthetically pleasing wind turbine installations, the video goes on to illustrate what the authors believe are examples of wind turbines successfully integrated into the landscape in a way that serves "as a source of inspiration for the future." They suggest that when assembled in distinct geometric patterns, wind turbines are acceptable as a supplement to an existing technical landscape, for example near an urban area or as an attractive demarcation of a harbor entrance. Wind turbines can also be tolerated when planted among hedgerows in a pastoral setting.[44] While advocates in the United States and Great Britain strive to expand the model of wind energy development to include more than just wind farms, the forestry agency wants Denmark, the nation that pioneered the installation of single turbines and small clusters, to turn toward larger wind power plants.

The Cemmaes Inquiry

One wind project where visual impact was *the* major issue was the Wind Energy Group's plan to build a small wind farm atop a broad plateau in mid-Wales. Although an extremely modest project by California standards, the proposal was seen as precedent setting in Britain. Objectors argued that there should be a presumption against the project because of its proximity to Snowdonia National Park and that approval would encourage development on any site not specifically within a national park or other area of statutory protection. Because of the controversy, the secretary of state for Wales called for a public inquiry into the project, and the resulting hearing record makes interesting reading for students of landscape aesthetics.

After a week of testimony, the hearing officer of the quasi-legal proceeding issued a report to the secretary of state, who subsequently approved the project, noting a national policy in favor of renewable energy; therefore, the presumption was for the project. The Cemmaes inquiry was important not so much for its outcome but for what it says about how different groups view wind turbines on the landscape.

Often, local communities bitterly oppose new developments because they must bear all the impacts and yet receive only some of the benefits. Cemmaes put an unusual twist on this situation. The local and regional community strongly supported the project, while those from afar opposed it. The Countryside Council for Wales and the Council for the Protection of Rural Wales, most of whose members reside outside the community, opposed WEG's plans and intervened to stop the project. The list of supporters would make an American wind developer green with envy: the Montgomeryshire District Council, the members of the Glantwymyn Community Council, the Llanbrynmair Community Council, the Farmers' Union of Wales, the National Farmers' Union, the Friends of the Earth, the Green Party's Energy Action Group, and the nearby Centre for Alternative Technology at Machynlleth.

In criticizing the intervention, Lord Stanley of Alderley charged that the Countryside Council did not weigh the difficulty of maintaining rural life for the benefit of "the occasional walker along a footpath." In a departure from most such conflicts, in which interveners typically demand that authorities thwart development out of respect for local opposition, the Countryside Council argued that broader issues were at stake and urged planning officials to disregard local desires.[45]

The project itself was simple enough: WEG planned to install 24 turbines on the 4-km (2.4-mile)-long plateau (Mynydd-y-Cemais) in Powys, the county with the lowest population density in England and Wales. If the site had been in the United States, developers would probably have installed two to three times the number of turbines in a densely packed array called a wind wall. Instead, WEG planned only one widely spaced string of turbines. Complications arose because some of the two-bladed 300-kW wind turbines would stand only

2.5 km (1.5 miles) from the nearest point of the park boundary. Tourists and residents alike would also see the turbines from points along the scenic Dovey (or Dyfi, in Welsh) valley and from public footpaths that skirt the plateau.

Snowdonia National Park charged that the project would produce an unacceptable impact on the park's viewshed, including a view from Cader Idris 16 km (10 miles) to the west. The Countryside Council followed suit, although they were more candid about their concern. Their warning that the turbines would introduce "an alien and man-made element into an otherwise remote landscape" forthrightly states the fundamental problem with siting wind turbines in pastoral settings anywhere: the introduction of man-made structures into a landscape where nothing of that type had previously been present, other contentions—that the turbines would subject hikers to their "unnatural" presence, for example—were peripheral to the main debate.[46]

The interveners trotted out the entire stable of objections heard in every similar case, whether for a proposed wind farm or a new park. Like their counterparts in the United States, they argued that "desecration of such a site" is unnecessary without a clearly demonstrable need, that is, if there is no other less environmentally damaging means for producing the same amount of energy. Following a textbook strategy seen time and again, opponents elevated the value of the status quo, denigrated the project's benefits, denied its need, challenged the breadth of support, amplified its impacts, and criticized farmers for their greed in endorsing the project.[47]

Objectors argued that there were alternative sites with less visual impact. "In identifying this ridge," said the Countryside Council, "the applicants could not have chosen a more beautiful part of Wales." The Countryside Council suggested instead that rolling farmland with a mix of trees and farm buildings would have a greater capacity to absorb the wind turbines visually than would the sweeping plateau.[48] But the secretary of state for Wales disagreed, noting that wind plants will be sited on a treeless, open expanse wherever they are located.[49]

The Centre for Alternative Technology's Roger Kelly notes the similarity between the conflicts involved in siting wind turbines at Cemmaes and those involved in reforestation as practiced in Great Britain. Kelly, who followed the inquiry because of its proximity and because of the Centre's long-standing advocacy of renewable energy, says several of those who testified found wind energy's impact on the landscape less damaging than that of reforestation. Yes, wind has an impact on the landscape, says Kelly, but determining to what degree it is harmful is subjective. Put another way, wind energy is conspicuous, yes, but not necessarily intrusive.

The hearing officer agreed. Although he found that the wind turbines would indeed intrude on the local landscape, he ruled that they would not "materially damage the many fine views and attractive landscapes . . . nor would they unduly harm the visual amenities of the nearby National Park." The wind turbines, he concluded, "will quickly become an accepted part of this landscape."[50]

In explaining his decision, the hearing officer made a statement reminiscent of Angus Duncan's assertion that wind energy is not a land use with which we are familiar today. "The proposed development," he said, "does not involve any of the activities or features commonly associated with industrial and commercial development," while the turbines themselves "may be man-made and contain machinery." The proposed wind farm "is a form of development of a somewhat exceptional nature which has to be treated as such." Unlike with other developments, for example, the turbines can be easily removed and the land restored to its original condition. Also, said the hearing officer, there was no need for the site to be fenced to protect people from the turbines.[51]

Further, he ruled that the project was in the interest of the nation, as it would diversify its sources of energy and offset pollution from fossil fuels. The hearing officer never says that wind energy is superior to energy conservation or that it should be pursued to the exclusion of other technologies, only that wind does contribute to diversity and therefore has value.[52]

"Planning permission should be granted," said the hearing officer, "unless the proposal would cause demonstrable harm to interests of acknowledged importance." Although the site does provide expansive views and is visible from nearby Snowdonia National Park, it was never included in that park nor in any other proposed park; the special protection of parks is not applicable merely because the project would be visible from the park. "It is wrong to assume," he said, "as shown in some of the reactions to this proposal, that merely because an object is visible it will necessarily harm" the landscape. "It is the turning blades," he goes on to say, "rather than the scale of the turbines, that will draw attention." The turbines, which will become indistinct beyond 6 km (3.6 miles), have clean lines with no extraneous features or visual clutter, he concluded, and "are considerably more pleasing than the large [electricity] pylons with which they are often compared." In particular, the hearing officer ruled, the project would have little impact on the landscape seen from the park or on visual amenities within the park. Visual impact, he found, is not the sole determinant of planning approval. It remains necessary to balance landscape protection with other considerations.[53]

As to the Countryside Council's concern about the impact on long-distance hikers, WEG's open array was seen as preserving the plateau's sense of expansiveness so essential to the hiking experience. The main interest of hikers on the footpaths that meander across the site, said the hearing officer, is the panoramic views from the plateau. These views will remain unchanged, "as will the sense of isolation due to the total lack of human activity on the site, other than the occasional farming and maintenance vehicle."[54]

Assimilation

Whether the public ultimately views wind turbines as monsters stalking through the garden or a part of the garden itself will be the result of efforts by wind turbine designers and wind plant developers to minimize the visual impact of wind energy and of the natural process of assimilation. The Eiffel Tower, for example, has become assimilated into western European culture.

In 1993, Parisian officials welcomed the 150 millionth visitor to the steps of the Eiffel Tower, now the most visited site in all France and possibly the world. Today, more than 100 years after its construction, the tower hosts 6 million visitors per year, twice the number visiting Versailles, and earns the city of Paris $4 million annually.[55] Time and the frequent incorporation of the Eiffel Tower as an image associated with Paris, France, and even "culture" in paintings, poems, novels, movies, and other literary forms have all but erased the tower's earlier image as a "black factory chimney." Seurat, Delaunay, and Chagall are only a few of the artists who have used the tower as part of a Parisian backdrop or as the central element of a painting.[56]

The Eiffel Tower is only one example of the many large man-made structures gracing the landscape that at one time stirred controversy: the Seattle Space Needle, the CN Tower in Toronto, the great British railway bridge across the Firth of Forth, or the many giant suspension bridges in the United States. More than a decade ago the Pompidou Center, with its external steel frame, opened in France to much the same criticism as that leveled at Eiffel's tower. Yet the Pompidou Center attracted more visitors during the first year of operation than did any other French museum.[57] Many of those visiting the Pompidou Center may indeed find its appearance displeasing, but nearly everyone has found it interesting.

There are important visual and cultural differences between the Eiffel Tower, wind turbines, and wind power plants. The appearance of modern wind turbines arrayed in a rural landscape is far different from that of a single structure in an urban setting. Yet the universal acceptance of the Eiffel Tower today obscures the similarities: the criticism of each on aesthetic grounds of what is or is not good taste, the powerful words and images used, and the vehemence of the emotions engendered. Most outside France know nothing about the tower's early controversy. They only know the "beautiful" Eiffel Tower and not its "odious" twin, the "factory chimney." During a television interview in the early 1990s, a reporter asked his guest whether people still objected to the "ugly towers" in Tehachapi. When the guest responded with a comparison to the Eiffel Tower, the reporter sarcastically interrupted and said, "You don't mean to say . . . ?" The reporter was blinded by the tower's image today. It was unimaginable to him that someone could object to the Eiffel Tower on the same grounds that they would object to wind turbines.

In an essay on bridges and viaducts, French author Jacques Lacarrière describes this process of assimilation: "Like pylons and water towers, they have for so long been a part of our country and mountain landscape that we now no longer even see them; at least we do not notice them, and are certainly no longer astonished at their existence. . . . They have become innocuous in becoming so familiar." Bridges may not have become as invisible as Lacarrière imagines, but they are an accepted, if not valued, part of the landscapes where they exist.[58]

Assimilation partly explains the growing field of industrial archeology that serves to document and preserve once-despised eighteenth- and nineteenth-century industrial structures in the United States and Great Britain. Guidebooks now go to some lengths to describe these structures and their significance. "Rather than an ugly blot [on the landscape]," says the *Blue Guide to England,* "[they are] an absorbing part of our history . . . and have begun to take their place beside the castle, church, and country house as significant monuments."[59] At one site, the National Trust seeks to preserve both a structure and wild lands, side by side. The key element in the view from one of the Trust's preserves is Brunel's suspension bridge across the Avon Gorge. Rather than decrying the despoilation of the view, the Trust calls the bridge "a miracle of engineering beauty" and seeks to protect it as well.[60]

There are similarities between wind turbines and bridges. Both are large utilitarian structures, and both can be seen as graceful additions to the landscape when designed with the landscape in mind. The link between the two was demonstrated in 1993, when the Kern Wind Energy Association's low-power radio station won a second-place award for its description of wind turbines in the Tehachapi Pass. Michigan's Department of Transportation won first place in the same category for its description of David Steinman's graceful suspension bridge crossing the Straits of Mackinac.[61]

At Oddesund in the northwestern corner of Denmark's Jutland peninsula, the visual connection between wind turbines and bridges is less an abstraction. There, wind turbines and bridges are so much a part of popular culture that they have become representative of the Thy district, for which a local beer is named (Figure 8.1). Thyholmer pilsner is brewed in the small district town of Thisted and depicts on its label a cluster of wind turbines adjacent to the well-known bridge across the sound. There are wind turbines at either approach to the bridge and the structures are visually linked on the label.

In explaining its interest in preserving technological artifacts such as Brunel's bridge, the National Trust notes that "[historical] windmills were the first to be rediscovered as things of beauty as well as utility." A windmill, says the Trust, is not just another industrial building but a machine that can be "admired for its combination of elegance and economy." The popularity of windmills, the Trust goes on, is "to some extent due to its place in English landscape painting." Constable, for example, prominently portrayed windmills in his works, popularizing an association between windmills and "idyllic rural scenes" celebrating the beauty of the English countryside.[62]

This assimilation also affects other prominent landscape features. Marcus Trinnick, of the English law firm Bond Pearce, observes that nineteenth-century lighthouses are often the objects of preservation movements, although they are found in highly prized coastal locations. Even old mines, the landscapes most ravaged by the industrial revolution, have found a degree of

Figure 8.1. Thyholmer Pilsner. Label depicting wind turbines and bridge across the Oddesund in northwestern Jutland. The Thy district, the commune on the northern side of the Oddesund, contains Denmark's largest concentration of wind turbines. The label appeals to the region's pride in their noteworthy structures. (Courtesy of Thisted Bryghus, Thisted, Denmark.)

acceptance. In California one mining landscape—and one entire mining town, Bodie—have become state parks. "Time softens the impact" of technology, says geographer Francaviglia, "both physically and psychologically"[63] (Figure 8.2).

Modern wind turbines are gradually becoming an accepted part of the cultural landscape. The process is well under way and may take place far faster than ever before because of today's instantaneous communication. Already, scenes showing wind turbines as visually interesting backgrounds have appeared in major Hollywood movies, in videos, in television commercials, and in print advertisements. As the artwork depicting the Eiffel Tower in a Parisian setting now denotes an interesting, vibrant, and sophisticated culture, these appearances of wind turbines in the artwork of our day will speed acceptance. Although some images of wind turbines, such as photos of warning signs posted at California wind projects, have been used to display the negative side of technology, the bulk of the images have portrayed wind turbines as symbols of a clean modern source of energy as opposed to polluting nineteenth-century technology, and a future built on sustainability rather than on the exploitation that characterized nineteenth-century industrialism.

Figure 8.2. Windmill tower. Wooden tower shelters storage tank for water-pumping windmill in Mendocino, a scenic village on California's north coast. Residents are proud of the structures, although they are no longer functional, and preserve them as part of the townscape.

Dutch surveys conclude that "wind turbines will only be appreciated in the landscape when they [through the process of assimilation] are seen as part of the landscape, and no longer as a result of a technological process."[64] Wind turbines, or more properly, windmills, have been an accepted part of our cultural as well as physical landscape for centuries. One of many examples illustrating how pervasive windmills have become occurred in early 1993, when millions of television viewers in the United States watched an episode of the popular game show "Wheel of Fortune." The set depicted a bucolic Dutch scene of contented cows, wooden shoes, and revolving windmills. The show played on the link between windmills and a pleasant pastoral life.

The image of scattered Dutch windmills on the plains of northern Europe is ingrained in our collective consciousness. In almost every household of western European descent, somewhere in a painting on the wall or among the bric-a-brac on a shelf there is a miniature Dutch windmill or the symbol of one. Many seventeenth-century paintings by the Dutch masters include a windmill somewhere in the scene.[65] Rembrandt van Rijn probably added windmills to his landscapes not only because they were prevalent in the countryside of his time but also because he was personally familiar with his father's windmill near Leiden.[66] Dutch windmills of old are seen as a natural part of the landscape today, after centuries of such associations. But it is useful to remember that at one time, Dutch windmills were seen as "technological" devices, being as they were the machines of their day. For some Dutch of the period, they were undoubtedly negative symbols of polders and land drainage.

The Netherlands' famous dikes are not universally loved even now. Half of the Netherlands is land taken from the sea. "It's artificial country," wrote an Italian travel writer in 1880. "The Dutch made it."[67] As with any major change in land use, whether in the 1990s or in the 1600s, some will gain, others will lose. When merchants drained a lake for new farmland, they threatened the livelihood of local fishermen, who sometimes fought back by sabotaging the dikes: Dutch counterparts to England's Luddites.[68] Windmills made polder drainage possible, and it is likely that those displaced by these land reclamation schemes thought none too kindly of windmills, regardless of what Rembrandt painted in Leiden or what potters fired in Delft. Despite the controversy around polder drainage and the role of the windmill in it, traditional Dutch windmills are a nearly universally positive symbol of the Netherlands today. In the lowlands of Germany and the Netherlands, windmills are not merely machines but part of the community: local landmarks, distinctively painted and individually named.

Windmills were an integral part of European urban, as well rural, life. Cities not on water courses required at least one windmill to grind grain for every 2000 people.[69] Up to the eighteenth century there were more than 30 windmills in Paris's hilly Montmartre district for grinding grain from the fields surrounding the city, pressing wine, and crushing stone. Many were transformed into popular cabarets at the end of the nineteenth century.[70] Windmills in such an urban setting are not an isolated occurrence, even in France. A similar example can be found in Avignon, a southern provincial capital. The windmills atop the *Rocher des Doms* (also known as *Rocher des Moulins,* Rock of the Mills) are shown in medieval paintings as being literally next door to the city's grand Papal Palace. Clearly, the powerful papal residents, who were known for their sumptuous lifestyle, would have quickly dismantled the windmills (and the millers) if they had found them unusually annoying.

Windmills are a common symbol of Provence, a region of France popular among both tourists and French elite. Windmills are found on pottery, in backdrops for crèches at Christmastime, and among the region's famed baked-clay santons. Their abandoned towers have become "the beacons of the rural [Provençal] landscape" according to Jean-Marie Homet in a description of local lore.[71]

Nearly every Provençal village bears testimony to the windmill's dominance on the landscape in its street and place names. There were nearly 20 windmills on the coastal buttes overlooking Marseille in the seventeenth century, nine on "windmill butte" alone.[72] There were 19 in the "mill quarter" of Arles, a city popular with Vincent van Gogh.[73]

The windmills of Montmartre, Avignon, and Provence were not Disney creations built for tourists nostalgic for a "kinder, gentler" era. Nor were they made appealing because they had been neutered by advancing technology and the passage of time. They were machines built to do a job: to grind grain. And it is doubtful that they went about it in a silent or dust-free manner. Yet there they were, essential parts of city life. Some Dutch in the province of Groningen still operate windmills in village settings, now grinding grain for tourists as well as townfolk (Figure 8.3).

Before the industrial revolution, grist mills were always close at hand because flour was difficult to transport. Grist mills, whether water- or wind-powered, may not have been loved by their neighbors, but they were commonly accepted. Everyone acknowledged that they were needed—not that everyone was agreeable during wind energy's golden age. There is historical evidence of disputes between some mill owners and their neighbors, one conflict involving the King of Prussia. Yet the weight of historical accounts points to a widespread acceptance of the windmill's role. Only in modern times has it become possible for us to divorce ourselves visually from our means of support by burying our sewers, water, power, and other utilities.

Similarly, American farm windmills did not proliferate on the Great Plains because they evoked a simpler, more rural America. They too were used to perform an essential task: to pump water, and like European windmills, they did so conspicuously. Farm windmills were not only utilitarian, quenching the thirst of sodbusters, they also became a symbol of sophistication. Farm

Figure 8.3. Urban windmill. Grain-grinding windmill "Grote Geert" operating in the heart of the Dutch village of Kantens northeast of Groningen.

windmills worked their way onto the manicured estates of the wealthy and, as electric lighting eventually did, became an essential part of the well-equipped manor, prominently pumping water to the first flush toilets. No fashionable Palm Springs resort of the day could have operated without them.

Farm windmills may not be needed today to prime our toilets, but many have been installed in the past few decades as oversized lawn ornaments, to give a home an air of rusticity. This could become a fad. Imagine Palm Springs' celebrities—the Fords, Hopes, and Bonos—clamoring to be the first on their block to erect a brightly galvanized Aermotor so that they could experience the sights and sounds of our past.

Farm windmills were, after all, just machines—one of the first machines truly adapted to American conditions, but still a machine—yet they became part of America's cultural heritage. In the United States, the image of the American farm windmill is nearly ubiquitous. It is rare to find a home that does not display somewhere a picture or image containing a farm windmill in a bucolic setting. American writers have often incorporated the windmill image to denote the rural life or the American West. Some have written longingly about the reassuring sounds of the farm windmill. "The clickety-clack, whirring, and turning sound of the old windmill just outside my window, comforted me as I drifted off to sleep each night," recalls V. Schurer of her childhood on the Great Plains. Schurer's description of rural tranquility might surprise those who have complained about modern wind turbine noise. She found the quiet unsettling during periods of calm winds. "What a wonderful sound it was to hear the windmill start its creaking and groaning again," after the wind returned.[74]

Just as the Eiffel Tower has become an integral part of Paris's cultural landscape, modern wind turbines are finding their way into the lives of the communities where they are located. The process is most advanced in Denmark, the Netherlands, and in Tehachapi, California. In Denmark and the Netherlands, wind turbines, like the traditional windmills before them, have already become part of the physical and cultural landscape. They appear on official tourist and topographic maps. Baedeker's guidebook on Denmark includes a whole section on wind turbines because of their importance in understanding the Danish countryside. In the Netherlands the Dutch touring association, ANWB (similar to the American Automobile Club in the United States and the Royal Automobile Club in Britain), designates on its maps wind plants alongside tulip beds and traditional windmills as sites worth visiting. Topographic maps in Britain, Denmark, and the Netherlands now routinely include newly coined symbols for modern wind turbines.

Of California's three wind major wind resource areas, only Tehachapi is an identifiable rural community. Palm Springs in the San Gorgonio Pass and Livermore (of nuclear bomb fame) in the Altamont Pass are suburban extensions of the large Los Angeles and San Francisco conurbations. If any community in the world is dominated by wind turbines on nearby hillsides, it is Tehachapi, a mountain hamlet of 6000 nestled in a scenic valley of the Tehachapi Mountains. During the late afternoon, the setting sun accents the white blades of more than 1000 turbines facing the village. With their towers nearly invisible from town at this time, the view is one of a spinning mass of three-bladed rotors. Many residents have likened the scene to white "flowers," "cross stitches," or the "glistening beads on a necklace." Of course, there have also been less flattering references, including the observation that "they stand upon the ridges like the bristles on a boar's back."

Yet Tehachapi has grown to accept the wind industry and its wind turbines. The local Rotary Club includes the symbol of a modern wind turbine along with a California poppy, apples, and a stylized mountain scene to represent those elements of their community they believe most noteworthy. The chamber of commerce touts the area with a brochure displaying a map of local sights, showing the wind plants alongside orchards and area resorts. During the annual summer parade, a float with children of wind company employees proudly proclaiming themselves "the Wind Generation" receives as many cheers as any other.

Children, in fact, have been the quickest to accept the new face of Tehachapi. To celebrate Kern County's new library, schoolchildren from each town made banners symbolically representing their community. The banner made by Tehachapi schoolchildren depicted scenes of Tehachapi's landscape: flowers on a mountain backdrop. On closer inspection, though, what look like flowers are stylized wind turbines. This association of modern wind turbines with pride in community portends growing acceptance elsewhere, among all generations.

Opinion Surveys

The preceding anecdotes are useful for gaining a sense of how people in a community view wind turbines on the landscape. But they are not helpful in determining how many in a community share various opinions. Fortunately, there have been several surveys in the United States and Europe that have examined the public's attitude toward wind energy. The California surveys differ from the European surveys in one important regard: the number of turbines respondents are asked to judge. Subjects interviewed in California were exposed, some daily, to thousands of wind turbines, whereas European subjects were asked to comment on at most tens of turbines and sometimes on only one experimental machine.

In general, the surveys on both sides of the Atlantic reveal that those who favor renewable energy are more likely to find wind's impact on the community acceptable, those who are neutral will accept wind turbines on the landscape if they know they are beneficial, and those who object to wind development on philosophical grounds will find it unacceptable regardless of mitigation measures.

National Surveys in the United States

Environmentalists have long advocated the "soft path" espoused by Amory Lovins and others. This path prominently features energy conservation and renewable resources such as wind and solar energy. Unlike photovoltaics and some other renewables, though, wind energy is now employed commercially on a utility scale. Some have speculated that the actual use of solar technology to such an extent would erode political support among environmentalists once the warts of these technologies—their environmental impacts—became more apparent.

Environmentalists have never given wind, or any other technology, an unqualified endorsement. Yet eight years after the first wind plants were built in California, support for wind energy among public-interest groups, including environmentalists, remained strong. In a survey by the League of Women Voters, 95% of the public-interest leaders polled supported expanding wind energy, whereas only 56% of those in the energy industry felt similarly. Wind's support ranked above all but solar heating (96%), conservation (96%), and energy efficiency (98%) among the public-interest groups surveyed.[75]

Another survey specifically targeted national environmental leaders. The study, by Phyllis Bosley at Towson State University, analyzed attitudes toward wind's role within the environmental community's desired mix of technologies. Bosley, who has conducted several surveys on how selected groups perceive California's wind industry, surveyed national environmental leaders representing 18 different groups. Subsequently, she surveyed national and regional leaders of the Sierra Club, because of Sierra Club California's highly publicized opposition to a specific wind project near Los Angeles.

The survey compared attitudes toward four energy technologies: solar, wind, fossil fuels, and nuclear power. There was general agreement that both wind and solar were environmentally superior to fossil fuels and nuclear power. Both national environmental leaders and the Sierra Club agreed that wind's greatest attribute is that it is a "clean" source of energy. Although all agreed that further wind development should be encouraged, they stressed that conserving energy is more important than finding new ways to produce it.[76]

Bosley found, somewhat surprisingly, that the Sierra Club held slightly more positive views toward wind energy than did environmental leaders in general. Sierra Club respondents selected wind as the "most environmentally superior" energy resource, ranking wind superior even to solar energy.[77]

Nationally, "solar was seen as the most cost-effective energy source, followed by wind," says Bosley. However, the Sierra Club ranked wind almost equally with solar in cost-effectiveness. Currently, wind energy is considerably less expensive than most other forms of solar energy, but Bosley believes that those surveyed defined *cost-effectiveness* as including price as well as environmental costs. Both groups, says Bosley, strongly believed that environmental costs should be factored into energy pricing. They agreed that the "benefits of clean energy sources should be included in the calculation of the prices paid for them" and that the environmental costs of conventional sources should be included in prices as well.[78]

Both groups identified visual pollution as wind's greatest drawback, although nearly 80% of government officials surveyed and 90% of the national environmental groups and the Sierra Club responded that wind energy was worth its environmental impact. But local activists, who live near California's wind plants, see the issue differently, says Bosley. Half of these activists, many whose sole *raison d'être* was stopping further wind development, saw no advantage to wind energy whatsoever. Only 22% of the local activists surveyed in 1988 believed wind energy worthwhile, in large part because they concluded that it "simply does not work."[79] That wind turbines must "work" to be worthwhile is a recurring theme in opinion surveys, whether in California or in Europe. To the industry's detriment, wind turbines in the United States seldom "worked" properly during the early 1980s, giving rise to a widespread perception in California that wind turbines and the developers who haphazardly installed them were both unreliable. It took wind farm operators nearly a decade of constant improvement to gradually dispel this image.

Altamont Pass

In a survey of public perceptions of the Altamont Pass, researchers from the University of California at Davis found that overall, people believe that wind energy symbolizes "progress," an "alternative to fossil fuels," and the "use of safe and natural energy." Those who liked wind turbines weighed their symbolic value heavily, whereas those who disliked them responded to more "basic visual attributes such as conspicuousness, clutter, and unattractiveness," says Robert Thayer, the study's principal author. The U.C.–Davis team also found that those favoring wind energy "were willing to forgive the visual intrusion of the turbines on the existing landscape for the presumably higher goals of the project, whereas dislikers were not."[80]

In the Altamont survey and the others that followed, Thayer found that it is this "visual intrusion" or the "loss of visual amenity" that elicits the greatest concern. Although wind plants create other environmental impacts, the principal impact is clearly visible for all to see. There are no containment buildings around wind plants to shield their inner workings from view.

Ironically, this is one of wind energy's principal assets: the costs associated with it are not obscured, buried, or shoved off onto future generations. We who use energy live with the consequences today, in our field of view. To some, the recognition that renewable energy is visible is not necessarily a drawback. In an editorial for a newsletter on renewable energy in Michigan's Leelanau peninsula, a hotbed of deep ecologists, John Richter wrote, "Seeing where your power comes from and relating the warmth of your home to the sunny skies provides a spiritual connection to your surroundings."[81] In commenting on proposed wind farms in the Angeles National Forest, landscape architects from California State Polytechnic at Pomona (CalPoly) noted that they "would show the public where some of their energy comes from, which would increase public awareness of energy generation" and the costs associated with it.[82]

This visibility, according to Thayer, is a double-edged sword. Wind turbines visually express their function and provide the viewer with immediate feedback on their operation. "They either spin, or they don't," says Thayer. In contrast to the visual clarity of wind turbines, the operation of nuclear power plants remains shrouded in mystery. "While looking deceptively static and benign up close," Thayer observes, nuclear power may be envisioned as a dangerous threat. Wind energy, on the other hand, may be envisioned as "benign and beneficial to society" yet appear unreliable up close, because seldom are all the turbines turning at one time. "When significant numbers of turbines do not turn when the wind is blowing," the simple expectation of the observer is violated, says Thayer.[83]

The effect of spinning and nonspinning turbines on the viewer's judgment of wind energy's "usefulness" is found in both Californian and Swedish studies. When the turbines are spinning, they are perceived as being useful and, therefore, beneficial. Observers are quicker to "forgive" the visual intrusion if the wind turbines serve a purpose; this they can do only when they are spinning. Even those opposed to wind energy often note that they would moderate their position if the turbines "worked" more often. Reviewing comments from respondents in his Altamont survey, Thayer found that "inoperative turbines equaled or exceeded siting, design, and scenic character in causing negative" responses.[84]

Wind turbines' visible operation and nonoperation is often perceived by viewers in terms of reliability. When the turbines are idle, for whatever reason, viewers see wind energy as "unreliable." The Altamont survey stated that "the disadvantage most often cited was the wind's day-to-day unreliability. Despoilation of the scenery was only the fourth most frequently mentioned disadvantage. Noise was not an issue."[85]

Consultants monitoring the wind industry near Palm Springs reached similar conclusions, observing that "the literature from the industrial revolution is filled with references to the beauty of moving machinery." Tierra Madre Consultants explained the relevance of operating wind turbines to public opinion in a report to Riverside County on a 1986 survey. "Two-thirds of those polled," said Tierra Madre, "supported wind development as long as the wind turbines produced appreciable energy." The survey found the wind plants "that appear to be working more attractive than those which do not." Motion, said Tierra Madre, "is equated with lower perceived visual impact."[86]

This problem of usefulness becomes increasingly acute as the number of turbines in a viewshed increases. Because the eye is adept at detecting motion, the absence of motion among a mass of wind turbines attracts disproportionate attention. And as the number of machines increases, there is greater likelihood that some will be inoperable, resulting in the "missing tooth" syndrome.

Just as we notice a missing tooth in an otherwise winning smile, observers of wind turbines invariably notice the inoperative turbines first. We expect to see even rows of perfectly formed teeth just as we expect to see wind turbines spinning in the wind. What jars us is the unexpected, the missing tooth, the inoperative wind turbine. The phenomenon has perplexed California's wind industry for more than a decade. An operator can show visitors a vista with thousands of spinning turbines. Yet the first question typically to arise is: "Why aren't they all turning?"

Thus California's wind industry inadvertently became its own worst enemy by deploying large numbers of poorly designed turbines in highly visible locations along heavily traveled highways. They could not have picked a worse place to begin than the Altamont Pass, where meteorological conditions exacerbate the problem. In the Altamont Pass, prime energy-producing winds occur principally during the summer months, when cool ocean air is drawn through the pass by convective heating of the San Joaquin valley. During much of the year, and especially during the morning commute, even the best turbines in the best locations stand idle. And all the early trouble-prone turbines were installed immediately adjacent to Interstate 580, one of the busiest highways in the state, with 36 million vehicles passing by per year.

The same effect was found in Tehachapi and near Palm Springs during the early to mid-1980s, when unreliable first-generation designs dominated California's wind turbine fleet. In Tehachapi, wind turbines border Highway 58, which 6 million vehicles per year use to pass through the Tehachapi Mountains; and in the San Gorgonio Pass, dense wind turbine arrays practically straddle Interstate 10 and the 18 million vehicles that use the road yearly.

Yet even with modern designs at the best of sites, wind turbines will operate only 50 to 75% of the time because of variations in the wind alone. By the nature of the wind resource, there will be substantial amounts of time when the turbines will be standing idle. This is compounded by the wide disparity in the wind resource from one site to the next within the same area. In low winds, for example, turbines on the crest of a hill may be operating, while those farther down the hillside may not be turning at all. Similarly, in winds above 20 m/s (45 mph), some hilltop sites may see gusts greater than 25 m/s (55 mph), at which speeds the wind turbines there begin to turn themselves off, while turbines in more sheltered sites will continue operating.

At the best-performing wind projects, 97% of the turbines will be in service at any one time. On a good day, with winds in the ideal operating range of 10 to 20 m/s (20 to 45 mph), only 3% or more of the turbines will be out of service awaiting routine maintenance. In an area such as the Altamont Pass, with a mix of old and new technology, 10 to 15% of the turbines may be out of service. With more than 6000 turbines, as many as 1000 turbines could be standing idle during ideal wind conditions. For those predisposed to find fault with a new technology, these nonspinning turbines amply reinforce the view that wind turbines are "unreliable" and fail to fulfill their function.

Somewhat surprisingly, nonoperating wind turbines have never become an issue in Europe. This may be due to the fact that the arrays found in Europe are smaller and more dispersed than those in California, and the visual impact on any view is more limited. Or it could result from the dominance of single turbines and small clusters of machines in Denmark, which has given Danes and other Europeans an opportunity to see a less threatening view of wind turbines than that presented in California. Or it could simply be due to the fact that Danish wind turbines during the early 1980s were more reliable than their American-built competition. It was not uncommon to see operating wind turbines on the Danish landscape in 1980. Even as late as 1983 it was rare to see large arrays of American-built wind turbines *operating* in California. Not until the large-scale introduction of Danish wind turbines and U.S. Windpower's model 56-100 in 1984 did the reliability of California wind plants begin to improve.

The need to prove that wind turbines are indeed useful led Thayer to suggest that wind companies make every effort to keep their turbines in operation.[87] California operators took some of Thayer's advice to heart and now try to show their turbines spinning in photographs. British developers discovered that the photographs of spinning wind turbines Californians found so successful had the opposite effect on the British public, and instead, raised fears that the machines would howl like buzz saws. Consequently, British photos typically show the turbines at rest. In England and Wales during the early to mid-1990s, the public had no prior experience with inoperative turbines as in California. As operators were installing state-of-the art technology, there was little concern about whether the turbines would be perceived as useful because they were likely to perform as expected.

San Gorgonio Pass

The San Gorgonio Pass near Palm Springs is the scene of the most protracted, bitter, and well-known controversy over the value of wind energy. The consequences have been far-reaching, not so much for the numbers of people involved but because of their celebrity status and political influence. The situation is somewhat reminiscent of the dispute surrounding the Eiffel Tower. Although no famous writers or artists are involved as in the Paris of 1889, they have been replaced by southern California's equivalent: retired entertainers (Bob Hope and Sonny Bono) and conservative ex-presidents (Gerald Ford).

During the early to mid-1980s, there was a nearly constant clamor in area newspapers and at public meetings about the wind energy scourge that was wreaking environmental havoc on the pristine desert surrounding the aging resort. From the flood of media attention and moves by local politicians to restrict wind development, an outside observer could easily be convinced that the entire community was ready to take up arms to stop this pillaging menace.

The wind industry certainly gave opponents ample opportunity to find fault. Development near Palm Springs proceeded at a faster and more haphazard pace than anywhere else in California. Wind turbines were erected with absolutely no regard for their collective aesthetic impact and seldom with any consideration of their impact on established desert neighborhoods. There is no worse example of wind energy than in the San Gorgonio Pass. Critics of wind energy in both the United States and Europe frequently use photographs of the San Gorgonio Pass to illustrate their concerns. But the wind turbines were only the latest form of development to sweep into the area, and they at least provided a net environmental benefit.

Palm Springs and its environs have appalled environmentalists for decades on many grounds. But nothing better epitomizes the conspicuous consumption for which it is known than the area's famous golf courses. Only massive infusions of chemicals, water, and electricity keep alive the so-called "Golf Capital of the World." These 100 or so bright green islands amidst the stark tan desert "are a gross violation of any person's notion of the sustainable carrying capacity of that desert region," dependent as they are on their umbilicals to more hospitable climes.[88]

With each course covering some 100 acres [40 hectares (ha)], golf has consumed nearly 10,000 acres (4000 ha) of the Sonoran Desert, home of the endangered fringe-toed lizard and desert tortoise. The recreation of a landscape resembling golf's Scottish birthplace required moving 3 to 6 million cubic yards (3 to 6 million cubic meters) of desert plants and soil to make way for the $\frac{5}{16}$ inch of Bent grass on each course.[89] The total soil moved for these courses would cover a 1 mi^2 (2.6 km^2) section of land one football field deep. Next to a hospital's intensive care ward there is no more artificial environment on earth.

Thus the stage was set for the most telling public opinion survey conducted to date. Conducted by contractors to Riverside County in 1986, the survey, because of its conclusions, remains controversial to this day. All of those surveyed were from Palm Springs and the small communities near the wind plants in the San Gorgonio Pass. Most (58%) lived within 2 miles (3 km) of the wind turbines; the remainder lived 2 to 5 miles (3 to 8 km) away. Of those nearest the wind plants, three-fourths could see the turbines.

The results shocked wind's critics as well as some elected officials who had staked careers on fighting wind turbine "blight." According to the study by Martin Pasqualetti, a geography professor at Arizona State University, and Edgar Butler, a professor of sociology at the University of California's Riverside campus, "the vocal opposition to the wind turbines . . . was not . . . borne out in fact." While the researchers acknowledged that there was "some opposition to the development of wind power at this site, particularly in terms of aesthetic degradation," they concluded that "the majority of respondents favored the development." Slightly more than half, 51%, considered the development of wind energy good, 21% thought it was not, and 23% were neutral. One of their more significant findings was that "two-thirds of this majority also believed further development should be encouraged." The authors, referring obliquely to Dolores Hope, wife of entertainer Bob Hope, Betty Ford, former first lady, and Sonny Bono, of the former pop duo Sonny & Cher, noted that "although some prominent Palm Springs residents openly advocated removing all the wind turbines, the survey indicated that the public at large generally did not hold such feelings. Only about 22% believed the wind turbines should be dismantled, while almost 50% thought they should not." The survey found residents (50%) ambivalent on whether the wind turbines were worth the environmental costs; 24% said they were not and 22% said they were. Nearly three-fourths said that the wind plants had not degraded the environment around their homes. On the question of aesthetics, "there was a fairly even distribution of opinion," said the authors: 36% thought they were attractive, and 45% thought they were not.[90] Because noise is a function of distance, those nearest the turbines should find them noisier than those who live some distance away. The survey concluded that this was indeed the case. Two-thirds said that noise from the turbines "did not disturb them, while 11% indicated that it did."[91]

Despite the controversy at the time, the study concluded that "overall, the public reaction to wind development in the San Gorgonio Pass has been positive, albeit at some recognized cost to local aesthetics." As expected, opposition to the wind turbines was most strongly held "by those who could see them from their houses." Thus the opinions of those surveyed contradicted the prevailing negative impression given by local opponents.[92]

By the early to mid-1990s, the tide had turned. Palm Springs' glamour had faded further and the city was desperately searching for revenue to revive its ailing downtown. For the first time, then-mayor Sonny Bono, who previously saw the wind turbines as a disgrace, now looked upon them as an asset to be courted. Overnight, wind energy went from pariah to savior. Since then, there has ensued an amusing dance between Riverside County and the city of Palm Springs, as each seeks to keep the wind industry in the tax base. Never more than half a dozen activists, the ranks of hard-core opponents of wind energy have withered without political support, and their protestations have diminished to an occasional whimper at public hearings.

Cornwall

The clearest example of a change in attitude toward wind energy, once a community has learned to live with the technology, occurred in Cornwall, where a survey measured the public's response before and after wind turbines were installed. The county of Cornwall extends from Devon in southwestern England to Land's End, where its exposed location takes the full brunt of North Atlantic storms. With open water on the west, the English Channel on the east, and a spine of nearly flat, treeless uplands stretching the length of the peninsula, Cornwall is an ideal area for wind prospecting.

As early as 1980, Cornish dairyman Peter Edwards considered installing wind turbines as an alternative to a nuclear power station then proposed for Cornwall. In 1983, he visited Denmark to explore the idea further, but Edwards held off until he was convinced that the technology and the market were ready. Edwards first sought planning approval in 1989, finally receiving permission two years later, with noise emissions strictly limited because of local fears. By late 1991, Edwards had commissioned Britain's first commercial wind power plant in the hamlet of Delabole a few kilometers northwest of the Cornish village of Camelford. On a clear day, which is rare, Tintagel Head—locally believed to be the site of King Arthur's round table—and the coast are visible to the west.

The project created controversy well out of proportion to its size. Neighbors feared that the noise from Delabole's ten turbines would drive them away. Owners of second homes and tourists from urban England, who frequent Cornwall, worried that the turbines would despoil their summertime destination. To gauge the effect of the controversy on the development of wind energy elsewhere in Britain, the Department of Trade and Industry sponsored the survey of the public's response to the Delabole wind farm.

During the summer of 1990, researchers polled nearby residents and residents of Exeter in Devon (the nearest major city) about their attitudes toward environmental issues, wind energy in general, and toward the Delabole project. The poll was repeated during the summer of 1992, six months after the ten 400-kW turbines were installed.[93]

During the first survey, two-thirds of those rural residents polled near Camelford identified themselves as "green," whereas three-fourths of those polled in urban Exeter so identified themselves. Opinion near Camelford was split on whether wind turbines should be built at Delabole, with two-thirds favoring the project and one-third opposed. Respondents in Exeter heavily favored the project. (The wind turbines were in someone else's backyard.) A majority of the Cornish group thought the wind turbines would spoil the scenery and a majority in Exeter thought they would not. More than 40% of the Cornish respondents thought the turbines would create a noise nuisance. Overall, the Cornish residents were less favorably disposed toward wind energy than were those living in Exeter.[94]

After installation, attitudes toward wind energy changed noticeably among those most affected: the residents near Camelford. Prior to installation, only 15% of those surveyed in Cornwall thought that noise would be insignificant. After the turbines were in service, says Brian Young, 80% of those surveyed in Cornwall found that noise was not a problem; 12% were unsure. Young, an Exeter consultant, found that among the Exeter group, which had not seen or heard the wind turbines, only 27% thought noise was not a problem, 19% still thought noise was a problem, and 54% were undecided. This led Young to conclude that noise is "less of a nuisance than people who had not lived near [wind turbines] would anticipate." In other words, noise is more of a concern for those who have never heard a wind turbine. [95]

Opinions toward Delabole's aesthetic impact also changed. After installation, about 28% of the Cornish sample still thought that the wind turbines spoiled the landscape, whereas more than half had thought so prior to the project. In Exeter, 29% thought wind turbines spoiled the scenery, agreeing with their

Cornish neighbors. Young determined that this group of aesthetic objectors in both samples believe that wind turbines spoil the scenery whether or not the respondents had ever seen wind turbines in Delabole or elsewhere. It was unnecessary to see them. The mere thought of wind turbines on the Cornish landscape was unacceptable. Still, nearly 60% of the Cornish sample found that wind turbines did not spoil the scenery, despite the fact that more than 90% thought preserving the scenic beauty of the countryside was important. Young determined that Cornish residents' negative opinion of wind energy's scenic impact prior to the actual project, like their concern about noise, had resulted from fear of the unknown.[96]

Overall, 85% of the respondents in Cornwall and Exeter approved of wind energy after completion of the Delabole project. Only 4% disapproved. Of the 289 surveyed, only 7 respondents continued to disapprove of wind energy after the turbines were installed, and only 3 switched from approval to disapproval after the project was completed. In contrast, says Young, 71 people, or about one-fourth of those polled, reversed their original opinion and approved of the project after completion.[97]

These results parallel those of Dutch wind developer Energy Connection, which has noted general approval of wind energy in the Netherlands until specific projects are proposed. These proposals elicit a negative reaction that dampens public support, but approval returns to near-normal levels once the project has been installed and communities have had time to adjust (Figure 8.4). "The reality of a wind farm," says British aerodynamicist Andrew Garrad, "is much better than the prospect of it." Acceptance grows as people learn that many of their misgivings were ill-founded.[98]

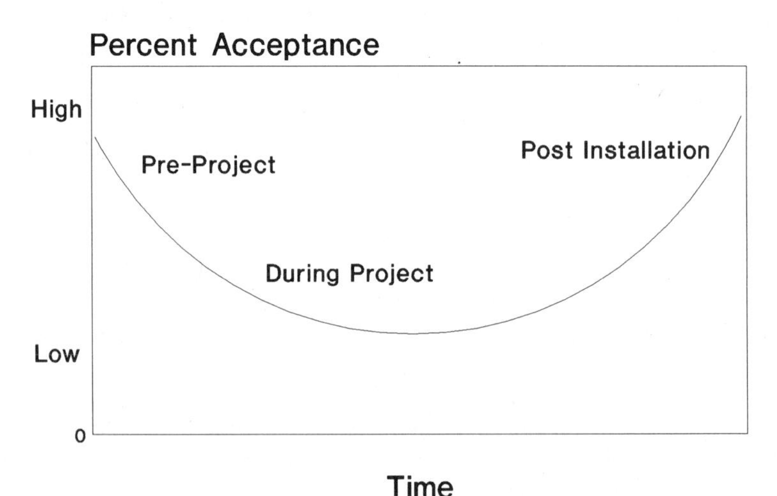

Figure 8.4. Acceptance of wind energy. Although wind energy has high approval ratings, Dutch developer Energy Connection has found that support erodes after proposal of a specific project. Acceptance returns to near pre-project levels once the community becomes accustomed to the wind turbines. (Courtesy of Energy Connection, Delft, The Netherlands.)

The survey found other interesting results as well. Substantial majorities in both the urban and rural groups agreed that wind energy was clean and unpolluting. This may shock some critics, who have made crusades charging that wind energy is not as "clean" as portrayed. Cornish residents also rejected other tired clichés about wind energy once they gained experience with the technology. Two-thirds of those in Cornwall found that wind turbines did not interfere with television reception, whereas in Exeter, two-thirds were still unsure because they had had no experience with wind turbines nearby. Again more than two-thirds of those in Cornwall found that the wind turbines did not disturb wildlife. Only 6% thought that they did, down substantially from the one-third who shared that opinion before the project was built.[99]

Wales

Subsequent surveys in Wales by the Department of Trade and Industry and the Countryside Council for Wales found results similar to those in Cornwall. Both surveys carried significant political import: The Trade and Industry study surveyed local opinions shortly after construction of WEG's hotly contested Cemmaes project, and the Countryside Council's publicized attempt to suppress the unexpected results of its survey embarrassed the quasi-public agency, which opposes expansion of wind energy on aesthetic grounds.

In the Trade and Industry poll, the surveyors conducted face-to-face interviews in the homes of residents living near Cemmaes both immediately after construction was completed and one year later. Three fourths of those interviewed saw the two-bladed WEG turbines every day. After one year of operation, 95% were "supportive" of wind energy. Only 16% thought that the wind turbines were "noisy." The overwhelming majority concluded that the turbines did not "spoil" the scenery, whereas 22% believed that they did. More than four-fifths of those surveyed (86%) favored the Cemmaes wind project after construction, and 11% were neutral. Only 3% objected. Two-thirds described the turbines as visually "interesting." More than 90% of the respondents said they were "not bothered at all" by the appearance of the turbines on the skyline. Upon more detailed questioning, the interviewers found that 54% of respondents held positive attitudes towards the appearance of the turbines, 27% were neutral, and 12% objected. Three-fourths of those who could see wind turbines from their homes expressed positive attitudes. As found in the Cornish study, there was less support among those with less exposure to the turbines. Of those who could not see the turbines, 63% expressed positive opinions. Of the 21 interviewed who could hear the turbines from their homes, 10 said the turbines were not noisy, four found it acceptable, and four found the noise "OK" after becoming accustomed to it. Only three of the 21 objected strongly to the noise. More than half of those interviewed supported further wind development locally, and 82% supported further wind development elsewhere. The latter response may reveal a subtle NIMBY phenomenon at work among residents, even though the majority have welcomed the WEG turbines.[100]

The Countryside Council conducted telephone surveys in four different areas of Wales, three in the vicinity of operating projects and one, used as a control, where there were no wind turbines. The surveys in late 1992 and early 1993 included residents near Llandinam, the site of Europe's largest wind plant, Penryddlan-Llidiartywaun or PnL. Expecting to find a groundswell of opposition to wind turbines, the Countryside Council was so shocked by its own survey that it attempted to quash the results. The effort backfired when Parliament intervened and ordered release of the publicly-funded study, prominently thrusting the results into the public arena.

Contrary to the Countryside Council's expectations, surveyors found that wind energy in Wales "had so far met with widespread approval, and even enthusiasm." They went on to note that "uncertainty and opposition emanate from a minority only," and are "closely associated with a lack of information and lack of experience of actual development." More specifically, three-fourths or more of those living near existing wind plants were willing "to see development take place within their own neighborhood." With the exception of Cemaes on Anglesey, where 17% said wind turbines should not be built in their neighborhood, only 3-8% objected to wind turbines at the three other locations. Between one-half and three-quarters of the respondents in areas with existing wind plants thought the wind turbines were in keeping with the landscape. As in Palm Springs, the survey found that 72-92% of respondents felt the wind turbines had caused little disruption "contrary to some press reports." Four out of five of those living near Llandinam and its 103 turbines said they would favor further development. About nine out of ten of those surveyed in all four areas agreed that wind energy was clean and would reduce pollution. The turbines were found neither noisy nor otherwise threatening. Only 11% of those surveyed at Llandinam found the turbines "noisy." Most damaging to the Countryside Council's opposition to wind energy was the realization that while nearly all those surveyed agreed that wind turbines were conspicuous, only "small numbers regarded them as intrusive." At Llandinam only 20% found the turbines visually objectionable.[101]

Elsewhere in Europe

Surveys elsewhere in Europe have reached similar conclusions, though not identical to those of Great Britain. In a survey for the European Community, Dutch researchers found that 80% favor wind energy, 5% oppose it, and 15% were neutral.[102]

In another survey, the former Central Electricity Generating Board (CEGB) polled the general public to gauge their probable response to a proposal for three demonstration projects. Few of those polled by the English utility knew what a modern wind turbine looked like. But after being shown a photograph,

54% said the machines would not "ruin the landscape." The results were little different for those who had some familiarity with wind turbines.[103] Two-thirds of those surveyed thought that the turbines would not spoil the landscape, and a majority believed that wind energy was an efficient use of rural land.[104]

Significantly for the future of massive California-style arrays, acceptance decreases with an increasing number of turbines. Projects with more than 50 turbines were acceptable to less than one-fifth of those surveyed in 1988.[105] Young discovered a similar result in Cornwall. About one-third of those who did not object to wind turbines at Delabole said that an acceptable number was "as many as possible," but a majority of the nonobjectors picked arrays of six to ten turbines.[106]

In an unusual finding, respondents to the CEGB survey suggested that the turbines should be scattered randomly across the landscape rather than placed in regular patterns.[107] This augers well for British support of single-turbine applications and small clusters, although other surveys have found an aesthetic preference for regular patterns of multiple turbines.

A British survey by Lee, Wren, and Hickman concluded that there is a consensus that wind plants "are an efficient use of rural land."[108] Pasqualetti uncovered a similar finding in his Palm Springs study. Despite the best efforts of opponents, 47% of respondents in the 1986 survey still thought that wind energy was a good use of the desert pass. "Vocal sentiments are not always representative of collective public opinion," says Pasqualletti, "even though a persistent minority may effect changes in political positions and development policy."[109]

The Dutch association of electricity producers found that those living near a small wind plant of 18 turbines at Sexbierum "considered wind energy to be 'harmless to people and the environment' and reasonably cost effective." The pre- and postconstruction poll of those within 3.5 km (2 miles) of the experimental project in Friesland, a northern province of the Netherlands, found that wind energy is "one of the most attractive ways of generating electricity." Coal and nuclear generation were the least attractive, especially after Chernobyl. Researchers noted that "the attitudes of the respondents toward 'windfarms-in-general' were, on average, positive, and very little difference was found after the project was installed." However, among inhabitants of a town where there are no wind plants, people responded less favorably to questions about "windfarms in the vicinity" than to those about "windfarms in general."[110]

Maarten Wolsink, who conducted the survey in Friesland and a parallel survey near a 1-MW experimental turbine in North Holland, concluded that "the knowledge that a turbine will actually exist in the near future seems to make people slightly less positive."[111] This is a nearly universal response, whether nearby residents speak Dutch, English, or American.

The results from early Swedish opinion surveys conducted by researcher Inga Carlman before and after installation of two multimegawatt turbines found that decision makers thought that aesthetic impact and noise were the greatest obstacles to using wind energy. But the public responded differently. It is "much more important to the public," says Carlman, that "the technology is useful," generally harmless, and beneficial. As Thayer learned in the Altamont surveys several years later, the public is more tolerant of wind's impacts if they perceive that there are countervailing benefits. Showing the importance that Swedes place on "usefulness" in mitigating the aesthetic impact, 81% of those surveyed by Carlman felt that wind turbines should be located where they would be most effective, not necessarily where they would be less visible. Rather than hide the turbines in sheltered sites as sometimes proposed by politicians fearful of public reaction, respondents in the Swedish survey expected that the turbines would work better and more often, thereby proving more beneficial, if installed in prominent locations such as along the coastline.[112]

Carlman also observed that attitudes were strongly influenced by the information available about the turbines' performance, again reflecting the public's desire to know that the turbines are useful. Support remained, even when the experimental turbines were not working properly, if the utility sponsors continued to inform the public about the turbines' status.[113] Whether as the result of Carlman's work or out of a sense of duty to the public for the use of the common visual resource, utility-developed wind plants throughout northern Europe include informational kiosks: some, as in Germany, are extremely elaborate; others, as in Denmark and the Netherlands, are simple and inexpensive (Figure 8.5).

Acceptability and NIMBY

Although overall public support for wind energy is high, controversy surrounding projects near Palm Springs and in Cornwall, Wales, and elsewhere illustrates that a phenomenon other than simple opposition to wind technology is at work. Most opponents to specific projects frequently note that in principle they have nothing against wind energy, "just don't put them here."

As diffuse sources of energy, renewables are likely to thrust more communities into energy decision making than before because of the tens of renewable projects necessary to accomplish the same result as that of a single conventional power plant. By their nature, renewables may stimulate a greater "Not in My Backyard" or NIMBY response than other technologies. As expected, researchers in California found that wind turbines produced the strongest response of this kind.

In a mail survey, the third in a series of studies on public response to wind energy by the Center for Design Research at the University of California's Davis campus, Robert Thayer examined public reaction to four energy technologies: biomass, nuclear, fossil, and wind. The researchers polled residents of Solano County, a once rural area east of San Francisco. Although they live 25 miles (40 km) north of the Altamont Pass, residents are familiar with wind turbines,

Figure 8.5. Information kiosk. Simple, effective kiosk at Tændpibe–Velling Mærsk explains how the wind turbines work, who is responsible for them, and what they accomplish. Informational kiosks are a common sight at European wind plants, reflecting the unspoken view that use of the public resource, the visual amenity, obligates the operator to justify what they do with it.

and more than half said they see Altamont's turbines often. This survey is revealing because of the light it sheds on the NIMBY response and where people would find these technologies most acceptable. The Davis team identified NIMBYs as those who found a technology "acceptable" in their county, although they would not accept it within 5 miles (8 km) of their home. Most of those surveyed found wind energy desirable—somewhere. Only 9% thought wind plants were completely unacceptable, whereas opinion was more polarized about nuclear and fossil fuels. One-fourth found fossil fuel–fired plants unacceptable in the county; nearly half found nuclear plants unacceptable. But wind drew the greatest NIMBY response[114] (Figure 8.6).

Among the results, some of which are surprising in light of charges by wind's critics, Thayer's team found wind plants the preferred power source of the four considered. People were willing to accept wind plants closer to their homes, within 2 to 5 miles (3 to 8 km), than any of the other technologies. In contrast, the minimal acceptable distance to a nuclear power plant was 20 to 100 miles (30 to 150 km)[115] (Figure 8.7).

Of the six factors Thayer measured—health and safety, reliability, environmental impacts, cost, dependence on foreign oil, and visual impacts—it was the latter, wind energy's Achilles' heel, that was the least important (Figure

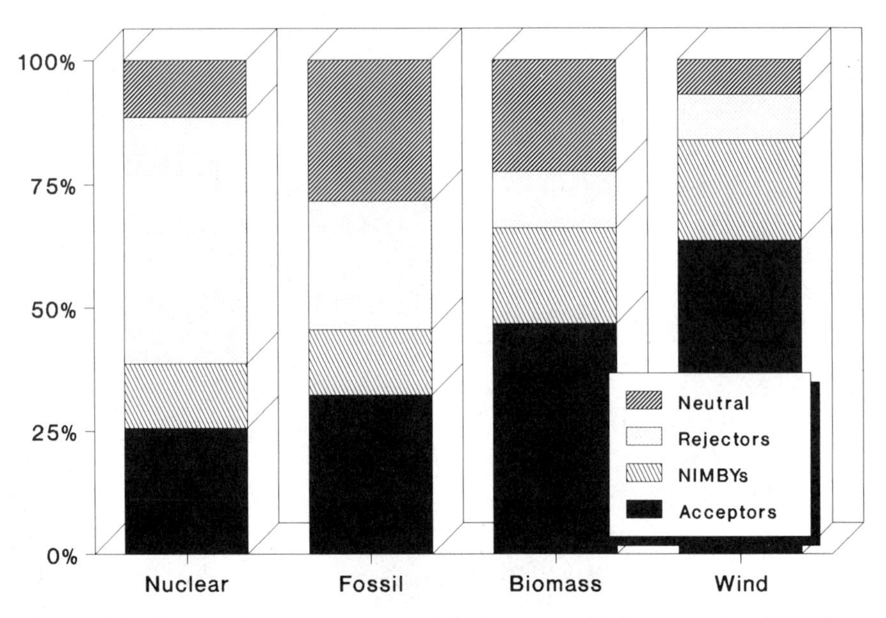

Figure 8.6. Power plant acceptance. Wind energy elicits a greater NIMBY response but fewer outright rejections than do other technologies. (Courtesy of U.C.–Davis.)

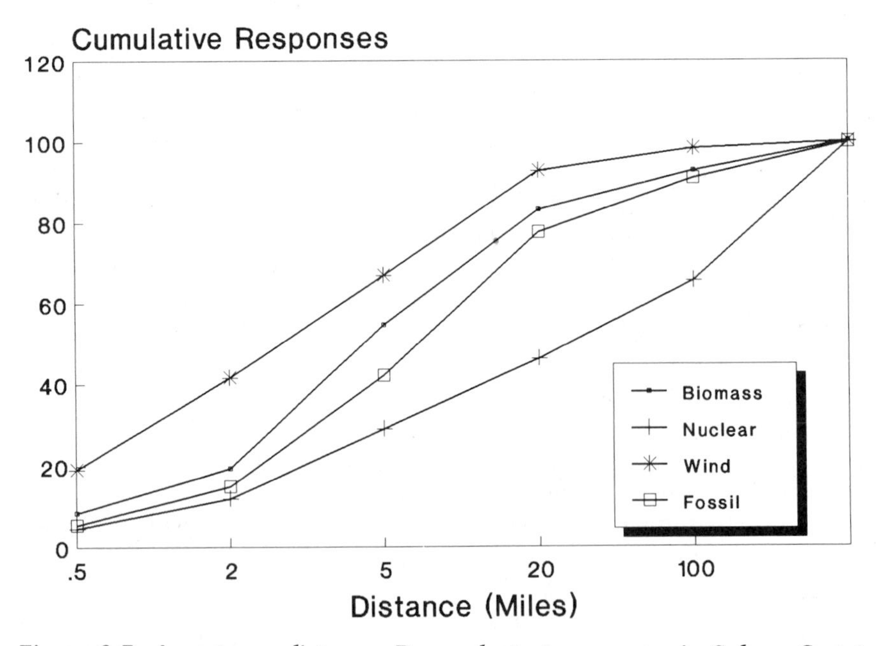

Figure 8.7. Acceptance distance. Respondents to a survey in Solano County were willing to live closer to a wind power plant than to other technologies. (Courtesy of U.C.–Davis.)

8.8). Although there was more disagreement about wind's visual costs than about those of the other technologies, wind still received the highest rating for visual acceptability of all the plants, including nuclear. This result puzzles researchers because nuclear plants are known for their clean lines and the absence of the cluttering conveyors and smokestacks common to coal-fired plants. Thayer suggests that the prominent cooling towers associated with nuclear plants have come to symbolize the controversy surrounding them, and this influences perceptions of the plants. Conversely, the Solano survey and others that Thayer has conducted in northern California reveal an ambivalence toward wind's aesthetic impact. On the one hand, wind is an energy technology preferred by most respondents, and its place on the landscape has some positive symbolic value. Respondents, for example, were more willing to view wind plants from highways, parks, and their offices than they were other technologies. Yet the visual impact, especially that of nonspinning turbines, troubles many.[116]

Thayer's research into the NIMBY phenomenon touched on the public's attitude toward electricity in general. He found that the utility system sends false signals to consumers. If local resources are unable to meet demand, more electricity is simply imported from outside the region. Thus there is no incentive for conservation, says Thayer rhetorically, because "we can just get more from somewhere else." As a result, consumers take electricity for granted, some viewing it as an inalienable, God-given "right." The consumer has never had to assume responsibility for supplying electricity. "That's the utility's problem, not ours," paraphrases Thayer.[117]

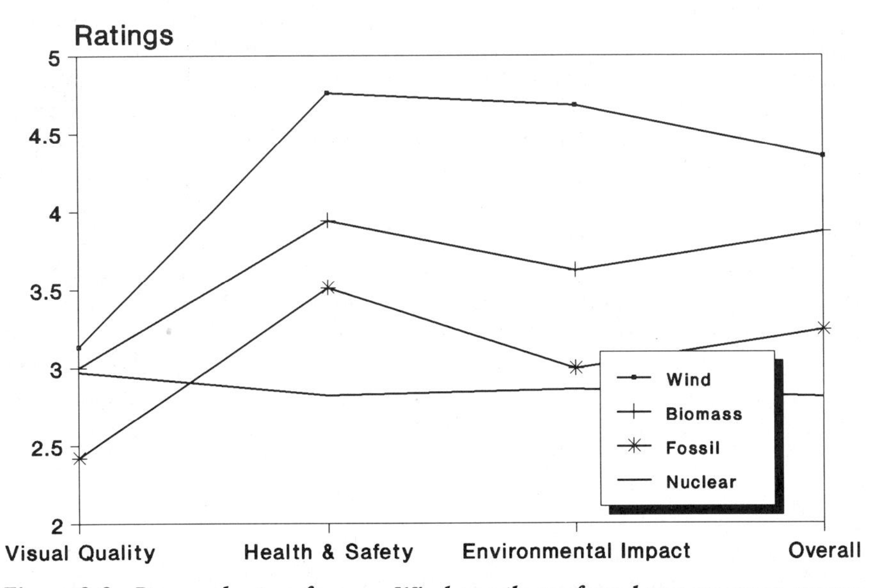

Figure 8.8. Power plant preference. Wind was the preferred power source among technologies surveyed in Solano County, even when its visual impact was considered. (Courtesy of U.C.–Davis.)

In a not unsurprising result, 85% wanted more control over energy siting decisions than they now have, yet nearly half felt little or no responsibility for the resulting impacts. "In general, the subjects felt little personal responsibility for the impacts of their own energy consumption," says Thayer, and they had "little concern about the sources which supply their electricity."[118] He is not alone in this observation. A *Newsweek* editorial ruefully commented that "if there's an energy crisis in the 1990s . . . it will involve electricity" because "we want cheap electricity without accepting the need for new plants."[119]

Given the choice, "most people would prefer no development," Thayer notes.[120] "With few exceptions, Americans tend to consider 'all' power plants objectionable. The result is a public with little feeling of personal responsibility for the environmental impacts of its own energy demands. In modern society we would rather not look at power plants. We have never really had to look at them—they have been located in other regions or even in other states."[121] But the public is not entirely divorced from reality either, says Thayer, because many consider power plants a necessary evil. The survey indicates "a public which is conditioned to the benefits of electricity but less eager to share the full costs."[122]

Like Thayer in California, Maarten Wolsink has observed the NIMBY phenomenon in the Netherlands. There Wolsink finds that a negative view of wind turbines on the landscape is the major factor determining opposition to wind energy. Other, though much less significant, factors are the disbelief that wind turbines will make a difference in improving air quality [usefulness], and the fear that the wind turbines will harm residents.[123]

Wolsink concludes that only a small number of those opposing specific wind projects are true NIMBYs and suggests that applying this demeaning label to all critics antagonizes the community and ultimately proves counterproductive.[124] Local opposition, despite generally favorable support of wind energy, is not always a result of NIMBY. Wind energy's impacts on the community become prominent in the public's mind only when a specific project is proposed. Wolsink's surveys find that the public's broad endorsement of wind energy seldom plays a role when people weigh the merits of specific projects.[125]

In the Netherlands, 90% of those surveyed reacted positively to wind energy. This support, says Wolsink, is tenuous and is limited by the distance between the respondent and the nearest turbine. The closer people live to proposed turbines, the less likely they are to endorse a proposed project. To further ward against complacency in the wind industry about the strength of its support, Wolsink warns that the other 10% are unsupportive from the start. Although only a small minority of residents ever consider taking action, Wolsink points out that it only takes one determined adversary to delay a project.[126]

Opposition is determined primarily by a negative reaction to seeing wind turbines on the landscape. But, says Wolsink, people unconsciously realize that opposition on aesthetic grounds is subjective and is, therefore, often dismissed by public officials. They then rationalize their opposition by citing concerns such as noise, shadow flicker, and birds, which can be evaluated objectively. But visual impact remains the primary cause of opposition.[127]

Local political support is crucial, says Wolsink, but not alone sufficient for success. Without political support, projects seldom proceed. With it, projects can still be stymied by vocal opposition. The manner in which projects are developed also determines the degree of opposition, he says. The "decide, announce, defend" approach commonly used in siting wind (and most other) projects antagonizes those in the community who fear the new development. This often leads to an escalating conflict as proponents become defensive and attack their critics.[128]

Visual Design

All three of California's wind resource areas are corridors not only for the wind but also for communication. Each straddles a major highway, unintentionally ensuring the maximum exposure possible for whatever visual response the wind turbines and their arrays produce—and they certainly generate a response. The view when first entering one of these mountain passes has variously been described as riveting, otherworldly, or shocking. Thousands of wind turbines in all shapes, sizes, and states of repair dot the landscape. In the Altamont Pass, and even more so in the San Gorgonio Pass, the turbines march right up to the roadway, welcoming, and yet at the same time overwhelming, the unwary motorist. They are an essential stop on any renewable energy itinerary of the United States. Yet for advocates and neutral observers alike, they create an uneasy feeling that it could have been done better, that the reality fails to meet the vision of what wind energy can be.

By the late 1980s, European wind companies were distancing themselves from the California model of wind development. When Birger Madsen of BTM Consult flashed photos of the San Gorgonio Pass on the screen at a wind energy conference in 1987 in Leeuwarden, the Netherlands, his message was clear. As his colleagues from California squirmed in their seats, Madsen told the audience that "never again" should wind farms proceed as they had in California. He assured them that no country in northern Europe would permit such haphazard development and that if they did there would be such a backlash that it would doom the industry. To demonstrate how a wind plant could be built artfully, he advanced to images of Tændpibe, the forerunner of Denmark's largest array of wind turbines, near Rinkøbing on the west coast of the Jutland peninsula. It was as if the Danes had set out to build a model wind power plant solely to counter the negative images from California (Figure 8.9).

Madsen's forecast has held true, at least through the mid-1990s. European wind plants do look different, and many would agree that they are more

Figure 8.9. Tændpibe–Velling Mærsk. One of the world's most aesthetically pleasing wind power plants. All 100 machines look alike, even though there are four different sizes of wind turbines in the array. Note the absence of visual clutter from power lines, transformers, substations, or roads. From 1987 to 1992, Tændpibe–Velling Mærsk was Europe's largest wind plant in number of turbines.

aesthetically pleasing than their California counterparts. Why is this? What observations can we make based on the contrast between the two models, and how can we use this to establish guidelines for maximizing the aesthetic acceptance of wind turbines?

First, the machines themselves differ. Early American designs were rickety, angular structures in comparison to the more proportioned lines of stout Danish machines. Second, the machines are sited differently. One of the most glaring differences between European and California wind plants is the visual clutter common to California. Clutter not just from inoperative wind turbines, but also buildings, power lines, roads, and scrap heaps or "bone piles" in California jargon.

Each resource area in California has its own unique aesthetic problems: hand-painted billboards in the Altamont, litter blowing from the freeway in San Gorgonio—both unrelated to wind energy—road scars and erosion gullies on the mountainsides in Tehachapi. Some of the clutter is the flotsam and jetsam found on the fringe of any major city in the United States. This is particularly true in the Altamont and San Gorgonio passes, where it is not uncommon to see abandoned cars, burned-out hulks, left in the ditch by urban gangs on a joyride—a little taste of the ghetto in California's golden hills. It is discouragingly reminiscent of the disrespect for the land around the strip mines of southern Indiana. In the days before the federal Surface Mining Act, the unrestored mines were a favorite dumping ground. Not until miners learned to respect the land in their care did their rural neighbors learn to respect mining and the land the miners used.

Like the miners of southern Indiana, wind companies will discover that the public has a voice in how natural resources are used, whether or not they are privately owned. And the public expects to be heard. Visual amenity is more a public resource than either coal, oil, or natural gas—a resource, like the land itself, which requires care and respect. Wind turbines cannot be placed in the abandoned V-2 missile silos at Peenemunde and brought to the surface only at night, a tongue-in-cheek proposal by Lothar Schulze of the German Wind Energy Association in response to aesthetic criticism.

Wind energy's principal environmental impact, and the principal concern of the public, is its visibility. Therefore, the aesthetic appearance of a wind turbine or the sight of a wind power plant must be weighted heavily in the design process, from inception through the life of the wind project.

This need not be an obstacle, just one more criterion in doing the job right. The Danes have proven it can be done at Tændpibe–Velling Mærsk, Noerrakær Enge, and other sites, the British in hilly Cornwall as well as in mountainous Wales, the Germans on the polder at Frederick Wilhelm Lubke Koog, and the Dutch on the dikes at Urk and Leystad.

Wind turbines can and should be built and sited in ways that accent—and do not detract from—their sublime role. Like a piano, a wind turbine is simply an artifact. The piano, a musical machine, can make great music only when played with skill. Similarly, a wind turbine can produce energy harmoniously with its environment only when designed, sited, and operated with skill. Even a critical study of wind's place on the landscape of the Angeles National Forest said that "if handled sensitively, a wind energy development can add excitement to a landscape." An array "can be a dramatic focal feature of the land."[129] Michael Layden describes the white wind turbines at Bellacorick in County Mayo as a "flash of white gulls" on the bleak Irish landscape. To Layden, the wind turbines have enhanced the deep greens and browns of the bog and offer contrast to the purple mountains in the distance. Just as important to Layden and other advocates, the array of turbines contrast symbolically with the nearby peat-burning power plant.

There may be no way to eliminate all objections to the appearance of wind turbines on the landscape, but there is some consensus on how to minimize these objections. The guidelines can be as simple as those of Energy Connection's Arkestiejn, who summarizes the lessons he has learned from developing projects in the Netherlands: Build an aesthetically attractive project, and keep the turbines turning.[130] Or as simple as that used by the Logstør district council in Denmark: All turbines should look alike, and they should all rotate the same way.[131]

Visual and Engineering Design

To Berkeley inventor Peter Sharp, the towers of Altamont's wind turbines look hard and mechanical against the soft curves of the rolling hills. Sharp, no unthinking critic, wonders if his middle-class concern for aesthetics is misplaced when it is his middle-class behavior (consumption) that underlies today's environmental dilemma. Perhaps, he continues, his concern reflects the same reverence for beauty and rejection of ugliness that underlies much of the environmental movement. "Maybe," he goes on, "wind turbines need to be regarded from an architectural and not just an engineering perspective."

Engineering, in the strict sense, is essential for wind turbines to compete effectively against fossil fuels. But when mechanical efficiency and economy become the sole driving force behind wind turbine design, that design—in its broadest sense—invariably suffers. Designers who fail to balance the three visual elements of a wind turbine (rotor, nacelle, and tower) or ignore appearance altogether miss a key component of design that will determine wind energy's ultimate success, its public acceptance, just as much as the efficiency of the airfoils used. Thoughtfully designed turbines increase acceptance.[132]

Many engineers equate good design with the aesthetic term *elegance.* They intuitively balance engineering utility, what it takes to do the job, with an often unarticulated appreciation of an aesthetically pleasing result.[133] The smooth curves of the Eiffel Tower illustrate the concept. There are no sharp demarcations from one shape to another as the eye ascends the tower. This is no accident. Eiffel could easily have opted for a more utilitarian shape than the compound taper, for which the tower is so famous, and accomplished the same goal. The tower's shape was an "elegant" Belle Epoque solution to Eiffel's challenge: how to use materials efficiently while distributing the tower's weight over a wide area of marshy ground. Some observers, engineers and artists alike, describe the result as "graceful."

Wind turbine towers can achieve the same end. They have a simple function: to elevate the wind turbine above the ground. Towers have a straightforward engineering solution. Whether the result is graceful or elegant depends on how the problem is approached. Unfortunately, towers are almost an afterthought to many wind turbine designers, who focus their creative energies on the rotor and drive train, mistakenly believing that the type of tower they ultimately use is immaterial. But aesthetically, the tower plays as important a role in the wind turbine design triad. as do the rotor and nacelle.

When designers put the tower last in the design process, they may find that they are limited by previous decisions to awkward solutions. This is most apparent in how the tower mates with the turbine at the juncture of the two shapes. When the nacelle is small relative to the diameter of the tower, the tower must neck down to meet it to avoid an ungainly mismatch between the nacelle and tower. The overhang of the rotor and the degree of blade flexure (toward the tower in upwind turbines) determine the clearance between the tower and the rotor and hence also limit the diameter of the tower through the region of the rotor disk. At the base of the tower, distributing the tower's load over a wide footprint lessens the depth and the amount of concrete needed to keep the tower from toppling over. Thus freestanding wind turbine towers narrow from a broad base to the nacelle. Designers that solve this requirement most elegantly, regardless of whether the tower is lattice or tubular, use a gentle, continuous taper. Those who do not use dramatic, awkward changes in shape.

A wind turbine and its tower need not be a box atop a cluttered mechanical pyramid, nor a "hotdog stand" on a stick, as Finn Hansen calls it (Figure 8.10). The turbine and its tower can be part of an aesthetic whole. Consider the Bonus, Vestas, and DanWin turbines shown in Figure 8.11. The nacelles and towers seem made for each other. They were.

Aesthetically, observers often prefer tubular towers to truss towers, however, engineers have also designed elegant lattice towers. Many of the truss towers supporting the Vestas turbines in Tehachapi employ a compound taper recalling the smooth curves of Eiffel's tower. As wind turbines increase in size, though, lattice towers become a more troublesome choice.

Figure 8.10. Box on a stick. Nacelle under- and overproportioned relative to the tower. The tower on the left lacks noticeable taper and is mounted by a stubby box of a nacelle. The nacelle on the right overwhelms its slender tower. Although aesthetically not the most pleasing Europe has to offer, these turbines have relatively clean uncluttered lines.

(a) (b) (c)

Figure 8.11. ***Aesthetically pleasing designs. Nacelles, nose cones, and towers form pleasing proportions on the Bonus Combi (a), Vestas (b), and DanWin (c) turbines.***

Lattice towers and their foundations are much less expensive than tubular towers, so are preferred where economy is of utmost importance. Their principal advantage is their ability to spread their legs over a wider footprint than those of tubular towers. This enables them to use three or four far-flung piers instead of a massive concrete pad. However, in larger turbines, those above 25 m (80 ft) in diameter, the clearance between the tip of the blade and the tower is limited. To provide sufficient clearance in the upper portion of the tower, U.S. lattice designs change shape markedly, and to add strength, increase the number of cross girts. This solves the problem technically but produces squat, angular towers with sharp edges and a dense clutter of angle iron (Figure 8.12).

Ironically, this technical and economic solution raises doubts among untrained observers about the structural integrity of the tower. Observers sense intuitively from the appearance of the resulting tower that something is amiss. Indeed, among some lattice towers in California, the radical changes in shape have concentrated stresses, leading to recurring structural problems.

By the mid-1990s, no one was installing medium-sized wind turbines on lattice towers in northern Europe because of the difficulty in obtaining planning approval. European regulators prefer the clean lines of most tubular designs over those of lattice towers. Yet designers have also produced ungainly tubular towers (Figure 8.10). Early Danish and Dutch machines were installed on tubular towers of two to three sections, like the stages of a rocket. The

Figure 8.12. Lattice tower shape. The three towers on the left have a sharp change in shape at the edge of the rotor disk. This and the linear taper detract from their appearance. The towers on the right, which use Eiffel's compound taper, have a more pleasing shape.

Dutch designs were particularly awkward, with an extremely thin "pencil neck" section where the tower mated with the nacelle.

For the design of the tower, the nacelle, or the entire assembly, management determines the design team's direction. Management is ultimately responsible for lending support to aesthetic design choices or for sacrificing aesthetics to utility. To optimize a design's acceptance, management must ensure that the mechanical engineer or aerodynamicist on the design team stands shoulder to shoulder with an industrial designer or someone else who has the entire picture in perspective: a view of aesthetically pleasing wind turbines in harmonius arrays on the landscape. Where management has had that perspective, it is reflected in their products.

Finn Hansen has that vision. The former managing director of Vestas, Hansen directed the family-owned firm during its successful foray into wind energy. He believed then, as he does now, that products in the public eye must be appealing to be successful. In 1979, his marketing director, Jonna, told him that this was even more critical for wind turbines than the other products Vestas produced. Hansen listened. He had to—he was married to her. The close-knit design team, which included Hansen's wife Jonna, Birger Madsen, and Per Krogsgaard provided immediate feedback to Hansen on design choices. For his part, Hansen made numerous sketches as he sought the right proportions among tower, nacelle, and rotor disk. The result, the Vestas V15, became the model for an entire line of wind turbines. To this day, the Danish manufacturer has continued to carry to each successive model the proportions Hansen derived.

One of the Vestas team's decisions concerned tower designs. Because of the cost, they resisted the move to tubular towers until Vestas began to lose ground to their competitors. When Vestas decided to offer a tubular tower their first decision was to produce a more aesthetically pleasing design than the three-stage "rocket" tower then popular in Denmark. To their surprise they found that their conservative Danish customers, after dickering for days on the price of the turbine with the cheaper lattice tower, would finally opt for the tubular tower. Ironically, price-conscious Danes so preferred the appearance of the tubular tower that they were willing to pay an 8 to 10% premium for it.

Per Lading, DanWin's chief designer, shared Hansen's foresight. Among his design criteria was the turbine's aesthetic appearance. To that end, Lading hired an industrial designer to produce the nacelle cover and tower for the DanWin turbine. The designer gave the nacelle its characteristic rounded corners, and like Hansen and the Vestas design team, rejected the rocket tower, opting instead for a two-piece tubular tower of constant slope. The designer then selected the tower's taper based on an intuitive sense of proportions. The joint between the two tower sections and hatch openings in the tower are intentionally concealed by careful design. In 1988, the Danish Design Council selected the DanWin 23 for its Industrial Design award, the country's highest honor for commercial products. Today, 91 DanWins stand amidst an array of 340 other turbines near Mojave, California. The soft, smooth DanWin towers and nacelles contrast sharply with the bulky angular lines of the surrounding turbines.

Like that of the Vestas and DanWin turbines, the appearance of the Bonus Combi was the result of a conscious design effort. Soren Vinther, Jens Veng, and Henrik Stiesdal, the Bonus Combi designers, produced numerous models in order to visualize the proportions among the tower, nacelle, rotor, and nose cone, all the while consulting with the company's staff, including the technicians, who would service the machine, and the marketing department, who would sell it. Like the others, they included a nose cone for purely aesthetic reasons.

The Danish design fraternity may quarrel politely among themselves about the merits of the rounded corners on the DanWin nacelle or the "mosquito-like" snout of early Bonus nose cones, but the characteristic most striking to outsiders is that they care, that these concerns are a subject of discussion, that aesthetics takes its place alongside more prosaic design considerations. Aesthetics are of such importance that Nordtank, another Danish manufacturer, hired the world-famous Danish industrial designer Jacob Jensen to examine their development of a wind turbine 60 m (200 ft) in diameter. Jensen applied his impressive design experience, including the design of Bang &

Olufsen's avant garde stereo systems, to convey "a feeling of modern clean technology."[134]

Nacelle Covers and Nose Cones

Technically, the nacelle encompasses the drivetrain, the bedplate on which it rests, and the nacelle cover, which protects components from the elements. On many medium-sized wind turbines, the nacelle cover shelters windsmiths who service the machine and is an integral part of the wind turbines' noise control. The nacelle cover also serves an important aesthetic function; it hides the drivetrain from view, wrapping the nacelle in a fiberglass shroud.

If the appearance of a wind turbine is critical to its public acceptance, and nacelle covers are instrumental in forming a pleasing design, then no nacelle is complete without its cover. Some contemporary manufacturers, notably the U.S. manufacturers Atlantic Orient and Northern Power Systems, have deliberately designed turbines for market without even a hint of screening the machinery from view, flaunting engineering economy over respect for the public's visual amenity.

Some California wind companies have operated wind turbines for years without nacelle covers, long after the covers were swept away by high winds. The absence of a nacelle cover in an otherwise uniform string of machines can be just as annoying to observers as an inoperative turbine. The disheveled turbine stands out among its peers. Just as parents ensure that their children dress appropriately every morning before they go off to school, wind turbine designers, manufacturers, and operators should vow never to let their wind turbines go out in public without proper attire.

Unlike nacelle covers, nose cones seldom provide any engineering utility. They are an aesthetic adornment. On some models, nose cones even obstruct servicing of the rotor. Ideally, the nose cone carries the horizontal lines of the nacelle past the blades to break up the vertical lines of the rotor. Wind turbines without prominent nose cones have a blunt, harsh functionality. All the turbines in Figures 8.10 and 8.11 employ nose cones, some better than others. In Figure 8.10, the turbines would appear even more "boxy" without their nose cones. Like the sleek, powerful snout of a P-40 Warhawk of Flying Tiger fame, the design would look incomplete without a large conical spinner capping the three-bladed propellers. Most engineers intuitively understand this and add nose cones as a matter of course. Unfortunately, some do not, and one major Tehachapi operator has gone so far as to remove and destroy the nose cones from thousands of its turbines in California in a shortsighted cost-cutting drive. The decision left the public to wonder why someone would chop the noses off so many windmills (Figure 8.13).

Whether a nose cone or even a nacelle cover influences the public's perception of a wind turbine depends largely on the distance from which the turbine

Figure 8.13. Wind turbine sans nose cone. Removing the nose cone destroyed the clean balanced lines of these turbines near Palm Springs, giving the nacelle a blunt, truncated finish.

is seen. The turbine's appearance will dominate foreground views, while the landscape and the turbines' placement on the landscape will dominate distant views.

Order and Chaos

For his research in the Altamont Pass, Thayer used a Swedish system for determining the various zones of visual influence. The turbine or an array of turbines will dominate the field of vision to a distance of about 10 times the turbine's height (Figure 8.14). The zone of "visibility," where the turbines can be seen but become part of the distant landscape extends to a distance of 400 times the turbine's height.[135] Although visible, they may not be visually intrusive.[136] Andrew Garrad, of the British consultants Garrad Hassan, describes visual intrusion in straightforward terms. "Can you see them?" he asks. "If so, how many can you see? And of these how much of them can be seen?"[137]

How we view wind turbines on the landscape, whether in the foreground or on a distant mountainside, strongly reflects our desire for visual "tidiness," a result of our need to create order out of chaos. Much of the aesthetic criticism of California wind farms is heard in terms of the "disorder, disarray, and clutter" of the turbines on the landscape (Figure 8.15). Maintaining order

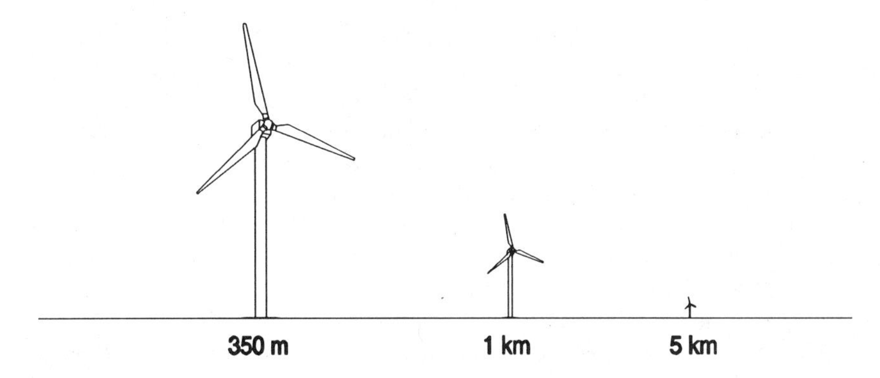

Figure 8.14. Visual influence zones. To-scale perspective of wind turbine 35 m (115 ft) in diameter on a 35-m tower. The turbine on the left dominates the foreground 350 m (1200 ft) from the observer. The middle turbine, 1 km (0.6 mile) from the observer, influences but does not dominate the middle distance. The turbine on the right remains visible at 5 km (3 miles) but becomes part of the landscape. (Courtesy of Renewable Energy Systems, Hemel Hempstead, Herts, England.)

and visual unity is the single most important means to lessening the visual impact of wind turbine arrays.

The Dutch are concerned about siting large numbers of wind turbines on their hard-won polder landscape. They, too, have found a strong desire for visual order. Their survey of before and after attitudes toward the country's

Figure 8.15. Visual clutter. Road cuts, erosion scars, and the jumble created by different types and sizes of wind turbines create visual chaos. (Courtesy of CalPoly.)

first wind plant concluded that people view the turbines as part of a technological process. Although most people "would rather not see technological elements in the landscape," they expect them to display geometric order.[138]

In response to the criticism leveled at wind development near Palm Springs during the mid-1980s, Riverside County supervised a series of studies to monitor wind energy's impacts and its compliance with county ordinances. The county's Wind Implementation Monitoring Program, which acquired the dubious acronym WIMP, was funded by a fee levied on each wind company.

Riverside County's WIMP reports are a gold mine for friend and foe of wind energy alike. The reports discuss visual impact at length. The first report, for example, examined ways to avoid the visual clutter caused by a jumble of different kinds of wind turbines on one site and concluded that "there should be clear separation of machine types" to establish a perception of order.[139] The first study also concluded that there was a high degree of visual contrast between the wind turbines and the open desert but that the cumulative impact was principally "marked by a lack of focus."

The second WIMP report found no visual center, that wind development seemed random and scattered. "Overall it's this high degree of visibility, but lack of central focus, which creates the cumulative visual impact" upon those driving through the San Gorgonio Pass.[140] For the core area on the Whitewater Wash, "the visual impact created by the density of development is also affected by a lack of visual unity" because of the different kinds of machines and towers used, and "patchy geographic placement." The effect, says the report, is "one of scattered disorder."[141]

Yet at medium distances, the sheer density of machines on the broad floor of the San Gorgonio Pass, where more than 1000 machines are concentrated, "creates its own peculiar unity." "Distance removes the cluttering effect of varied turbine types, instead substituting the visual unity of many white blades."[142]

Not all the turbines in the pass are located on the floodplain of the Whitewater River. Large arrays are also found on Whitewater Hill, near Painted Hills, and on Alta Mesa, a prominent plateau within the pass.

The second WIMP study found that on clear afternoons, the turbines atop Alta Mesa can be seen from 7 miles (11 km) away. "Viewers," the report said, "are split on its impact. Some strongly object to its high visual contrast, while others find it attractive" because of its visual unity.[143] Unlike other areas of the pass, which were developed by different companies, all 460 Vestas turbines on Alta Mesa were installed by Zond Systems, using identical turbines. Alta Mesa was distinctive because of the visual uniformity among the three-bladed turbines on open lattice towers.

The 1986 WIMP report warned that haphazardly installing other projects on the mesa could destroy the effect of uniformity, especially if different turbines were used.[144] The warning went unheeded, and 117 larger DanWin turbines on tubular towers were installed by a competing firm. The DanWin turbines themselves are pleasing, but in the context of the existing Vestas turbines on lattice towers, they detracted from the mesa's previous harmony.

Similar warnings in the WIMP reports—that large turbines or dense arrays of smaller machines could overwhelm the geographic features of Whitewater Hill—were ignored and the site is now fully developed.[145] The U.S. Bureau of Land Management relaxed their aesthetic standards partly because of preexisting visual disruption on the site from power lines and communication towers. Despite their reputation as the most productive wind machines in the world, the turbines on Whitewater Hill have served as a lightning rod for critics of wind energy in nearby desert communities and among Palm Springs officialdom due to their prominent visual impact on the pass.[146]

Changing weather and lighting also affect the view of desert landscapes such as that found in the San Gorgonio Pass. For example, a row of turbines "which stand out starkly when highlighted by the setting sun" may "nearly disappear when viewed through wind-raised dust" in the heat of high noon. Under certain conditions, the white blades on many of the turbines "can be seen flashing for miles."[147]

Whitewater Hill welcomes eastbound travelers on I-10 at the portal to the broad expanse of the Coachella Valley with a clutter of various machines of different types and sizes towering above the highway. Yet Whitewater Hill, where at least all the machines are installed atop similar tubular towers, is just a taste of what motorists find a few miles farther east, near Indian Avenue.

Motorists on I-10 just west of Indian Avenue encounter an otherworldly sight, which critics such as John Warner call deplorable. Wind turbines of every conceivable description crowd the highway. There are some wind turbines with two blades, others with three. Some turbines are small, others big. There are turbines on lattice towers, others on tubular towers. Some spin gently, others frantically, and some not at all. There are turbines laying on the ground, and there are towers without turbines. It is truly American in scale, and on a windy day, it is alive with action.

Motion

Because the natural landscape is nearly motionless, the human eye can detect movement at great distances. Motion powerfully attracts and holds the observer's attention, providing contrast with the stationary landscape. Even when wind turbines are well designed and placed optimally on the landscape, "the movement of the blades is certain to catch the eye," say landscape architects who studied the visual impact of wind turbines in Wales.[148]

The effect of motion is striking on the desert landscape near Palm Springs. From major roadways, the motorist sees a single mass of moving blades on the Whitewater floodplain, where the densest concentration of wind turbines in the world is to be found. The spinning rotors meld together into pure motion. The eye largely ignores the towers because of the mass of motion.

Motion extends the visual zone influenced by the wind turbines. At distances of 2.5 km (1.5 miles), it is the motion rather than the size of the turbine that draws the eye.[149] Motion continues to highlight the turbines at distances of

8 km (5 miles) or more. "Were the blades not moving," says the third and final WIMP report, "the machines would be impossible to see." While motion makes the turbines visible, "the visual contrast in these distant views remains low," says the report. The spinning turbines appear as a shimmering haze in the distance. "A shimmering haze is certainly visible, but it scarcely constitutes a substantial visual impact."[150]

This motion makes the wind turbines a visually interesting addition to the landscape. Viewers may or may not find distant wind power plants beautiful or attractive, but they frequently label them interesting. "When perceived abstractly, says landscape architect Thayer, "the turbines are simply looked upon as moving visual forms contrasting with a smooth landscape. In this mode," he says, "the turbines are perceived somewhat like abstract sculptures, arousing interest with their novel, unfamiliar forms and animation."[151] Visual interest can be a powerful force, as demonstrated by the millions of tourists who visit Pisa's Leaning Tower or San Francisco's sinuous Lombard Street. Motion also signifies usefulness: the wind turbines are working, doing what they are meant to do.

Use Them or Lose Them

There are few more convincing demonstrations of how well wind energy works than to see Zond's "wind wall" in Tehachapi on a blustery day. If watching 400 turbines dancing on the hillside is unconvincing, a drive over the pass to SeaWest's Mojave site can be instructive. There on the gently sloping flank of Cameron Ridge stand another 1060 turbines, nearly every one of them in operation. Unfortunately, there are hundreds, if not thousands, of wind turbines in California that are less reliable, less well maintained, and less well sited than these. Some simply do not work.

These inoperative turbines stir controversy far out of proportion to their numbers. "Abandoned or derelict turbines," consultants to Riverside County advised, "have become one of the unexpected major sources of visual impact in the San Gorgonio Pass."[152] The public's concern about usefulness and the importance they placed on whether the turbines operate properly also surprised the U.C.–Davis researchers in their study of Altamont pass. Thayer concluded from this that the single most significant action that California wind companies could take to boost public acceptance of wind energy was to keep their wind turbines operating as much as possible by quickly fixing broken turbines and removing those that were unrepairable.[153] Yet by 1991, there were still enough derelict turbines near Palm Springs alone for the Edison Electric Institute's Charles Linderman to plead with attendees of AWEA's annual conference: "Please get those inoperative machines down, to avoid the misinterpretation that wind still doesn't work." Few listened.

Only U.S. Windpower and SeaWest have had the foresight to heed such advice and remove arrays of inoperative turbines. U.S. Windpower removed several hundred of their first-generation turbines, the 56-50s, and additionally cleaned up dozens more of another company's machines at one of the Altamont's most notorious sites alongside I-580, doing themselves and the industry a favor. SeaWest removed turbines from its abandoned site east of San Diego.

The problem is not unique to California. Dutch planners worry that inoperative turbines will make acceptance of new installations more difficult. "Support wanes quickly when people see rotors at a standstill," says Gerrit de Long, a planner for Het Bildt, a small Dutch village. Long explains that "people say that if they [wind turbines] are not working they are just polluting the horizon." To avoid this, the village conditions the turbine's environmental permit on removal of all turbines that have stood idle for one year.[154]

Because removal may incur modest costs or trigger new legal obligations to restore the site, most firms with inoperative turbines choose to sit and wait. Riverside County was forced to remove several hundred abandoned wind turbines on the Maeva site near Palm Springs at its own expense. Near Tehachapi, a half dozen twisted hulks were removed only after the local Sierra Club filed a nuisance complaint because the turbines stood near the Pacific Crest Trail. One major Tehachapi operator has turbines on its site that have stood inoperative since 1985. At a small project nearby another firm operates only half its 80 turbines, and towers for the other half stand in mute testimony to a flawed design. These 100 or so turbines account for only a fraction of the 5100 in the Tehachapi Pass. Nevertheless, to draw the public's ire, only a few inoperative turbines are needed.

It was the fear that scam artists riding on tax-credit fervor would litter scenic hillsides with thousands of broken turbines that stirred early opposition to wind development in Tehachapi and the nearby desert crossroads town of Mojave. Critics demanded that the county require removal of inoperative turbines and that each turbine be bonded to pay for removal if the operator failed to do so. Kern County's wind energy ordinance never included such provisions. But even counties with removal ordinances, such as Alameda and Riverside, find it hard to remove derelict machines.

One of the worst examples is Altamont's Fayette machines. Fatally flawed from its inception in the mountains of western Pennsylvania, the Fayette design has embarrassed the wind industry for more than a decade. Of the 1500 turbines installed during the early 1980s, most bordering I-580, only half were in service by 1993, and many of these were seldom in operation. An untold number lay where they had fallen, visible to all passing along the nearby highway.

Regulators found a significant obstacle to quick removal: inoperative wind turbines are not necessarily abandoned, in the legal sense. Some wind machines encounter design defects that are difficult and time-consuming to correct. This occurred frequently during California's early days, when many of the first-generation turbines were inoperative much of the time. Well-meaning regulators gave operators ample time to make repairs. Technical problems plagued the Fayette design, for example, and Alameda regulators allowed more than enough time for Fayette to correct the problems.

Even when the turbines are operable, repairable, or slated for removal, legal disputes can delay resolution for years. Still, wind companies and their regulators have an obligation to ensure that wind turbines are either operable or are removed if they are not. Wind turbines are a modular technology. They can be quickly installed and just as easily dismantled. With the exception of the concrete foundation and any deep road cuts, the site can be restored quickly once the turbines have been removed.

Keep Them Spinning

The desire to see wind turbines spinning also affects design choices. Designers can ensure that the turbines are spinning as much as possible by choosing self-starting wind turbines over those that motor up to speed. Self-starting turbines, like most European designs, begin freewheeling in light winds of 3.5 m/s (9 mph); that is, the wind spins the rotor even though the generators are not producing electricity. Once moving, the rotors will coast in wind speeds down to 2.5 m/s (6 mph). Designs such as FloWind's Darrieus turbines or the ESI, park their rotors until the wind speed is sufficiently strong to generate electricity. At wind speeds of 4.5 m/s (10 mph) or greater, the wind turbine's generator, acting as a motor, spins the turbine up to its operating speed. Thus freewheeling turbines have a wider operating range than the Darrieus turbine, for example, and will operate more often.

This is critical in light winds, because some of the turbines will begin operating while others remain idle, due to subtle differences in wind speed. The lower startup speed of the freewheeling machines will ensure that a greater percentage of the turbines in one string will be spinning at any one time than in a row of Darrieus turbines.

As with aesthetic turbine design, nacelle covers, and nose cones, freewheeling rotors offer no engineering utility. Freewheeling rotors generate little more energy than those that park the rotor, but they do garner more acceptance. To save a few dollars on transformer losses (as transformers consume energy even when they are idle), one Tehachapi operator restricts operation of their freewheeling turbines by delaying release of the brake until generating wind speeds are reached. This penny-wise-pound-foolish approach defeats the advantage of a freewheeling rotor and is counter to what should be every wind plant operator's maxim: keep them spinning.

Visual Density

The most striking difference between European wind plants and those in California is the scale. The largest European projects, North Friesland in Germany (35), Urk in the Netherlands (50), Tændpibe–Velling Mærsk (100) in Denmark, EcoGen's PnL site in mid-Wales (103), and PESUR in Spain (184), deploy 35 to 200 turbines each. However, these arrays are unrepresentative of most European wind development. By 1993 British and Dutch wind plants averaged 10 to 20 machines; Danish arrays, 10 to 35 turbines; and German clusters, 4 to 15. Most turbines on the continent are sited individually or in small clusters of two to four machines. Of Denmark's 3500 wind turbines operating in 1993, only one-fourth were in wind plant arrays. In contrast, U.S. Windpower manages 600 turbines at its Solano County site alone, and Zond operates 342 turbines at Sky River.

Nowhere in Europe, not even in the South of Spain near Gibraltar, is there the density of wind turbines like that found in California's principal wind resource areas. From any prospect in the Altamont, San Gorgonio, and Tehachapi passes, thousands of wind turbines can be seen marching off into the distance.

As more turbines are added to an array, the influence zone expands. For example, more than 1000 turbines are clearly visible from the town of Tehachapi 4 miles (7 km) distant. "Such a high level of visibility, says U.C.–Davis's Thayer, "virtually guarantees that large wind power developments will invoke strong reactions among viewers."[155] Europeans have encountered the same response. Marcus Trinnick of the British law firm Bond Pearce has found in his practice that clusters of 15 turbines are much easier to site than bigger projects.

The public appears better able to digest wind turbine arrays in distinct visual portions of uniform density. Organizing wind turbines into distinct groups provides a sense of order absent when the machines are seen sprawling hither and yon across the countryside.[156] Consciously using the topography, such as hills, drainage courses, or ridge lines to separate machines visually into distinct clusters would enable wind development on the scale of Altamont or Tehachapi while minimizing the visual impact on the landscape.

Disregarding the disparity in number between European and Californian developments, there is another visible difference as well: the density of the arrays (Figure 8.16). For various reasons, California developers packed their turbines much closer together than did their European colleagues. A spacing between turbines of 2 by 6 rotor diameters is not uncommon in California. In some California projects, developers placed the turbines so close together across the wind that the blades of one turbine nearly meet the blades of the next (1 diameter spacing). At Tændpibe–Velling Mærsk, designers specified a more open, less dense array of a 4 by 7 diameter spacing. It is this openness that makes European wind plants more welcoming than the dense California arrays, where the turbines stand cheek by jowl like the slats in a picket fence.

The density of development, says British aerodynamicist Andrew Garrad, is a trade-off between capacity and energy. Packing the turbines closely together maximizes the installed capacity on a site but cuts energy production as interference among turbines increases. Decreasing the visual density by spreading the turbines farther apart improves energy capture and lessens the visual impact, but it increases electrical losses from the longer wire runs between

Figure 8.16. Visual Density. Kirkby Moor is a good example of the open uncluttered wind plants developed in Great Britain. The 12 turbines rise above the heather in the Cumbrian Mountains of England's scenic Lake District. They overlook five turbines at the former Haverigg air base on the Irish Sea.

turbines. Net electrical losses within a wind project are not inconsequential, averaging 2 to 2.5%, and designers seek to balance the losses against greater energy capture.[157] As with other elements of wind turbine and wind farm design, in the planning process, the visual impact of the array's density should be weighed equally alongside energy capture, wire losses, and economics.

Wales Landscape Study

An example of weighing both number and density of wind turbines on the British landscape is contained in a 1992 report by consulting landscape architects for Dyfed County. They considered the scale, topography, and "grain" of land cover (woods, hedgerows, roads, field boundaries, and buildings) in determining the character of the landscape.[158] The architects found dispersed dwellings punctuating a patchwork of fields on the gently undulating plateaus typical of the countryside of southwestern Wales. They decided that individual turbines were acceptable when associated with other rural or agricultural features, such as farmhouses and outbuildings.[159]

Because the landscape lacks distinctive natural features, said the architects, it is susceptible to domination by large-scale structures such as California-style wind development. Although planners classify Dyfed's visual quality as "average" by British standards, the architects felt that large arrays of 50 to 100 turbines would overwhelm the landscape. They questioned whether even the typical British wind plant comprising only 25 turbines was appropriate and concluded that small clusters of wind turbines would ease "absorption" of the turbines into the landscape.[160]

They proposed grouping three clusters of 5 to 10 turbines each within close proximity. This would provide visual distinctness and limit the scale of any one cluster while permitting the use of common services such as electrical substations. With a separation distance between clusters of 750 to 1250 m ($\frac{1}{2}$ to $\frac{3}{4}$ mile), observers would still be able to see all three groups of turbines, but their scale and spacing would minimize their visual impact. The architects felt that this arrangement best reflected the plateau landscape of southwestern Wales.[161] They suggested that a distance of 3 to 5 km (2 to 3 miles) separate each grouping to ensure that the wind turbines would became part of the landscape rather than dominate it.[162]

Visual Uniformity

Next to keeping the wind turbines spinning, the most significant measure for improving public acceptance is visual uniformity. Even when large numbers of turbines are concentrated in a single array or there are several large arrays in one locale, visual uniformity can create harmony in an otherwise disturbing vista. Visual uniformity simply means that the rotor, nacelle, and tower of each machine look alike, forming one visual unit (Figure 8.17).

Providing visual "unity" in type of turbine, tower, and spacing should be the single most important consideration in the design of a wind plant, said a team of landscape architects in a study commissioned by the Angeles National Forest. CalPoly's landscape architects recommended that developers use only one kind of turbine in each project, to reduce the visual clutter they found at California sites.[163] The intent is to encourage the eye to follow across a line of machines without abruptly halting at a visual interruption; examples of such an interruption are turbines spinning clockwise in a field of turbines spinning counterclockwise, two-bladed turbines interspersed among three-bladed machines, and turbines on tubular towers in an array of turbines on truss towers.[164]

The arrays should also have uniform density and spacing. The CalPoly team shares the British architects' view of the need for visually distinct groupings. Long lines of turbines, for example, should "be separated by open zones to create orderly visual units," they reported[165] (Figure 8.18).

Riverside County's WIMP studies reached similar conclusions when weighing the aesthetic impact of repowering projects near Palm Springs with newer, larger turbines. Acknowledging that replacing the existing turbines with fewer, more reliable machines would greatly improve public acceptance,

Figure 8.17. Linear uniformity. The 50 wind turbines north of Urk in the Noordoost polder comprise the world's longest linear array. The 6-km (3.5-mile) string of 300-kW turbines borders a dike (on the left) separating the polder from the Ijsselmeer, the Netherlands' great inland sea. Note tourists, grazing sheep, and visitors center (lower right). These wind turbines provide 12% of the electricity used by the inhabitants of the polder.

the study warned against "piecemeal replacement with differing turbines" and "extensive mixed arrays."[166] If a project uses a three-bladed turbine, all turbines installed later should also sport three blades. If the original turbines use a truss tower, all later turbines should use a truss tower. If the nacelles of the turbines first installed have a distinctive shape, all subsequent turbines should use a similar nacelle. In addition, all the turbines should spin in the same direction.

It is equally important that all towers be of consistent height, unless part of a whole as in a "wind wall." The CalPoly team suggests that turbines on towers of varying height are not only acceptable when in a uniform pattern but can also add visual interest to an array.[167] Towers of seemingly random heights destroy any uniformity that otherwise might exist in an array (Figure 8.19). The uniformity among many of the turbines on the Whitewater Wash

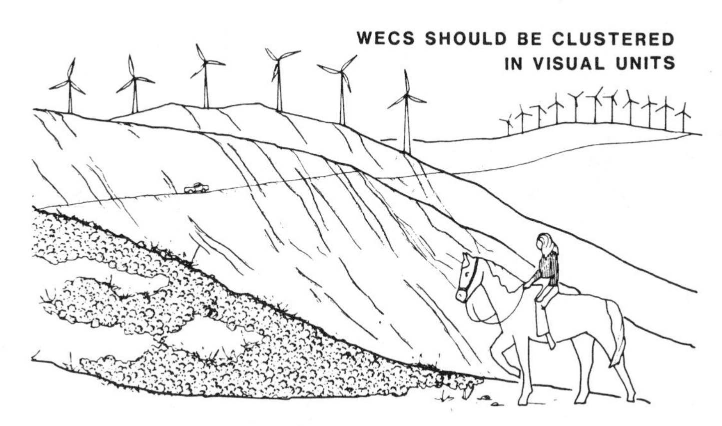

Figure 8.18. Visual units. Arrays of turbines of similar design and similar size clustered into distinct visual units eliminates clutter on the landscape. (Courtesy of CalPoly.)

(most of which use three-bladed rotors on tubular towers of equal height), coupled with the great density, creates a unity of its own. "The sole distraction from the horizontal mass" of machines is the array of Carter turbines, which stand out on guyed towers twice the height of those around them.[168]

Some argue that tubular towers are preferable to truss towers on aesthetic grounds. But truss towers, although less pleasing in foreground views, can be a "good choice," said the team of CalPoly architects. When the turbines are seen at a distance, the towers "fade away into the landscape."[169] One good

Figure 8.19. Nonuniform tower height. Juxtaposition of towers of different heights adds to the visual clutter on the Whitewater Wash near Palm Springs.

example of this is the Los Angeles Department of Water & Power's series of truss towers carrying hydroelectricity across the Mojave Desert from its water projects on the eastern side of the Sierra Nevada. During the late afternoon, the towers become nearly invisible against the desert landscape. Another is the view from Tehachapi during late afternoon, when nearly 1000 turbines are seen as waving white flowers suspended on the hillside. All sit atop truss towers that have vanished into the landscape. Tubular towers bordering these machines remain visible throughout the day, most notably when silhouetted against the sky.

Two or Three Blades

In the debate as to whether two-bladed turbines are as aesthetically appealing as three-bladed designs, engineering utility often has a too powerful voice. Designers from the influential aerospace field invariably opt for two blades. The teetering hub that two blades permit, and its ability to shed loads, holds these engineers spellbound. In a sometimes blind drive to cut costs and lift efficiency, these designers forget that people will have to live with the machines they build. Whether the public favors two-bladed wind turbines has never entered the design equation.

Some proponents of two-bladed turbines callously suggest that the American public will tolerate the design, as the machines will be deployed at remote sites on the Great Plains where few will ever see them. The residents of Wyoming, Montana, and other plains states may disagree and instead demand that wind turbines sited in their midst meet the same aesthetic requirements as wind turbines sited in populous California or northern Europe.

The two-versus-three-blade dispute touches several sensitive issues in the wind industry: U.S. government–sponsored research versus private-sector research and Danish versus American design philosophy. The U.S. Department of Energy has placed its bets on two blades. The Danish private sector has put its money on three blades.

There is mounting evidence that the public prefers three-bladed designs.[170] In testimony at the Cemmaes inquiry, the Countryside Council attacked National Windpower's choice of the two-bladed WEG turbine, saying in effect that if National Windpower were to install wind turbines on the plateau, they would, on visual grounds, prefer three-bladed turbines. An opinion survey carried out by Dutch developer Energy Connection found three blades preferred in the Netherlands as well.[171]

Dutch surveys have also found that "the tranquility of the landscape is disturbed by high rotor speeds." This observation can affect the preferred number of blades, as two-bladed turbines spin faster than comparable three-bladed turbines.[172] To some observers, the high rotor speeds of two-bladed turbines gives them a more frenzied appearance, especially at rotor speeds above 50 rpm.[173]

Two-bladed rotors also produce unusual optical effects that observers find distracting. According to Alexi Clarke of Britain's Open University, the speed of rotation appears to vary as the blades of a two-bladed turbine pass the tower. Clarke finds three-bladed turbines more pleasing to the eye because "there appears to be a succession of tumbling rotor arms wheeling overhead, [and] there is more continuity of motion."[174] Others describe the rotor disk of a two-bladed turbine as changing shape whenever one of the blades passes the tower. Just as three blades furnish dynamic balance, they also deliver visual balance not found on two-bladed rotors.[175] Roger Kelly, director of the Centre for Alternative Technology in Wales and an advocate of renewable energy, finds the absence of symmetry bothersome when two-bladed turbines are at rest. Like others, he finds two-bladed turbines make him ill at ease when they are operating.

The Wind Energy Group's John Armstrong questions whether it is simply a matter of familiarity (what people are used to) and rotor speed. Most two-bladed turbines, like the old ESI and Carter designs, operate at high rotor speeds. It is the frenetic pace of these machines that disturbs viewers most, says Armstrong. He argues that the slower speeds of WEG's two-bladed machines will reassure the public that two blades are an acceptable choice. Armstrong may have a point.

Most are familiar with rotors using three and four blades. Most modern wind turbines, like many of the most famous World War II aircraft, use three blades. The traditional Dutch windmill uses four blades.[176] But few have the opportunity to see rotors with two blades. Only small private planes use two-blade propellers today, so few in the public at large would be familiar with the image of two blades.

Related to the three-versus-two-blade debate is how the rotor is parked. Because turbines are seen frequently with their rotors at rest, it is important to consider their appearance when parked, particularly that of two-bladed machines. "Arrays of machines that stop in synchronized positions are much more attractive in the landscape," say the CalPoly architects, "than ones that stop at random, and form haphazard lines."[177] Turbines with three-bladed rotors, because of their inherent symmetry, are generally more pleasing than two-bladed machines at any parked position.

Because of the lines formed by the rotor and the tower, a two-bladed rotor should only be parked on the horizontal. Only the horizontal provides visual harmony from one two-bladed turbine to the next. Part of the disarray characteristic of the turbines on the Whitewater Wash near Palm Springs is the random resting position of rotors on Carter, ESI, and Windtech turbines, all two-bladed machines. For practical as well as aesthetic reasons, both WEG and R. Lynette & Associates park their two-bladed rotors on the horizontal.

Materials and Color

Fearful of wind energy's aesthetic impact, regulators or political leaders unfamiliar with the technology often seize on the false hope that there is a high-tech paint that will make wind turbines disappear. Wind turbines are unavoidably visible; they can be designed and sited in a manner that pleases as many people as possible, but no amount of paint or camouflage will make wind turbines invisible. Whatever their color, the turbines and towers will always appear black in silhouette.

Although color and construction materials offer no panaceas for concealing the wind turbine, they should not be overlooked. In the desert and rangeland of the southwestern United States, regulators prefer beige or tan color schemes to lessen the contrast between the turbines, their towers, and the landscape. Elsewhere, off-white appears best. In England, planning guides suggest light gray or white matte, as these blend into a range of backgrounds and the often cloudy weather common to the British Isles. The Dyfed study in Wales recommends off-white or pale gray and advises against green and brown unless the turbines will always be seen below the skyline. They also suggest a matte gel coat on the blades to reduce reflection.[178]

Bright flashes of light reflected from a glistening gel coat or from a flat panel of galvanized metal on a nacelle have antagonized residents near Palm Springs. These "sparklies," as residents call them, are largely self-correcting. The gel coat, the topmost finish on fiberglass and composite blades, checks over time from attack by the elements, eventually losing its luster. Nacelles, on the other hand, should be covered with nonreflecting fiberglass; where galvanizing is used, it should be buffed to a dull finish to eliminate reflections.

Some observers find that painted tubular towers look less "industrial" than brightly galvanized towers. In more extreme cases, some have demanded the use of "natural" materials. The latter follows from the widespread acceptance of the Dutch windmill. A 1977 survey of 300 people in several areas across the United States first noticed this phenomenon. When questioned about the aesthetic impact of wind energy, respondents invariably preferred traditional "Dutch" windmills to modern wind turbines. In this early survey, the margin of preference was three to four times that for modern wind turbines.[179]

Armin Keuper of the German Wind Energy Institute was shocked to discover the same effect in northern Germany during the early 1990s. Nature protection societies, says Keuper, want the turbines to "look like the land where they're placed," that is, to use thatch or stone as building materials rather than steel or fiberglass.

The dilemma is not unlike that which faced the British post office when it sought to introduce the now familiar (and recently abandoned) red phone booths. The Council for Rural England opposed introduction of the boxes as inappropriate in color and materials for use in the countryside. The British post office replied that if they were to make all modern amenities of native materials so that they might blend in with rural surroundings, they would have to thatch intercity buses. For inhabitants and tourists alike, the red phone booths eventually became symbolic of England and an expected part of both the urban and rural landscape.

Structures other than the wind turbines themselves should be made of native materials to harmonize with traditional buildings already existing on the landscape. But wind turbines must use modern materials to compete against conventional fuels.

Color, along with line, form, and texture, is one of the criteria for determining contrast used by the U.S. Bureau of Land Management in screening visual impact. For its domain in the western United States, the BLM evaluates the color of development activities such as roads, not just the structures on the landscape. The impact of roads is principally the introduction of a dramatic color contrast between disturbed soil and undisturbed vegetation. In California, it is not unusual for the wind turbines themselves to nearly disappear into the landscape, while the scars from service roads remain distinctly visible for great distances.

Verticality

Apart from its motion, the most striking visual feature of a wind turbine is its vertical lines. On many landscapes this verticality forms a distinct contrast with horizontal landforms. This is particularly evident on the lowlands of northern Europe, where the tallest objects on the landscape were historically church steeples and traditional windmills. Both pierced the skyline, extending above the treetops and cityscape. German groups, says the Wind Energy Institute's Keuper, want to limit wind turbines from protruding above rooftops to preserve the polder landscape's dominant horizontal line (Figure 8.20).

But German wind energy advocates point to other structures already on the landscape that extend above village roof lines. The modern landscape includes grain silos, not unlike those found in the midwestern United States, and ubiquitous television towers. They also note that deforestation and eighteenth-century construction of dikes across the North German Plain substantially altered the then-existing landscape. The horizontal lines of this artificial landscape obstructed views of distant church steeples, the otherwise dominant vertical features of the polder lands. The addition of wind turbines

Figure 8.20. Iconography of North German Plain. Windmills and church steeples pierce the traditional skyline of Jever in Niedersachsen. Lowland communities from the Netherlands on the west to Denmark in the northeast associate windmills, church steeples, and dikes with the flat polder landscape. (Courtesy of Jever Tourist Board, Jever, Germany.)

to the landscape may alter the visual balance between vertical and horizontal again, but to no greater extent than past actions taken in the common interest.

This balance will be an important visual element of wind development on the American Great Plains, where only the occasional water tower, radio mast, or church steeple pierces the monotonously flat landscape. Midwestern homesteads, with their barns, windmills, and grain silos, are islands of verticality in a sea of gentle horizontal swales.

Wind turbines are of necessity vertical structures. Just as there is no magic paint that will make the turbines invisible, there are few means of lessening the contrast between the vertical lines of a wind turbine and the horizon. However, there are ways to lessen the contrast of structures associated with wind turbines.

Ancillary Structures

The architectural study for Dyfed County recommended placing substations, transformers, and other ancillary structures off the horizon line of the hilly sites found in Wales, and screening them with existing features common throughout Britain, such as hedges and stone fences. In general, the study found that to reduce clutter, no support structure should pierce the skyline, and buildings should use materials in harmony with the surrounding landscape[180] (Figure 8.21).

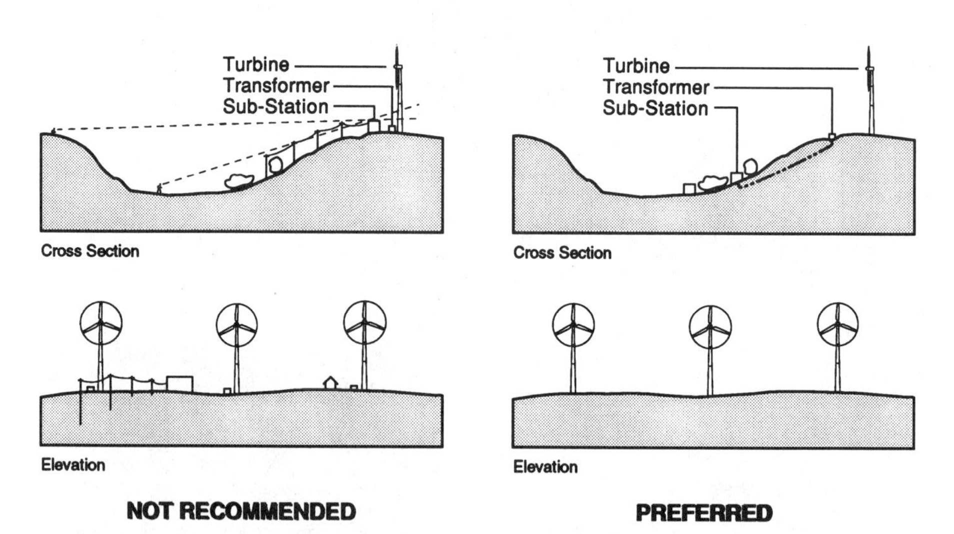

Figure 8.21. Ancillary structures. Removing all structures other than the wind turbines from hilltop sites reduces visual clutter. The visibility of structures relocated on lower slopes can be reduced by vegetative screening. (Adapted from Landscape Impact *by Chris Blandford Associates.)*

From his surveys in California, Robert Thayer recommends that developers bury all power lines and integrate extraneous equipment, such as transformers, into the turbines themselves.[181] The latter is now possible with the advent of larger turbines. When used with tubular towers, the transformers and control panels can be installed inside the towers, as is done on offshore and harbor breakwater installations. British architects agree with Thayer.

These measures have become common practice at British wind plants, where developers bury power lines and construct ancillary buildings of local materials. Both Renewable Energy Systems at Carland Cross in Cornwall and EcoGen at New Town in Wales use native stone for the façades and slate for the roofing of their substations, to match traditional building styles.

Both medium- and low-voltage power lines are carried off the sites underground until they are among other structures or natural features that can screen them from view.[182] In some cases it may be less damaging to the terrain to carry lines overhead down steep slopes than to trench. Where overhead lines are necessary, designers can sometimes route them away from the most popular viewsheds.

Taken together with the more open, less dense arrays of European wind plants, the absence of other intruding structures among the wind turbines provides an orderliness not found on any California wind farm (Figure 8.22).

Figure 8.22. Visually pleasing hilltop array. The clean uncluttered lines of the wind plant at St. Breock Downs coexist harmoniously with nearby Celtic standing stones. The turbines are clearly visible from the Cornish town of Wadebridge (foreground).

In an effort to further reduce visual clutter, European designers must even consider the number and placement of meteorological masts in the planning process. It is this thoroughness and the high value placed on the public's acceptance of the finished product that have resulted in some of the world's premier wind projects.

In a provision unheard of in the United States, British architects urged Wales' Dyfed County to prohibit fencing because of the resulting visual clutter. All wind turbines and their equipment, say the landscape architects, should be designed to withstand vandals without fencing.[183]

Fencing

Wind plants occupy large tracts of land. Fencing the perimeter unnecessarily removes the tract, both visually and physically, from public use. It sends a negative signal to the public that wind turbines are dangerous, threatening machines.

Only one European wind plant restricts access by fencing: the 23 turbines at the LNG terminal at Zeebrugge, Belgium. There the fencing prohibits entry to the LNG terminal, to protect the tankers from terrorists. There are fences at other wind plants in Europe as well. These fences too are for purposes other than preventing access to the wind turbines. In the Netherlands, for example, fences prevent vehicle access to the dikes. (Imagine the dike manager's worst nightmare: "Joyride on off-road vehicle floods Holland.") Pedestrians and cyclists have unrestricted access through gates or turnstiles designed for the purpose.

One of the most pleasing attributes of European wind plants is their openness. Nearly all are accessible. Even at sites where gates prevent vehicular access, there is usually a public footpath or hiking trail nearby that can be used to reach the turbines.

In the midwestern United States, where fencing of fields is common, wind turbines in an array will be bound by the existing field pattern and existing fences. The fences are a part of the existing landscape. There is no need to add fences for the turbines themselves.

Another intrusive aspect of fences surrounding California wind plants is the existence of the blaring signs posted on them: "Danger—Wind Turbines." The signs are a misguided attempt to forestall negligence suits against county governments by trespassers who injure themselves climbing a wind turbine. The Dutch deal with the same problem in a more subdued manner. A single sign at the gate says: "Walking and cycling are permitted at your own risk. Don't bother the sheep." It says what needs to be said and puts responsibility for safety on the user.

Designers, operators, and regulators of wind power plants should avoid needlessly scaring the public about wind turbine hazards. Modern well-designed wind turbines are not dangerous. Climbing them is. No amount of preventive measures will divert a determined vandal or someone bent on self-destruction. The fencing issue had bedeviled the wind industry in the United States since the mid-1970s, occasionally reaching absurd proportions. Some local authorities have required owners of small wind turbines costing less than $10,000 to install fences costing $1000 or more.

Yet the tracks and platforms at railroad passenger stations are not fenced, nor are those at metro rail stations in urban areas. It is assumed that users, including parents with children, have enough common sense to stand clear of the train as it moves into the station. Imagine the confusion during rush hour if thousands of passengers had to squeeze through gates in a fence "protecting" them from the train they wanted to board. In many places in the United States, there are no platforms or grade separations between trains and passengers ready to board. Passengers wait at track level. Again, there are no fences.

Fortunately, there are simple and inexpensive means for discouraging people from climbing wind turbines. For medium-size machines, the most effective methods are simply locking the access door on tubular towers, or removing the climbing rungs for the first 3 m (10 ft) on lattice towers.[184]

Occasionally, common sense prevails—even in litigious California. Kern County granted Zond Systems a variance from the fencing requirement on Zond's remote Sky River project. Zond bars unauthorized traffic to the site, but access to hikers and wildlife is unrestricted.

Roads

As mentioned in the discussion of color, the most visible aspects of a wind power plant in hilly terrain, apart from the wind turbines themselves, are the service roads cut into the slopes (Figure 8.23). To some California environmentalists, such as Howard Wilshire of the U.S. Geological Survey, roads and the erosion they cause are the principal environmental impact of wind development.[185] The CalPoly team of architects also found terracing for service roads visually disruptive.[186]

There is no simpler way to minimize this impact than to minimize the construction of roads or to eliminate them altogether. The ideal location of a wind plant is adjacent to the existing roads and power lines found in urban areas or near the tilled land found in the midwestern United States or in northern Europe.

The architect's report for Dyfed County recommended that developers use existing farm tracks wherever possible, justify the need for all new roads, and limit the width of permanent roads to 3 to 3.5 m (10 to 12 ft). These roads, said the landscape architects, should follow field boundaries as much as possible to minimize visual impact.[187]

In their study of California wind plants, the CalPoly team suggests that after construction is complete, developers promptly reseed the graded areas to enhance the site visually and reduce erosion. They further recommend that the surface of infrequently used roads be revegetated to lower visual contrast

Figure 8.23. Scarring from road cut. Scarring from the bench detracts from the visual uniformity of this "wind wall" in the Tehachapi Pass. The color of the rock in the cut slope contrasts sharply with the preexisting vegetation, enhancing the bench's visibility. The fill slope, below the turbines on the bench, supports only sparse vegetation 10 years after construction. The debris cascading down the hill also draws attention to the scar. The turbines on the bench are far less productive than those on the ridge. In retrospect, the environment would have been better served if the developer had restricted the turbines to the ridgetop. Despite the bench, this view from Highway 58 is an impressive example of wind energy at work. During prevailing winds, nearly all of the 400-turbine array will be in motion.

between the road and undisturbed terrain.[188] This is a common practice on wind farms in England and Wales, where turf is expected to cover the road up to the wheel tracks.

The need for roads and the degree of their subsequent impact increase with the steepness of the terrain. In steep terrain, the cut-and-fill slopes left by road construction scar hillsides permanently, detracting from what could be an otherwise pleasing wind turbine array. For this reason the CalPoly architects advise against slopes greater than 40% and conclude that sites with slopes of 25 to 40% can be used only with the greatest care.[189]

Logging for access to sites in forested terrain can produce the same contrast in color and form as that produced by road construction. The CalPoly architects recommend sites in open grasslands, treeless plateaus, or over forested sites with low vegetation because of the felling that would be necessary to build the access roads into forested areas.[190]

Housekeeping

A long list of items that can be used to reduce the visual clutter and disorder typical of California wind plants falls under the rubric of general housekeeping. Some, such as visual density or the preferences for three-bladed turbines, are unique to wind energy. Most are not. They are the prosaic prescriptions for living in a civilized world that our parents teach us as children. We learn to pick up after ourselves and generally to consider the effects our actions have on others.

For managers of wind plants, this translates into respect for the environment and the community of which they are a part. Fastidious site managers care enough to ensure that the turbines, where numbered, are identified with a crisp, legible stencil rather than a slovenly spray-painted scrawl. They care enough to require technicians to pack their litter out with them at the end of the day instead of allowing it to blow across the countryside. And they are never too busy watching the bottom line to notice the day-to-day details that govern how the public views them and wind energy.

Paving service road entrances is one "good neighbor" practice. This prevents tracking of mud from service roads onto paved highways, reduces fugitive dust, and minimizes nuisance complaints from other users of nearby roadways. Anyone who has had their windshield cracked by a stone thrown from a heavy truck will appreciate this care. Another is to eliminate eyesores such as the "bone yards" or scrap heaps of abandoned machinery, that litter some California sites.

Responsible managers and wind turbine designers alike also ensure that nacelles contain all oil or fluids that are likely to leak. If they do leak, these managers promptly clean the turbine and tower, returning the site to its pristine condition. They know that no manager at a nuclear power plant or an auto assembly line in a Western country would long keep their job if they permitted oil to pool on the shop floor. A wind plant is no different. Operators and employees alike understand that the public intuitively judges management by how they execute such simple chores as housekeeping, for if the company shows little concern for the obvious, how much do they care about less visible tasks, such as the safe disposal of hazardous wastes?

Lights

No wind turbine should call attention to itself with flashing lights like some garish billboard along the Las Vegas strip. Unfortunately, when the total height of a wind turbine in the United States reaches 200 ft (60 m), the Federal Aviation Administration requires that owners of the "obstruction," in FAA parlance, provide flashing lights or paint the structure in contrasting bands of red and white. Similar requirements to protect aircraft exist in other countries. The giant MAN wind turbine on the island of Helgoland in Germany and the multimegawatt machines in the United States have all fallen prey to this regulation.

Such flashing lights are particularly annoying at night, as is the bright "security" lighting common at wind plant substations in California. Wind turbines and their supporting structures are conspicuous enough during daylight hours—they need not be so at night, especially in the rural areas where wind turbines are often located. For many rural residents, nightfall is a time of tranquility. Flashing strobe lights atop wind turbines or security lighting will exacerbate any annoyance the turbines' presence causes residents.

The gaudy stripes and flashing lights of wind turbines taller than 200 ft (60 m) should be avoided at all costs. This requirement may limit the ultimate size of wind turbines and their towers. Similarly, security lighting should be used only when necessary. At substations in residential neighborhoods, for example, Southern California Edison uses outdoor security lighting only when workers are on the premises.

Aesthetically pleasing design is the result of conscious intent. Where aesthetics have played a part in the design process, whether on the drawing boards of Danish wind turbine manufacturers or in the boardrooms of British wind farm developers, the results speak highly for those involved. Unfortunately, the opposite is also true. California's wind resource areas are, according to CalPoly's landscape architects, "examples of wind energy developments based on economics with little regard to visual quality."[191]

Some of the excesses found in California are excusable because of the ignorance of manufacturers and developers. Few knew, even fewer cared, how the public would react to wind turbines on the landscape. These men and women were intent on creating an industry from scratch overnight. After much toil, they succeeded. But the manner in which they did so is unacceptable in the 1990s. Ignorance is no longer an alibi.

There is ample information on what makes wind turbines and wind power plants most acceptable to the greatest number (Table 8.1). To ignore this

Table 8.1
Aesthetic Guidelines for Wind Plants

Ensure visual uniformity (direction of rotation, type of turbine and tower, and height)
Avoid fencing
Minimize or eliminate roads
Bury intraproject power lines
Limit or remove ancillary structures from site
Remove inoperative turbines
Avoid steep slopes
Control erosion and promptly revegetate
Remove litter and scrap
Clean dirty turbines and towers

knowledge today, gained at such great cost to the wind industry and the public alike, would be folly. Ideally, aesthetics, like workplace safety, will become institutionalized. Every MBA learns about the costs of a lax safety program but they learn nothing about aesthetic design. Improvements in safety have resulted less from the fear of regulatory retribution than from the certainty that accidents will lead to higher insurance premiums. When designers and developers realize that wind will never deliver its full economic potential until aesthetic concerns are addressed, visual design will find its niche in the planning process.

The public holds wind energy to a higher standard than other technologies. Wind must compete economically, yet be environmentally benign. The public does not question whether or not a conventional power plant works or why oil-field pump jacks stand idle most of the time. But they do ask such questions of wind turbines. For the same reason, those in the wind industry must meet more stringent environmental standards than those of conventional power plants, because the public expects them to do so.

As Danish industrial designer Jacob Jensen advises, for ultimate success the goal of manufacturers and developers alike must be that wind energy represents in the public eye "a beautiful human manifestation." When well-designed wind turbines are sited with sensitivity, many agree with Robert Thayer at U.C.–Davis, who believes that "wind energy could achieve a serene, utilitarian beauty common to other working landscapes."[192]

Community Acceptance

Wind energy advocates would prefer that everyone like wind turbines as much as they like traditional windmills: to listen for their creaking and groaning with anticipation, not hostility. But that is, if not unrealistic, certainly naive. Acceptance should be the goal instead. Acceptance implies a weighing, sometimes conscious but often not, of the costs against the benefits wind energy offers. As with automobiles, they may not be universally loved, but everyone uses them, and most people (sometimes begrudgingly) accept them. Wind turbines too can become a universally accepted technology.

Opposition Is Normal

As wind energy becomes more widespread, proponents will soon realize that opposition by some in the community is normal and that they should expect it. There will always be some opposition to any project, whether it is a synagogue in a suburb of Atlanta, Georgia, a new school in Palmdale, California, or a national park in France. No matter how beneficial or worthy the land use, there will always be some portion of the local population that objects if for no other reason than fear of change: a tendency to prefer "the devil we know to the devil we don't know."

Like traditional windmills, parks are universally appreciated. Yet even parks can fall victim to local opposition. In northern Indiana, neighbors blocked creation of a new park because the site lacked adequate trees, they argued, and the trees that would be planted would not grow sufficiently in the neighbors' lifetimes to shield the park from view. The state estimated that the park would have drawn 50,000 visitors per year. According to news accounts, opposition killed two other park proposals in the same area.[193]

In a more celebrated case, inhabitants of France's Massif Central waged a decade-long guerrilla campaign to stop the government from creating a new national park in their midst. They feared that the park would impose draconian restrictions on how residents used their land in the sparsely populated region, especially on the right to hunt. Pastoral uses, commerce, and habitation—although strictly controlled—are permitted in French parks, in contrast to what is permitted in North American national parks. Evidence of the conflict can still be seen 20 years after the park was first proposed, in graffiti declaring "Non au parc" painted on road signs and walls. Ultimately, the Parc National des Cévennes became an international model for integrating the concerns of residents into park planning. The park sees its mission as preserving not only the flora and fauna but also the rural population and their pastoral heritage.

That there will always be some opposition does not absolve project proponents from addressing community concerns, nor does it justify a confrontational approach when it inevitably appears. Ignoring antagonists may save the developer headaches in the short run, but concerted opposition can easily prevent a project from proceeding to fruition. Even in Britain, where wind companies are responsive to local issues, only 50% of proposals win planning approval.

British surveys find that it is not only people's attitudes toward wind energy that determine their acceptance or rejection of particular projects, but also the value they place on the landscape in question and their expectation of how wind turbines will affect that value. One finding that appears universal: those opposed on visual grounds will raise other objections. Moreover, as Wolsink observed, the visual response can be a surrogate for other fears or objections about proposed wind turbines that are more difficult to articulate.[194]

Pace of Development

The pace of development alone can generate as much opposition as the manner of development. California's wind rush stirred some Tehachapi and Mojave residents, who feared that the turbines and their profit-driven developers would overrun them, into a fury of activity. This phenomenon can also be seen in Britain, where EcoGen's Tim Kirby warns: "We must not run so fast that we create a bow wave," or backlash. Kirby's boating analogy describes the challenge facing the budding British wind industry: how to grow while giving the public enough time to adjust to the technology. This is difficult for the wind industry to control because wind plants can be built so quickly.

The operator of a bed and breakfast learned firsthand how quickly change can come. Looking out across Anglesey's Cemaes Bay at the concentrated power of the Wylfa nuclear power station, the woman sighed that one day there was only Welsh farmland and the next there were windmills among the hedgerows (Figure 8.24). The countryside with which she was so familiar had changed literally overnight into a landscape with wind turbines. Not that she opposed wind energy—to the contrary, after Chernobyl she had marched in protest against expansion of the Wylfa nuclear plant. The Ukrainian disaster struck the area hard, not only with the nearby reactor's threat to the community, but with the threat from reactors in distant lands. When area farmers were ordered to keep their sheep indoors, the risks hit home. It is just that she realized after the turbines sprang out of the ground that there was a price to pay for wind energy too. She was weighing the cost anew, since it would take thousands of windmills to equal the output of Wylfa's single—but ominous—structure. She still believes that wind energy makes sense, but finding them in her garden makes her less sanguine about its promise.

Speaking at the 1993 British Wind Energy Association conference in York, Godfrey Bevin, of the Department of Trade and Industry's Renewable Energy

Figure 8.24. Rhyd–y–Groes, Wales. Bonus wind turbines at Rhyd–y–Groes frame the Wylfa nuclear power station on Anglesey. (Telephoto lens foreshortens the distance.)

Branch, chided the industry that it must build quieter turbines, reduce visual intrusion, and allow the public time for familiarization. Bevin also urged the industry to consider building community-owned or cooperative wind farms as a means of winning local support, a refrain heard increasingly in England and Wales.

Sharing the Benefits

The costs and benefits of development, whether a wind farm or a new shopping center, are distributed unequally. Society as a whole shares the benefits accruing from power plants: the electricity. But the costs of the pollution from the plant are borne by those who live nearby. They share the benefits but bear all the costs.

Although the impacts of wind plants may be minor in comparison to those of a coal-fired power plant, they are no less real to those living nearby. Seldom does the comparison between the impacts of wind energy and conventional resources carry much weight with those affected. It is unlikely that any other kind of energy development would ever be built in their midst. Wind plants are usually sited in inhospitable locales supporting a sparse population at best, typically areas with few competing resources.

The people who choose to live in such locations do so primarily because the land is unsuitable for other urban uses. They reasonably expect that the area will remain rural and undeveloped. Thus, disregarding any real or imagined environmental impacts, the mere act of erecting the wind plants affects local residents by changing the character of the rural landscape around them. Those most vocal in their opposition to wind projects "chose their home sites because of the scenery and the solitude that the windy conditions provide. They never expected any substantial, non-residential use of the land near them," says Martin Pasqualetti at Arizona State University. "These are now the people affected most directly by the visual and audio impacts of the wind turbines."[195]

In a study of the controversy surrounding wind development near Palm Springs, James Throgmorton faulted developers, planners, and community leaders alike for missing an opportunity to creatively resolve concerns about the rapidly expanding wind industry in the mid-1980s. Unlike supportive councilman Ben Austin, who wanted Tehachapi to build its own wind farm, Riverside County political leaders were openly hostile to wind energy and wanted the turbines removed. Both proposals fell short. Just as others lacked Austin's vision and Tehachapi's municipal wind plant was never built, Riverside County's troublesome turbines remained. The Palm Springs conflict served only to appease a mere handful of critics and enrich a few attorneys.

Throgmorton, an urban planner, challenged all involved, particularly planners, to envision a way of distributing some of the projects' benefits to those most directly affected. His findings suggest that some residents interpreted change, the addition of the wind turbines, as degradation of their community "because outside developers would obtain most of the benefits" while they would bear most of the costs.[196]

Planners, developers, and the community needed to explore alternative mechanisms, says Throgmorton, that would have allocated the risks and benefits of wind development in a manner that local residents considered fair. "The Palm Springs case suggests that local residents define as a NIMBY any energy development that exposes them to risks that exceed the benefits they might accrue."[197] Phyllis Bosley, in her survey of antiwind activists, found identical results: the "local risks of a wind power plant are often perceived to outweigh any possible benefits."[198]

This observation expands on the insight of Thayer and Carlman about usefulness: the public accepts wind energy's impacts when the turbines are perceived as useful. Not only must wind energy be useful in the broader sense—that the turbines must be seen working—but wind energy must be of some use or benefit locally. Residents near Palm Springs said "time and again that if their electric power rates were reduced, they would be happy to allow more wind energy development." The benefits need not be solely in the form of energy. Throgmorton suggests that "direct payments, guarantees of property values, insurance, and in-kind improvements," such as paving the dirt roads common in the desert near Palm Springs, could have mitigated concerns by nearby residents that they were absorbing all the costs without gaining any of the benefits.[199]

Thayer reached similar conclusions about a proposed project in the Cordelia Hills of Solano County east of San Francisco. The project failed because of local opposition: "Had they [Cordelia Hills] possessed even a degree of symbolic ownership or investment in the turbines, the entire visual meaning of the landscape might have changed, and the outcome might have been different. "To succeed," says Thayer, "future wind-power plants must somehow enfranchise their 'visual consumers'—those neighboring residents who must look at the wind turbines in their landscape." Thayers observation reflects the outlook that visual resources belong to the public and their use implies an obligation to use the public resource wisely. Like Wolsink, Thayer believes that enfranchisement requires community involvement from the beginning of the planning process rather than after decisions have already been made, and where appropriate, that enfranchisement could include "compensation (such as financial return or amenity development) for the scenic 'damage' wrought by the turbines."[200]

Surprisingly, the French nuclear industry pioneered community compensation; its ability to build reactors is partially attributable to the success of that program. At Cattenom, a village in northern France, Electricité de France's largess supports one-fifth of the municipal budget and contributed to the construction of Cattenom's community center, sports stadium, town hall, and tennis courts.[201]

Compensation is no panacea. It is fraught with risks: chiefly that it can be perceived as a form of bribery. In some domains, such as under English planning law, the manner and amount of compensation may be strictly limited. While justified, impact compensation will not placate all opposition to siting wind turbines, because of other political and social currents moving through the community.

There is certainly no guarantee that a responsible developer who provides local benefits will be welcomed with open arms. In one noteworthy case, no amount of effort by the developer to sincerely meet local needs was sufficient. When Jim Dehlsen, the founder and then chief executive of Zond Systems, proposed to fund the Meritus Foundation with a royalty on gross earnings from their Gorman project, he was met with derision by antiwind activists. Through an independent board of directors drawn from the communities of Gorman and Frazier Park, the foundation would have controlled a budget of nearly $250,000 per year. Dehlsen stipulated only that the funds be used for environmental protection. Instead of applauding Dehlsen's sensitivity and Zond's corporate responsiveness, opponents publicized the offer as a payoff. Following the humiliating rebuff and eventual defeat of Zond's Gorman proposal, the company has taken an aggressive and confrontational stance in subsequent encounters with local activists. Everyone lost.

A similarly intractable conflict between public benefit and private cost continues to simmer in the Pyrenees, where farmers in the Aspe valley have staged a long-running battle against preserving the last bears in France. The farmers fear that measures designed to protect the less than a dozen remaining bears will forever change the way they live. They charge that the bears startle their sheep and cattle into leaping off mountain cliffs to their deaths. But righting the balance by compensating the farmers for their real or imagined losses has been insufficient to eliminate the visceral hatred the farmers have for the bears and what they represent. In the same way, compensation proposed to ranchers for losses of cattle from reintroduction of wolves into Yellowstone National Park has not eliminated opposition. Like their rural counterparts in the Pyrenees, the independent-minded ranchers of Montana and Wyoming resent the dictates and influence of a distant urban center, be it Washington, DC or Paris. Although compensated financially, they must still bear the social cost of change for the benefit of urban dwellers with newly won environmental consciousness.[202]

Cooperatives

Another means of sharing the benefits of wind energy with the community in which the turbines are located—"enfranchisement," as Thayer calls it—is through ownership. Wind energy has succeeded in Denmark largely because of its origin at the local level. Two-thirds of the turbines are owned by individuals or local cooperatives. This personal financial stake that thousands of Danes hold in their own communities accelerated the technology's broad acceptance. Wind farms built by utilities and projects sponsored by investors from outside the community have played a far lesser role.

Obviously, if landowners install turbines on their own property, they gain the financial benefits. Not everyone is so fortunately situated, even in Denmark. Cooperatives are a successful Danish vehicle for encouraging participation of those who otherwise could never use wind energy. In the Netherlands, where cooperatives are less plentiful than in Denmark, one Dutch coop alone has 750 members to operate six turbines.[203]

Individuals can also participate indirectly through a municipal government that owns turbines. Such is the case for residents of Ebeltoft, which installed a cluster of Nordtank turbines on a breakwater to pay for harbor improvements (Figure 8.25). Unlike a municipally owned project, a cooperative directs benefits back to its investors, individual members of the community, rather than to the community at large. Cooperative ownership ensures that individuals see the benefits as they pass through their hands; municipal ownership can obscure benefits to individuals. Whether owned individually, by a cooperative,

Figure 8.25. Ebeltoft Harbor. Municipally owned Nordtank turbines on harbor breakwater. The pathway beneath the turbines is open to the public.

or by a municipality, many Danish wind turbines are identified as a product of local effort that benefits local individuals or the community.[204]

Communities favor local ownership over the imposition of projects from outside by either developers or utilities.[205] Some Welsh criticize private wind farms as a form of eco-colonialism because the owners are in "the City," London's financial center, and the developers are all from England, that is, from outside Wales. Unlike Danish cooperatives, British wind projects are structured more traditionally, with development and financing from outside the community. The exception is Delabole, where innovative Cornish farmer Peter Edwards built his own wind farm.

During the early 1980s, Tehachapi residents directed the same resentment toward real estate speculators from Newport Beach, a wealthy suburb of Los Angeles akin to Beverly Hills, who would reap quick rewards and then flee to beach-front mansions in their BMWs. Even customer-owned Danish utilities have fallen prey to local resentment of "outsiders" profiting at local expense.

Winkra's Henning Holst witnessed this effect at Germany's Friedrich-Wilhelm-Lubke-Koog near the Danish border, where the 1300 ha (3200-acre) polder adjoins the famous Hindenburg railway bridge to the resort island of Sylt. The polder was diked in the 1950s, and the first generation of pioneer settlers and their immediate descendants, people willing to take chances, still live there. When Hamburg investors built one of Germany's largest wind plants in their midst, they felt cheated out of their own resource and decided to take matters into their own hands. But rather than bemoan the newly constructed turbines, the residents chose to build a second 22 million DM ($36 million) wind plant, their own. Of the 50 to 60 households in the polder, 43 became shareholders.

Steve Wade, of the small British company Wind & Sun, believes that what he calls *community-based development* allows use of the greatest number of potential sites. The larger the pool of acceptable sites, the greater the flexibility in siting. Community ownership, he says, may also permit use of less environmentally sensitive sites than may more conventionally financed projects. A cooperative, for example, may accept a lower return on its wind investment than will a bank. A portion of the cooperative's "earnings" appear in the form of the project's "green" value and its local control. Wade and others believe that this approach could alleviate objections to wind projects perceived as "foreign" intrusions into small communities by faceless corporations in distant cities.

Unlike their counterparts in Britain, where NFFO bidding discourages community ownership, German regulators foster it. By the mid-1990s, two-thirds of the applications for the German federal subsidy program were from individuals.[206] German planners were pushing applicants to pool their efforts into cooperatives. This, say the planners, will ensure that the wind turbines are sited to the community's best advantage. Holst of Winkra, Germany's largest independent wind developer, believes pressure from environmental groups will lead to even more local ownership in the future.

Occasionally, residents of urban areas want to play a part too. The benefits of cooperative ownership in Denmark has been limited to residents in the rural areas adjacent to those where the turbines are predominately located. This effectively excluded the 2 million people living in Copenhagen from participation until the Avedøre Vindkraft cooperative built the city's first wind plant. The project, the first wind farm within a major metropolitan area, is also an unusual example of cooperation between Greens, who founded the cooperative, and conservative executives from the local utility. Potentially a model for similar joint ventures elsewhere, the cooperative initiates planning; the utility builds and operates the wind plant (Figure 8.26).

The utility NESA finds that working with a cooperative, although difficult at times, enables them to better meet their obligations under the Danish 100-MW program. Planning approvals are easier to obtain and there is less public resistance. The utility brings to the table its project development skills and financial clout. NESA buys any turbines in the project not sold by the coop to its members. The venture's joint decision making is time consuming but "is a reasonable price to pay," declares NESA, "when sites are scarce." According to NESA, the paucity of sites in urban areas demands close collaboration between Danish utilities and local groups such as coops.

Figure 8.26. Cooperative membership meeting. Annual meeting of Avedøre Vindkraft cooperative. The 720 members of the cooperative own turbines on the outskirts of Copenhagen. (Courtesy of Jens Larsen, Københavns Miljø- og Energikontor, Copenhagen, Denmark.)

Coop members like those of Avedøre Vindkraft come from all walks of Danish life. The coop owning three turbines at Vederso includes a headmaster, a janitor, a clerk, a grocer, a mechanic, a doctor, a plumber, an electrician, and farmers. Although a business—the local bank takes payments from the utility and automatically credits each coop member's account—annual shareholder meetings call for a celebration like that at a family reunion.[207]

Planning

Enfranchisement also means including the community in the planning process. Two California developers stand out for their attempts to work with local communities near their proposed wind plants. Both SeaWest and U.S. Windpower have practiced textbook examples of how to deal with their prospective neighbors: Seek community input early, and often.

This is the policy followed by most European companies. Some, such as Britain's Renewable Energy Systems make frequent use of computer-aided graphics to provide visual simulations of how the turbines will appear on the landscape for those who will live nearby. These are not simple photo-montages or crude line-of-sight views but full-color renditions of the actual site from several perspectives, with the proposed wind turbines in place.

In Wales, WEG and its contractor Taylor Woodrow did everything possible to accommodate local interests near the Cemmaes site. For example, in a much-appreciated courtesy, WEG issued a regular newsletter during construction to keep neighbors informed of its activities and to head off rumors. WEG's community support paid dividends during Cemmaes' long public inquiry. Unfortunately, such practices are not universal. One bumbling California firm antagonizes its neighbors at every turn.

Planning, rather than being an impediment to development, can further wind energy by resolving conflicts before they escalate to the political and legal arenas. Early debate helps all parties. Through it, interest groups cannot only determine where wind turbines will be unacceptable but also where they will be permissible. Both Denmark and the Netherlands have undertaken nationwide planning surveys to do just that. Ruud de Bruijne of NOVEM, the Netherlands agency for energy and the environment, notes that the Dutch ministry for physical planning reached agreement with the provinces for installing 1000 MW of wind capacity, and through consultation with environmental and other interest groups has already identified sites for 600 MW of that.[208]

Planning can also provide protection once the turbines are in place. Tehachapi planner Chris Grimes proposed adding provisions to the city's general plan aimed at safeguarding local wind plants from suburban encroachment. Across the Atlantic, British planners were thinking along the same lines as Grimes. After ensuring that wind companies complied with strict guidelines, British planners prohibited construction of buildings or other structures nearby to prevent upstarts from robbing wind projects of their wind.[209]

One nettlesome question that planning addresses is whether the visual intrusion is worth the potential generation. When a small wind plant or an individual turbine is compared to a nuclear plant, the wind generation appears minuscule. But upon closer examination, the visual impact of wind energy may also be proportionally less than that of a nuclear plant and its attendant transmission lines. The planning process and the debate surrounding specific projects should be educational, enabling the community to put wind energy into context with other land uses, including other sources of energy, and guiding developers in how best to minimize their impact.

Wind as Alternative

As Wolsink found in the Netherlands, wind energy as an alternative to conventional sources seldom carries much clout with those most directly affected by proposed projects. The fact that wind plants consume fewer visual resources than the high walls and spoil banks left by strip mining does little to console those who live in the vicinity of a proposed wind plant and who object to this or any other form of development. To them it is not a question of trading off the visual impact of a wind plant for that of a coal mine, because the coal mine has always been somewhere else.

However, it would be wrong to dismiss entirely wind's value as an alternative source of energy in encouraging community acceptance. Brigitt Gubbins, a British energy activist, found several cases where wind plants have been built on sites previously closed to nuclear development. At Nojsomheds Odde in Denmark, Gubbins found 24 turbines on what was once a hotly contested site for a nuclear plant. Farmers on whose land the turbines are located initially opposed the wind project. It was not that they wanted the turbines to go the way of the nuclear plant; it was just that their compensation was too miserly. They told Gubbins that no amount of compensation would have been sufficient for the nuclear plant. Although they clearly hear the turbines 300 m (1000 ft) away, noise is not a problem.[210]

An architect with a district council near Ålborg, the site of a large coal plant, worried about the effect the Danish wind program would have on the countryside, birds, and wildlife. Yet, Gubbins notes, in almost the same breath he put his concerns in perspective when he added: "If they refuse this [permit], they will be shoveling in more coal."[211]

For many environmental groups who actively campaign against new coal and nuclear plants, wind energy's chief advantage is that it is an alternative to technologies they find unacceptable. For example, the Izaak Walton League applauded the installation of the small municipal wind farm at Marshall, Minnesota as part of its drive to thwart new coal and nuclear plants in the upper Midwest.[212] Alone, the five turbines are no substitute for a giant coal plant, but they are representative of potentially thousands of such machines forming a sustainable energy supply.

Sustainability

Wind energy can be symbolic of what Thayer calls higher concepts, such as stewardship, or it can be seen as symbolic of "ugly technology."[213] He poses a question as to whether "wind power landscapes will be publicly interpreted as an indicator of man's callousness toward scenic beauty or a symbol of society's conservation ethic and wise use of a renewable energy resource." Wind energy, among its advocates at least, is not so much another technological assault on the land as it is a harbinger of more responsible land use, comparable to environmentally conscious farming. Thayer's surveys in the Altamont reveal that most agree and view the turbines as "positive symbols of a responsible energy policy."[214]

Many find the agricultural setting of tidy farms among tilled fields stretching to the horizon pleasing, despite our awareness that there are hidden environmental impacts. Farming imposes massive changes on the land. Yet these changes are more readily acceptable to us than, say, the massive changes brought about by urban sprawl or other "modern uses." This may be due to our long association of farming with the necessity for raising food. We unconsciously identify the benefits that farming produces, and willingly accept the changes to the land that result. In time, wind power plants will be viewed similarly, as a necessary use of the rural landscape, implying an unconscious acceptance of its benefits and costs.

Our sense of what constitutes a wise land use changes over time. Until recently, for example, when talking of fall plowing, Minnesota farmers wanted to "turn her black," a highly destructive practice of leaving freshly plowed fields bare throughout the winter. Farmers looked at the newly turned soil with pride. They considered the black fields beautiful. In contrast, the "crop residue left in the field was looked upon as trash."[215]

Naturalists who grapple with the question of what is "sustainable" suggest that agriculture—to become truly sustainable—requires a more mature land ethic than is prevalent today. We must learn not to "turn her black" because of a misplaced sense of aesthetic order, they say. Now, more conservation-minded farmers "have come to see the crop residue as beautiful" because it represents better land stewardship than turning the fields black.[216]

In time, those who object to wind development may come to see wind turbines, like the windmills before them, as an acceptable land use. Like Minnesota farmers who have learned to appreciate the value of stubble in the fields, one-time critics can learn to accept wind turbines because of the wise stewardship they represent rather than being repelled by an ill-founded aesthetic sensitivity that the machines are ugly.

In a discussion of living within the bounds ot seasonal rhythms, writer and environmental activist Stephanie Mills lovingly describes the setting for an interview with friends in Michigan's Leelanau peninsula: "The background music to our conservation was the little tikka-tikka-tikka–sewing machine noise from their wind generator twirling merrily in the sunny breezes." Mills used the small battery-charging turbine to paint a picture of ecologists living in harmony with the land around them, a picture of sustainability.[217]

Ringkøbing: A Model for the 1990s

One example of a community striving toward a sustainable energy supply is Ringkøbing. The commune or district spans 40 km^2 (15 mi^2) on the west coast of Denmark's Jutland peninsula. Ringkøbing has cut the per capita coal consumption of its 17,000 mostly rural residents to one-eighth the 1986 level and expects to eliminate coal altogether during the mid-1990s. The district has reduced its overall fossil fuel use by 25%.

Ringkøbing operates two combined heat and power (cogeneration) plants fueled with natural gas. The district plans to add a third by the mid-1990s, eliminating any further need to burn coal. When the third plant is complete, Ringkøbing will get 90% of its heating and 40% of its electricity from cogeneration. Wind turbines will generate the remainder of the electricity needed.

The 160 wind turbines already operating in the district, most in two nearby wind farms, generate 45% of the 100 GWh consumed by Ringkøbing. The wind power plant at Tændpibe–Velling Mærsk alone meets the needs of Ringkøbing village, the commune's district seat and largest city.

Siting has not proven a problem, even though wind turbines are prominent throughout the district and are visible from such popular vacation spots as the Holmsland Klit, a narrow sandy spit on the west side of Ringkøbing Fjord. Wind turbines can be seen from all the approaches to Ringkøbing, and one turbine operates within the city limits, at the town's sewage treatment plant. On a summer day, tourists can be seen cycling the popular bike path by Tændpibe–Velling Mærsk. Some occasionally stop at the visitors' kiosk to satisfy their curiosity, then ride on.

There is strong local support for the area's two wind farms, especially Tændpibe–Velling Mærsk, where 35 of the 100 turbines are cooperatively owned by 500 area residents. Ringkøbing's success is due to a combination of local initiative, local ownership, market incentives, and an environmental awareness almost unique to Denmark.

Beckoning Flowers

Most accept wind energy and support further development when it is deemed useful, that is, when the wind turbines are perceived as working reliably. Some have opposed specific wind projects on various grounds, principally wind's aesthetic impact. Because wind turbines use a public resource, visual amenity, proponents must be sensitive to these community concerns.

In the mid-1980s, Dutch researcher Maarten Wolsink began warning manufacturers and developers alike that they must take aesthetics and community impact seriously or face growing resistance in the years ahead. The perceived aesthetic and social impact of wind turbines on the immediate community, not wind's contribution to global environmental quality, will determine public acceptance, says Wolsink. Wind's attributes will not be sufficient to counter the damage done by ugly turbines, improperly chosen sites, or poorly developed projects. It is only through the careful siting, design, and installation of wind turbines, as well as the involvement of the community, that the wind industry will turn general support into widespread acceptance.[218]

Wind's advocates find Wolsink's observations frustrating. Many work with wind energy because they believe that it is an environmentally superior alternative to fossil fuels and nuclear power. The Cemmaes conflict in Wales arose because the Countryside Council and the other interveners saw wind turbines on the landscape in the same way that the Romantics had viewed the railroad. Wind turbines need not be the pounding, steaming, sulfurous, whistling creatures that locomotives were in the nineteenth century. They can be, however, and some wind turbines have been: gaunt structures with their arms flailing frantically at the wind with a maddening thump-thump. Some residents, like John Warner near Palm Springs, awoke one day to find such unwelcome machines in their gardens. Warner, unknowingly mimicking Hawthorne's metaphor for the horrors of the industrial revolution, likened them to a thundering locomotive that never passed by.

Others, such as tourist Sandy Lloyd or French poet Jacques Lacarrière, see them differently. Lacarrière found that California's wind turbines resembled "giant flowers which, rather than practicing photosynthesis like plants, practice aeolosynthesis, taking their life from the strength of the wind."[219] Thus wind turbines need be neither brutish intruders imposing their will upon a virginal landscape nor roaring, clattering, greasy monsters that send mothers fleeing with their children. They can and should be soft, beckoning flowers with their whirring petals murmuring like a distant brook, both a symbol and an artifact of a people in harmony with their environment.

Part II and Chapter 8 Endnotes

1. Alphonse Daudet, *Lettres de mon moulin* (Vanves, France: Hachette Collection Lecture Facile, 1993), Le secret de maître Cornille, p. 17. The quote is from a condensed version of the original . . . moi, je travaille avec le mistral et la tramontane, qui sont la respiration du bon Dieu. "As for me, I work with the mistral and the tramontana, which are the breath of the good lord." Alphonse Daudet, *Lettres de mon moulin* (Paris: Fasquelle, 1965), Le secret de maître Cornille, p. 23.
2. Ibid. The quote is from a condensed version of the original N'allez pas là-bas, disait-il; ces brigands-là, pour faire le pain, se servent de la vapeur, qui est une invention du diable. "Don't go there, he said; to make their bread, those crooks use steam, which is an invention of the devil."

3. Thomas Lippman, "A Breath of Fresh Air for Wind Power," *Washington Post,* November 25, 1991.
4. Steve Ginsberg, "The Wind Power Panacea: Is There Snake Oil in Paradise?" *Audubon Imprint,* Santa Monica (Calif.) Bay Audubon, 17:1, 1993, pp. 1–5.
5. *RENEW 85,* September/October 1993, p. 7.
6. Bonnie Allen, posting on Econet, an international electronic bulletin board for environmentalists, October 9, 1991.
7. Amélie Chazelles, *La Tour Eiffel: vue par les Peintres* (Lausanne, Switzerland: Edita, 1988), p. 9.
8. Aline Vidal, "Eiffel's Towering Achievement" *France,* La Maison Française, Washington, DC, Fall 1988, pp.12–13.
9. "La Tour Eiffel fait voir rouge," *Journal Français D'Amerique,* 15:3, January 22–February 4 1993.
10. Ronald W. Clark, *Works of Man* (New York: Viking Penguin, 1985, p. 280.
11. Nicole Claveloux, *La Livre de la tour Eiffel* (Paris: Gallimard, 1983), p. 89.
12. Pierre Salinger, "High and Mighty," *American Way,* April 15, 1989, pp. 83–85.
13. "La Tour Eiffel fait voir rouge."
14. George Will, "1789's Echo in Steel," *Newsweek,* July 17, 1989, p. 62.
15. "Windmills Around Haworth," letter to the editor, *The Times Literary Supplement,* February 18, 1994.
16. Chazelles, *La Tour Eiffel,* p. 12.
17. Claveloux, *La Livre de la tour Eiffel,* p. 47.
18. Chazelles, *La Tour Eiffel,* dust jacket blurb.
19. Ibid., p. 10.
20. Ibid., p. 9.
21. Leo Marx, *The Machine in the Garden* (New York: Oxford University Press, 1964), pp. 18–22.
22. Ibid., p. 28.
23. *The National Trust Guide to England, Wales, and Northern Ireland* (New York: W.W. Norton, 1977), p. 421.
24. Richard Francaviglia, *Hard Places* (Iowa City, IA: University of Iowa Press, 1991), p. 66.
25. Robert Thayer and Carla Freeman, "Altamont: Public Perceptions of a Wind Energy Landscape," Davis, CA: Center for Design Research, Department of Environmental Design, University of California, February 2, 1987, p. 6.
26. Francaviglia, *Hard Places,* p. 9.
27. *The National Trust Guide,* pp. 419–420. See the introduction to Chapter VII, "Coast and Country," for why the Trust values certain landscapes.
28. Paul Shepard, *Man in the Landscape* (New York: Alfred A. Knopf, 1967).
29. Marx, *The Machine in the Garden,* pp. 18–22.
30. H. Martin Edge, "The Contribution of the Social Sciences to Wind Power Research: Environmental Ideology and Public Perception," Robert Gordon Institute of Technology, Aberdeen, Scotland, undated.
31. Roderick Nash, "Problems in Paradise," *Environment,* 20:6, July/August 1979, pp. 25–40.
32. Ibid. At the time, Nash was imagining the use of DOE's multimegawatt wind turbines, which truly were "enormous." They stood two to three times the height of medium-sized wind turbines used commercially during the mid-1990s.
33. Ibid.

34. Ibid.
35. Dave Elliott, "Windfarm Planning and the Environment," *RENEW,* Natta Newsletter, 80, November/December 1992, pp. 14–16.
36. Ibid.
37. *The National Trust Guide,* p. 423.
38. Ibid., pp. 423–424.
39. Thayer and Freeman, "Altamont," p. 29.
40. Elliott, "Windfarm Planning and the Environment."
41. Alexi Clarke, "Windfarm Location and Environmental Impact," Natta, Open University, Milton Keynes, England, June 1988, p. 64.
42. "Catching the Wind," video by Danish Ministry of the Environment, the National Forest and Nature Agency, Copenhagen, 1992.
43. Ibid.
44. Ibid.
45. D. Sheers, "Report on an Application for 24 Wind Turbine Generators at Mynydd-y-Cemais, Machynlleth," Montgomeryshire District Council, Powys, Wales, April 1991, p. 36.
46. Ibid., p. 34.
47. Ibid., p. 38. F. Mattick charged that those he interviewed did not support the project; household energy consumption is 4000 kWh/yr, not the 3215 kWh/yr in the report; the energy used to make the turbines was not considered in evaluating their benefits; noise will be greater at night than projected; and farmers are neither altruistic nor environmentally friendly.
48. Ibid., p. 36.
49. "Landscape Impact Assessment for Wind Turbine Development in Dyfed" (Cardiff, Wales: Chris Blandford Associates, February 1992), p. 4.
50. Sheers, "Report on an Application," pp. 39–50.
51. Ibid.
52. Ibid.
53. Ibid.
54. Ibid.
55. "La Tour Eiffel: Sept million de visiterus avant 1'an 2000," *Journal Français D'Amerique,* October 1–14, 1993, p. 7; also, Champs-Elysées, 12:11, May 1994, p. 1.
56. Vidal, "Eiffel's Towering Achievement."
57. Salinger, "High and Mighty."
58. Jacques Lacarrière, *Ce Bel aujourd'hui,* chapter on "Viaducs" (Paris: J. C. Lattès, 1989), pp. 198–200.
59. Ian Ousby, *Blue Guide to England* (New York: W.W. Norton, 1989), p. 417. See the introduction to the North Midlands, England's industrial heartland.
60. *The National Trust Guide,* pp. 429–430.
61. The station, WNYD 244, was the only station west of the Rocky Mountains to win an award by the National Travelers Information Radio Exchange in its Travelers Information Message or "TIMMY" contest. Entrants included national parks, national forests, state highway agencies, area tourism offices, and universities from across the United States.
62. *The National Trust Guide,* p. 360.
63. Francaviglia, *Hard Places,* p. 207.
64. F. Lubbers, "Research Program Concerning the Social and Environmental Aspects Related to the Windfarm Project of the Dutch Electricity Generating Board," SEP (the Dutch Electricity Board), Arnhem, The Netherlands, undated, pp. 10–11.

65. Helen Colijn, *Backroads of Holland* (San Francisco: Bicycle Books, 1992), p. 127.
66. Helen Colijn, *Of Dutch Ways* (New York: Harper & Row, 1984), p. 29.
67. Ibid., pp. 11–12.
68. Ibid., p. 40.
69. Frederick Stokhuyzen, *The Dutch Windmill,* trans. Carry Dikshoorn (New York: Universe Books, 1963), p. 12.
70. Francoise Fix and Philippe Fix, *Le Livre de Paris* (Paris: Gallimard, 1990), p. 54.
71. Jean-Marie Homet, *Provence des moulins a vent* (Aix-en-Provence, France: Edisud, 1984), p. 8.
72. Ibid., p. 12.
73. Ibid., p. 32.
74. V. Schurer, "Windmills," *Country,* premier issue, 1988.
75. Kathleen Smith and David Loveland, "U.S. Energy Policy: The 1990's and Beyond," The League of Women Voters Education Fund, Washington, DC, 1989.
76. Phyllis Bosley, "A Study of Energy Resources and Issues: Perceptions and Attitudes Held by National Environmental Thought Leaders," Towson State University, Towson, MD, 1989.
77. Ibid.
78. Ibid.
79. Phyllis Bosley, "California Wind Energy Development: Environmental Support and Opposition," *Energy & Environment,* 1:2, 1990, pp. 171–182.
80. Thayer and Freeman, "Altamont," pp. 14, 24.
81. John Richter, *Energy Paths,* newsletter of the Great Lakes Renewable Energy Association, Maple City, MI, Summer 1993.
82. R. Fulton, K. Koch, and C. Moffat, "Wind Energy Study, Angeles National Forest," Graduate Studies in Landscape Architecture, California State Polytechnic University, Pomona, CA, June 1984, p. 15.
83. Thayer and Freeman, "Altamont," pp. 25–26.
84. Robert Thayer and Heather Hansen, "Wind on the Land," *Landscape Architecture,* March 1988, pp. 68–73.
85. Ibid.
86. WIMP, Phase III, Wind Implementation Monitoring Program, Draft Report, Riverside County, Riverside, CA, October, 1987, p. C-9; see the Visual Element.
87. Thayer and Freeman, "Altamont," pp. 25–26.
88. Robert Thayer, "Technophobia and Topophilia: The Dynamic Meanings of Technology in the Landscape," paper presented to the conference of the Society on the Social Implications of Technology, Los Angeles, October 1989.
89. M. McGehee, "California's Booming Desert: Where Golf Never Stops," *Golf Course Management,* November 1986.
90. M. Pasqualetti and E. Butler, "Public Reaction to Wind Development in California," *International Journal of Ambient Energy,* 8:2, August 1987, pp. 83–90.
91. Ibid.
92. Ibid.
93. Brian Young, "Attitudes Towards Wind Power: A Survey of Opinion in Cornwall and Devon," Energy Technology Support Unit, Department of Trade and Industry, Harwell, Berkshire, England, 1993, pp. 1–3. The second poll used a different set of respondents.
94. Ibid., pp. 9–11.
95. Ibid., pp. 44–45.
96. Ibid., pp. 43–44.

97. Ibid., p. 40.
98. Andrew Garrad, "Wind Energy in Europe: Time for Action!" European Wind Energy Association Strategy Document, August 31, 1990, p. 6.
99. Young, "Attitudes Towards Wind Power," pp. 46, 47.
100. Estas Esselmont, "Cemmaes Windfarm Sociological Impact Study," Energy Technology Support Unit, Harwell, United Kingdom, March, 1994, Summary of Findings.
101. Chris Blandford Associates, "Wind Turbine Power Station Construction Monitoring Study," in association with the University of Wales, Bangor, February 1994, pp. 42–55. The study was commissioned by the Countryside Council for Wales.
102. C. Westra and L. Arkesteijn, "Physical Planning, Incentives, and Constraints in Denmark, Germany, and the Netherlands," paper presented at "The Potential of Wind Farms," European Wind Energy Association's special topic conference, Herning, Denmark, September 8–11, 1992.
103. J. M. Brown, letter to R. Thayer, University of California at Davis, February 13, 1989.
104. Alison Tasker, "Future Perfect," *Review*, Department of Energy, United Kingdom, 13, Autumn 1990, pp. 12–13.
105. Ibid.
106. Young, "Attitudes Towards Wind Power," p. 37.
107. Tasker, "Future Perfect."
108. "Landscape Impact Assessment," p. 5.
109. Pasqualetti and Butler, "Public Reaction."
110. Lubbers, "Research Program," p. 12.
111. Maarten Wolsink, "Public Acceptance of Large WECS in the Netherlands," Department of Environmental Sciences, University of Amsterdam, The Netherlands, undated.
112. Inga Carlman, "Public Opinion on the Use of Wind Power in Sweden," paper presented at the European Wind Energy Association conference, Rome, October 1986.
113. Ibid.
114. Robert Thayer and Heather Hansen, "Consumer Attitude and Choice in Local Energy Development," Department of Environmental Design, University of California–Davis, May 1989, pp. 12–19.
115. Ibid., p. 16.
116. Ibid., p. 20.
117. Ibid., p. 23.
118. Ibid., p. 8–10.
119. R. Samuelson, "1990s Energy Crisis," *Newsweek*, 1989.
120. Thayer and Hansen, "Consumer Attitude," p. 23.
121. Thayer and Hansen, "Wind on the Land."
122. Thayer and Hansen, "Consumer Attitude," p. 23.
123. Maarten Wolsink, "The Siting Problem: Wind Power as a Social Dilemma," Department of Environmental Science, University of Amsterdam, The Netherlands, undated.
124. Ibid.
125. Ibid.
126. Ibid.
127. Maarten Wolsink, "Attitudes and Expectancies About Wind Turbines and Wind Farms," Wind Engineering, 13:4, 1989, pp. 196–206.
128. Wolsink, "The Siting Problem."
129. Fulton, Koch, and Moffat, "Wind Energy Study," p. 128.
130. Westra and Arkesteijn, "Physical Planning."
131. Bridgett Gubbins, "Living with Windfarms in Denmark and The Netherlands," North Energy Associates, Northumberland, England, September 1992, p. 7.

132. L. Arkesteijn and R. Havinga, "Wind Farms and Planning: Practical Experiences in the Netherlands," paper presented at "The Potential of Windfarms," European Wind Energy Association's special topic conference, Herning, Denmark, September 8–11, 1992.
133. For an engaging discussion of engineering intuition, see Eugene S. Ferguson, *Engineering and the Mind's Eye* (Cambridge, Mass. MIT Press, 1992).
134. Timothy Jensen, in a letter describing their design approach, Jacob Jensen Design, Højslev, Denmark, March 21, 1994.
135. Thayer and Freeman, "Altamont," p. 5.
136. A. J. Robotham, "Visual Impact," in short course on "Principles of Wind Energy Conversion," Imperial College of Science, Technology, and Medicine, London, July 1993.
137. Andrew Garrad, "An Analytical Approach to Wind Farm Design," paper presented at the British Wind Energy Association's annual conference, York, October 6–8, 1993.
138. Lubbers, "Research Program," p. 10.
139. WIMP, Phase II, "Wind Implemetation Monitoring Program," Riverside County, Riverside, CA, Tierra Madre Consultants, August 1986, p. B-6.
140. Ibid., p. B-19.
141. Ibid., p. B-20.
142. Ibid., p. B-15.
143. WIMP, Phase III, "Wind Implementation Monitoring Program," p. C-6.
144. WIMP, Phase II, "Wind Implementation Monitoring Program," p. B-11.
145. Ibid., pp. B-12–B-13.
146. Ibid., p. B-14.
147. Ibid., p. B-3.
148. "Landscape Impact Assessment," p. 6.
149. Robotham, "Visual Impact."
150. WIMP, Phase III, "Wind Implementation Monitoring Program," p. C-3.
151. Thayer and Hansen, "Wind on the Land."
152. WIMP, Phase III, "Wind Implementation Monitoring Program," p. C-15.
153. Thayer and Hansen, "Wind on the Land."
154. "No Idle Turbines Allowed," *Windpower Monthly,* 10:4, April 1994, p. 14.
155. Thayer and Freeman, "Altamont," p. 5.
156. Fulton, Koch, and Moffat, "Wind Energy Study," p. 129.
157. Garrad, "An Analytical Approach."
158. "Landscape Impact Assessment," p. 11.
159. Ibid., p. a1.
160. Ibid., p. i.
161. Ibid., p. 11.
162. Ibid., p. 17.
163. Fulton, Koch, and Moffat, "Wind Energy Study," p. 64.
164. WIMP, Phase III, "Wind Implementation Monitoring Program," p. C-10.
165. Fulton, Koch, and Moffat, "Wind Energy Study," 1984, p. 64.
166. WIMP, Phase III, "Wind Implementation Monitoring Program," pp. C-12–C-15.
167. Fulton, Koch, and Moffat, "Wind Energy Study," p. 64.
168. WIMP, Phase III, "Wind Implementation Monitoring Program," p. C-4.
169. Fulton, Koch, and Moffat, "Wind Energy Study," p. 128.
170. Robotham, "Visual Impact."
171. Arkesteijn and Havinga, "Wind Farms and Planning."
172. Lubbers, "Research Program," p. 8.

173. Robotham, "Visual Impact."
174. Clarke, Windfarm Location.
175. Ruud de Bruijne, NOVEM, oral comments, European Wind Energy Association's special topic conference. "The Potential of Wind Farms," Herning, Denmark, September 8–11, 1992.
176. Although most British mills used four blades, some used five, six and eight blades. J. Kenneth Major and Martin Watts, *Victorian and Edwardian Windmills and Watermills from Old Photographs* (London: Fitzhouse Books, 1977).
177. Fulton, Koch, and Moffat, "Wind Energy Study," p. 130.
178. "Landscape Impact Assessment," p. 9.
179. Robert Ferber, "Public Reactions to Wind Energy and Windmill Designs," *Proceedings of the 3rd Wind Energy Workshop,* U.S. Department of Energy, Washington, DC, September 1977, pp. 413–419.
180. "Landscape Impact Assessment," pp. a2, 10.
181. Thayer and Hansen, "Wind on the Land."
182. "Landscape Impact Assessment," p. 10.
183. Ibid.
184. For further suggestions, see Paul Gipe, *Wind Power for Home & Business* (Post Mills, VT: Chelsea Green Publishing, 1993).
185. Howard Wilshire and Douglas Prose, "Wind Energy Development in California, USA," Environmental Management, 10:6, 1986.
186. Fulton, Koch, and Moffat, "Wind Energy Study," p. 130.
187. "Landscape Impact Assessment," p. 10.
188. Fulton, Koch, and Moffat, "Wind Energy Study," pp. 62, 130.
189. Ibid., p. 60.
190. Ibid., p. 61.
191. Ibid., p. 127.
192. Thayer and Hansen, "Wind on the Land."
193. J. Swiatek, "Proposed Park Site Is Short of Trees, Long on Controversy," *Indianapolis Star,* September 10, 1989.
194. Robotham, "Visual Impact."
195. Pasqualetti and Butler, "Public Reaction."
196. James A. Throgmorton, "Community Energy Planning: Winds of Change from the San Gorgonio Pass," *APA Journal,* Summer 1987, pp. 358–367.
197. Ibid.
198. Bosley, "California Wind Energy Development."
199. Throgmorton, "Community Energy Planning."
200. Thayer and Hansen, "Wind on the Land."
201. "A Comeback for Nuclear Power," *National Geographic,* August 1991, pp. 73–75.
202. Douglas Day, "A Little Bear Laissez Faire," *American Way,* December 1, 1993, pp. 84–125.
203. Gubbins, "Living with Windfarms," p. 21.
204. Jamie Chapman, "European Wind Technology," Electric Power Research Institute, Palo Alto, CA, March 1993, pp. 1–9.
205. Robotham, "Visual Impact."
206. Sara Knight, "Portrait of a Booming Market," *Windpower Monthly,* 9:3, March 1993, pp. 26–31.
207. Gubbins, "Living with Windfarms," p. 6.
208. de Bruijne, NOVEM, oral comments.

209. "Renouvelables: inventaire et état de l'art," *Systémes Solaires,* 94/95, 1992, p. 22.
210. Gubbins, "Living with Windfarms, pp. 13–15.
211. Ibid., p. 9.
212. William Grant, "Power to Spare," *Outdoor America,* 58:1, Winter 1993, pp. 31–33.
213. Thayer and Hansen, "Wind on the Land."
214. Ibid.
215. J. Paddock, N. Paddock, and Carol Bly, *Soil and Survival* (San Francisco: Sierra Club Books, 1986) Chapter 6, "The Uses of Beauty: Farmland Aesthetics in the Twentieth Century," pp. 64–71.
216. Ibid.
217. Stephanie Mills, *Whatever Happened to Ecology* (San Francisco: Sierra Club Books, 1989), p. 214.
218. Maarten Wolsink, "The Social Impact of a Large Wind Turbine," Environmental Impact Assessment Review, 8, 1988, pp. 323–334.
219. Lacarrière, *Ce Bel aujourd'ui,* p. 132.

9

Impact on Flora and Fauna

The impact of wind energy on plants and wildlife results primarily from road grading and the disturbance of habitat. (The impact on birds is discussed separately). In their study of the San Gorgonio Pass for Riverside County, Tierra Madre Consultants found that few small animals were killed or injured directly through the installation and servicing of the wind turbines, and estimated that there was only a modest incremental loss of habitat. They did find severe impacts on rattlesnake populations. This is more an indictment of our culture's attitudes toward venomous snakes than of wind development per se. Despite concerns that fences and human activity would disrupt predator–prey relationships, predatory animals were still found to frequent the wind plants. Human activity may discourage mammals from denning on the site, but it will not discourage them from foraging. Shy animals such as bobcats may suffer more disruption than animals lower on the food chain. Noting that there were slight reductions in the populations of native plants and animals due directly to loss of individuals and indirectly through loss of habitat, principally from grading, Tierra Madre summed up their study by saying that there was a quantitative loss of biological resources but no qualitative loss. "We noted no obvious effect on [biological] diversity," they said. Although the sites are no longer "pristine," Tierra Madre considered the cumulative loss of wildlife insignificant and much less severe than that caused by residential development.[1]

Many jurisdictions in the United States require that wind plants fence their project boundaries and prohibit unauthorized entry. Some biologists assert that fencing will create additional impacts by restricting migratory patterns. But an early WIMP study examined fencing in detail; it found "no major negative effects," and concluded that fencing has reduced the impact on the Coachella Valley fringe-toed lizard (an endangered species) by reducing off-road vehicle traffic.[2] Tierra Madre recommended that Riverside County discontinue further study of fencing impacts and that the county require fencing in areas of "critical environmental concern" to discourage use by off-road vehicles.

Planners and wind plant operators must strike a balance between fencing for the prevention of vandalism, for the protection of threatened or endangered plants, and for the protection of the public from the hazards that do exist at a wind plant (high-voltage electricity, for example) while still allowing access to the public and permitting wildlife unrestricted movement.

Impact on Birds

Cynthia Struzik strode purposely to the blackboard, where she wrote in bold letters, "Thou shalt not kill," then added, "Acceptable mortality level is zero!" Struzik, a special agent of the U.S. Fish and Wildlife Service, made her point. It was not lost on the others gathered at Pacific Gas & Electric's research center in San Ramon, California to discuss the conflict between birds and the wind turbines in the Altamont Pass. A groan of disbelief swept across industry representatives in the room. But it was only the beginning. By the end of the day, Struzik, her colleagues from the California Department of Fish and Game, and representatives of the Sierra Club's Mother Lode chapter were openly mocking attempts by the industry and biologists on that December day in 1992 to understand what was happening in the pass just a few miles away.

Wind turbines and wind plants have little or no impact on most plants and animals, but they have killed some birds, notably in the Altamont Pass and near the straits of Gibraltar. No single environmental issue has caused more consternation among wind energy advocates and environmentalists alike than the existing or potential effects that wind turbines have on birds. It is the kind of "hot button" issue that elicits strong emotional responses, one that could, if not addressed, derail plans for expansion of wind energy not only in California, but elsewhere.

That some wind turbines kill birds some of the time should come as no surprise. Most tall structures kill birds to some degree, as do most sources of energy. Although undesirable, this revelation reflects no differently on wind energy than the realization that the swoosh of the blades or the whine of their generators annoys some people some of the time. This issue, like all others, must be considered in context.

The "bird problem," as the wind industry calls it, came to light over a three-year period in the late 1980s when the California Energy Commission tallied reports that as many as 160 birds had been killed or had died in the vicinity of the state's wind power plants, including a protected and highly valued species: the golden eagle. After surveying the Fish and Wildlife Service, the California Department of Fish and Game, wildlife rehabilitation centers, and wind plant operators, the CEC learned that 99 dead birds had been recovered in the Altamont, 9 in Tehachapi, and 40 in the San Gorgonio Pass from 1984 to 1988. These birds had been killed either by the wind turbines or by the transmission lines serving the wind plants, or else they had died from some unknown natural cause.[3]

Nearly 70% of the incidents were attributed to collisions with the turbines or their associated transmission lines. Golden eagles and red-tailed hawks were the species most often affected by collisions. Despite the data, the CEC was uncertain as to whether regional populations of raptors were being harmed, but they did conclude that local breeding populations could be reduced by the collisions, and this warranted concern.[4]

The deaths also worried the Department of Fish and Game because of its attempt to stabilize the state's raptor population. California's raptors, or birds of prey, are fast losing their habitat to an exploding human population. In the San Joaquin Valley alone, more than 95% of wildlife habitat has already been converted to other uses. Consequently, wildlife becomes increasingly dependent on the remaining "islands" of undeveloped land. Some of this land remains undeveloped because high winds make it hostile to human habitation. Thus there is the potential for increasing competition between raptors and wind development for the same resource.[5]

In California, golden eagles are a "species of special concern." This designation mandates that Fish and Game protect them. Federal law in the United States also prohibits the "taking" of golden eagles under the Bald Eagle Protection Act and the Migratory Bird Treaty Act. *Taking* is a euphemism for killing, derived from the days when the predecessor of the Fish and Wildlife Service efficiently managed the extermination of the wolf and other predators.[6] The agency still has teeth. Anyone who "knowingly or with wanton disregard" kills bald or golden eagles commits a felony in the United States, punishable by two years in prison and a fine of up to $250,000.[7] When Struzik wrote her commandments on the board in San Ramon, she was simply stating federal law.

To resolve this simmering conflict, the CEC and the counties of Solano, Alameda, and Contra Costa, sponsored the most extensive study of bird deaths near wind farms ever undertaken. Begun in early 1989, the study to determine the exact number of birds being killed, why, and what mitigation measures may be required, took two years of field work to complete. The conclusions were not pleasing.

The study found 182 dead birds, two-thirds of them raptors. Red-tailed hawks headed the list, followed by American kestrels and golden eagles. BioSystems, the consortium's contractor, estimated that wind turbines and related facilities in the Altamont pass were killing 160 to 400 birds per year, most of which were birds of prey, including up to 40 golden eagles per year.

BioSystems' study raised as many intriguing questions as it answered. Why, for example, are more raptors killed at end-row turbines than elsewhere? Why are more killed at lattice towers than tubular towers? Does a difference in foraging behavior explain why turkey vultures and ravens are not killed at the same rate as raptors? BioSystems speculated on the reasons for the collisions but offered no conclusions. For example, the birds may habituate to the presence of the turbines, says the report, and essentially become careless and inattentive when searching for prey, much like a driver too busy reading a billboard to notice a car stopped ahead, and slam into the wind turbines.[8] Biologists also question why the problem appears so much more severe in the Altamont Pass than in San Gorgonio, Tehachapi, or northern Europe.

Even more damaging to wind energy's reputation as a relatively benign technology than the numbers of birds being killed is the manner in which they died. Two-thirds of the golden eagles were killed after colliding with wind turbines or their towers. The powerful visual imagery of graceful hawks soaring into the "Cuisinarts of the air," as one former Sierra Club lobbyist termed wind turbines, lends itself to blaring headlines and self-styled investigative reports revealing the "true story" behind one green technology. The wind industry's principal trade magazine, *Windpower Monthly,* graphically drew attention to the problem in its February 1994 issue with a gruesome cover photo of a disemboweled raptor beneath wind turbines at Tarifa, Spain.

European environmental groups are just as concerned about wildlife as their American counterparts. Questions from "nature protection societies," as they are called, about the impact of wind turbines on wildlife in general and birds in particular, have led to several similar field surveys in northern Europe.

European Bird Studies

Europeans appear more concerned about the loss of habitat than about the number of birds that wind turbines may kill, even where studies have shown that wind turbines have killed birds. This concern led the Dutch Institute for Forestry and Nature Research in 1984 to begin a before-and-after study of the Netherlands' first wind plant. Designers for the Dutch association of electricity producers, SEP, arrayed 18 experimental turbines on 55 ha of drained farmland near Sexbierum in the northern province of Friesland. The 30-m-diameter 300-kW variable-speed turbines were only 3 to 4 km (about 2 miles) from the Wadden Sea, an important wildlife area.

Although the project is small by California standards, the Dutch were actually able to watch birds pass near the wind turbines both during the day and at night with radar, passive image intensifiers, and infrared spotlights. They found that one of the 14 birds trying to cross the rotor disks during daylight hours collided with a turbine. Of the 51 birds trying to cross the rotor disk during the night, 14 collided. Not all collisions were fatal. Four birds recovered and continued their flight. In six of the collisions, the birds did not strike the turbine or tower but were swept to the ground in the rotor's wake when flying with a tailwind. Three of these birds recovered and continued their flight. Only 5% of the collisions were fatal. Small songbirds were particularly susceptible. Overall, biologist J. E. Winkelman estimates that 68 birds were killed during the seven days and nights of her observations.[9]

Winkelman suspected that there was too little noise from the wind farm for noise to influence the breeding of meadow birds, and test results showed little effect on breeding. She did detect an overall effect on feeding and resting birds up to a distance of 500 m (1600 ft) from the turbines, but found that most of the impact was contained within 100 to 250 m (300 to 800 ft). The wind farm also affected migratory birds, some of whom were found to avoid the wind farm after its completion.[10]

In a survey of European studies, Dutch consultants found negligible impact on breeding birds and concluded that the number of birds killed per turbine per year appeared acceptable. The results of the studies were often conflicting they noted, indicating on the one hand that there could be serious impacts at sites with large numbers of birds; yet in other areas, where significant kills were expected, such incidents have not materialized.[11] They also found that wind turbines will disturb resting and migrating birds to some extent.[12] But, the greatest impact on birds in Europe appears to be the loss of habitat when untilled soil is tilled, whether for row crops or wind turbines.

Danish conservationists hold diverse views about wind energy. There is both strong support for wind energy and serious concern about its potential impact on wildlife. Yet wind turbines have been erected near or alongside nature reserves without serious opposition. Concern about the potential impact on birds at a reserve near Tjæreborg did lead to a detailed bird study before and after construction of a large 2-MW wind turbine.[13]

Because of the Wadden Sea's importance to waterfowl, Danish conservationists criticized construction of the experimental turbine on the west coast of Jutland near Esbjerg. As a result, the National Environmental Research Institute conducted a four-year study of the wind turbine's effect on nearby birds. Regardless of whether the turbine was operating, the study found birds changing course when passing nearby. Radar observations indicated that birds in general were able to detect and avoid the turbine. Researchers did find 15 dead birds at the site. Of these, four had hit the wind turbine, the remainder had collided with the meteorological mast. Nevertheless, the turbine creates a "vacuum effect" that prevents birds from using the area closest to the turbine for breeding and nesting. The polder's habitat, which has been diminished extensively by farming, is further reduced for breeding, staging, and foraging birds by the turbine's disturbance.[14]

In 1983, the Game Biology Station at Rønde began a study to determine potential conflicts between birds and intermediate-sized wind turbines at Koldby and Nibe in northern Jutland. The latter is the site of two large early turbines on the Limfjord west of Ålborg. The biologists, who were principally concerned with collisions and loss of habitat, found that migratory birds used routes along the Limfjord, parallel to the shore. On occasion, birds were observed changing course as they neared the turbines. No dead birds attributable to collisions with the wind turbines were found, despite the fencing of the site at Nibe to discourage scavenging of carcasses. The Biology Station tentatively concluded that the turbines caused little impact but suggested that collisions could occur under special weather conditions.[15]

Another Danish study examined 135 wind turbines at nine sites. The researchers counted nearly 2000 birds passing within 150 m (500 ft) of the turbines. Of these, 17% altered course. Two birds may have been killed by the turbines. The ornithologists concluded that the risk of collisions seemed negligible. They also suggested that local birds may habituate to the presence of the machines. To avoid conflicts, they sensibly recommended that wind turbine development should avoid the migratory routes and staging areas of sensitive species.[16]

Avoiding a problem is always easier than trying to treat it after the fact. This appears to be the case in a burgeoning international problem near Gibraltar, in a major migratory flyway between Europe and Africa. The nearby windsurfing center of Tarifa has become famous as home to Spain's largest concentration of wind turbines. The 270 wind machines have also become notorious among Spain's environmental groups for killing raptors staging above the arid ridgetops before soaring across the straits. The turbines were killing protected birds such as griffon vultures and storks, according to local observers. By the mid-1990s, the severity of the problem was unknown, but studies were under way.[17]

German environmentalists are equally concerned about wind turbines near breeding and staging areas on the North Sea. The North German Academy for Nature Protection examined 11 sites with wind turbines and found 32 dead birds in total. Responding to concerns that the wind turbines could be killing birds indirectly by destroying their food supply, researchers measured, with classic Teutonic thoroughness, the density of insect splatter on the blades and even determined the species composition. They concluded that wind energy's effect on insect populations was negligible.[18]

The Germans are not alone in spending public resources to study whether wind turbines will slap too many insects from the skies. The U.S. Department of Energy (DOE) filmed the release of honeybees and blowflies to determine their interaction with the experimental Mod-0 turbine near Sandusky, Ohio in the mid-1970s. They reached the same conclusion as the Germans 15 years later: there was little observable impact.

To be fair, insects were not at the top of DOE's list of concerns. Initially, they feared that birds migrating at low altitudes during the night would collide with the turbine. DOE's biologists concluded that even on nights of high migration, the single turbine would not be lethal to a significant number of birds, but they did observe a number of birds changing course to avoid the turbine.[19]

San Gorgonio Pass

Because the San Gorgonio Pass is one migration corridor along the Pacific Flyway, biologists are anxious to avoid a major bird kill there. Mass bird kills are not unknown, although they have yet to occur at any wind plants in California or abroad. Millions of birds fly through the San Gorgonio Pass during peak migration, mostly at night at low elevations: conditions thought most likely for birds to fly into wind turbines or transmission lines. The first study by Riverside County's Wind Implementation Monitoring Program concluded that the potential for collisions was minor. Nevertheless, a subsequent study for WIMP II in Painted Hills, an area with wind turbines, found a high scavenging rate and concluded that only one or two birds out of every three killed would be seen. No dead birds were found during a study of spring migration.[20]

Tierra Madre Consultants warned in WIMP II that there remained the potential for a "catastrophic collision event" should a flock of migrating birds such as the Salton Sea's white pelicans descend through the wind turbines to the water recharge ponds on the Whitewater Wash. Tierra Madre considered this a possible, though unlikely event. Overall, WIMP II concluded that if collisions were taking place, only a small number of birds were involved. The turbines, WIMP II reported, constituted only a minor threat to migrating birds.[21]

The 32 to 37 million birds that migrate through the San Gorgonio Pass each spring and fall also concern Southern California Edison, the regional utility, because they operate several high-voltage transmission lines there that feed the Los Angeles basin. Possible bird collisions with the area's growing population of wind turbines also interested the utility, the purchaser of the wind-generated electricity. In a 1985 field survey of area wind plants, the utility found 38 dead birds, nearly all songbirds and no sensitive or endangered species. From this small sample, SCE's biologists estimated that 3900 to 6900 birds could be killed each year. Although the number is large, the mortality rate is extremely small (0.006 to 0.009%) and SCE's biologists consider it insignificant compared to the large number of songbirds migrating through the pass. The biologists observed that many of those birds migrating at night flew low enough to hit the turbines but that most avoided them.[22]

Solano County

When U.S. Windpower proposed building a 60-MW wind plant just north of the confluence of the San Joaquin and Sacramento rivers, their permit was opposed by an unlikely but influential group—duck hunters. The 600 turbines that USW wanted to install were between two wildlife sanctuaries: the Grizzly Island Wildlife Area to the northwest and the Lower Sherman Island Wildlife Area to the southeast. Hunters feared that the wind turbines would "take" the area's waterfowl before they did.

To address the hunters' concerns, Solano County required U.S. Windpower to survey bird populations before and after construction. During their four-year study, biologist Judd Howell and his team observed nearly 15,000 birds while monitoring 237 turbines. They found 22 dead birds, of which half were raptors, primarily red-tailed hawks. The mortality rate was similar to that of the Altamont (Table 9.1).

Table 9.1
Estimated Number of Birds Killed by Wind Turbines in Northern California

	Altamont Raptors		*Solano* All Birds		*Solano* Raptors	
	Low	High	Low	High	Low	High
Birds/yr	164	403	17	44	11	2
Turbines	6800	6800	600	600	600	60
Birds/turbine/yr	0.024	0.059	0.029	0.074	0.018	0.04
MW	700	700	60	60	60	6
Birds/MW/yr	0.23	0.58	0.29	0.74	0.18	0.4

Source: For Altamont, Susan Orloff and Ann Flannery, California Energy Commission, 1992; for Solano, Judd Howell and Jennifer Noone, Solano County, Fairfield, California, 1992.

Given that both red-tailed hawks and American kestrels are abundant in California, Howell concluded that the Solano project should have no biologically significant impact on regional or state populations, despite a mortality rate similar to that of the Altamont. He says that the early concerns of hunters were ill-founded, as waterfowl flew along the watercourses some distance from the wind turbines. The turbines would have no effect on the number of ducks the hunters could shoot.[23]

Possibly most important was Howell's observation that the mortality of the most abundant species in the area, turkey vultures, was "practically zero."[24] This may be a key to understanding what is happening to Altamont's raptors because it mirrors BioSystems' findings in their Altamont study. They calculated that golden eagles, red-tailed hawks, and American kestrels were nine times more likely than vultures to collide with wind turbines.[25] Why, no one knows. But the answer may help biologists and wind turbine designers craft a response that protects all birds.

Tehachapi

Once the extent of the problem in the Altamont became apparent, the CEC contracted with BioSystems to examine the Tehachapi area to determine if a similar problem existed there. No in-depth studies of bird populations had been conducted in the Tehachapi Pass prior to the CEC-funded project.[26]

Dead birds had been found previously in the vicinity of Tehachapi's wind plants. The CEC reported that nine raptors had been killed in Tehachapi from 1984 to 1988, according to accounts from rehabilitation centers, wildlife agencies, and wind companies themselves.[27] From 1988 to 1991, Marty Smith, the local game warden, reported that four owls had been killed near the wind farms.

BioSystems expected to find a higher number of birds killed in Tehachapi relative to those in the Altamont, because Tehachapi had several characteristics they believed contributed to killing birds. Tehachapi is more mountainous than the Altamont and has a higher average elevation. Because of its steep terrain, Tehachapi has more end-row turbines. BioSystems had linked all three factors to bird deaths in the Altamont. Despite their expectations, BioSystems could find "no dead or injured birds" during their brief survey. The scavenging rate of bird carcasses in Tehachapi was no different than that of the Altamont, ruling out the possibility that dead birds were being removed before they could be observed. BioSystems reluctantly concluded that the mortality of all birds in Tehachapi was in fact significantly lower than in the Altamont, stating that "the lower mortality we found is particularly noteworthy."[28]

At a loss to explain their results fully, BioSystems suggests that fewer raptors live in the Tehachapi area than in the Altamont and that those that do fly higher and perch less often on the turbines. There are also fewer ground squirrels than in the Altamont. Although the Tehachapi Pass is a migration corridor, few raptors use the route. Those raptors that do use the pass fly higher than those in the Altamont, according to BioSystems' field observations. This may remove them from the zone where collisions with the wind turbines are more likely.

BioSystems' results were buttressed by a two-year study of wind turbines near Mojave, on the lower east-facing slopes of the Tehachapi Mountains. Bakersfield biologist Diane Mitchell could also not find any dead raptors during a two-year period. She did find 20 dead songbirds, and SeaWest employees found one raven near a transmission line outside her survey area.[29]

The study was required as part of an agreement between wildlife agencies, Kern County, and SeaWest for a permit to install 300 Mitsubishi wind turbines in the early 1990s. Mitchell conducted the study for SeaWest under Kern County's Mitigation Monitoring Program. She followed the approach pioneered by BioSystems in the Altamont Pass and monitored 10 sample plots, comprising 17% of the 300 turbines, for eight seasons.[30]

In a memo to the Fish and Wildlife Service, the California Department of Fish and Game, and the CEC, Mitchell concluded that few raptors used the site and saw no need for continuing her study. She suggested that other habitats, including higher elevations and ridgetops, may have more raptor activity, and she cautioned against extrapolating her results to the entire Tehachapi Pass.

Other Structures That Kill Birds

Most large structures (television towers, smokestacks, lighthouses, transmission lines, monuments, and so on) kill birds to some extent. In early environmental reports, the U.S. Department of Energy cites studies finding that 2700 birds were killed annually over 11 years at a television tower in Florida, that 800 to 1400 birds per season were killed over a five-year period at a radio tower in North Dakota, that a cooling tower at a nuclear plant on Lake Erie is believed to kill 40 to 100 birds per season, and that 50,000 birds were killed one night at an airport in Georgia.[31] In its report on the Altamont Pass, BioSystems tried to put the 160 to 400 deaths per year there in perspective by noting that 5 to 80 million birds die annually in the United States from collisions.[32]

All energy sources kill birds to some degree. The extent varies. The *Exxon Valdez* oil spill alone killed from 375,000 to more than 500,000 birds, far more than the *Torrey Canyon* (30,000) or the *Amoco Cadiz* (20,000) spills.[33] Wildlife biologists estimate that the *Valdez* spill killed 900 bald eagles, 11% of the local population.[34] Biologists at the Fish and Wildlife Service's Alaska Research Center say that it may take 20 to 70 years for bird populations to recover in Prince William Sound.[35]

To put the situation in perspective, California wind plants offset 14 times the oil spilled by the *Exxon Valdez* in primary energy every year, nearly a third of that being generated in the Altamont Pass. If BioSystems' estimates are correct, it will take wind turbines in the Altamont Pass 500 to 1000 years to kill as many birds as the *Exxon Valdez.*

Other solar-electric technologies also kill birds. Southern California Edison estimated that its solar concentrator near Barstow, Solar One, killed 70 songbirds during one study. Biologists estimated that the 10-MW solar plant killed 2 birds per week or about 100 per year. The birds died primarily from collisions with the picture-window-like surface of the heliostats.[36] There have also been reports of collisions with photovoltaic panels, for similar reasons. Although the heliostats at Solar One may kill 10 times the number of birds per megawatt as the Altamont Pass wind turbines, the species killed are of lesser concern and certainly of lesser symbolic value than the golden eagles and other raptors killed in the Altamont Pass.

Birds are killed not only in the production of electricity, but also through its transmission and distribution. Birds die by striking power lines (or even telephone lines) or by electrocution. Most deaths by electrocution are avoidable and can be prevented by modifying transmission-line towers. The University of Minnesota's Raptor Research Foundation suggests techniques for preventing raptors from contacting the energized cables. Simple techniques for adding safe perches or antiperching devices can be found in a Foundation report, Suggested Practices for Raptor Protection on Power Lines, and subsequent updates.[37] Provisions for preventing electrocutions should be standard practice within California wind plants and elsewhere. Alameda county now requires that all intraproject power lines must be "raptor-proofed." Other counties will probably follow suit at the request of wildlife agencies.

In California, power lines themselves should not reduce raptor populations significantly. However, birds of special concern are killed by power lines often enough to worry biologists. Within one month during the summer of 1993, three condors worth $1 million each were killed by transmission lines near Los Angeles. (Another condor died after drinking antifreeze.)

Raptor collisions with utility lines (as opposed to electrocutions) will always remain a problem. Peregrine falcons are the species most susceptible to power-line collisions, and they do collide with power lines in California. Yet no peregrine falcons have been reported killed near wind plants.[38]

The Danish Game Biology Station examined a transmission line in the same vicinity as the wind turbines they were studying in northern Jutland. They found 14 to 15 dead birds per 2 km (1 mile) of high-voltage line per visit.[39] This cursory examination echoes a three-year study by the French equivalent of the National Audubon Society, the *Ligue de Protection des Oiseaux,* which found 700 carcasses over a 300-km (180-mile) stretch of high-voltage transmission line. Most of the birds, 85%, had been electrocuted, and the remainder had died from collisions. Altogether, one-third of those killed were raptors, including six eagles. At the power line studied, 8 birds were killed every 10 km (6 miles). The Ligue calls the 70,000 km (40,000 miles) of high-voltage lines in France a "grave menace."[40]

What Does It Mean?

Although the death of any bird is unfortunate and should be avoided wherever possible, that alone is insufficient to condemn a technology and the people who make it work. Rather than focusing on the number of individual birds killed, it is important to place the issue in the context of the total population, says Tom Cade, founder of the Peregrine Fund and director of the World Center for Birds of Prey. Regrettable as they are, the deaths "may really have no biological significance."[41]

The number of birds killed in the Altamont Pass could be significant for protected species such as the golden eagle. The CEC believes that enough golden eagles are being killed near the Altamont Pass to disrupt the resident population. This is particularly significant with a species such as the golden eagle which has suffered population declines throughout its range in California, due to urban encroachment. The CEC suggests that raptor populations could decline further in all three passes from urban growth alone.[42]

Not everyone agrees with the CEC's belief that the regional golden eagle population has a low reproductive rate in the Altamont Pass. Until the regional population is known with more certainty, no one can determine if wind plants are killing so many golden eagles that they will eventually reduce the population below a sustainable level. For this reason, Cade says that ornithologists need to study the regional golden eagle population. Then they can determine how the number of deaths due to wind turbines affects this population. Only then can they answer the crucial question: Are Altamont's wind plants threatening the population's stability.[43]

This is the question that Rich Ferguson wants answered. How many dead birds, specifically eagles, are too many? Ferguson, energy chair for the national Sierra Club, is the environmental community's point man on the bird–wind energy issue. He is trying to mediate an internal debate within the club's powerful California contingent, which is fueled by charges from a local activist that a 50-MW wind project in Solano County proposed by the Sacramento Municipal Utility District will kill golden eagles and should therefore be stopped. Ferguson, who calls the situation in the Altamont Pass "tragic and unacceptable," nevertheless believes that the issue is less than black and white.[44]

Ferguson wants to avert a split within the nation's largest environment group like that of the deep division over nuclear power that rent the Sierra Club during the 1960s, when an antinuclear faction gained supremacy over a group led by famed photographer Ansel Adams. Ferguson, a former physics professor, hopes to head off just such an all-or-nothing battle over wind energy, which would certainly damage the wind industry but could potentially damage the Sierra Club's authority as a proponent of renewable energy as well.

To foster dialogue and explore the concerns of all parties, Ferguson has led meetings among activists, the wind industry, and biologists. At one such meeting in August 1992, Judd Howell, the biologist conducting U.S. Windpower's field survey in Solano County, presented his findings. Despite Howell's generally polite reception, two members from Sacramento's Mother Lode Chapter were clearly impatient with the slow pace of research. They queried Howell closely, particularly on the extent of U.S. Windpower's experiments with possible mitigation measures. After Howell explained that USW had painted five sets of five turbines each as a means of warning raptors, one skeptically exclaimed, "Are you telling me they tested 25 turbines out of 7000?" Howell responded that it is not easy to paint stripes on the rotor of a wind turbine, and that it was only an experiment, after all—an experiment with inconclusive results, at that.

Howell regained rapport with his critics after he explained that the Solano study was commissioned because duck hunters feared that the wind turbines would kill waterfowl from nearby Suisun Marsh. The irony of U.S. Windpower being forced to fund a study to ensure that hunters had enough birds to shoot was not lost on the antihunting audience. He further convinced the group of U.S. Windpower's good intentions by recounting a move by the company only a few days earlier.

Just prior to the Sierra Club briefing, U.S. Windpower had convened the first meeting of what Howell calls the "Mount Olympus of ornithology," comprising five of the leading authorities on birds in the United States. The Avian Advisory Task Force includes Tom Cade, founder of the Peregrine Fund and the Fish and Wildlife Foundation,[45] Mark Fuller, director of the Raptor Research and Technical Assistance Center, Melvin Kreithen at the University of Pittsburgh, Vance Tucker at Duke University, and Charles Walcott from Cornell's laboratory of ornithology.

They were called together, said Howell, to find out why a small number of birds "held in high esteem" had been killed in the Altamont Pass. "Flying is hazardous," Howell said to nods of agreement from the audience; "the first year, mortality among raptors [learning to fly] is 30%." The job of the task force is to make flying less hazardous around wind turbines. One intriguing idea, Howell told the Sierra Club group, is painting eyespots on the nacelles. Eyespots are universally recognized by raptors, who have no desire to become prey themselves. The eyespots could put the birds on alert, thus possibly helping them avoid the turbines. The task force will explore this and many other avenues in its research.

Bird–Wind Turbine Research

One idea proposed by the task force, says Howell, is an experiment releasing homing pigeons around wind turbines to study their behavior. Both Kreithen and Walcott are experts on pigeons. The experiment could shed light on how raptors are being killed by the turbines. Kreithen's team is studying collision avoidance behavior. With the exception of an observation made by Sheila Byrne, a biologist at Pacific Gas & Electric, and one made by Cornell's Kreithen during his pigeon releases, few have seen birds of any kind strike a wind turbine. Ornithologists can only speculate on what happens as birds fly near the turbine—thus the need for direct observations. "It's a tricky business to be a fast-flying animal at low altitude," says Kreithen. "They make mistakes," he adds. His goal is to find out why birds, particularly raptors, make such mistakes around wind turbines.

Releasing trained pigeons is also a necessary control for later judging the success—or failure—of proposed mitigation measures, says Howell. "The pigeons are the bellwether species," explains Kreithen, as "they assure the safety of the more valuable birds." Eventually, the task force will also fly trained raptors around the turbines.[46]

The Raptor Research Center in Boise, Idaho will study the visual acuity of raptors and their ability to recognize objects, shapes, and colors. Despite the old adage of an "eagle eye," biologists do not know how well eagles really see. The Center will also study avoidance behavior. Raptors may see the turbines, for example, but they may choose to ignore the warning signs. "Overall, the numbers of birds striking the turbines are few compared to all the birds that are out there," says Cade, "but why do those few individuals make mistakes and get killed?"[47]

Meanwhile, biologists will continue the field work begun by BioSystems in the Altamont Pass. They will perform necropsies on the carcasses they find to try to determine how the birds were killed, including the measurement of any residual poisons that may predispose birds to collisions. Poisons intended for ground squirrels or coyotes could become concentrated as they move through the food chain. In the Altamont, raptors perch atop the food pyramid and could suffer from eating prey tainted with poison or lead pellets from gunshots.

Ultimately, the task force will study how to enhance the natural avoidance characteristics of raptors. Noise could be useful, not in scaring a bird away but in making it more aware of the turbine's presence. Birds, like other animals, quickly habituate to measures intended to startle them. As frustrated fruit growers have discovered, gunshots are ineffective over the long term in driving away hungry birds. But biologists say that other auditory or visual stimuli may provide the recognition and response desired. The noise could be in the nonaudible range, like that of silent dog whistles, so as not to annoy neighbors. Similarly, there may be visual patterns or colors that birds recognize but humans ignore.

With the creation of the task force and its study of Solano County as well as its own studies in the Altamont Pass, U.S. Windpower has visibly taken the lead on this issue. Through 1993, USW spent more than $1 million on its aggressive program to find solutions in the Altamont Pass. Darryl Gray, Alameda county's land-use planner responsible for wind energy, emphasizes that "some [companies] are doing more than their share to solve the problem, not just the minimum to get by."

U.S. Windpower's immediate concern is the collison of raptors, specifically golden eagles, in the Altamont Pass. But as scientists, the task force wants to derive answers or solutions that will apply broadly to different places, species, and types of wind turbines. The task force could find that there are no simple solutions and that wind plants will continue to kill some number of birds per year, including golden eagles. If this is the case, California wind companies will violate the letter of the law. There are ample examples, however, of wildlife agencies accepting "off-site mitigation" as compensation. To protect the species elsewhere, the violator donates suitable habitat to the responsible wildlife agency in some proportion to that removed from the species' former territory. Likened by critics as "wildlife extortion," and frowned upon by purists such as Fish and Wildlife's Struzik, off-site mitigation is an alternative.

Fortunately, no rare or endangered birds such as bald eagles, peregrine falcons, or California condors are known to have been killed by wind turbines or their power lines anywhere in California. In an understatement, BioSystems' Sue Orloff warns that "the death of even one condor would be a significant event."[48] The mere threat of injury to the condor killed the Gorman wind project and could forestall others.

Struzik Strikes

As part of its raptor studies, U.S. Windpower proposed replacing 160 of its old model 56–100 turbines with half as many of its new, larger turbines. They argued that studies should begin immediately on the next generation of wind turbines, the ones likely to replace those now in the Altamont Pass. This was the opportunity Fish and Wildlife's Struzik had been waiting for; she objected to Alameda County that this was merely a ruse for doubling the capacity of the site. If USW wants to experiment, that is acceptable, as long as there is no net increase in the hazard to golden eagles, that is, as long as there is no net increase in the wind capacity of the site.

In an unusual turn of events, Sierra Club California sprang to U.S. Windpower's defense. "The [Altamont] study took place during a period of severe drought," said Rich Ferguson in testimony to the county, "thus the deaths may or may not have been typical." The species killed are not under stress in California or the western United States, added Ferguson, although local populations may be affected. True, "raptor mortality at Altamont is unacceptably high," and for wind energy to fulfill its potential in the United States, "the problem must be solved quickly." Nevertheless, said Ferguson, the project should be approved.[49]

Despite Struzik's stridency, Ferguson said that shutting the turbines off was unwarranted. "Although the negative impacts are significant, it is also clear that wind energy has distinct benefits to our society," he said in his testimony. Ferguson specifically challenged Struzik's effort to limit the number of new turbines. Reducing the number would do more harm than good, he argued. Reducing the number of new turbines installed would double the time necessary to produce statistically significant results. Struzik's position would, Ferguson conjectured, kill 400 more raptors per year during 15 years of operation than under the original proposal.[50] Nevertheless, Struzik prevailed.

The Consequences for Wind Energy

Most observers believe that particularly with raptors, the problem is unique to the Altamont and sites on major flyways. Sites outside California may have conditions far different than those in the Golden State. In the Midwest, for example, waterfowl migrate along the principal watercourses. Wind plants will tend toward upland locations with the highest elevations, such as atop Buffalo Ridge in southwestern Minnesota, avoiding most conflicts. But the American public perceives that the problem is more widespread, and perception is reality in politics.

Wind energy's chief attribute is its environmental benefits. When opposing Zond's Gorman project in 1989 on grounds that it could kill condors, even the National Audubon Society, reiterated their long-standing support of wind-energy because of wind's environmental benefits.[51] Without these benefits, wind is just another energy technology to be hindered at every turn. The fact that wind power plants can kill birds could turn widespread environmental support of wind energy into hostility literally overnight. Many in the environmental community are sensitive to the dilemma. Don Aitken, a senior scientist with the Union of Concerned Scientists and a vocal proponent of renewables, advises the industry to face the problem squarely and handle it with sensitivity.

Yet no wind company other than U.S. Windpower has contributed directly to solving the problem. (All have willingly cooperated with researchers, notes Alameda County's Darryl Gray). Nearly half of the wind turbines in the Altamont Pass are operated by other companies, so this concerns Audubon and some biologists. If no solution is found quickly, Audubon will call for a moratorium on all projects in migration corridors as a means of pressuring the industry for action.[52] To Audubon, the definition of a migration corridor can be quite broad.

Opponents have been quick to seize upon the issue. According to a Dutch study for the European Community, Sweden's "nuclear lobby" has begun using the bird issue to discredit wind energy.[53] Groups as diverse as Montana ranchers and the West Virginia Coal Association have publicly aligned themselves with opponents of local wind projects on grounds that wind turbines kill birds. If unresolved, the issue could eventually pit one environmental group against another. Conceivably, the Union of Concerned Scientists could stand in one corner of the ring, urging sensitive siting, while Audubon could stand in another, urging a moratorium on all new development and dismantlement of existing plants.

Wind companies must avoid the fortress mentality that these possibilities create. To the dismay of environmentalists and industry leaders alike, some companies are responding by trying to control the damage instead of trying to solve the problem. As Exxon found with the *Valdez,* this "damage control" may cause as much damage to the company's interests as the disaster itself.

The realization that wind turbines or their ancillary structures have killed birds, including raptors, pains wind energy advocates. The goal of wind energy's proponents is to provide a sustainable source of electricity in as environmentally benign a manner as possible. UCS's Aitken says that the number of birds killed so far is extremely small in relation to the number killed by other human activities that we have come to accept. Conservationists interviewed in Sweden expressed continued support for wind energy despite the number of birds killed. Ralph Cavanagh of the Natural Resource Defense Council maintains that wind's overall impact on birds is minor but that more research to resolve the dilemma is urgently needed.[54]

Although the overall impact may be slight, the fact that there is an impact at all illustrates, once again, that there are costs to all energy choices. Kevin Cousineau, an electronics engineer for Zond Systems, asks rhetorically: "Have you ever watched TV, listened to the radio, made a long-distance telephone call, or used an electric appliance?" Certainly, everyone has. "If so, then you too have contributed to killing birds."

Some birds, including eagles, will continue to fly into wind turbines, regardless of mitigation measures, just as they fly into power lines and other structures. An unpleasant thought, yes—but unavoidable. Those who think otherwise are deluding themselves. "Zero kill?" says Tom Cade of the Peregrine Fund. "That's not ever going to happen."[55]

Chapter 9 Endnotes

1. WIMP, Phase II, "Wind Implementation Monitoring Program," Tierra Madre Consultants, Riverside County, Riverside, CA, August 1986, pp. C23–C24.
2. Ibid., p. C11.
3. James A. Estep, "Avian Mortality at Large Wind Energy Facilities in California: Identification of a Problem" California Energy Commission, Sacramento, CA, 1989, pp. 10–12.
4. Ibid., pp. xi, 9, xii.
5. Robert Haussler, "Avian Mortality at Wind Turbine Facilities in California" California Energy Commission, Sacramento, CA, 1988.
6. See Alston Chase, *Playing God in Yellowstone: The Destruction of America's First National Park* (San Diego: Harcourt Brace Jovanovich, 1986), p. 128. The Fish and Wildlife Service is the direct descendent of the U.S. Biological Survey, whose principal function was "predator control." It was not uncommon in the western United States through the mid-1970s for ranchers to kill raptors, including eagles, under the guise of controlling predators suspected of preying on their livestock.
7. Theresa Tamkins, "Tilting at Windpower," *Audubon,* September/October 1993, pp. 24–27.
8. Susan Orloff and Anne Flannery, "Wind Turbine Effects on Avian Activity, Habitat Use, and Mortality in Altamont Pass and Solano County Wind Resource Areas" California Energy Commission, Sacramento, CA, March 1992.
9. J. E. Winkelman, "Nachtelijke Aanvaringskansen voor Vogels in de SEP-Proefwindcentrale te Oosterbierum" Arnhem, The Netherlands: Rijksinstituut voor Natuurbeheer, 1990, pp. 127–130; and J. E. Winkelman, "De Invloed van de SEP-Proefwindcentrale te Oosterbierum: Nachtelijke Aanvaringskansen" Institut voor Bos en Naturronderzoek, Arnhem, The Netherlands, 1990, pp. 118–120.
10. J. E. Winkelman, "De Invloed van de SEP-Proefwindcentrale te Oosterbierum op Vogels: 4. Verstoring" Institut voor Bos en Natuuronderzoek, Arnhem, The Netherlands, 1990, pp. 103–106.
11. J. H. B. Benner, J. C. Berkhuizen, et al., "Impact of Wind Turbines on Birdlife: An Overview of Existing Data and Lacks in Knowledge in Order of the European Community" Consultants on Energy and the Environment (CEA), Rotterdam, The Netherlands, July 1992, pp. 22–23.
12. Ibid., p. 1.
13. J. C. Berkuizen and A. D. Postma, "Impact of Wind Turbines on Birdlife," Consultants on Energy and the Environment, paper presented at the European Wind Energy Association's special topic conference, "The Potential of Windfarms," Herning, Denmark, September 8–11, 1992.
14. Michael Brinch Pedersen and Erik Poulsen, "Impact of a 90m/2MW Wind Turbine on Birds: Avian Responses to the implementation of the Tjæreborg Wind Turbine at the Danish Wadden Sea," Danmarks Miljoeuindersoegelser, Afdeling for Flora og Faunaoekologi, Rønde, Denmark, 1991, English summary, pp. 34–36.
15. Niels Walter Moeller and Erik Poulsen, *Vindmøller og Fugle* (Rønde, Denmark: Vildtbiologisk Station, 1984), p. 37.
16. B. Pedersen and H. Nohr, "Consequences of Minor Windmills for Bird Fauna," Ornis Consult, Denmark, June 1989, English summary.
17. Anthony Luke, Alicia Hosmer, and Lyn Harrison, "Bird Deaths Prompt Rethink on Wind Farming in Spain," *Windpower Monthly,* 10:2, February 1994, pp. 14–16.
18. Gottfried Vauk et al., "Bioligisch-ökologische Begleitunterschungen zum Bau und

Betrieb von Windkraftanlagen" Norddeutsche Naturschutzakademie, Schneverdingen, Germany, 1990.
19. Sharron Rogers, Barney Cornaby, et al., "Environmental Studies Related to the Operation of Wind Energy Conversion Systems" U.S. Department of Energy, Washington, DC, December 1977, pp. xiii, 47.
20. WIMP, Phase II, "Wind Implementation Monitoring Program," pp. C2–C3.
21. Ibid.
22. M. D. McCrary et al., "Summary of Southern California Edison's Bird Monitoring Studies in the San Gorgonio Pass," unpublished, Rosemead, CA, 1993.
23. Judd Howell and Jennifer Noone, "Examination of Avian Use and Mortality at a U.S. Windpower, Wind Energy Development Site, Montezuma Hills, Solano County, California," Solano County, Fairfield, CA, September 10, 1992, p. 15. At the time of the study, Howell also worked for the National Park Service's Golden Gate National Recreation Area.
24. Ibid., p. 23.
25. Orloff and Flannery, "Wind Turbine Effects," p. xi.
26. Susan Orloff, "Tehachapi Wind Resource Area Avian Collision Baseline Study," Biosystems Analysis, Tiburon, CA, January 1992.
27. Estep, "Avian Mortality," p. 10.
28. Orloff, "Tehachapi Wind Resource Area," pp. 4–6.
29. Diane Mitchell, "SeaWest Raptor Monitoring Program: SeaWest 90 Project," Kern County Building Services, Bakersfield, CA, April 5, 1993.
30. Biologists consider the number of turbines surveyed sufficient to characterize the entire group. Field observations are extremely labor-intensive and it is difficult to survey an entire wind plant site.
31. Rogers, et al., "Environmental Studies," p. 49.
32. Orloff and Flannery, "Wind Turbine Effects," pp. 1–4.
33. Biologist John Piatt, who counted 37,000 actual carcasses after the spill, estimates that 375,000 birds were killed, based on his research and that of others. Michael Fry, a consultant to the oil spill trustees, estimates that more than 500,000 birds may have been killed. John Piatt, "The Oil Spill and Seabirds: Three Years Later," *Alaska's Wildlife,* 25:1, January/February 1993, p. 11; Michael Fry, *Exxon Valdez Oil Spill Symposium,* Anchorage, AK, February 2–5, 1993, pp. 30–33.
34. Timothy Bowman, "Bald Eagles: After the Spill," *Alaska's Wildlife,* 25:1, January/February 1993, p. 13.
35. John Piatt and Calvin Lensink, "*Exxon Valdez* Bird Toll," *Nature,* 342, December 21–28, 1989 pp. 865–866.
36. Michael McCrary, Robert McKernan, et al., "Avian Mortality at a Solar Energy Power Plant," *Journal of Field Ornithology,* 57:2, Spring 1986, pp. 135–140.
37. Richard R. Olendorf, A. Dean Miller, and Robert N. Lehman, *Suggested Practices for Raptor Protection on Power Lines: The State of the Art in 1981* (St. Paul, MN: Raptor Research Foundation, University of Minnesota, 1981). This pioneering work was being updated in 1994 by biologists at Pacific Gas & Electric and Southern California Edison, among others, who had experience with wind turbines and bird electrocution hazards.
38. Richard Olendorff and Robert Lehman, "Raptor Collisions with Utility Lines: An Analysis Using Subjective Field Observations" Pacific Gas & Electric Co., San Ramon, CA, February 1986.
39. Moeller and Poulsen, "Vindmøller og Fugle," pp. 56–58.
40. "La mort sur le fil: des oiseaux victimes des lines èlectriques," *Le Figaro,* July 1, 1991.

41. Transcript, Avian Research Task Force convened by U.S. Windpower, April 20, 1993, p. 39. Comments by Tom Cade of the Peregrine Fund, Boise, Idaho.
42. Estep, "Avian Mortality," p. 14.
43. Transcript, Avian Research Task Force, p. 39. Comments by Tom Cade of the Peregrine Fund, Boise, Idaho. Initial reports from a study of the golden eagle population in the Altamont Pass indicate that it may have a far larger population than anyone thought possible.
44. Rich Ferguson, Sierra Club California, testimony before Alameda County Planning Commission on U.S. Windpower's proposal to replace some of its 56–100 turbines with 80 larger turbines, March 31, 1993.
45. The Fish and Wildlife Foundation is not affiliated with the U.S. Fish and Wildlife Service.
46. Transcript, Avian Research Task Force, p. 27. Comments by Melvin Kreithen, University of Pittsburgh.
47. Transcript, Avian Research Task Force, p. 12. Comments by Tom Cade of the Peregrine Fund, Boise, Idaho.
48. Orloff, "Tehachapi Wind Resource Area," pp. 4–7.
49. Rich Ferguson, "Birds and Wind Turbines: Can They Coexist?" Coalition Energy News, Coalition for Energy Efficiency and Renewable Technologies, Spring 1993, pp. 9–10.
50. Ferguson, testimony before Alameda County Planning Commission.
51. Linda Blum, "National Audubon Cites Threat to Condor and Other Raptors in Opposing Wind Farm Location," press release, National Audubon Society, New York, July 14, 1989.
52. Tamkins, "Tilting at Windpower."
53. Benner et al., "Impact of Wind Turbines on Birdlife" pp. 73–75.
54. Ibid.
55. Transcript, Avian Research Task Force, p. 49. Comments by Tom Cade of the Peregrine Fund, Boise, Idaho.

10

Impact on People

Public Safety and Setbacks

"Danger—Wind Turbines" proclaim signs prominently posted every several hundred feet along the perimeter of California's wind plants. They make great fodder for opponents of wind energy, who like to point to them and say "See, we told you they were dangerous, malicious monsters devouring our landscape."[1] The signs imply a degree of risk that is nonexistent; they are an outgrowth of California's litigious legal environment, and nothing more. They respond to county regulations governing what planning officers and lawyers describe as "attractive nuisances," anything that attracts trespassers, who may subsequently injure themselves. Europeans find the concept hard to grasp: that someone who climbs to the top of a tower and injures himself in the process has a legal claim against the owner. Europeans still have the innocent view that people are at least partially responsible for their own actions.

Despite the signs, and despite the rare catastrophic destruction of a wind turbine, no member of the public has ever been injured by a wind turbine. To Swiss analyst Andrew Fritzsche, the operation of wind turbines is "practically risk-free for the public." In contrast to other energy sources, renewables "have practically no potential for severe accidents" that would endanger the public.[2]

An incident in the spring of 1981 illustrates the difference between energy technologies, regarding risk to the public. California's Governor Jerry Brown planned to launch what would become the world's largest wind industry at a conference in Palm Springs. The featured event of the conference was Alcoa's demonstration of its new Darrieus wind turbine, operating at Southern California Edison's nearby test center. Just prior to the conference, a malfunction in the turbine's controller led to the turbine's destruction. Grim-faced but retaining his sense of humor, Alcoa's program manager, Paul Vosburgh, addressed the gathered dignitaries and announced: "I have some good news and some bad news. The bad news is that our wind turbine has destroyed itself. However, the good news is that we did not have to evacuate Los Angeles."

His reminder of the 1979 nuclear accident at Three Mile Island near Harrisburg skillfully turned attention from the pubic relations fiasco to wind's promise as a more benign technology.

During tumultuous public hearings in Tehachapi in the early 1980s, some fearful residents charged that the unreliable wind turbines then in use would throw parts long distances, thus endangering them and their families. Wind turbines, notably during the technology's early days, have indeed self-destructed, and falling components have damaged trucks, trailers, and even nearby wind turbines. Some machines, such as the Storm Master, have thrown parts several hundred meters. In Denmark, where the turbines are often located close to inhabited buildings, broken blades have landed near the homes of their owners. Yet no one has been injured, either in Denmark or in California, and damage has been limited to the area near the turbine. After examining several wind turbine failures in Europe and the United States, Alexi Clarke concluded that the risk of being hit within 210 m (700 ft) of a wind turbine is comparable to the risk of dying from a lightning strike; beyond 210 m, the risk is even lower.[3]

Setbacks from habitation of 150 to 300 m (500 to 1000 ft), sufficient to reduce the noise and visual impacts of commercial wind plants, are more than ample to protect public safety. Noise setbacks are derived from projections of noise impact and can vary with the noise strength and the number of wind turbines proposed. The noise setback from a single low-noise turbine falls at the lower end of the range. Noisier machines require a greater separation distance. Such setbacks are unwarranted on safety grounds alone.

In a study for Dyfed County, Wales, Derek Taylor concluded that setbacks of 120 to 170 m (400 to 600 ft) were sufficient to protect habitations, and that only 50 to 100 m (150 to 350 ft) was needed to protect the public using nearby roadways. To minimize complaints from neighbors about noise and shadow flicker, Taylor recommends a 300-m setback from nearby dwellings, twice that needed to protect public safety.[4]

The perceived risk from wind turbines has led some communities, mostly in California, to establish setbacks or buffer zones between the wind turbines and their neighbors (Table 10.1). The need for these buffer zones to separate wind turbines from homes and public facilities remains controversial, even though it has been 15 years since the question was first raised. Intended to compensate for the involuntary risk of real or imagined hazards that are borne by the public, setbacks impose a standard on wind turbines not applied to all other technologies or comparable land uses.

In California, setbacks are sometimes given in multiples of the wind turbine's height. For example, in Kern County, which administers California's Tehachapi Pass, wind turbines must be set back from roads and trails a distance of $1\frac{1}{2}$ times the total height of the turbine (hub height plus rotor radius). For a 40-m turbine on a 40-m tower, the setback is 90 m (300 ft). In Palm Springs, where local officials have been less receptive to wind energy than in Kern County, setbacks are greater. The buffer zone of six times the total height

Table 10.1
Safety Setbacks

	Kern County		*Palm Springs*		*Taylor & Rand (UK)*	
	ft	m	ft	m	ft	m
Off-site residences	1000	305	6 × height		400–600	120–170
On-site residences	$1\frac{1}{2}$ height					
Roads and trails	$1\frac{1}{2}$ height		500	150	150–350	50–100

Note: Kern County reduces setback from off-site residences to $1\frac{1}{2}$ times the height of the wind turbine, with the written permission of the owner. Palm Springs applies setback to the lot line. If the lot is undeveloped, the setback is reduced to three times the height.

from a parcel with a habitation prohibits the installation of the 40-m turbine in the example closer than 360 m (1200 ft) from the lot line.

Safety setbacks are unnecessary in some cases, such as with individual wind turbines, and should never impinge on those people who voluntarily choose to accept the risk: those who want to install wind turbines near their own homes or who grant easements for others to do so. If the setbacks in Table 10.1 had been applied to individual wind turbines, not just to California wind plants, most of the wind turbines in Denmark and nearly all of the 5000 small wind turbines in the United States would never have been installed. These machines were deliberately located by their owners near their homes or businesses for their own benefit, and few had to comply with setback standards such as those in Table 10.1.

In the United States, safety buffers from some individual turbines at suburban locations have been required. Often, this is equivalent to the height of the turbine, so that if the tower collapsed, it would not fall across the property line.

Planning officers must have sufficient flexibility to exclude individual wind turbines or small clusters of turbines from restrictive setback requirements; otherwise, they discriminate against wind energy compared with other technologies. Wind turbines are no more dangerous to their owners, or to the public, than other sources of energy that course through our homes and workplaces every day. Natural gas, fuel oil, and electricity all impose risks that we willingly accept. We also willingly accept the public risk imposed by our neighbors' use of these technologies. The same standards should be applied to wind energy: nothing more, nothing less.

Setbacks on aesthetic grounds may be appropriate for commercial wind power plants comprising tens of machines. For example, Palm Springs requires a scenic setback of 500 to 3500 ft (150 to 1000 m) from designated roads leading into the resort city nestled at the base of Mount San Jacinto. But the impact of hundreds of wind turbines on the landscape differs markedly from that of a single turbine. Most landscapes can easily assimilate a few scattered wind turbines. Thus setbacks for single wind turbines on grounds of either safety or aesthetics are inappropriate.

Europeans are more tolerant and less fearful than Americans of wind turbines in their communities, possibly because of their prior experience with the technology. In at least three officially sanctioned settings in the Netherlands, visitors can stroll unhindered among operating windmills: at Arnheim's open air museum, at the seventeenth-century "windfarm" of Kinderdijk, and at the re-created Dutch village at Zaanse Schans. In each case the windmills are fully accessible to visitors. Tourists can also visit hundreds of other operating windmills throughout the country.

Modern wind turbines are equally accessible. In 1992 a restaurant on the autobahn near Aachen installed a 150-kW wind machine; a McDonald's restaurant near the north German port of Kiel installed a Ventis wind turbine (dubbed "McVentis" by local journalists) in its parking lot. It is the same across northern Europe. An 80-kW Lagerwey turbine operates prominently at a lock on the northern end of the Afsluitdijk, the dam closing the former Zuider Zee. The turbine stands within 50 m (150 ft) of the heavily traveled route, in plain view of traffic which frequently stops for the lock. Nearby, several Lagerweys spin merrily near the parking lot of a restaurant just off the motorway. In Montabaur, provincial capital of Germany's Westerwald, a small wind turbine whirring away on a rooftop in the city center stirs little interest from pedestrians on the street below. These and countless examples in Denmark prove that the public need not fear wind turbines and that setbacks are sometimes inappropriate.

Occupational Safety

The capture and concentration of energy—in any form—is inherently dangerous. Wind energy exposes those who work with it to hazards similar to those encountered in heavy construction, oil extraction, and the electric power industry. It also exposes workers to hazards unique to wind energy: high winds, heights, rotating machinery, and the large spinning mass of the wind turbine rotor. Like its aesthetic impacts, wind energy's hazards are apparent to anyone who wishes to see them. There are no latent diseases, no black lung, no radiation-induced cancers. When wind kills, it does so directly and with gruesome effect.

Death in the maw of a wind machine is not new. There is at least one historical account of a miller becoming ensnared in his own machinery. In *The Story of Sprowston Mill,* H. C. Harrison recounts how his great-grandfather, Robert Robertson, was killed in 1842 after becoming entangled in the sack hoist.[5]

Since its rebirth in the 1970s, wind energy has killed 14 men worldwide directly or indirectly and seriously injured or crippled three men and one woman in the United States. Most have died during construction or construction-related accidents. Five died during operation or maintenance of the turbines. These accidents are not simply statistics. They happened to living, breathing human beings who left behind grieving families.

The first was Tim McCartney, who fell to his death near Conrad, Montana, in the early 1980s while trying to salvage a 1930s-era windcharger. There are few details on McCartney's death other than that his broken body was found near the tower. News reports said simply that he fell during high winds. McCartney was followed by Terry Mehrkam, a pioneering Pennsylvania designer and manufacturer of wind turbines. Mehrkam was killed in late 1981 near Boulevard, California. A few years later, a man was crushed to death unloading a container of tubular towers in the Altamont Pass. Canadian Eric Wright rode an experimental Darrieus wind turbine to his death when it fell over during installation near Palm Springs in 1983. Not far away in the following year, Art Gomez was killed while servicing a crane.

There was a long hiatus until a series of accidents struck during the late 1980s and early 1990s. The simple medical description of John Donnelly's death found in the files of his company's insurer fails to describe the horror of his fate. Death by "multiple amputations" sanitizes a truly grisly accident, a nightmare witnessed by his partner, who watched helplessly as Donnelly was drawn inexorably into the nacelle's slowly spinning machinery. What made Donnelly's accident even more terrifying for windsmiths throughout California is how it happened: Donnelly's safety lanyard, a device designed to prevent injury, became entangled on the revolving main shaft and dragged him to his death.

Not long after Donnelly's accident near Palm Springs, Dutch homeowner Dick Hozeman was killed in the same manner. Against professional advice, he climbed to the nacelle of his turbine in a vain attempt to stop the runaway rotor from spinning in a violent winter storm. Tragically, the turbine had been inoperative for two years. After squeezing into the small nacelle Hoseman got caught on the turning main shaft. Rescue crews retrieved his body the next day, after the wind subsided.

The same year, two men were killed in a single accident on the Danish island of Lolland when the rotor they were servicing unexpectedly began to turn. A third man dangled from his lanyard 30 m (100 ft) above ground until rescued. Then in 1991, Thomas Swan, a crane operator, was electrocuted near Tehachapi when the boom on his crane snared a 66,000-volt power line. Richard Zawlocki fell to his death in 1992 while descending a ladder inside a tubular tower near Palm Springs. Robert Skarski died in 1993 while installing a small wind turbine. Most recently, Mark (Eddie) Ketterling was killed after a chunk of ice knocked him off the interior ladder of a tubular tower in Minnesota.

These deaths should serve to warn the industry. managers and staff alike, that hazards exist and that they are deadly. Yet these men need not have died in vain if the industry can learn from them how to prevent such accidents in the future. Industry practice, for example, and what some would argue to be common sense, suggest that McCartney, Mehrkam, and Zawlocki all made the same fatal mistake: they did not use a work belt and lanyard, which is called in the jargon of the trade a fall-restraint system. The lanyard connects the work belt (or body harness) to a sturdy attachment on the nacelle or tower. The nylon lanyard, kept as short as possible, prevents the windsmith from striking the ground after a fall. Falling from the tower is the single most apparent occupational hazard of working with wind energy.[6]

Little is known about how McCartney died. He obviously was not using a safety belt and lanyard. But Mehrkam's death was investigated by California's Department of Occupational Safety and Health. They concluded that Mehrkam climbed to the top of the tower without a safety belt and either fell or was thrown off the tower to his death. Unlike Mehrkam, Zawlocki was wearing his work belt when he fell while descending the ladder. However, he was not using his lanyard. It was later found atop the tower attached to the nacelle cover, holding the cover open. Like their workmate, the two Danes killed on Lolland might have survived had they been attached securely to the top of the tower.

Zawlocki's death is more troublesome for what it says about the human factor in all accidents. It is evident, that he was aware of the risk of working on the tower because he was wearing his work belt. But he failed to use it—we will never know why. We do know that when safety equipment is inconvenient or uncomfortable, there is a tendency to avoid using it.

Nearly all medium-sized wind turbines in California and northern Europe include a fall-restraint system designed specifically to prevent the kind of accident that killed Zawlocki. It uses a metal sleeve that slides along a steel cable from the base of the tower to the top. When ascending or descending the tower, windsmiths attach their work belts to this sleeve. Should they slip, the sleeve grips the cable, stopping their fall. The tower Zawlocki was descending had only recently been installed and lacked this cable-and-lock system. As an alternative, technicians were instructed to attach their lanyards to the ladder when climbing the tower. This entails climbing a few rungs, removing, and then reattaching the lanyard. (This is recommended practice for climbing towers of small wind turbines, which seldom use cable and sleeve systems.) However, it is inconvenient, and in combination with complacency, is more honored in the breech than observed in practice.

In the day-to-day operation of a wind plant, crews that have serviced hundreds and possibly thousands of machines without incident become complacent about the risk of falling. After the fear has waned, crews are tempted to climb the towers without attaching their lanyards. Tubular towers may breed more complacency than lattice towers, which are open to the elements. When one is climbing a lattice tower, the sense of danger is more apparent. Tubular towers may create a false sense of security because they are sheltered from the wind and the height above ground is more difficult to discern. It is also more difficult for managers to supervise technicians once the latter are ensconced inside the tower.

On lattice towers, a superviser may simply drive by and quickly determine whether a windsmith is using a safety lanyard. Not so with tubular towers. Managers are dependent on the cooperation of the technicians, who are hidden from view. Observers have noted that when technicians scale a tubular tower with both internal and external ladders (on older Danish turbines, a portion of the tower must be scaled outside to reach the nacelle), they often climb unprotected on the internal ladder, then clip on when switching to the external section that leads to the top of the turbine, where they might be seen. To prevent this, wind companies must constantly inculcate their staff with the absolute necessity of taking precautions, especially against falls.

The second lesson to be learned from these accidents is that no one should ever work atop the nacelle when the wind turbine is spinning. Technicians must lock the rotor in place before servicing components in the nacelle. How this is done varies from turbine to turbine, but it typically involves placing a bolt or pin through a rotating component of the drivetrain. This would have prevented the accident in Denmark, where the brake was inadvertently released, allowing the rotor to begin turning. It would also have prevented John Donnelly's death in Palm Springs and Dick Hoseman's in the Netherlands. During a seemingly calm day, Donnelly climbed the turbine to repair a damaged brake. A slight breeze started the rotor turning, catching Donnelly offguard. By the time he realized that his lanyard was snagged, it was too late. One manufacturer, the Wind Energy Group, considers the rotor locking pin of sufficient importance to prospective users that it notes prominently in its technical specifications the provision for such a pin.

Mehrkam died when he tried to stop the rotor on one of his wind turbines after the brakes failed in high winds. This is the worst scenario imaginable for working atop the tower. The rotor is not simply spinning; in the case of Mehrkam's machine under these conditions, the rotor would have been spinning so fast that it was just a blur. As insane as it seems now, it was Terry's practice with "runaways" such as this to climb the tower and manually brake the rotor to a halt by wedging a crowbar into the brake calipers.[7]

Terry Mehrkam made two mistakes: he climbed the tower, and he did so without a safety belt and lanyard. He must have been frantic to save his machine, because he had used a belt and lanyard in the past. The belt would have restrained his fall, and although he would have been injured, it would have saved his life. It was utter foolhardiness to mount the turbine and straddle the nacelle like Slim Pickins riding a nuclear bomb to its target in the movie *Dr. Strangelove.* He should have used other means to control the rotor, from a safe distance, or he should have simply walked away, cleared the site, and waited for the wind to subside.

Others have also been seriously injured by falls when not using their safety belts. They lived only because they were near the ground when they fell. Some of these men have been permanently crippled.

The other hazards encountered in building and operating a wind power plant are not unique to wind energy. The hazard of working with tubular tower sections is similar to that of working with wood poles in the electric utility industry, and the precautions are similar.

Next to falls, the most serious hazard to windsmiths is working around electricity. The most common form of serious accidents in California wind plants is injuries from electrical burns. In one case, a Tehachapi woman was maimed when she touched energized equipment inside a transformer cabinet. Again, the hazards are similar to those in the electric utility industry, and the precautions developed during the past century for generating and transmitting electricity safely are applicable; if they had been followed, it would have prevented this accident and others like it.

These morbid figures, while alarming, may not be statistically representative. The statistics include several men killed during the technology's formative days that may not accurately reflect what can be expected from a mature industry. Nine of those killed died in construction-related accidents. McCartney was dismantling a turbine, and Mehrkam was trying to rescue a turbine he had recently installed. Swan was driving his crane to move a turbine to a better site, and Zawlocki was completing the reinstallation of a turbine that had been moved from a site in the Altamont pass to Palm Springs. Skarski was installing a small turbine and Ketterling was installing a medium-sized turbine when they were killed.

Construction is not an ongoing activity, and the risk associated with it normally occurs once in the life of the wind turbine. Wind turbines worldwide generated about 6 TWh per year in 1993. During the expected 20-year life of the machines already in place, they will generate a total of 120 TWh. This gives a mortality rate of 0.08 death per terawatt-hour from construction.

Since the first wind turbines were installed in California and Denmark during the early 1980s, five men have died during ongoing maintenance activities. Wind turbines during the period from 1980 to 1994 generated a total of 33 TWh, for a mortality rate of 0.15 death per terawatt-hour. The total mortality rate, admittedly based on scanty data from a young technology, is 0.23 death per terawatt-hour. Two deaths in the single Danish accident skew the data substantially.

The occupational death rate from wind energy is far below that of the accidental death rate of the general population. The general risk to the population of accidental death is 450 deaths per million, half of which are automobile-related.[8] A population of 1 million Americans consumes 2.5 to 5 TWh per year. Thus the rate of accidental death is 450 deaths per 2.5 to 5 TWh per year, or 90 to 180 deaths per terawatt-hour. Nevertheless, the wind industry's occupational death rate is disturbingly similar to that of the coal industry and far higher than expected by most risk analysts.[9]

In a study of the external costs of various energy sources, British researcher David Pearce reports 1.1 occupational deaths per gigawatt-year from coal. Pearce's work is based largely on that of Andrew Fritzsche for the Swiss Federal Energy Office. The latter estimates that the mining, processing, and transport of coal lead to 1.1 to 2 deaths per gigawatt-year, including latent deaths from respiratory diseases such as pneumoconiosis or black lung. This is equivalent to 0.13 to 0.26 death per terawatt-hour. Samuel Morris at Brookhaven National Laboratories estimates that there are 1.3 deaths per gigawatt-year from the average fuel cycle of coal in the United States. There are up to 3.16 occupational deaths per gigawatt-year when coal is mined underground and transported by truck. From these sources the death rate from coal is 0.15 to 0.36 death per terawatt-hour.[10]

This is surprising. In 1989, Fritzsche wrote that wind energy would have a mortality rate similar to that of coal, but for other reasons. He believed that the deaths would accumulate through the manufacture of the materials used to build the wind turbines. There is no evidence to support his contention. Modern wind turbines are less material-intensive than the megawatt-sized wind turbines once believed optimum. But generating electricity with wind energy is more labor-intensive than with other generating sources, and this may be a factor. There may be more exposure to common hazards than that found in other energy industries. Or the data may simply be unrepresentative of a mature industry.

Still, the wind industry, particularly in the United States, must do better. No one should ever fall to his or her death from a wind turbine. Falling is the one hazard that is apparent to everyone all the time. It should always top the list of hazards to avoid.

Unlike environmental standards, safety is strictly regulated, both by insurers, who want to avoid paying accident claims, and by occupational and safety agencies. Managers have made safety consciousness a part of day-to-day business in California's wind plants. There is a designated safety officer in each of California's wind companies. This person is responsible for ensuring compliance with the company's safety practices. In Tehachapi, the Kern Wind Energy Association sponsors a safety committee among its member companies, and participants meet regularly to exchange information and discuss thorny issues such as whether or not windsmiths should work in pairs, as well as more pragmatic questions such as which work belt offers the most protection. SeaWest and Zond Systems have worked with the county fire department to devise emergency procedures for jointly rescuing injured technicians from the cramped confines of wind turbine nacelles. If maintained, these efforts should eventually bear fruit and reduce the rate of injuries and deaths from working with the wind.

Wind energy's principal advantage over coal has always been the reduced risk to public health. This substantial benefit remains unchanged by the data on occupational risks and is discussed in Chapter 12.

Impact on Fire and Police Services

Unlike residential development, wind power plants require few public services. In California, they are a net generator of revenue for the counties in which they are located. Riverside County found during the WIMP studies that wind plants in the San Gorgonio Pass seldom used fire and police services. Further, they found that revenues from property taxes surpassed any costs incurred for these and other social services.[11] Wind plant operators in California do occasionally call on police to apprehend trespassers or vandals, and fires sweeping through California wind farms sometimes require costly control efforts.

Wildfires, earthquakes, and landslides are the triple scourges of life in the Golden State. Wildfire is the most frequent threat to life and property, and the annual fire season is a time of dread. Above-average rainfall in 1992 produced a bountiful crop of grass the following summer. After responding to several wild fires among wind turbines in 1993, Kern County fire captain Greg Black commented that it was "not necessarily a bad year for safety, but a good year for grass."

Wind turbines in arid lands are not immune from the hazard of fire. Indeed, their very existence puts them at risk: they are installed in the midst of the fuel supply, there are often strong winds to fan the flames, and there are many ways in which a wind turbine can ignite a wildfire. Electrical short circuits, an overheated bearing, downed electrical cables, welding splatter from technicians servicing the turbines, or even the catalytic converter on service vehicles can start a conflagration. An overheated generator on an Enertech wind turbine in the Altamont Pass started a wildfire that nearly killed one technician, disabling him for life.

Jurisdictions in California have the right to recover firefighting costs when the fire was due to negligence. An occasional fire from overheated bearings or electrical short circuits is no cause for concern, but frequent fires draw regulators' attention. When a power line blew down on Cameron Ridge in the Tehachapi Pass and started a fire, Kern County firefighters responded. And when it was out, they sent the wind company a bill. They reasoned: "It is a wind farm, after all. The power lines should be able to withstand the wind."

The number of fires in the Altamont Pass doubled during 1992, leading Alameda County to examine fire prevention practices among the area's wind companies. As in Tehachapi, they found that the service roads provided numerous fire breaks and offered ready access for firefighters, but that high winds, lack of water, and the danger from electrical lines made firefighting difficult. They concluded that wind companies needed to monitor their turbines more carefully, ensure that the turbines with pendant power-supply cables were not capable of twisting them off, and make certain that all technicians carry some firefighting equipment and be sufficiently trained to handle minor fires before they get out of control. Most important, firefighters wanted to see more fire breaks around power lines and other trouble spots, including fire breaks on adjoining property.

Wildfires from wind turbines certainly pose no problem in humid climates such as those of the British Isles, northern Europe, and the eastern United States. But they could be of concern in Mediterranean climates—in Spain, southern France, and Greece, for example.

Shadow Flicker

To North Americans, Europeans are inordinately concerned about an unusual phenomenon caused by wind turbines called *shadow flicker*. Most Americans have never heard of it. Shadow flicker may be more of a problem in northern Europe than elsewhere, because of its lattitude and the low angle of the sun in the winter sky, or because of the closer proximity between wind turbines and inhabited buildings than in North America.

Flicker occurs when the blades of the rotor cast shadows that move rapidly across the ground and nearby structures. This can create a disturbance when the shadow falls across occupied buildings, especially when windows open onto a turbine turning in front of the sun.[12] Europeans fret that the flickering shadow could disorient or trigger seizures in the 2% of the population who are epileptic.[13]

Near Flensburg in Schleswig-Holstein, German researchers examined the effect and found that flicker, under worst-case conditions, would affect neighboring residents a total of 100 minutes per year. Under normal circumstances, the turbine in question would produce a flickering shadow only 20 minutes per year. Psychologists are now studying the issue of whether neighbors can reasonably be expected to accept this impact.[14]

There are few recorded occurrences of such concern in the United States, although Ruth Gerath notes that the flickering shadows from the turbines on Cameron Ridge near Tehachapi have startled her horse and those of others in the local equestrian club when riding this section of the Pacific Crest Trail. Except for the flickering shadows, she says that the turbines seem to have no effect on the horses. The shadows simply cause the horses to stop briefly until their riders urge them on.

Noise

Next to aesthetic impact, no aspect of wind energy creates more alarm or more debate than noise. Whether wind turbines are "noisy" is as much a subjective determination as whether wind machines appear "beautiful" or "ugly" on the landscape. However, unlike aesthetics, noise is measurable, and some researchers have noted that because it is measurable, neighbors will "transfer" their concern about wind energy's aesthetic intrusion to the increase in background noise attributable to the wind turbines.[15]

Wind turbines are not silent. They are audible. All wind turbines create unwanted sound, that is, noise. Some do so to a greater degree than others. And the sounds they produce—the swish of blades through the air, the whir of gears inside the transmission, and the hum of the generator—are typically foreign to the rural settings where wind turbines are most often used. These sounds are not physiologically unhealthful; they do not damage hearing, for example. Nor do they interfere with normal activities, such as talking quietly to one's neighbor, any more than do the sounds common in any suburban setting. But the sounds are new, and they are different.

Those who live in the rural settings where wind turbines are suitable do so because they prefer the peaceful lifestyle of the country to that of the city. Long-time residents are accustomed to the relative quiet of rural life. They are familiar with the noises that exist and have learned to live with them or even to find them desirable: the wind in the trees, the chirping of birds, the creaking of a nearby farm windmill, the hum of the neighbor's tractor. Rather than being nuisances, these sounds reinforce the bucolic sensation of living in the country.

The addition of new sounds, which most residents have had little or no part in creating and from which they receive no direct benefit, can be disturbing. No matter how insignificant they may be in a technical sense, these new sounds signify an outsider's intrusion. The effect is magnified when the source, such as a wind turbine, is also highly visible.

Where wind turbines have been seen as an intrusion on an otherwise rural setting, some nearby residents have objected to them on the grounds of their noise impact. "The perception and reaction to noise is related to level of acceptance and personal preference. For example, people often prefer to camp or live near running water or crashing waves" despite the noise, says geographer Martin Pasqualetti. "An intrusion rather than a personal choice produces different opinions about noise."[16] If wind turbines are unwanted for other reasons, such as their impact on the landscape, noise serves as the lightning rod for disaffection.

On a field trip to examine new sites for wind turbines near the Tehachapi Pass, area resident Sandy Hare complained that existing wind turbines on Pajuela Peak were "terribly noisy," even though she lived more than a mile from the site overlooking Sand Canyon. Hare, a resident of the remote canyon and vocal opponent of wind projects moving into the mountains surrounding her rural homestead, said she was "shocked" to hear the turbines, even though they were hidden from her view.

She should not have been. If Hare had listened carefully, she could also have heard the murmur of trucks on Highway 58, the freeway carrying traffic through the Tehachapi Pass, and the rumble of 48 trains per day on the adjoining Southern Pacific tracks. But Hare had no reason to fear that the freeway or the railroad threatened her tranquility. She had long ago learned to tolerate, if not accept, their presence. The wind turbines are different. They are relatively new (those on Pajuela Peak were installed only a decade ago) and they were marching in her direction.

Nor should Hare have been surprised that sounds carry long distances in the mountains. The situation in Sand Canyon is not unusual. The problem is the same in mountainous Wales or in the rolling hills of England's Pennines. Low ambient noise levels in sheltered valleys, and refraction of the turbines' noise emissions—which bends the sound toward the valley floor, particularly under nighttime temperature inversions—enable residents to "hear" the wind turbines, or any other noise source, for great distances. The sound of the distant turbines is discernible, but not "noisy," in technical terms. It would hardly deflect the needle on a sound-level meter.

In Hare's case, hearing the turbines makes her aware of their presence and reminds her that the wind turbines have trespassed on what she views as her private back-country domain. In this situation and others like it, the issue is more one of learning to tolerate a new, but permitted, use of the land than it is one of noise emissions. Hare must learn to coexist peacefully with her new neighbors, just as wind companies must learn to minimize their intrusion onto Hare's peaceful rural setting. Both share a responsibility for becoming good neighbors. In this regard, the wind industry's record in California has been less than sterling.

During the early 1980s, some developers installed a cluster of wind turbines literally across the street from a residential neighborhood in North Palm Springs. Sandy Hare's problem pales in comparison. These early wind turbines were particularly noisy, and their pounding and throbbing in close proximity to the bedroom windows of nearby desert dwellers stirred up a hornet's nest. Led by John Warner, a retiree and capable organizer, the residents' complaints reached the ears of local politicians and nearly brought a halt to further wind development. Warner is one of those rare people who is not content to sit back and simply complain; he went on to spend countless hours working with the American Wind Energy Association's noise committee, trying to derive standards for the young industry. In the process, he hoped to find a resolution to his dilemma and prevent such unpleasant encounters from happening again.

The eventual solution for Warner and his neighbors involves removing the offending machines and replacing them with fewer wind turbines, which although more powerful, are also quieter. The *repowering,* as it is called, will also create a greater buffer between the residents and the machines. These measures will cut the noise to a legally permitted level but will not return the neighborhood to its pre–wind turbine repose. The turbines will remain visible and audible. Because the wind turbines will still be heard, repowering is unacceptable to some residents. To them, the only acceptable solution is complete removal.

In an opinion survey near Palm Springs, Pasqualetti found that two-thirds of those surveyed said noise from the wind turbines did not disturb them; 11%, most of whom live within 2 miles (3 km) of the machines, said that it did. Of those surveyed, 46% thought the noise was no worse than that of the nearby freeway. But 9% claimed that the turbines disturbed their sleep. Of these, two-thirds could see the turbines from their homes; for most of them, reducing the noise from the turbines would not make the machines any more tolerable. "The people living closer would still object," says Pasqualetti, "even if the turbines were less noisy."[17] Thus noise remains an issue for those who will continue to hear the turbines.

Sandy Hare would still find the wind turbines on Pajuela Peak near Tehachapi disturbing, even if they were quieter than those there now. Occasionally, she would still hear them, and that would be sufficient for her to associate them with all the environmental ills befalling the industrialized world.

Acousticians can describe noise at great length in technical terms, such as decibels, that have little meaning for most people. To evaluate just how noisy wind turbines are, each person must listen to them at various times of the day from various positions and make a personal determination. When visitors ask the Kern Wind Energy Association, "Just how noisy are they?", the trade group tells them to judge for themselves and takes them to the nearest clump of machines, and that is just what a reporter for the French solar energy magazine *Systèmes Solaires* decided to do. After hearing of complaints about noise, Jean-Paul Louineau went to England to hear them for himself. Like visitors elsewhere, whether at Delabole in Cornwall or Tehachapi in California, he noticed that the machines were not as noisy as he had been led to believe. It was hard for him to understand what the fuss was about.[18]

"Considering that 20,000 wind turbines have been installed worldwide," say industry analysts at the Open University in England, "the number of complaints about noise has been very small."[19] Of the 3500 turbines in Denmark, less than 2% have caused noise complaints.[20] Nearly all of these are less than 225 m (700 ft) from the complaining neighbor. Only one wind plant, at Kyndby, has encountered serious noise problems that required extensive mitigation.

Noise from an 80-kW Lagerwey turbine does not bother Dutch farmer Lolle Hylkema, even though the turbine is close to his house. Bridgett Gubbins quotes him as explaining, "We say in Friesland, your own pigs don't smell."[21] Countless similar tales can be told by older farmers of the American Great Plains, who grew up with creaking water-pumping windmills outside their bedroom windows. Yet six pages of the 26-page British wind energy planning guide are devoted to noise.[22]

Noise will always haunt wind energy, as do aesthetics, because the costs and benefits are distributed unequally. Not everyone can have or would even want a wind turbine in his or her backyard like Dutch farmer Hylkema. For some, exposure to wind turbine noise is involuntary; and they may well wish that it would simply go away.

Nevertheless, some associate wind energy with tranquility. Honda launched its new Concerto automobile in England by directly associating its "1.6 litres of quiet power" with California wind turbines, noting that its car was "capable of generating power without noise."

Noise Principles

Noise is measured in decibels (dB). The decibel scale spans the range from the threshold of hearing to the threshold of pain. Further, the scale is logarithmic (not linear), crudely attempting to match how human hearing responds to sound. Therein lies a serious problem. The logarithmic or exponential relationship between levels on the decibel scale causes more confusion about noise than any other aspect. Doubling the power of the noise source—for example, by installing two wind turbines instead of one—increases the noise level only 3 dB, the smallest change between levels that most people can detect. Doubling the noise content or acoustic energy from a group of wind turbines may thus appear horrific on paper, yet be barely perceptible to nearby residents. Increasing acoustic energy by 26%, for example, raises the noise level only 1 dB, an insignificant amount relative to human hearing. Tripling acoustic energy increases sound level 5 dB, an increase that is clearly noticeable. It takes 10 times the acoustic energy to raise the noise level 10 dB and double its intensity—sound twice as loud.[23]

For most discrete sources such as wind machines, the distance from the listener is just as important as the noise level of the source (Table 10.2). Whenever noise is presented as sound pressure levels, the location is always specified, or implied, because sound levels decrease with increasing distance. For example, Danish wind turbine manufacturers estimate that the noise from a typical medium-sized wind turbine will drop to 45 dB(A) within 150 m

Table 10.2
Typical Sound Pressure Levels

Source	Distance from the Source		SPL
	ft	m	dB(A)
Threshold of pain			140
Ship siren	100	30	130
Jet engine	200	61	120
Jackhammer			100
Inside sports car			80
Freight train	100	30	70
Vacuum cleaner	10	3	70
Freeway	100	30	70
Small (10-kW) wind turbine	120	37	57
Large transformer	200	61	55
Small (10-kW) wind turbine	323	100	55
Wind in trees	40	12	55
Light traffic	100	30	50
Average home			50
300-kW wind turbine	400	150	45
USW 56-100	800	243	45
30- to 300-kW wind turbines	1800	550	45
Soft whisper	5	2	30
Sound studio/quiet bedroom			20
Threshold of hearing			0

Source: Handbook of Noise Measurement, General Radio; Bergey Windpower Co.; and U.S. Windpower.

(500 ft) (Figure 10.1). The aggregate noise from a small wind power plant of 30 such turbines will drop to 45 dB(A) within 500 m (1800 ft).

The noise footprint of a wind turbine is not perfectly symmetrical: Noise is greatest downwind of the tower (Figure 10.2). The contours of equal loudness drawn by acoustic engineers around a wind turbine in noise assessments seldom resemble ever-expanding concentric circles.

The perceived loudness varies not only with the sound level but also with the frequency, or pitch. Human hearing detects high-pitched sounds more readily than it hears those low in pitch. The sound of a complex machine such as a wind turbine is composed of sounds from many sources, including the swoosh of the wind over the blades and the whir of the generator. Each source has a characteristic pitch, giving the composite sound a characteristic tonal quality. When measuring noise, we try to take into account the way the human ear perceives pitch by using a scale weighted for those frequencies we hear best. The A scale is most commonly used. This scale ignores inaudible frequencies and emphasizes those that are most noticeable.

Impulsive sounds, those that rise sharply and fall just as quickly, like a sonic boom, elicit a greater response than sounds at a constant level over time. Wind machines using two blades spinning downwind of the tower emit a characteristic "whop-whop" as the blades pass through the turbulent wake behind the tower. This impulsive sound and its effect on those nearby may be missed by standard A-weighted measurements. Many of the complaints about wind turbine noise near Palm Springs have been directed at the impulsive noise from two-bladed, downwind turbines, requiring special measurements of low-frequency sounds.[24]

Thus noise containing pure tones or impulsive sounds is perceived as being louder than broadband noise. Broadband noise, such as the aerodynamic noise from the wind rushing over a turbine's blades, is composed of sounds across

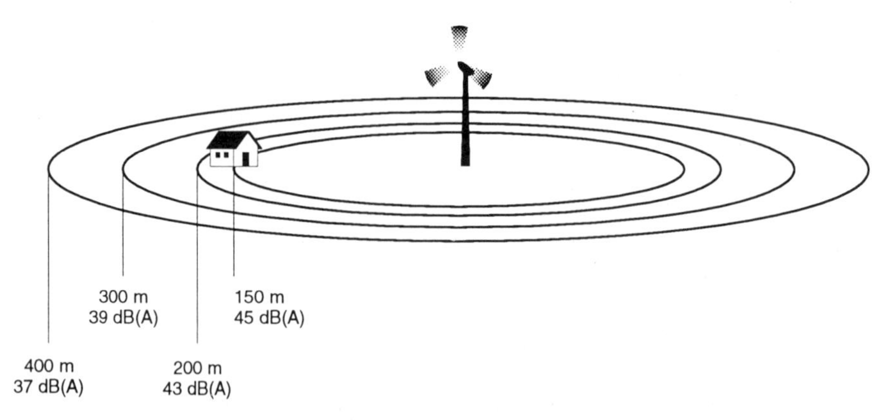

Figure 10.1. Contours of equal noise level around a hypothetical Danish wind turbine. Noise levels decrease with increasing distance. (Courtesy of Foreningen Danske Vindmøllefabrikanter, Copenhagen, Denmark.)

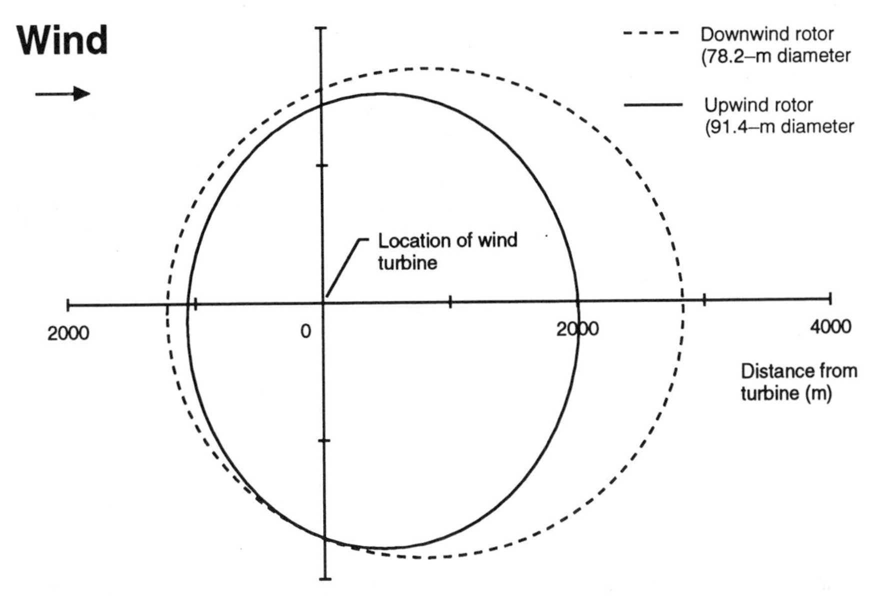

***Figure 10.2.** Noise footprint. NASA measurements of noise threshold from large wind turbines. The contours of equal sound level are typically elongated downwind of the turbines, more so for downwind rotors.*

the spectrum of human frequency response. It is less intrusive than either impulsive noise or noise with distinct tonal components.

Another component of noise is time. Simpler noise ordinances specify a maximum noise level that must not be exceeded. Others weigh the amount of time the noise occurs at various levels. This complicates the task of estimating a wind turbine's noise impact. Unlike the noise of trains or airplanes, which emit high levels infrequently throughout the day, a wind turbine may emit far less noise but do so continuously for days on end. Some find this trait of wind energy noise more annoying than any other. In windy regions, the sound may appear incessant. The literature of life on the Great Plains is full of references to the ever-present sound of the wind. In a classic 1928 film, the oppressive wind drives silent-screen star Lillian Gish mad in her role as a sod-busting pioneer. Appropriately, the film was titled simply *The Wind.*

This time weighting is expressed as the *noise exceedance level,* the amount of time the noise exceeds a specified value. For example, L_{10} is the noise level exceeded 10% of the time; L_{50} is the average noise level, meaning that half the time the noise is greater, half the time less; L_{90}, the noise level exceeded 90% of the time; and L_{eq}, the continuous sound pressure level which gives the same energy as a varying sound level. A noise standard of 45 dB(A) L_{90} is stricter than a standard of L_{10}, because 90% of the time the noise must be below 45 dB(A) instead of only 10% of the time. Highway noise standards in the United States are often in L_{10}, because the noise is intermittent. In contrast, wind turbine noise is nearly continuous when the turbine is in operation. A wind turbine at a good site may emit noise as much as 6000 hours per year.

Because of the logarithmic scale, noise levels are not simply additive. If two wind turbines producing 45 dB(A) are installed side by side, which doubles the acoustic energy, the decibel sum is 3 dB greater than the highest level, or 45 dB(A) + 3 dB(A) = 48 dB(A) (Table 10.3). However, if one of the wind turbines is 6 dB noisier, say 51 dB(A), the sum is 51 dB(A) + 1 dB(A) = 52 dB(A). If two noise levels are more than 15 dB apart—for example, the fall of a pin onto a noisy street—the lower level has a negligible effect on the total. The noise of the street masks the sound of the falling pin.

Noise Propagation

Noise levels decrease with increasing distance as the sound propagates away from the source. Under ideal conditions, sound radiates spherically from a point source, such as a helicopter, and for every doubling of distance the noise level decreases 6 dB. However, wind turbines seldom hover high above the ground like a balloon. They are earthbound, and their noise emissions spread outward hemispherically. Over a flat reflective surface such as a lake, noise decays 3 to 6 dB per doubling of distance. The atmosphere and objects on the landscape also absorb some of the noise energy, further attenuating the noise over distance. The International Energy Agency (IEA) assumes hemispherical spreading in its commonly used noise propagation model. They also incorporate a modest amount of atmospheric absorption.[25]

Danish acousticians have found that the IEA method overestimates the noise level 1 to 6 dB from wind turbines at appreciable distances.[26] The IEA model does not adequately account for ground cover or meteorological effects. Both can greatly influence noise levels. Temperature and wind shear, for

Table 10.3
Decibel Addition

Difference Between Two Levels (dB)	Add to Highest Level (dB)
0	3.0
1	2.5
2	2.1
3	1.8
4	1.5
5	1.2
6	1.0
7	0.8
8	0.6
9	0.5
10	0.4

example, refract or bend sound waves from those expected, and vegetation attenuates or absorbs more sound than the model assumes.

The rate at which noise decays increases with increasing atmospheric absorption. Relatively close to the tower, within 100 to 200 m (300 to 600 ft), atmospheric absorption has little effect. As distance increases, for example from 200 to 400 m (600 to 1300 ft), the decay rate with absorption increases to 7 dB with every doubling of distance. Thus the noise attenuated by atmospheric absorption can be important in projecting noise levels surrounding a wind turbine.

Unfortunately, meteorological effects vary with the season, weather patterns, and time of day. Vegetation may vary seasonally as well. Row crops may be tilled in the fall when deciduous trees also lose their leaves, which removes much of the vegetation that dampens noise from nearby turbines. Moreover, nighttime temperature inversions refract sound waves, bending them back to earth, increasing the noise level over that from simple propagation models. Valley inversions during the fall and winter produce a similar effect. Anyone living alongside a lake or river has experienced sound carrying great distances during wintertime inversions.[27] There is also little or no atmospheric absorption of extremely low frequency sound. For these reasons, engineers are hesitant to incorporate greater atmospheric absorption into their noise propagation models. Thus the models remain conservative.

Multiple wind turbines complicate matters further. From relatively long distances, an assembly of machines appear as a point source, and doubling the number of turbines simply doubles the acoustic power, increasing noise levels 3 dB. Closer to the turbines, they begin to act as a line source.[28] The decay rate for line sources is 3 dB per doubling of distance, not 6 dB for true spherical propagation.

Even the wind itself will influence noise propagation. Noise levels are typically higher downwind of a turbine than upwind, and more so for downwind turbines (Figure 10.2). Thus projecting the noise emitted by a single wind turbine or a large array is no simple matter and is fraught with uncertainty. Although unlike aesthetic impact, noise is quantifiable, interpreting the results of field measurements and mathematical projections requires almost as much subjective judgment as it does objective analysis.

Ambient Noise

The total perceived noise is the logarithmic sum of the ambient or background noise and the projected wind turbine noise. Thus the noise generated by a wind turbine must always be placed within the context of other noises around it. Wind turbines near airports or busy highways will hardly create a problem, no matter how noisy they are. Conversely, wind turbines, no matter how quiet, may be heard above the background at great distances in the stillness of a sheltered mountain cove.

At the top of a windswept hill, the noise of the wind will overshadow that of the turbine. The wind itself often masks wind turbine noise by raising the ambient noise level. At exposed locations, there will always be noise from the wind whenever the wind machine is operating because the wind rustles the leaves in nearby trees or sets power lines whistling. Studies of the ambient or background noise created by wind in trees have found noise levels of 51 to 53 dB(A) at 40 ft (12 m) in winds of 15 mph (7 m/s). At this level, the noise of nearby trees can mask the noise from a wind turbine operating in the same winds. For example, consider the problem faced by Bergey Windpower when asked to measure the noise produced by one of their small wind turbines. They measured an ambient noise level of 53 dB(A) at wind speeds of 25 mph (11 m/s) near Norman, Oklahoma. At 323 ft (100 m) from the turbine, they measured 54 to 55 dB(A). At 600 ft (200 m), the wind turbine noise of 53 to 54 dB(A) had approached that of the background.

Despite the masking effect of high winds, a wind turbine will still be audible to those nearby, particularly when they are sheltered from the wind. The sounds emitted by these machines are easily distinguishable from those of the wind. The generator or transmission may produce a noticeable whine, for example, or the passage of the blades may generate more discrete sounds. The "swish-swish-swish" aerodynamic noise from three-bladed rotors is a common wind turbine sound. These sounds may not be objectionable, but they are detectable. The whir of the compressor in a refrigerator is audible, for example, but few find the sound objectionable. Some have compared this situation to that of a leaky faucet. Once recognized, the noise is hard to ignore.

Where the background noise level is low, as in a mountain hollow, a new noise may be considered intrusive, particularly at night when few other man-made sounds are present or a nighttime temperature inversion has brought a deathly hush to the valley. Whether or not a noise is intrusive depends on the nature of the noise: that is, its tonal or impulse character, the perception of the noise source (whether the wind turbines are loved, despised, or merely tolerated), the distance from the source, and the activity (for example, whether one is sleeping inside with the windows closed, or conversing with a neighbor in the yard).[29] But no wind turbine, no matter how quiet, can do better than the ambient noise. It is the difference between ambient noise and wind turbine noise that determines most people's response.

Noise complaints in Wales are similar to those first encountered by the Mod-1 in the Appalachian Mountains during the 1970s and those encountered in southern California in the 1980s. These are areas where topographic features shelter residents from the wind and there is no wind noise to mask the noise from the turbines on the exposed heights. At Delabole in Cornwall, some neighbors complained that the occasional aerodynamic noise created as the turbine blades changed pitch during high winds, and the turbines' mechanical noise, were most noticeable when the neighbors were sheltered from the wind. From a sheltered vantage point, they found the noise more noticeable. Others elsewhere in Wales and California have complained about placing a continuous mechanical noise into a quiet rural environment.

Community Noise Standards

On the European continent, fixed noise limits are the norm. In Denmark, for example, noise is limited to 45 dB(A) L_{eq} outdoors at the nearest habitation in rural areas, and 40 dB(A) in residential areas and other noise-sensitive locations, such as schools and hospitals. The Netherlands employs a stricter limit at night when it reduces daytime limits 10 dB(A) (Table 10.4). In Britain, a host of criteria have been employed.[30]

Community noise standards are even less consistent in the United States than in Britain. California's Kern County, for example, limits wind turbine noise to 45 dB(A) at $L_{8.3}$ for sensitive receptors. ($L_{8.3}$ is the noise level exceeded for 5 minutes out of every hour.) This appears lenient, but the same county regulations also prohibit any noise exceeding 50 dB(A), an exceedence level of L_{100}, unless the ambient noise is greater. On closer examination, the Kern County ordinance is stricter than those of Riverside County or the city of Palm Springs. Where noise levels from a proposed project exceed these limits in Kern County, the developer can receive a waiver by gaining the assent of affected property owners and filing a permanent noise easement with the county.[31]

All community noise standards incorporate a penalty for pure tones, typically 5 dB. If a wind turbine meets a 45-dB noise standard, for example, but produces an annoying generator whine, planning officers dock the offending turbine, 5 dB. The operator must then lower the turbine's overall noise level 5 dB or eliminate the whine.

Table 10.4
Noise Limits [dB(A)] of Sound Pressure Levels

	Commercial	Mixed	Residential	Rural
Germany				
Day	65	60	55	50
Night	50	45	40	35
Netherlands				
Day (L_{eq})		50	45	40
Night		40	35	30
Denmark[a] (L_{eq})			40	45
England[b]				
High speed (L_{50})				45
Low speed (L_{50})				40
Kern County[c] ($L_{8.3}$)			45	45
Riverside County (L_{90})			45	
Palm Springs[d] (L_{90})			50	60

[a] Proposal to limit noise to 45 dB(A) 400 m from wind turbine.

[b] L_{50} approximately 350 m from the nearest turbine.

[c] Not to exceed 50 dB(A).

[d] 50 dB(A) if lot is actually used for dwelling.

Source: M. Trinnick and I. Page, Kern and Riverside counties, city of Palm Springs.

Despite compliance with community noise standards, operators of wind turbines still run the risk of annoying their neighbors. Whenever any wind turbine noise exceeds the threshold of perception, there is the potential for complaints[32] (Table 10.5). Fluctuations in background noise and variations in the quality or tonal component complicate determining whether wind turbine noise will exceed the perception threshold and stimulate complaints. Table 10.5 was derived for noise sources other than wind turbines, and neighbors could be either more or less sensitive to wind turbine noise than that indicated by the table.

In complex terrain, ambient noise can vary dramatically from one site to the next. A wind-swept moorland site now occupied by wind turbines can overlook a secluded valley where the background noise level L_{90} is as low as 20 dB(A) at night.[33] Thus a noise level of 30 dB(A) of a wind turbine 600 m (2000 ft) distant may still stimulate complaints, such as those of Sandy Hare, even though it is equivalent to that of a human whisper. Rough terrain also complicates modeling noise propagation.

Sound Pressure and Sound Power

Local noise ordinances typically state the acceptable sound pressure level in dB(A) at the property line or nearest residence, because noise measurements are commonly made in these units. Wind turbine manufacturers in the United States report noise measurements in sound pressure levels at a specified distance from the wind turbine. The norm is 100 m (328 ft) downwind from the nacelle, but it can be any distance. For example, U.S. Windpower reports the noise from its model 56-100 as 45 dB(A) at 800 ft (243 m) from the turbine (Table 10.2).

However, the mathematical models used to project noise levels surrounding a wind turbine use the acoustic energy created by the machine, not sound pressure directly. Acousticians use field measurements of the sound pressure level, L_p, to calculate the sound power levels, L_W, emitted from the wind

Table 10.5
Community Response to Noise from Sources Other Than Wind Turbines

Amount by Which Rated Noise Exceeds Background Level (dB)	*Estimated Community Response*	
	Category	Description
0	None	No observed reaction
5	Little	Sporadic complaints
10	Medium	Widespread complaints
15	Strong	Threats of action
20	Very strong	Vigorous action

Source: Harvey Hubbard and Kevin Shepherd, NASA, 1990.

turbine. As if the similar-sounding names were not confusing enough, both noise measures use the same units, dB(A). While sound pressure levels will always be specified at some distance from the turbine, the sound power level will always be presented at the source: the wind turbine itself. Engineers use the sound power level to project the noise at various distances from the wind turbine.[34] The distinction is important. The sound power level of most commercial wind turbines varies from 95 dB(A) to more than 100 dB(A). Yet a wind turbine emitting a sound power level of 100 dB(A) can meet a 45-dB(A) noise limit in sound pressure level given sufficient distance.

European wind turbine manufacturers report noise emissions in sound power levels. Thus, the noise data presented in product literature from European manufacturers cannot be compared directly to noise data in product literature from U.S. manufacturers, unless both are presented in sound power levels. Ideally, all manufacturers would report noise emissions in sound power levels to simplify comparisons.

To recapitulate: acousticians measure wind turbine noise in sound pressure levels. They then use these measurements to calculate the strength of the noise source, the wind turbine, in sound power. From this they are then able to project the noise impact in sound pressure levels at any distance from the machine. By reporting the emission strength in sound power at the source—ground zero—the noise from one wind turbine can be compared directly with that of another and there are substantial differences in noise emissions from one wind turbine to the next.

Wind Turbine Noise

There are two sources of wind turbine noise: aerodynamic and mechanical. Aerodynamic noise is produced by the flow of the wind over the blades. Mechanical noise results from the meshing of the gears in the transmission and the high-speed whir of the generator.

Unless there is a whistling effect from slots or holes in the blades, aerodynamic noise, like white noise, occurs across the frequency spectrum. It is principally a function of tip speed and shape, but aerodynamic noise is also influenced by trailing-edge thickness and blade surface finish. The number of blades is also a factor. Neil Kelley, a researcher at the National Renewable Energy Laboratory, finds that the aerodynamic noise of two-bladed wind turbines is greater than that of three-bladed machines, all else being equal, because the two-bladed turbines place higher loads on each blade for an equivalent output. Further, the type of rotor control, whether fixed or variable pitch, affects aerodynamic noise. On rotors with fixed-pitch blades, noise increases when the blades enter stall during high winds. But rotor diameter and speed are the primary determinants of aerodynamic noise. High tip speeds create greater tip vortices, which are believed to be the dominant emission source.

Noise also increases with increasing wind speed and power (Figure 10.3). This is especially true of variable speed turbines that operate at faster tip speeds in strong winds than conventional constant-speed turbines. The source strength typically increases 1 dB(A) for every 1 m/s increase in wind speed until the turbine nears its rated capacity.[35]

However, the influence of wind noise on the microphone of sound level meters reduces the accuracy of noise measurements at higher wind speeds. Because of this effect, acousticians have standardized noise calculations at wind speeds below rated capacity where wind noise is less prominent and noise from the wind turbine is more noticeable. Europeans report noise measurements for a reference wind speed of 8 m/s. In the United States the reference wind speed is 10 m/s.

Research by Nico van der Borg at ECN found that the source power of noise from wind turbines can be approximated using a formula based on rotor diameter, where larger wind turbines generate more noise than do smaller machines.[36] The model is derived from data on wind turbines designed in the 1970s and early 1980s, a number of them large experimental turbines in government-sponsored programs (Figure 10.3). Many of these machines operated at much higher tip speeds than are commonly found today in commercial products (Table 10.6). Modern medium-sized wind turbines, such as those built by Danish manufacturer Bonus, emit considerably less noise—about 7 dB less—than did their predecessors.

A measure of this progress can be seen when we consider that both the

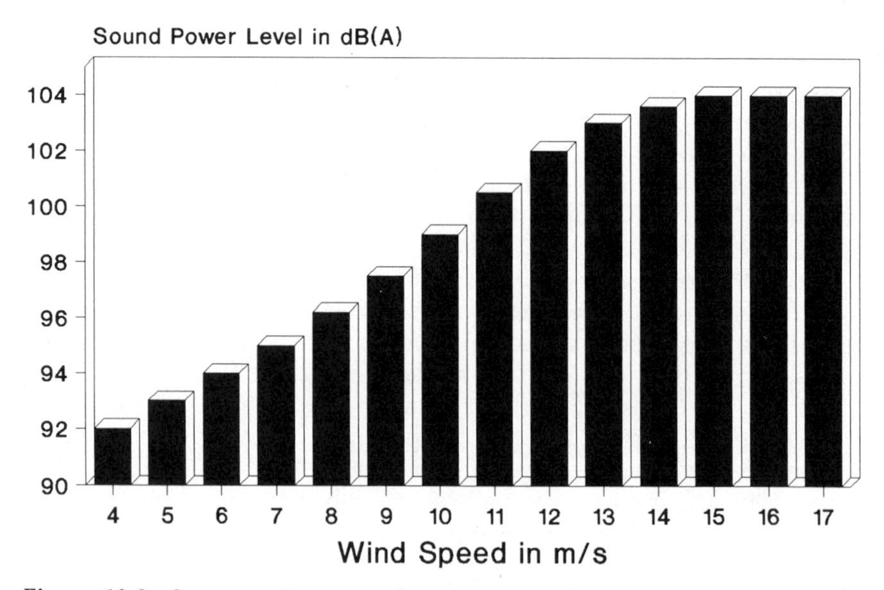

Figure 10.3. Source noise strength increases with increasing wind speed and power for the Bonus Combi.

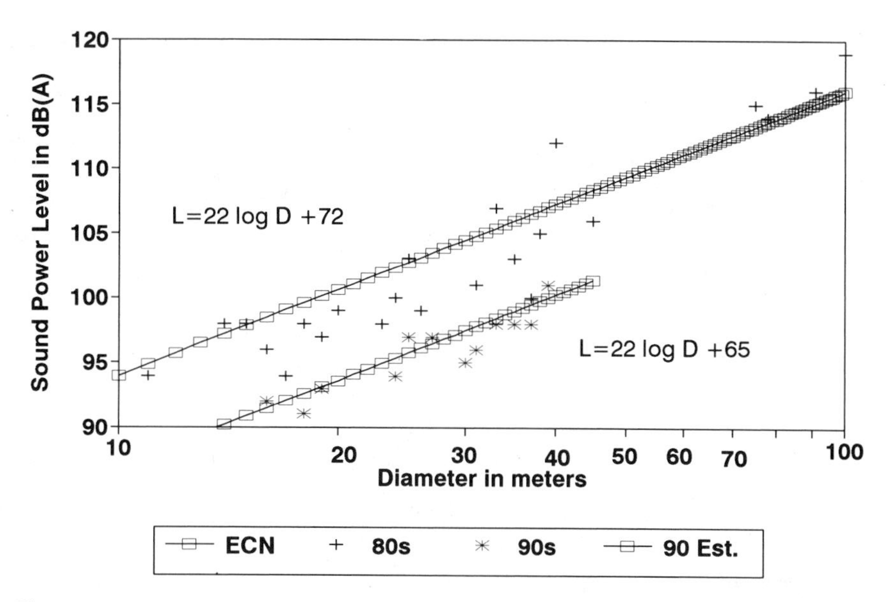

Figure 10.4. Source sound power level. Modern medium-sized wind turbines emit less noise than wind turbines designed in the 1970s and early 1980s.

Bonus and Vestas 55 kW (Table 10.6) were designed during the late 1970s and introduced to the market during the early 1980s. These turbines emitted 98 to 104 dB at tip speeds of about 40 m/s. The Vestas V27, introduced during the late 1980s, emits less noise while operating at slightly higher tip speeds and intercepting three times more wind energy than did the earlier machines.

Table 10.6
Sound Power Level for Selected Wind Turbines

Turbine	Diameter m	ft	Area (m^2)	Capacity (kW)	Wind Speed (m/s)	Blade Tip Speed (m/s)	Sound Power Level [dB(A)]
Bonus	15	50	181	55	10	38	98
Vestas	15	51	186	55	10	40	104
USW 56-100	18	58	243	100		67	98
Lagerwey	18	59	254	80	7	61	91
WindMaster	25	82	490	300	7	68	105
Vestas	27	89	572	225	8	61	97
Bonus Combi	31	102	754	300	8	50	96
Enercon	33	108	855	330	7	43	98
WEG	37	121	1060	400		52	98
Bonus	37	121	1075	450	8	58	97
Mod-0A	38	125	1140	200		80	105
Vestas	39	128	1194	500	8	61	101
Nibe A,B	40	131	1256	630	10	71	112

Note: For a more complete list, see Appendix B.

Source: ECN and product literature.

The Bonus Combi, introduced in the early 1990s, also operates at slightly higher tip speeds than those of the early Danish machines, yet it produces less noise while capturing more than four times more wind energy.

All medium-sized wind turbines with modest noise emissions operate at tip speeds of 50 to 60 m/s. For example, the 80-kW Lagerwey was introduced in the early 1990s. At a tip speed of 61 m/s, it generates less noise than does U.S. Windpower's model 56-100, introduced during the early to mid-1980s, which operates at a slightly higher speed.

Advances in airfoils and reductions in tip speeds have essentially decoupled noise emissions from the size of the wind turbine. The Vestas V39 (introduced in the mid-1990s) sweeps the same area as NASA-DOE's Mod-0A of the mid-1970s and Denmark's Nibe A&B turbines installed in 1980. Yet the V39, which operates at three-fourths the tip speed of the Mod-0A, emits half the noise energy (4 dB) and the V39 emits only half the loudness (10 dB) of the Nibe designs.

Cutting Wind Turbine Noise

Building quieter turbines not only makes wind energy a better neighbor, it also makes good business sense. In Europe, where competition is fierce, manufacturers find that quieter turbines give them an edge over their rivals. Manufacturers with quieter turbines can site them in areas where planning officials would prohibit other turbines. Quieter turbines also ensure that there are fewer headaches after installation and less bad press eroding support for wind energy.

In the densely populated Netherlands, the government chose to accelerate the progress toward quieter turbines by targeting a subsidy for noise reduction. To encourage development of quieter turbines during the late 1980s and early 1990s, the Dutch environment ministry gave a bonus of NLG 50 ($28) per kilowatt for designs emitting 3 dB(A) less than the average. In the case of the Dutch manufacturer Holec, the noise incentive influenced a turbine's design. Holec opted for three blades on its 500-kW prototype, partly to reduce tip speed and hence to keep noise below the average of machines its size. Holec recouped the cost of the extra blade through the noise reduction incentive.[37]

The most direct route for lowering noise emissions is to attack rotor speed. Tests by Britain's Energy Technology Support Unit illustrate the effect the reduction of rotor speed has on cutting aerodynamic noise. Slowing the rotor speed on ETSU's test turbine from 60 rpm to 40 rpm, and thereby proportionally reducing tip speed, cut noise emissions by 7 dB(A).[38] British manufacturer WEG demonstrated the relationship commercially when they upgraded their MS-3 model. In the 400-kW version of the original 300-kW design, WEG stretched the rotor from 33 m to nearly 37 m in diameter, increasing the swept area by 24%. To cut noise emissions, WEG reduced rotor speed from an aerodynamic optimum of 43 rpm for the stretched rotor to 40 rpm. They then designed a compliant drivetrain, carefully feathered the trailing edge, and added an unusual swept tip. Altogether these measures reduced noise emissions by 2 dB(A), WEG realized that these efforts are for nought if the blades are damaged during shipping or installation. As a final step to minimize noise, WEG buffs imperfections out of the blades' trailing edge before installation (Figure 10.5).

Another means for cutting aerodynamic noise by lowering rotor speed is to operate the turbine at dual speeds. This permits operating the turbine at a lower rotor speed in light winds, when there is less wind noise to mask noise from the turbine. Variable-speed operation is also effective, enabling designers to program operation for lower speeds at night, when noise sensitivity is greatest.

The use of dual-speed operation on the WindMaster turbines on a breakwater at Blyth, on England's northeastern coast, clearly demonstrates its value for noise reduction compared to WindMaster's standard single-speed design. Noise was an early concern in the installation of the nine WindMaster turbines, because of the proximity of a nearby neighborhood. Daytime noise levels can be high as 60 dB(A) L_{90} as ships load and unload in the industrial harbor. As harbor noises subside at nightfall, ambient noise falls to 43 dB(A) L_{90}. Local officials consequently limited wind turbine noise to no more than the ambient nighttime levels. Dual-speed operation reduced the noise emissions from the WindMaster turbines, one of the noisiest designs on the international market,

Figure 10.5. Treating trailing edge. Buffing nicks in trailing edge on WEG MS-3 turbine prior to installation reduces noise near Cemmaes, Wales.

from 101 dB(A) to 94 dB(A) in low winds. In high winds, the dual-speed turbines are inexplicably quieter than are the single-speed machines.[39]

Mechanical noise often has tonal components. These can be reduced by redesigning the gearbox and by adding resilient couplings in the drivetrain to isolate vibrations, as WEG did in its 400-kW version of the MS-3. Sandbags can be added selectively for further vibration dampening, and acoustic insulation can be installed inside the nacelle to reduce mechanical noise.[40] At Delabole, for example, Vestas dampened a mechanical resonance in the tubular towers by fitting a sand jacket to the top 2 m (6 ft) of the tower, says Peter Edwards. An aerodynamic squeak in the blades was also readily cured, he says, by taping a small section of the trailing edge on each blade.[41]

The most severe problem with mechanical noise was encountered by Danish manufacturer DanWin at Kyndby. Because of the proximity of one residence, DanWin took special precautions when building and installing the 21 turbines. They mounted the 180-kW nacelles on rubber dampers, sharpened the trailing edges of blades on the eight nearest turbines, mounted sand-dampening chambers on four towers, and reduced generator speed to 1000 rpm from the typical 1200 rpm. (By comparison, most U.S. generators operate at 1800 rpm.) Despite these precautions, the noise at the nearest neighbor, a farmhouse 220 m (720 ft) away, was 48 dB(A), including a pure tone component from the gearbox. After four years of work and 4.5 million DKK ($750,000), the turbines' noise emissions were reduced from 97 to 102 dB(A) to 95 dB(A), resulting in an acceptable noise level of 44 dB(A) at the dwelling. DanWin's successors achieved this level by redesigning the gear teeth and adding further noise treatment. The engineers found that they could gain 4 dB(A) on three test turbines by sharpening the trailing edges of each blade, providing one of the most convincing demonstrations that trailing-edge thickness is a significant factor in aerodynamic noise.[42]

As seen with Holec, WEG, WindMaster, and DanWin, concern about noise can influence the design of the turbine. Noise impact also affects the configuration of multiple-turbine arrays near dwellings. And in the Netherlands, noise emissions may also determine the size and number of machines chosen for particular projects. For linear arrays along dikes, Dutch officials prefer as large a wind turbine as possible, and they have pushed manufacturers toward giant megawatt machines. Large wind turbines have historically generated more noise than do commercially available medium-sized turbines. Where noise is a concern, it may be easier to comply with community noise limits by installing more medium-sized machines rather than fewer multimegawatt turbines, even though this may not optimize use of the wind resource along a dike. Along with aesthetics, designing and installing quieter wind turbines must be given equal weight with engineering efficiency.

The Bonus Combi

During the early 1990s, Danish manufacturer Bonus set out to design a medium-sized wind turbine that was both pleasing to the eye and as quiet as today's technology permitted. Bonus earned industry plaudits when they succeeded in developing the quietest wind turbine in its class: the Bonus Combi, a 300-kW turbine driven by a rotor 31 m (100 ft) in diameter. Designer Henrik Stiesdal maintains that "there are no secrets" to the techniques he used, just the application of what is already known about how to reduce wind turbine noise. "It's a reasonably quiet turbine, but one is never satisfied," he says modestly.[43]

The rotor dominates the sound power emitted by a wind turbine, so Stiesdal first addressed the rotor's aerodynamic noise. Bonus has always run its turbines at low rotor speeds, and the low noise emissions from the Bonus Combi can be attributed partly to its low tip speed of 50 m/s (112 mph), which is among the lowest in its class. The blade tip is often the dominant source of aerodynamic noise for modern, high-performance blades. To reduce tip noise, Stiesdal had a choice: return to the old elliptically shaped tips used on earlier blades and suffer a 5% performance penalty, or design a new blade tip. He opted for the latter, and progressively developed, by trial and error, what he calls a "tip torpedo," which helps shape the vortex found at the tip of the airfoil (Figure 10.6) Sharpening the blade's trailing edge to a thickness of about 1 mm (0.04 inch) on the outboard half of the blade also helps measurably. However, Stiesdal found that the razor-thin trailing edge made the blades more difficult to handle without damage in the field.[44]

On the Combi, Bonus's noise reduction efforts cut noise 2 to 3 dB, and Bonus's tip torpedo has been incorporated into other models of the Bonus line. For stall-controlled turbines such as those built by Bonus, stall can be a significant source of noise at high wind speeds. To reduce noise during stall, Stiesdal added a turbulator strip on the leading edge of the blade near the tip. The material they chose, says Stiesdal, was as low-tech as imaginable. It is cheap and can be found in any hardware store on either side of the Atlantic: adhesive antislip strips used in bathtubs.

The thumping noise of the blades passing through turbulence eddies is more difficult to control. Unlike the swirling eddies created downwind of the tower, turbulence eddies in the wind stream itself will affect upwind turbines

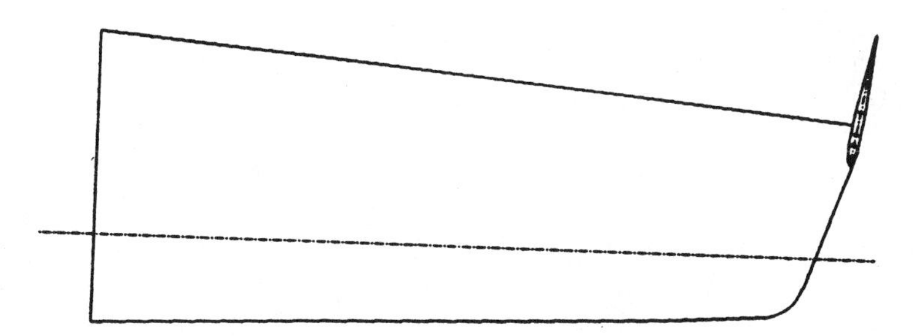

Figure 10.6. Bonus tip torpedo. Modification of the blade tip on the Bonus Combi, designed to shed vortices and reduce noise. (Courtesy of Bonus Energy, Brande, Denmark.)

at exposed sites. At a sheltered position at the bottom of a valley in hilly terrain, this thumping can be the dominant noise.[45]

Stiesdal also addressed common mechanical noise sources such as generators and transmissions. The drivetrain determines any problematic tonal noise, and the gearbox's high-speed shaft is the most critical component, says Stiesdal. By a mix of measures, including specialized grinding, and hardening of the gears, suspension of the gearbox and generator on rubber dampers, and the addition of an insulating cover on the gearbox housing, Bonus reduced gearbox noise an impressive 5 dB and nearly eliminated any pure tones. The unusually long main shaft on Bonus machines (2 m between bearings on the Bonus Combi) dampens much of the remaining metal-to-metal contact in the drivetrain. He also lined the nacelle cover with noise-absorbing materials.

Stiesdal is adamant about totally enclosing the drivetrain and sealing the nacelle canopy. Even ventilation louvers must be carefully designed as sound baffles, he says, or a significant part of the turbine's machinery noise, especially noise at higher frequencies, will escape the nacelle.[46] This is one reason all wind turbines should be housed inside a tight fitting nacelle cover or canopy.

While quieting the Bonus Combi is a significant accomplishment, Stiesdal is the first to admit that there is a 1- to 2-dB uncertainty in measurements of sound power emissions of wind turbines. The noise emissions of other machines (Vestas's V27, WEG's 400 model, the Enercon 33) in its class easily fall within this range of uncertainty. To paraphrase Stiesdal, Bonus and its competitors "should never be satisfied."

Noise from the Bonus Combi, with a source strength of 96 dB(A) will decay to 45 dB(A) within 125 m (400 ft) from the nacelle, according to the IEA model (Figure 10.7). A wind turbine emitting 100 dB(A) will require up to 200 m (650 ft) before reaching the 45-dB(A) level. This gives the Bonus Combi a 75-m (250-ft) siting advantage over a turbine emitting 100 dB.

It is the objective of planning officials to limit the increase in total noise. They seldom demand the impossible by requiring a wind turbine to meet a standard that is lower than the ambient noise level. To determine the potential impact, a noise assessment will first determine the ambient noise level. The acousticians will then project the noise from the wind turbine, using the IEA model for example, evaluate the acceptability of that noise, and estimate the level of annoyance.

The Danish windmill owners' association takes a strong stand on wind turbine noise. The association's members are not only the chief advocates of wind energy in Denmark, they are also those who own wind turbines; many literally can see wind machines outside their windows. They can speak with authority as those who want wind energy and those who demand that it be a good neighbor. Their position is clear: noisy turbines are unacceptable. Noisy machines should either be soundproofed or moved. If not, the turbines should at least be stopped in light winds, when the noise is most annoying. The goal of the owners' association, one that should be the goal of all manufacturers, is to avoid the problem from the start. They have found, as has the

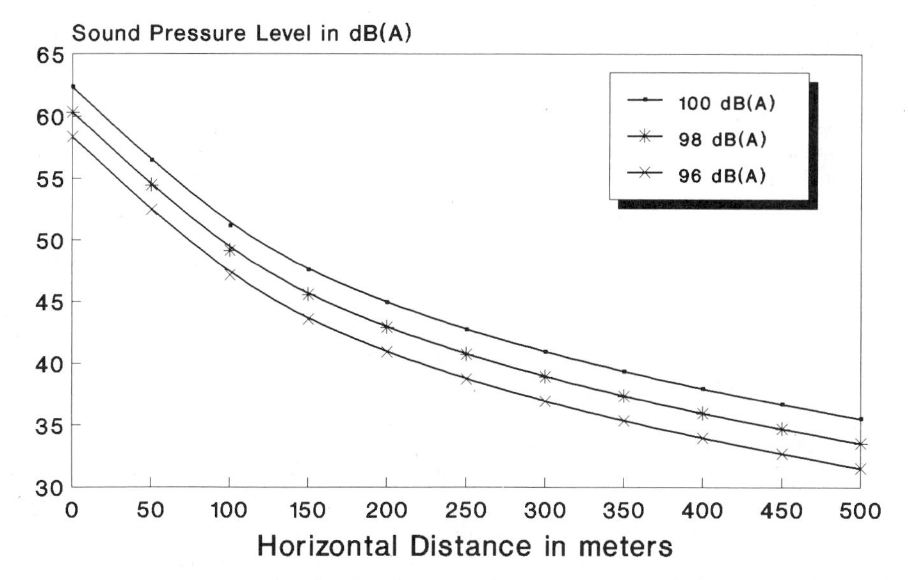

Figure 10.7. Projected noise level using the IEA noise model for a single wind turbine on a 30 m tower. A wind turbine emitting 96 dB(A) of sound power will produce 45 dB(A) of sound pressure 125 meters from the source.

Danish environment ministry, that once people have been bothered by noise, they remain disturbed, even after the noise has been abated.

Though manufacturers and operators should continually strive to quiet their machines, they will never silence them. Wind turbines will always remain audible to those who choose to hear them.

Chapter 10 Endnotes

1. See the cover photo for the September 1993 issue of the Santa Monica Bay Audubon newsletter. The quote summarizes both the content and the tone of the article. Steve Ginsberg, "The Wind Power Panacea: Is There Snake Oil in Paradise?" *Audubon Imprint,* Santa Monica (Calif.) Bay Audubon, 17:1, September 1993, pp. 1–5.
2. Andrew F. Fritzsche, "The Health Risks of Energy Production," *Risk Analysis,* 9:4, 1989, pp. 565–577.
3. Alexi Clarke, "Windfarm Location and Environmental Impact," Open University, Milton Keynes, England, June 1988, pp. 55–57.
4. Derek Taylor and Marcus Rand, "How to Plan the Nuisance Out of Wind Energy," *Town and Country Planning,* May 1991, pp. 152–155.
5. H. C. Harrison, *The Story of Sprowston Mill* (London: Phoenix House, 1949), p. 13.
6. For more information about fall protection systems and how to use them, read the chapter on safety in Paul Gipe, *Wind Power for Home & Business* White River Business (White River Junction, VT: Chelsea Green Publishing, 1993).
7. I listened, dumbfounded, as Mehrkam calmly described this procedure during an

interview in 1980. It occurred to me that some day I might write his obituary. I did, only a year later.

8. L. D. Hamilton, "Health and Environmental Risks of Energy Systems," *Proceedings of an International Symposium on the Risks and Benefits of Energy Systems,* International Atomic Energy Agency, Vienna, 1984, p. 30.
9. Michael Rowe, "Health Risks in Perspective: Judging Health Risks of Energy Technologies" (Upton, NY: Brookhaven National Laboratory, September, 18, 1992), p. 105. Rowe cites "The Environmental Impacts of Production and Use of Energy," Part IV: "The Comparative Assessment of the Environmental Impacts of Energy Sources" Nairobi, Kenya: United Nations Environment Program, January 1985. The rate of occupational deaths for wind energy is 0.000125 per terawatthour; for solar thermal, 0.000958 per terawatthour; for photovoltaics, 0.000053 per terawatthour; for geothermal, 0.000490 per terawatthour.
10. "Some Externalities Come Home," *RENEW,* 83, May/June 1993, pp. 9–11. For the number of deaths per gigawatt-year, the article cites "The Social Costs of Fuel Cycles," by David Pearce, University College London, Centre for Social and Economic Research on the Global Environment (London: Her Majesty's Stationery Office, September 1992). The study used mortality data from Andrew Fritzsche of the Swiss Federal Energy Office; Fritzsche, "The Health Risks of Energy Production;" Samuel C. Morris, "Health Risks of Coal Energy Technology," in *Health Risks of Energy Technologies,* Curtis C. Travis and Elizabeth L. Etnier, eds. (Boulder, CO: Westview Press, 1983), Chapter 4, p. 128.
11. WIMP, Phase II, "Wind Implementation Monitoring Program," Riverside County, Riverside, CA, Tierra Madre Consultants, August 1986, pp. A4–A7.
12. "Landscape Impact Assessment for Wind Turbine Development in Dyfed," Chris Blandford Associates, Cardiff, Wales, February 1992, p. 6.
13. Clarke, Windfarm Location, p. 48.
14. Sara Knight, "Study on Flickering Shadow Nuisance from Rotor Blades," *Windpower Monthly,* 8:10, p. 15.
15. J. H. Bass, "WECS Environmental Aspects: Acoustic Noise," in course notes for "Principles of Wind Energy Conversion," L. L. Freris, ed., Imperial College of Science, Technology and Medicine, London, July 1993.
16. Martin Pasqualetti and Edgar Butler, "Public Reaction to Wind Development in California," *International Journal of Ambient Energy,* 8:2, April 1987, pp. 83–90.
17. Ibid.
18. "Renouvelables: inventaire et état de l'art," *Systèmes Solaires,* 94/95, 1992, p. 21.
19. Taylor and Rand, "How to Plan."
20. Poul Nielsen, "Development of Wind Energy in Denmark," DEFU The Association of Danish Utilities, paper presented at the American Wind Energy Association's Annual Conference, Windpower '93, San Francisco, July 1993.
21. Bridgett Gubbins, "Living with Windfarms in Denmark and the Netherlands," North Energy Associates, Northumberland, England, September 1992, pp. 16–17.
22. "Renouvelables'" Planning Policy Guidance, Renewable Energy, Wind Energy Annex, December 5, 1991.
23. Lyle Yerges, Sound, *Noise and Vibration Control,* 2nd ed. (New York: Van Nostrand Reinhold, 1978), p. 7.
24. Neil Kelley, "A Proposed Metric for Assessing the Potential of Community Annoyance from Wind Turbine Low-Frequency Noise Emissions," Solar Energy Research Institute, Golden, CO, November 1987.
25. The IEA model for sound pressure level is $L_p = L_w - 10 \log (2\pi R^2) - \alpha R$, where L_W

is the source, strength, and α is the atmospheric absorption rate, which for broadband noise is 0.005 dB(A) per meter. Atmospheric absorption varies from 0 to 0.0054 dB(A) per meter from the source.

26. Bent Andersen, "Noise from Wind Farms: Measurement and Prediction," European Wind Energy Association: special topic conference, "The Potential of Wind Farms," Herning, Denmark, September 8–11, 1992.
27. Bass, "WECS Environmental Aspects."
28. Harvey Hubbard and Kevin Shepherd, "Wind Turbine Acoustics," NASA Technical Paper 3057, December 1990, p. 6.
29. Bass, "WECS Environmental Aspects."
30. Ten-minute average 40 dB(A) L_{50} at any dwelling 380 m (1250 ft or $\frac{1}{4}$ mile) or greater during low-speed operation, and 45 dB(A) during high-speed operation; the ambient L_{50} plus 5 dB(a) at sensitive receivers; a 10-minute average L_{90} at any dwelling 350 m from the nearest turbine but limited to raising the ambient noise by no more than 7.5 dB(A). Ian Page, "The Development of Wind Farms in England and Wales," European Wind Energy Association's special topic conference, "The Potential of Wind Farms," Herning," Denmark, September 8–11, 1992.
31. Kern County Ordinance, 19.64.140j, Development Standards and Conditions, Bakersfield, CA, pp. 322–323.
32. Hubbard and Shepherd, Wind Turbine Acoustics, p. 31.
33. D. I. Page, M. L. Legerton, and H. G. Parkinson, "The Development of Wind Energy in England and Wales," paper presented at the American Wind Energy Association's annual conference, Windpower '93, San Francisco, July 1993.
34. Sound power is $L_w = L_p + 10 \log (4\pi R^2)$, where L_p is the measured sound pressure level; and R is the slant distance from the nacelle. If the sound pressure level was measured on a reflective panel, 6 dB must be deducted from the L_p.
35. Bass, "WECS Environmental Aspects."
36. N. C. J. M. van der Borg and W. J. Stam, "Acoustic Noise Measurements of Wind Turbines," Energieonderzoek Centrum Nederland (ECN), Petten, The Netherlands, June 1989, p. 2.
37. Jos van Beek, "Noise," *Windpower Monthly,* August 1990, pp. 12–19.
38. M. Legerton, "Wind Turbine Noise in the DTI Wind Energy Programme," paper presented at the British Wind Energy Association's annual conference, York, October 6–8, 1993.
39. W. Grainger, D. Still, N. Rogers, and A. Gammidge, "A Wind Farm in a Mixed Industrial and Residential Area," paper presented at the British Wind Energy Association's annual conference, York, October 6–8, 1993.
40. Bass, "WECS Environmental Aspects."
41. Peter Edwards, "The Performance and Problems of and the Public Attitude to the Delabole Windfarm," paper presented at the British Wind Energy Association's annual conference, York, October 6–8, 1993.
42. Carsten Skamris and Soren Pedersen, "Structural Noise at Kyndby Windfarm," paper presented at the European Wind Energy Association's special topic conference, "The Potential of Wind Farms," Herning, Denmark, September 8–11, 1992.
43. H. Stiesdal and E. Kristensen, "Noise Control on the Bonus 300 kW Wind Turbine," paper presented at the British Wind Energy Association's annual conference, York, October 6–8, 1993.
44. Ibid.
45. Ibid.
46. Ibid.

11

Impact on Land

Opponents of wind energy charge that wind plants require more land than do conventional power plants because the energy in the wind is more diffuse than that, say, in enriched uranium. Gilles Ménage, president of Électricité de France, for example, downplays wind's potential, saying that it would require a "carpet" of 90,000 wind turbines stretching 7 miles inland from Dunkerque to Biarritz to equal the output of the Gravelines reactors. Indeed, it would take thousands of wind turbines to replace Gravelines, whose six 910-MW reactors near Dunkerque comprise one of the world's largest concentrations of nuclear power outside Chernobyl. But to drive the point home, EDF resorts to the common tactic of understating wind's productivity and overstating wind energy's land requirements.

The same criticism has often been leveled at other sources of solar energy: they are too land-intensive for even a land-rich country such as the United States. However, a study for the U.S. Department of Energy concluded that contrary to widespread belief, solar-electric technologies consume no more land than do coal-fired power plants after accounting for the entire fuel cycle, and possibly no more than nuclear plants if enrichment and waste storage are included.[1] Germany's Öko Institut finds that small wind turbines use about the same amount of land as bituminous coal and nuclear power, yet only one-tenth the land of plants fired by brown coal.[2]

The distinction lies in the amount of land occupied by the wind plant versus the amount actually used by wind turbines, roads, and power lines. Most of the land in a wind plant remains untouched and usable for other purposes. Consider, for example, the land requirement of a typical coal-fired power plant and that of a comparable wind plant.

In *Energy and Ecology,* David Gates at the University of Michigan calculates the land required for a 1000-MW coal-fired power plant. He estimates that the plant itself requires 140 hectares (350 acres) for coal delivery, handling and storage, buildings, ash and sludge ponds, roads and parking, switchyards, and landscaping. Dedicating a railroad right-of-way to carry coal from mines on the Great Plains to midwestern power plants could add another 1500 h (3700 acres) to the total.[3] This land will remain in use for the life of the plant, typically 30 years. However, most of the land used by a coal-fired power plant is the land disturbed by mining. Consultants to the U.S. Department of Energy estimate that a typical bituminous surface mine feeding such a plant consumes about 240 h (600 acres, nearly 1 mi^2) of land per year[4] (Table 11.1).

A 1000-MW plant operating at a capacity factor of 68% will generate about 6 TWh per year. To generate an equivalent amount of electricity from wind turbines capturing 1000 kWh/m^2 of swept area per year in an open array 5 by 10 rotor diameters apart requires 16 km^2. This array is considerably less dense than those now found in California and resembles the spacing at European wind plants. Such a hypothetical array has a specific land requirement of nearly 16 ha/MW (40 acres/MW).

The wind farm *occupies* four to five times more land than the coal plant, mine, and railroad after completion of the full 30-year life cycle. Yet the wind turbines and their roads use or consume only 5% of the land area. The surface mine consumes 100% of the mining site every year for the life of the power plant. Thus wind energy *uses* from one-fourth to one-third the amount of land used by coal over 30 years.

Coal mined from thicker seams will consume less land than in this example. Some strip mines in Wyoming are essentially coal quarries, mining seams hundreds of feet thick. Land must also be "reclaimed" after mining ceases. However, it is difficult to restore mined lands to pre-mining levels of productivity. Mining also permanently disturbs groundwater aquifers and often unleashes acidic drainage after reclamation is completed. Thus mining is a consumptive use of the land. On the other hand, the wind plant in this example could be located at a windier site than that assumed and could employ denser packing to optimize its use of land and infrastructure.

Table 11.1
Land-Use Requirements (ha) for Coal-Fired and Wind-Driven Power Plants to Generate 6 TWh/yr for 30 Years

	Coal, Deep Mine	Coal, Surface Mine	Wind: 5 × 10 Array at 1000 kWh/m^2
Mine/yr	120	240	
Power plant/yr	5	5	38,200
Railroad right-of-way/yr	50	50	
Land/yr	175	295	38,200
% land used	100	100	5
Land used/yr	175	295	1,910
After 30 years of mining:			
Total land	5,240	8,840	38,200
Total land used	5,240	8,840	1,910
Total used/TWh	873	1,473	318

Source: Data on deep-mined coal from David Gates, *Energy and Ecology* (Sunderland, Mass.: Sinauer Associates, 1985). Data on surface-mined coal from Meridian Corporation, Energy System Emissions and Material Requirements, U.S. Department of Energy, Washington, D.C., 1989.

Wind plant designers array turbines in geometric patterns to minimize the interference from upwind turbines on those downwind. These patterns and the spacing between turbines depends on the terrain and the prevailing winds. In areas where the winds are omnidirectional, designers space the turbines equidistant from one another: for example, 8 to 10 rotor diameters apart. Where winds are unidirectional or bidirectional, as they are in California's mountain passes, designers tightly pack turbines 1.5 to 3 rotor diameters apart across the prevailing wind while spacing the machines 8 to 10 rotor diameters apart downwind.

Contrary to the implication of early studies, the land required for geometric arrays is, paradoxically, independent of size. As turbines become larger, they occupy more land per unit because of the greater spacing required between them. However, the relative spacing between turbines, governed by their rotor diameter, remains the same. Only in linear arrays will increased size increase land utilization rates.

Wind plants occupy from as little as 10 acres (4 ha) per megawatt in linear arrays to as much as 80 acres (32 hectares) per megawatt in a widely-spaced 10 by 10 geometric array for omnidirectional winds. In a highly critical analysis of wind energy, the U.S. Geological Survey estimated that California wind plants occupy up to 30 acres/MW (12 ha/MW) on steep terrain.[5]

Based on field experience, California's dense arrays occupy 15 to 18 acres/MW (6 to 7 ha/MW). The actual amount of land will depend on the project's boundaries and how much land within the boundaries is suitable for wind development. European wind plants occupy proportionally more land. At the 2.6-MW Tændpibe project, Vestas installed 35 turbines in a 6.5 by 9 array on 44 ha (109 acres) of an ancient lake bed, an average of 17 ha/MW (40 acres/MW). The combined 12.6-MW Tændpibe–Velling Mærsk plant occupies 153 ha (380 acres), an average of 12 ha/MW (30 acres/MW). In Germany, Winkra's Henning Holst installed a 27-MW project requiring 300 ha (740 acres) on the island of Fehmarn, an average of 11 ha/MW (27 acres/MW). The European Wind Energy Association estimates that wind plants under typical European conditions will require 13 to 20 ha/MW (30 to 50 acres/MW) of installed capacity.[6]

Land Disturbed

Development and continued use disturb only a small portion of the land occupied by wind plants. The amount of land affected depends on the terrain, the size and number of roads, mounting pads, buildings, and other structures (Figure 11.1).

Wind projects disturb more land in hilly terrain than on level ground. Grading a road on steep slopes demands that the bulldozer operator make a cut on the uphill side while pushing the loosened soil downslope. These cut-

Figure 11.1. Land disturbed. View from Mt. San Jacinto onto the floor of the San Gorgonio Pass. The wind plant in the lower center straddles the Whitewater Wash. The wind plant in the upper right-hand corner adjoins Painted Hills. Both illustrate the amount of land occupied and used by wind plants. Far more land than was needed to install and service the turbines was graded at both projects.

and-fill slopes extend the disturbance beyond the immediate vicinity of the road or tower pad. On level terrain, grading disturbs less soil because there are no cut-and-fill slopes.

After examining two projects on steep terrain in southern California, one operator found that 7 to 10% of the soil of the total project area was disturbed for road and crane pad construction. According to an inspector formerly responsible for the Bureau of Land Management's oversight of the Alta Mesa site near Palm Springs, at most 45 acres (10%) were disturbed during construction of the 440-acre project, based on actual site observation.[7] In the hilly Altamost Pass, most of the lease agreements between landowners and wind plant operators stipulate that no more than 5% of the land can be removed from grazing. Yet where operators have shown little regard for the impact of grading, far more land has been disturbed than needed.

In a detailed evaluation for Riverside County as part of its Wind Implementation Monitoring Program, Tierra Madre Consultants found on average that 0.47 acre (0.2 ha) of soil was disturbed per turbine near Palm Springs, almost twice the amount of disturbed land anticipated, because developers graded wider roads and more of them than were necessary. They calculated that these projects disturbed a shocking 23% of the area occupied by the turbines. Several sites, such as those on Whitewater Hill, contributed substantially more than their share to excessive grading. Fortunately, Tierra Madre offered simple remedies: limit the amount of grading, and place stricter enforcement on these limitations.[8]

European wind plants disturb less land than plants in California. On at least two European projects, the land is tilled up to the base of the tower, and no allowance is made for access roads or even surface expressions of the foundation. When access is needed for heavy equipment, temporary roads are laid over the tilled soil. Wind turbines, roads, parking, and a small control building occupy only 3% of the flat Tændpibe–Velling Mærsk site. Most of the roads were preexisting. British wind engineers estimate that at sites such as Delabole in Cornwall, where the turbines are placed among the hedgerows, development permanently disturbs less than 1% of the land area.

Nonexclusive Land Use and Compatibility

Critics also charge that development of wind energy precludes use of the land for other purposes. To the contrary, wind turbines are compatible with most land uses. In a study intended specifically to address wind energy's land-use impact, landscape architects at CalPoly note that "it is important to emphasize that WECS [Wind Energy Conversion Systems] do not constitute a singular use of the land, as most other energy generation sources do. Grazing, recreation and numerous other activities can occur in and around wind farms."[9] In fact, they recommend deliberately choosing "sites that offer the possibility of multiple use" in preference to using sites where only wind energy is suitable. Grasslands, for example, allow both wind turbines and grazing. The CalPoly architects go on to propose that "locations of dispersed ORV [off-road vehicle] use and hiking are also good choices."[10]

Grazing is the most common example of a land use compatible with wind development. In the Altamont Pass and to some extent in the Tehachapi Pass, land leases for wind development specifically restrict the amount of land taken out of pasturage for cattle. Sheep graze peacefully beneath wind turbines along dikes in the Netherlands and on sites throughout Britain. To a comment about the numerous sheep grazing on the site during construction of the Cemmaes project in Wales, Taylor Woodrow's Tony Burton responded "Of course!" as if to say "Why wouldn't sheep be grazing here during construction?" To Burton it is the norm, and he thought no more about it.

However, grazing is but one example. In northern Europe the land beneath some wind power plants is tilled to the base of each tower. At others, the turbines are planted within the hedgerows and field boundaries. Some wind turbines in Denmark, machines the size used in California wind plants, are erected in their owners' backyards. Businesses in England, the Netherlands, Germany, and Denmark install wind turbines near their establishments. Outside Kiel, one McDonald's has humorously dubbed the Ventis wind turbine in its parking lot "McVentis" (Figure 11.2). At the port of Lauwersoog on the Wadden Sea in the Netherlands, a Lagerwey wind turbine stands near the entrance of a small business. In England, a Vestas wind turbine powers a veterinary clinic. Across the breadth of the United States, some 5000 small wind turbines are used in residential applications. In Europe it is not uncommon to see wind turbines operating in public places such as schools and sporting centers. On the popular strand at Dunkerque, mothers push their prams unconcernedly beneath a WindMaster turbine. Fishermen sit beneath the Nordtank turbines on the breakwater at Ebeltoft, while children play around the wind turbine at the ferry terminal's nearby visitors' center.

Helen Colijn, in *The Backroads of Holland,* suggests biking to the wind turbines on the dike at Urk. She describes how cyclists can find wind farms on Dutch tourist maps, reassuring them that cycling is permitted near the wind turbines just as it is on bikeways at dikes throughout the country.[11] Once there, cyclists will find signposted directions to sights along the route (Figure 11.3). Tourists cycling the scenic bike path on the eastern side of Denmark's Ringkøbing Fjord are tempted to detour into the heart of the adjacent Tændpibe–Velling Mærsk wind plant and visit the information kiosk

Figure 11.2. McVentis. Wind turbine at fast-food restaurant off autobahn near Kiel. (Courtesy of Winkra, Husum, Germany.)

(a)

(b)

Figure 11.3. (a) Dike Walk. The sign suggests sights to see on a walk or ride along the dike bordering the Ijsselmeer. Like most European wind projects, the WindMaster turbines at Urk are accessible to the public. Dogs (which bother the sheep), camping (which damages the dike), and motorcycles (which disturb the tranquility) are prohibited. (b), Marlies Kort studies her map before continuing her ride beneath the turbines at Urk. Kort coauthored a cycling guide to Dutch wind turbines. (Courtesy of Chris Westra Produktie, Amsterdam.)

before continuing on their way. Farther south, across the German border, the Wiedingharde Tourism Association may lead an evening walk along the dikes to the wind farm at Friedrick-Wilhelm-Lubke-Koog.

In practical terms, nearly all land uses except hunting and wilderness preservation can be compatible with wind development. By definition, wilderness requires the absence of man, and wind development is, after all, development. Hunting, though, is not inherently incompatible: California wind turbines have suffered surprisingly little damage from gunfire. Wind turbines do not hinder hunters, but hunting endangers those who service wind turbines.

For this reason alone, hunting is prohibited at all California wind plants. However, hunting from blinds with a controlled field of fire may be compatible.

Some charge that wind development effectively closes public lands to public use. This is not the case in either California or Europe. In the San Gorgonio Pass, for example, hikers on the Pacific Crest Trail have full access to the trail near 460 turbines on a Bureau of Land Management (BLM) lease atop Alta Mesa.

Coal Clough in England's South Pennines practically sits astride a footpath across the "long causeway" above the historic dale of Calder. A public car park solely for the use of hikers is just across the road from the turbines. The footpath itself crosses the wind farm at several points via gates installed for the purpose. As at Coal Clough, most wind projects in England and Wales can be reached by signposted public footpaths that existed long before the wind turbines were in place. Most turbines on public land, and many on private land, can be reached by public walking or cycling paths throughout northern Europe.

Normally, private land in the United States is closed to the public. Wind plants in the Tehachapi Pass have enabled access to previously restricted private land where the Pacific Crest Trail crosses BLM and private leases. The trail, which stretches from Canada to Mexico, is the longest footpath in the United States. Through the vehicle of easements, both private and public lands are now open to hikers along the Pacific Crest Trail (PCT) through the heart of Tehachapi's wind development. To publicize the trail and the public's free access to it, the Kern-Kaweah Chapter of the Sierra Club and the Kern Wind Energy Association jointly sponsor an annual spring hike among the wind turbines. They have been doing so since 1986. In 1990, nearly 100 people took the hike (Figure 11.4).

Clearly, the experience of hiking among wind turbines is different from that of hiking in a verdant forest. The mere thought of hiking among wind turbines enrages some purists. A local Audubon activist, Steve Ginsberg, not one to forsake hyperbole when it suits his purpose, proclaims that "Any hiker with the slightest sense of aesthetics or ecology [sic] might question such an absurd suggestion" that hiking among the wind turbines in the Tehachapi Pass is a pleasing recreational activity. Yet even he begrudgingly admits that "hundreds have taken the hikes," apparently responding to the "absurd suggestion."[12]

Ginsberg's diatribe serves only to illuminate his limited outdoor experience. Hiking across Tehachapi's Cameron Ridge on the PCT resembles the type of hiking found throughout Europe or in the eastern United States, for example on the Appalachian Trail, where the works of man are often present. Cameron Ridge is not a wilderness, park, or preserve, nor is it wooded. The sense of space, so important to the hiking experience in the western United States, remains. Hikers can still enjoy the spectacular views of the Mojave Desert, the Garlock Fault, and other prominent features. In short, the experience is

Figure 11.4. Lunch break on PCT. Hike on the Pacific Crest Trail across Cameron Ridge near Tehachapi, California. The local wind energy association and the local chapter of the Sierra Club sponsor the annual spring hike to promote both the trail and renewable energy.

different from walking through a coniferous forest, but many of the attributes of hiking on the desert remain.

Rural Preservation

The concern about how much land wind energy requires reflects the fear that the technology will consume valued open space. But rather than contribute to urban sprawl, wind plants help preserve rural land for rural uses. They do so by enabling those who live on the land to resist the relentless pressure to abandon the land and subdivide it for residential development. This is particularly important in California, where real-estate speculators hunger after the state's remaining open land. (Large-scale real-estate projects of 5000 to 10,000 homes each are not uncommon in California.) Once the land has been lost to suburbia, it is lost permanently to all rural uses, including cattle grazing, farming, wind generation, or simply open space.

Ranchers and farmers can protect their way of life by raising a new cash crop, the wind. Because the revenue from wind resources is steadier than that from agriculture, this new source of income helps them to better weather the farm economy's boom-and-bust cycles.

Some farmers are able to install wind turbines for their own use. But most wind plants will be built, owned, and operated by third parties who will pay the farmers royalties for the use of their land and the wind that flows over it. The royalties enable those who live on the land to supplement their income while continuing to do what they have done before: farming or ranching. Royalties from wind generation in the Altamont Pass have insulated area ranches from large-scale suburban development through several years of serious drought. "It absolutely saved the Walker family bacon," said Hugh Walker in an *Oakland Tribune* article. Walker was referring to the 700 wind turbines on his Altamont property. "It was a godsend in a time of need. We probably would have been forced to sell out" without the wind turbines, he concluded.[13]

California landowners typically receive 2 to 5% of gross revenues as royalties. These royalties can be a significant boost to rural economies. In many areas of the United States, the royalty revenue from wind energy can be several times that earned by grazing or tilling the land. For example, a 17-m 100-kW turbine can produce 200,000 kWh per year at good sites in California. At $0.10 per kilowatt-hour, the turbine will earn $20,000 in annual gross revenues, producing $400 to $1000 per year in royalties for the landowner. Assuming a turbine density of 30 acres/MW (12 ha/MW), each turbine requires 3 acres (1 hectare). As a result, the rancher could earn from $130 to $330 per acre ($500 to 1200 per hectare) per year—four to ten times the amount earned from grazing on California's arid range lands.

Owners earn these royalties over the course of the year, helping farmers bridge the gap between the expense of planting and the income from the harvest. In the midwestern United States, says the Union of Concerned Scientists' Michael Tennis, most of the royalties will occur during the winter months, when they are needed most.[14]

Some royalty contracts in California increase payments after 10 years. One lease increases the royalty to 10% of gross revenues after the first decade of operation. Wind developers also sometimes offer a modest one-time, up-front payment to secure the contract.

Royalties in Britain are similar to those in California and average 2 to 2.5% of gross revenues prorated among property owners. German royalties are comparable. Enercon, for example, says that one of their 500-kW turbines will earn a German farmer about 6000 DM ($3500) per year. One 500-kW Enercon turbine occupies about 6 ha (15 acres), for royalties of 1000 DM ($600) per hectare ($230 per acre) per year. However, Dutch and Danish landowners receive less. Turbine owners pay only 1000 to 1400 Ecu ($1100 to $1600) per turbine in the Netherlands; thus the landowner receives $180 to $270 per hectare ($70 to $100 per acre). Twenty-year leases in Denmark cost 25,000 to 50,000 DKK ($4000 to $8000) per turbine, according to Vestas. When built in the mid-1980s, Tændpibe–Velling Mærsk paid landowners a one-time fee of 20,000 DKK ($3300) per turbine, in addition to which they pay a prorated payment of about 13,000 DKK ($2100) per year in crop compensation for all 100 turbines. Vestas' V25 at Velling Maersk will generate about

$60,000 per year in revenues. Farmers receive the equivalent of $200 to $400 per turbine per year, 0.3 to 0.6% royalty plus 0.3% in payment for crop damage. Altogether, Danish farmers receive royalties from one-fifth to one-third those paid in California and Germany.

Planners in several north German districts have set aside locations where wind turbines are acceptable. As it has in Denmark, this action raised the price of land leases. Germany's Winkra finds royalties on land with planning permission of 15 to 20 DM per kilowatt. For a 500-kW turbine this is about 7500 to 10,000 DM ($4500 to $6000), equivalent to a 3 to 7% royalty on gross revenues.

The pattern of land ownership in the midwestern United States differs from that in California. There are far more parcels and owners to deal with in Minnesota, typically four owners per square mile (1.5 per square kilometer), than in California. Settlement patterns in the Midwest more closely resemble those in Europe than those in California.[15]

European developers have encountered an unexpected dilemma as a result of the large number of neighbors at each site. Down at the local pub in Cemaes Bay, for example, neighbors of EcoGen's Rhy-d-Groes project grumble: "Why couldn't they have put them on my land?" The farmers say they could use the money. Under the current system, adjoining landowners must endure the impact without gaining any of the direct benefits. Winkra's Henning Holst faces the same problem in Germany. It is possible, be believes, to construct a royalty system on impact zones. Those with turbines on their land would receive compensation as they do today. But turbine owners could also pay peripheral landowners more modest sums based on the parcel's distance from the project. Holst envisions compensation based on a series of concentric zones around the turbines. Those nearest the turbines would receive more than those farther away.

As Europe and North America rush to reduce farm supports in the face of new trade treaties such as GATT and NAFTA, there could be a vast conversion of agricultural land now in production. In early 1994, Mexican farmers so feared they would lose control of their staple crop—corn—to efficient Yankee farmers in El Norte that they were near rebellion. Such treaties will have far-reaching effects on rural populations, making it even more difficult for those living on the land to remain there. For those who find themselves growing an uncompetitive crop, there are few alternatives. Biomass crops for liquid fuels are one possibility; wind energy is another.

Turbine Density

Ultimately, the question about wind energy's land requirement becomes "How many wind turbines will people accept on the landscape?" It is that question to which EDF's president, quoted at the beginning of this section. refers when he speaks of "carpeting" the French countryside with wind turbines. Ménage might be surprised to learn that long before the advent of the French nuclear program, there were thousands of windmills "carpeting" parts of France.

In 1694, French military engineer Vauban estimated that there were 16,000 windmills in what was then France. Historians believe that during the eighteenth century, as many as 20,000 windmills ground grain or pumped water throughout the country.[16] These would most certainly have been concentrated along the coasts, the drainage divide of the Loire valley, and on hilltops in proximity to major cities such as on Montmartre in Paris. There were 2000 windmills along the English Channel in Nord-Pas-de-Calais alone.[17] In Germany during the nineteenth century, there were some 20,000 windmills, concentrated primarily in the northern coastal states.[18]

Jens-Peter Molly, director of the German Wind Energy Institute, estimates that before their decline in the late nineteenth century there were 100,000 windmills operating in northern Europe.[19] That is a lot of windmills. Unlike the spindly water-pumping windmills dotting North America, these were conspicuous machines. Many were driven by rotors 25 m (80 ft) in diameter and stood atop towers of equal height. They were as big as today's medium-sized wind turbines. In some regions, such as the lowlands of northern Europe and central England, they were a significant feature of the landscape, and residents lived peacefully in their presence. The large number of windmills, and occasional local concentrations, seem to have had no ill effects.

With 10,000 windmills on the Netherlands' landscape in the eighteenth century, a country of much less land area than that of the modern nation we know, these machines were literally everywhere. Some were even clustered along drainage canals much like the rows of wind turbines in a modern-day wind plant. The 19 drainage windmills at Kinderdijk, probably the world's oldest wind farm, operated until 1950 (Table 11.2).

During the late eighteenth century, windmills were nearly eight times more common on the Dutch landscape than modern wind turbines are in California. There was a windmill on nearly every square mile of the Netherlands (or one every 3 km^2). The visual image of these turbines would have been far different from that of the concentrated arrays in California. Nevertheless, in Denmark and the Netherlands during this period, windmills would have been visible in every direction.

Table 11.2
Density of Windmills and Medium-Sized Wind Turbines on the Landscape

	Turbines	Units/mi^2	Units/km^2
California, 1993	16,500	0.10	0.04
Denmark, 1993	4,000	0.24	0.09
Denmark, nineteenth century	3,000	0.18	0.07
Netherlands, eighteenth century	10,000	0.77	0.30
England and Wales, nineteenth century	8,000	0.16	0.06
Northern Germany, nineteenth century[a]	18,000	0.74	0.29
Nord-Pas-de-Calais	2,000	0.42	0.16

[a] Lower Saxony and Schleswig-Holstein.

Denmark is the only nation where the density of modern wind turbines rivals that of the nineteenth century. Today, as it must have been then, it is difficult to go anywhere in Denmark without seeing a wind turbine somewhere on the horizon. During the 1920s there were even more. Danish researchers Bent Rasmussen and Flemming Øster estimate that there were 30,000 small windmills in use across the country. These machines were a cross between the American water-pumping windmill and the traditional European mill. With a rotor 5 meters (16 feet) in diameter, they were bigger than most American farm windmills but much smaller than the traditional windmill.[20] A traveler could find almost two of these windmills every square mile (about one per square kilometer), roughly equivalent to the density of farm windmills on the American Great Plains.

British policymakers are striving toward a "landscape with wind turbines," with which they are familiar, and want to avoid the "wind turbine landscape" they see in California. Academician, inventor, and wind farm developer Peter Musgrove calculates that 2000 MW of wind generating capacity—7000 wind turbines—could be installed in Britain by 2005. These machines would be similar in size to traditional windmills, although on taller, more slender towers, and would represent about the same density on the landscape as in the nineteenth century.[21]

Comparative Land Use: Wind and Solar

Earlier in this section we saw how wind energy's specific land intensity compares with that of the fuel cycle for a coal-fired power plant. Wind energy also compares favorably to other solar-electric technologies. When used in power plant applications, most solar-electric technologies share similar land requirements, although there are some important exceptions.

The California Energy Commission estimates that photovoltaic power plants occupy about 9 acres/MW (4 ha/MW) of installed capacity.[22] This is verified by several installations, including those of the Sacramento Municipal Utility District, Pacific Gas & Electric, and the Mont Soleil project in Switzerland. Southern California Edison found a similar land requirement for Solar One, its idle 10-MW central receiver. The CEC estimates that a new central receiver will require up to 19 acres/MW (8 ha/MW). Because their efficiency is higher than that of photovoltaics, solar thermal technologies using arrays of parabolic troughs, as at the Luz plants, occupy 4 to 5 acres/MW (2 ha/MW), about half the land required by photovoltaic plants (Table 11.3).

However, some photovoltaic plants occupy more land than is commonly realized. The slightly greater requirement for the Chronar module at Pacific Gas & Electric's PV-USA (Photovoltaics for Utility Scale Applications) demonstration plant results from the cells' low efficiency. Arco's Carrizo Plain and Lugo plants are not necessarily representative of typical photovoltaic installa-

Table 11.3
Comparative Land Use: Wind and Solar

	Track Axis	*Area*			*Area Occupied*		*Area Used*		Used/ Occup.
		acres	ha	MW	acres/MW	ha/MW	acres/MW	ha/MW	
Wind, 5 × 10 array					38.6	16	1.9	0.8	0.05
PV-CEC					8.9	4	2.1	0.9	0.24
Mont Soleil, China	1	5	2	0.5	9.9	4	2.3	0.9	0.23
PV RWE-Kobern, Germany	F	12	5	0.34	36.3	15	2.6	1.1	0.072
PV-Pellworm, Germany		4	1.6	0.3	13.2	5	3.2	1.3	0.24
Solar thermal-CEC					4.4	2	4.4	1.8	1.00
Luz solar thermal	1	380	154	80	4.8	2	4.8	1.9	1.00
Carrizo Plains PV	2	170	69	6.5	26.2	11	6.3	2.5	0.24
Coal (30 yr, 68% cf)					8.0	3	8.0	3.3	1.0
SMUD PV	1	18	7	2	9.0	4	9.0	3.6	1.00
Solar One	2	90	36	10	9.0	4	9.0	3.6	1.00
PV-USA, IPS	1	2	1	0.2	9.2	4	9.2	3.7	1.00
Central Recvr-CEC					18.5	7	9.3	3.7	0.5
PV-USA, Arco	1	2	1	0.2	11.1	4	11.1	4.5	1.00
PV-USA, Chronar	F	5	2	0.4	13.2	5	13.2	5.3	1.00
Lugo PV	2	20	8	1	20.0	8	20.0	8.1	1.00

tions. The plants occupy two to three times the land of most other photovoltaic arrays, due primarily to an unnecessary buffer required by county planners. The 0.5-MW German plant at Kobern-Gondorf west of Koblenz is also atypical. There the arrays are managed as part of ecologic units adapted to the terrain. This creates a pleasing aesthetic effect but occupies more land than a rectangular array. Interestingly, the photovoltaic units at Kobern-Gondorf have nearly the same amount of specific land requirement as the wind plant used in the previous comparison with a 1000-MW coal plant.

Wind energy occupies far more land than do most other solar technologies. However, as in the example for coal, the amount of land used by wind plants is comparable to that of solar central stations. The solar plants in Table 11.3 use a portion of their sites for roads and buildings. This is land unavailable for other uses. The land "used" by photovoltaic arrays varies substantially, depending on how the site is managed. The impacts may not be directly comparable with those from wind projects. For example, the impact on vegetation of shading from solar arrays differs markedly from the impact of a bulldozer blade scraping the surface bare for a road. The actual land consumed by solar plants is not well understood.

Of the solar plants in Table 11.3, Arco's plant at Carrizo and the German plant at Kobern-Gondorf most resemble a wind power plant, where much of the land on the site remains undisturbed. At one time, sheep even grazed on the Arco site—until a few bucks butted the mirrors, causing minor damage.[23]

In contrast to the sites at Carrizo and Kobern-Gondorf, the entire site of the 2-MW plant operated by the Sacramento Municipal Utility District at the defunct Rancho Seco nuclear power station is cleared to bare earth as a precaution against fire. On one of the two projects at the site, SMUD laid down gravel to prevent growth of vegetation; on the other, they doused the area with a herbicide.[24] Similarly, the entire area beneath Luz's Solar Energy Generating Stations is cleared, in part to build a berm against flash flooding, a common desert hazard.[25]

Southern California Edison faced an unusual dilemma at its Solar One central receiver. Initially, SCE graded the site, then allowed native vegetation to return. However, after finding sidewinders (venomous desert snakes akin to rattlesnakes) taking up residence in the junction boxes beneath the heliostats, SCE sprayed the site with a herbicide.[26]

Thus at the SMUD, Luz, Lugo, and Solar One plants, 100% of each site is used, leaving the site devoid of vegetation. These practices need not necessarily be the pattern for all future solar plants. Panels could be elevated sufficiently to allow cattle and sheep to graze beneath them without interference. Grazing, in turn, could limit the fire hazard, obviating the need to clear the site of vegetation with herbicides.

PG&E is grappling with these land-use issues at its PV-USA demonstration project. PG&E chose to build the initial stages of PV-USA atop an abandoned sewage disposal pond near Davis, California, turning previously unused, and unusable, land to a productive purpose. According to Steve Hester, PV-USA's project leader, PG&E hopes to control dust and weeds without resorting to herbicides. Ideally, Hester would like to find crops that will prosper in the partial shade of the collectors.[27] Management practices determine the extent of land disturbed by each technology, and these management practices can be changed when minimizing soil disturbance becomes a priority.

Wind and photovoltaics are also not mutually exclusive on the same site; they are compatible technologies. Some have suggested that in windy areas with intense sunlight, such as on the fringes of the Mojave Desert near Tehachapi and on the valley floor of the San Gorgonio Pass near Palm Springs, photovoltaic arrays could be built among the wind turbines already in service. Such an array of wind turbines and photovoltaic modules would use land more efficiently than would either technology alone, reducing the total land area disturbed by each technology through joint use of roads and other infrastructure. For the time being, this is just a dream. Plans to build such projects in the past have always faltered.

When developed with care, solar technologies use no more land than do coal-fired power plants and their mines. This finding debunks a myth that solar-electric technologies, wind in particular, are land-intensive.[28]

Roads and Road Construction

Next to the turbines themselves, it is roads that determine how the public sees the wind industry. Roads consume land, and roads are visually prominent. It is the roads—their number, width, and location—that are the most visible manifestations of how developers and site managers view the land and their responsibility for protecting it from abuse.

Environmentalists cringe at the thought of a bulldozer blade slicing into the earth, a part of the living biosphere. Out of reverence for the land, their credo is "to go lightly." Only with great reluctance do they agree that the land can be "worked," the soil tilled for crops, a trail hacked into a hillside for hikers, or a pad bulldozed for a park visitor center—and only then, if efforts are made to minimize the damage. Their green consciousness can be heard pleading: "If there must be wind turbines, and they must have roads, then please minimize the scraping and gouging as much as is humanly possible."

In the United States, where roads may use up to 5% of a site, roads account for nearly all the land disturbed by wind power plants. But in northern Europe, where almost every aspect of a proposed wind plant is examined more critically than in North America, the need for roads is often called into question. Where roads are used in Europe, there are far fewer of them, and they are considerably narrower, than those crisscrossing American projects.

Some European projects have been built without adding new roads. Vestas and the Danish utility Vestkraft installed 100 turbines at Tændpibe–Velling Mærsk using only those roads already present. The turbines stand in tilled fields. Should the turbines need service requiring a crane, a temporary road of steel panels is laid, and the panels are later removed. According to Vestas, the operator, the absence of roads to individual turbines has not created any undue hardship, although a smaller project would incur additional costs. Tændpibe–Velling Mærsk is large enough, says Vestas, to justify purchase of the special all-terrain vehicles used for routine access (Figure 11.5).

Germany's first wind plant at Kaiser-Wilhelm-Koog, on the north bank of the Elbe estuary, also uses only existing roads for access. Neighboring farmers till the polder to the base of the towers. Across the English Channel, Peter Edwards was determined to use wind energy to supplement his farming income. He built Delabole, Britain's first wind farm, without roads, using bulldozers and all-terrain cranes to maneuver around his muddy site. Not willing to part with his prime agricultural land, Edwards installed his 10 turbines within the farm's hedgerows (Figure 11.6).

Similarly, EcoGen installed the 24 turbines at Rhyd-y-Groes on Anglesey within the hedgerows to minimize the impact on local landowners. This is common practice in Denmark as well, where individual turbines or small clusters of machines often stand along fencerows or in the border between fields.

At Carland Cross, Britain's second wind plant, Renewable Energy Systems used only existing farm tracks, and restored temporary construction roads to

*Figure 11.5. Wind energy and cropland. At Tændpibe–Velling Mærsk (*a*) and Kaiser-Wilhelm-Koog (*b*), farmers till the soil to the towers. Servicing the turbines at Velling Mærsk requires all-terrain vehicles.*

pasture. There, as at Kaiser-Wilhelm-Koog and Velling Mærsk, there are no surface expressions of the foundations. The tubular towers seem to plunge directly into the ground without the large concrete pad often seen at the base of such towers. RES achieves this effect by burying the concrete pads for all 15 machines 1 to 2 m (3 to 6 ft) below the surface, thus allowing the farmer to plow up to the base of the towers.

Most of National Windpower's 21 sites on the moorland at Cold Northcott in Cornwall are reachable by existing farm tracks, says Mike Hitner of constructor Taylor Woodrow. The boggy ground unavoidably required some temporary roads, he says, but Taylor Woodrow tried to minimize their number to reduce the loss of cropland. Taylor Woodrow will remove these roads after construction. As has been done elsewhere in Britain, Taylor Woodrow laid temporary roads over moorland with geotextiles, a tough fabric mat that can literally be peeled off the ground after construction (Figure 11.7).

Renewable Energy Systems notes that if the foundations are buried and no new roads are built, British wind plants use only 0.01% of the site's land area; where roads are necessary, about 1% of the site is taken out of agriculture.[29]

Not all Europeans agree that wind turbines can be installed and serviced without roads. "They're machines, after all, and you do have to work on them," says Manfred Luhrs, of Enercon's Marne office east of Kaiser-Wilhelm-Koog. Enercon, says Luhrs, typically uses roads 4 m (13 ft) wide. Some wind plant roads in California are twice that width.

Figure 11.6. Delabole hedgerows. Farmer Peter Edwards planted his wind turbines among existing hedgerows to minimize the amount of land used.

California wind companies are products of their culture. It is not common to see roads of four lanes, each 12 ft (3.7 m) wide, serving new housing tracts of only a dozen homes. In a region accustomed to explosive growth, the roads are needed "to accommodate future traffic" says one city planner. Such practices are found throughout the vast reaches of the western United States, where land has long been viewed as a disposable commodity.

California wind developers say wide roads speed construction by enabling two-way traffic of heavy vehicles to move at high speed. These roads met the need of the frantic year-end construction schedules typical of California's tax-credit era. Yet even during the mid-1980s, one BLM staffer called a halt to such practices on some federal land near Palm Springs. The BLM insisted that the developer of Alta Mesa use narrower roads than previously permitted and allow only one-way traffic. The developer complied, and the project was built on time despite more limited access.

SeaWest, another California wind company, operates Ecogen's PnL site in mid-Wales. Despite their previous experience with wide roads, SeaWest found the 2.5-m (8-ft) roads on the PnL site to be entirely adequate. If they had to do it over, says SeaWest's Mark Haller, they would widen the roads slightly at the base of each tower. This would enable the erection crane to extend its outriggers from the road safely without construction of the separate crane pads found at a few British projects and on most California sites.

Renewable Energy Systems uses this technique at Coal Clough in the Pennines, where it links each turbine in a "daisy chain" along a winding road.

Figure 11.7. British road construction. Yorkshire Water Co.'s Keith Pitcher and Angela Willis examine the crushed stone laid atop a geotextile mat on a road at Chelker Reservoir in the Pennines. The width of the road, 3 to 4 m (10 to 13 ft), is typical of that used for access to wind farms in Britain. The geotextile mat permits removal of temporary roads after construction is complete. (Photo by Guzelian; courtesy of Yorkshire Water, Bradford, Yorkshire, England.)

The boggy upland soils at both Coal Clough and nearby Ovenden Moor required the construction of roadbeds raised slightly above the mire. This alters surface features but less dramatically than the benches created by the erection of turbines on steep slopes in California. The Dutch use a similar technique in the polderlands of Noord Holland. At a NedWind site in the Wieringerwaard, the Dutch utility PEN paved an access road about 3 m wide over a raised right-of-way 10 to 12 m in width.

British developers expect that within a year, moorland grasses will cover newly constructed roads, reducing the visual contrast between the roads and the surrounding terrain and softening the roads' intrusive effect. After completion of the WEG project on the Mynydd-y-Cemais plateau in Wales, Taylor Woodrow quickly seeded the 3.75-m (12-ft) wide roads. Taylor Woodrow's Tony Burton says that grass will eventually cover most of the road surface, leaving only a 2.5-m (8-ft) path for service vehicles. The roads amid the moor's sedges and peat will then resemble local farm tracks (Figure 11-8). Rather than grade wide roads, Taylor Woodrow provided occasional passing bays to ease traffic flow during construction. These too were seeded.

Of the hundreds of miles of roads weaving among California's wind turbines, those built by U.S. Windpower come closest to resembling those in Europe. As with other California firms, the principal access roads to U.S. Windpower

Figure 11.8. Farm track. A graveled farm track is used to service these Nordtank turbines at Norrekær Enge in Denmark.

sites are wide thoroughfares. U.S. Windpower's roads differ from other California companies, though, where the roads branch off to reach distant strings of turbines. "We don't have a bare-earth policy" toward grading roads, says USW's Joanie Stewart. U.S. Windpower grades its roads to a width of 12 feet (3.7 meters), substantially less than other California firms, and they grade less often than other companies, allowing some grass to revegetate the road surface. Some of the roads Stewart refers to have reverted to mere farm tracks.

At its Solano County site, for example, U.S. Windpower graded roads to each turbine during construction of its 600-machine project. They regrade the roads periodically during the summer months, to reduce the fire hazard. (The roads act as fire breaks). But the access roads are seldom visible from nearby highways, and the turbines appear to spring out of the golden wheat fields that surround them. This provides a far more pastoral setting than elsewhere in the state, where wind turbines rest on benches cut deeply into hillsides. These "tread lightly" practices effectively make USW's Solano project one of California's premier wind plants.

Roads and Erosion

In the spring of 1992, Tehachapi-area wind plants were ill-prepared for a deluge. After seven years of drought, runoff-absorbing vegetation was sparse. Bankruptcies and lax management had taken their toll on maintenance of erosion-control structures. When heavy rains struck, runoff surged along roadcuts to cascade down steep slopes, gouging deep gullies into the mountainsides and leaving some wind turbines standing precariously on exposed foundations.

Although the Tehachapi area is susceptible to cloudbursts typical of the world's desert regions and critics had warned developers during the mid-1980s that they were cutting roads in erodable soils too close to the edge of steep escarpments, the industry appeared dumbfounded when it awoke to the damage soon afterward.

To some wind companies, erosion is a natural process. "So what if it's a little more than usual?" they say. "There's no harm done." Environmentalists see the issue differently. Erosion is indeed natural. There are many barren, highly eroded slopes in the Tehachapi Mountains not unlike those found in the badlands of South Dakota. However, accelerated erosion, caused by artificially baring the earth to the impact of rainfall and concentrating the resulting runoff along roads, is not "natural."

Environmentalists' distaste of erosion, apart from the increased siltation of streambeds, alteration of stream courses, and increased flooding that accompany it, results from the scars it leaves on the land. The rill-and-gully erosion seen in Tehachapi cuts deep into the surface of the landscape. More galling than the erosion itself is the abuse of the resource it represents, because accelerated erosion is unnecessary and can be avoided. The industry must control erosion or it will certainly suffer further at the pen of activists such as Audubon's Steve Ginsberg, to whom erosion "is just one of many egregious examples of how wind energy is ripping up the Tehachapis, and its [the industry's] lack of true environmental concern"[30] (Figure 11.9).

Wind companies can reduce the risk of serious erosion by minimizing the amount of earth disturbed during construction, principally by eliminating unnecessary roads, avoiding construction on steep slopes, allowing buffers of undisturbed soil near drainages and at the edge of plateaus, assuring revegetation of disturbed soils, and designing erosion-control structures adequate to the task (Figure 11.10).

Figure 11.9. Unnecessary grading. Zephyr site near Tehachapi is the wind industry's most notorious example of unnecessary grading and of a site where wind energy was clearly unsuitable. Since this photo was taken in the mid-1980s, all the turbines have been removed. The environmental damage was for nought and there is little that can be done now to heal these scars on the hillside.

The single most reliable technique for limiting erosion is to avoid grading roads in the first place. Glenn Harris, a biologist for BLM's Ridgecrest office, suggests that driving overland rather than grading roads to install and service turbines will significantly lessen erosion damage in arid lands. He admits that the tires or treads of vehicles crush the surface of perennial plants such as bunch grasses and shrubs, but that they leave the root systems intact. "Ninety percent of the plant is still there," says Harris. "They're root sprouters," because they have adapted to frequent fires. The plants will continue to live and bind the soil in place. A bulldozer blade, in contrast, scrapes the earth bare, removing plants, root systems and all. If traffic frequently follows the same overland route, the path can be "hardened," Harris adds, to prevent ruts from concentrating runoff. Hardening could entail selective placement of open paving blocks like those used in parking lots outside Copenhagen, which support the weight of the vehicle while allowing plants to grow between spaces in the blocks.

SeaWest's Jeff Ghilardi emphasizes that maintaining erosion-control structures is essential to fighting erosion. It is possible that the original practices on Tehachapi's Cameron Ridge, where the erosion is most serious, were

Figure 11.10. Pleasing ridgetop array. Although the lattice tower is angular and somewhat cluttered, U.S. Windpower's model 56-100 is among the most aesthetically pleasing American-designed wind turbines. The California manufacturer and developer has also taken pains to minimize road construction and attendant scarring. U.S. Windpower also relies entirely on its own product, giving a uniformity to the more than 4000 turbines it operates in the Altamont Pass.

sufficient, he says, but that over time, funds were diverted to other activities. With an extended drought, erosion control became secondary.

Once gullies form, they need treatment immediately or erosion will accelerate exponentially during subsequent storms. Although rills on moderate slopes can be graded and revegetated, gullies on steep slopes become a permanent feature of the landscape.

Although the wind industry is clearly at fault, its sins should be seen in perspective. Nearby, 120 years after the Southern Pacific punched its way through the Tehachapi Mountains, the railroads' bare embankments continue to erode without treatment, and erosion from housing construction on steep hillsides overlooking Highway 58 in Tehachapi digs deeper gullies year by year. Even the Tejon Ranch, which brags about its environmental attentiveness, saw sheet-and-rill erosion wash silt down its steep slopes bordering Interstate 5 during the same heavy rains that afflicted Tehachapi.

Still, the industry must do better. The public expects it. To gauge the amount the wind industry should be willing to pay for site restoration and erosion control, consider that the average coal mine in Maryland spends $4500 per acre ($11,000 per hectare) for reclamation. Certainly, wind companies can do as least as much as the coal industry.

Roads and Fugitive Dust

It is hard for Cornish farmer Peter Edwards, up to his knees in mud from 60 inches (1.5 m) of rain per year, to imagine anyone concerned about dust. But dust from unpaved roads in arid regions of the United States is a major source of air pollution. Dust could eventually force closure of two important military bases in Kern County: China Lake Naval Weapons Center, and Edwards Air Force Base. (The latter is where the space shuttle often lands.)

Dust has become a concern of the wind industry because of regulations that may affect the miles of unpaved roads used on California wind farms. These regulations could limit traffic on the roads or require regular watering or chemical treatment. In the San Gorgonio Pass, Riverside County regulations require operators to reduce speeds on unpaved roads in high winds. Regulators do not rule out the possibility of more stringent action in the future.

Reducing speeds on dirt roads is as much a question of neighborliness as it is a need to meet an abstract regulation. Although the wind farms are seemingly remote, people live in the vicinity of wind plants in all three of California's resource areas. In the Tehachapi Pass, for example, prevailing winds carry dust from roads servicing the 1000 turbines on the SeaWest site over the Homestead tract and on toward the town of Mojave. Dust is a part of life on the desert, where the landscape is dotted with Joshua trees, saltbush, creosote bush, and other desert plants adapted to scant rainfall. Despite their willingness to live under harsh conditions, Mojave residents have occasionally complained about dust from the wind plants.

Rather than dismiss the residents' complaints as a price of living on the desert, SeaWest's site manager Sean Roberts listens. He lives downwind in the Homestead area himself. These are his neighbors. Roberts takes a personal interest in dust plumes rising from SeaWest's service roads. He prominently posts a 25-mph speed limit, and instructs his employees to obey it. He also regularly waters main thoroughfares to hold down dust. Roberts also prohibits employees from indiscriminately driving vehicles overland and damaging fragile desert vegetation that shields the soil from the wind. Violators are subject to dismissal. Roberts' personal attention to this management detail gets results. More so than that at any other site in Tehachapi, the mile-long entrance road to the SeaWest site often has a hard-packed surface from frequent watering that is nearly pavement-like in preventing dust. As Roberts demonstrates, the best treatment for fugitive dust, as for erosion, is to avoid the problem in the first place by minimizing bare land exposed to the elements.

To summarize, wind energy uses far less land than commonly believed. With careful project design such as that at Tændpibe–Velling Mærsk and sites in Britain, and attentive managers such as SeaWest's Roberts, the impact on the land that wind plants do use can be minimized.

Chapter 11 Endnotes

1. Meridian Corporation, "Energy System Emissions and Material Requirements," U.S. Department of Energy, Washington, DC, February 1989, p. 22.
2. U. Fritsche, L. Rausch, and K. Simon, Environmental Analysis of Energy Systems: The Total-Emission-Model for Integrated Systems, Summary and Major Findings, Institut fur Angewandte Ökologie, Darmstadt, Germany, September 1989, p. 17.
3. David M. Gates, *Energy and Ecology* (Sunderland, Mass.: Sinauer Associates, 1985), pp. 249–250. Gates estimates the land requirement for both surface and deep mines to serve a 1000-MW plant per year. However, he overstates the land requirement by failing to account for data which assume 30 years of mining. The data for the deep mine have been corrected.
4. Meridian Corporation, Energy System Emissions, pp. 8, 12, 22. Meridian uses slightly different assumptions than those in the example. They estimated that a 500-MW plant would require 365 acres per year to generate 3.5 TWh, about 100 acres/TWh. Generating 6 TWh would consume 600 acres (242 ha).
5. Howard Wilshire and Douglas Prose, "Wind Energy Development in California, USA," Environmental Management, 10:6, 1986.
6. Ezio Sesto, "L'Energie eolienne en Europe: il est temps d'Agir," in Energie Eolienne, Séminaire de Contractants, Lastours, France, September 1991, pp. 173–174.
7. Leslie Cone, U.S. Department of the Interior, Bureau of Land Management, Indio Resource Area, Riverside, CA, personal communication, November 1985.
8. WIMP, Phase II, "Wind Implementation Monitoring Program," Riverside County, Riverside, CA, Tierra Madre Consultants, August 1986, pp. C11–C16.
9. R. Fulton, K. Koch, and C. Moffat, "Wind Energy Study, Angeles National Forest," Graduate Studies in Landscape Architecture, California State Polytechnic University, Pomona, CA, June 1984, p. 14.
10. Ibid., p. 64.
11. Helen Colijn, *The Backroads of Holland: Scenic Excursions by Bicycle, Car, Train, or Boat* (San Francisco: Bicycle Books, 1992), pp. 71–72. See also the cycling itineraries in *A Closer Look at Wind Energy* (Amsterdam: Chris Westra Produktie, 1993).
12. Steve Ginsberg, "The Wind Power Panacea: Is There Snake Oil in Paradise?" *Audubon Imprint,* Santa Monica, CA Bay Audubon, 17:1, 19, pp. 1–5.
13. J. Miller, "Windfall: Altamont Wind Farms Play Integral Energy Role," *Oakland Tribune,* Oakland, CA, July 30, 1989.
14. Michael Tennis, oral comments at Windpower '94, annual conference of the American Wind Energy Association, Minneapolis, MN, May 10, 1994.
15. Leslie Lamarre, "A Growth Market in Wind Power," *EPRI Journal,* December, 1992, pp. 4–15.
16. Jean-Claude Debeir, Jean-Paul Deléage, and Daniel Hémery, *In the Servitude of Power: Energy and Civilization Through the Ages,* trans. John Barzman (London: Zed Books, 1991), p. 90.

17. "Les Verts du Nord-Pas-de-Calais," *Systèmes Solaires,* 79, 1992.
18. Wolfgang Frode, *Windmühlen* (Hamburg: Ellert & Richter, 1987), p. 7.
19. Jens-Peter Molly, *Windenergie: Theorie, Anwendung, Messung* (Karlsruhe: Verlag C. F. Müller, 1990), p. 56.
20. Bent Rasmussen and Flemming Øster, "Power from the Wind," in *Wind Energy in Denmark: Research and Technological Development,* F. Øster and H. Andersen, eds. (Copenhagen) Danish Energy Agency 1990), pp. 7–11.
21. Peter Musgrove, National Windpower, oral comments, 2nd World Renewable Energy Congress, Reading, England, September 1992.
22. "Technology Characterization Final Report," Sacramento, CA: California Energy Commission, November 22, 1991, p. E16-6.
23. M. Stimson, Arco (now Siemens) Solar, Camarillo, CA, personal communication, October 1989.
24. D. Collier, Sacramento Municipal Utility District, Sacramento, CA, personal communication, November 1989.
25. A. Erickson, Luz International Ltd., Los Angeles, personal communication, October 1989.
26. C. Lopez, Southern California Edison Co., Rosemead, CA, personal communication, November 1989.
27. S. Hester, Pacific Gas & Electric Co., PV-USA, San Ramon, CA, personal communication, November 1989.
28. Meridian Corporation, "Energy System Emission," p. 22. Central station PV requires 0.08 acre/GWh; coal, 0.09 acre/GWh.
29. P. M. Quilleash, "Wind Farm Design I: Principles of Wind Energy Conversion," a short course, Imperial College of Science, Technology and Medicine, London, July 1993.
30. Ginsberg, "The Wind Power Panacea."

12

Benefits

"Vindkraft? Ja tak!—Wind turbines? Yes, thanks!" Danish spin on antinuclear bumper sticker popular among German Greens, "Atomkraft? Nein Danke!—Nuclear plants? No thanks!"

This chapter skirts a comprehensive examination of wind energy's benefits. Other advocates of the "soft path," such as Amory Lovins, have previously detailed the benefits of wind and solar energy at length. However, new information on the energy balance of wind turbines, the social costs of competing sources, and the relationship of wind energy to global warming that have come to light during the past decade warrant discussion. Several other topics that have generated interest among public officials and energy policy analysts are also introduced.

Energy Balance

The energy generated by wind turbines pays for the materials used in their construction within a matter of months. Yet the question as to whether they do, thought by industry analysts to have been effectively answered during the 1970s, is continually raised by critics of wind energy. It was the first avenue that desert activist Howard Wilshire sought in his quest to find a magic bullet that would kill the wind energy monster for all time.

The question possibly arises from reports about the poor energy balance of photovoltaics. Early solar cells consumed more energy than they produced, according to Jos Beurskens, manager of Renewable Energy for the Netherlands' Energy Research Foundation, ECN. He says that contemporary products perform far better and pay back the energy contained in their manufacture within 10 years. As the performance of photovoltaics continues to improve, so will their energy balance. In contrast to photovoltaics, notes Beurskens, wind turbines pay for themselves quickly, despite the use of such seemingly energy-intensive materials as steel and fiberglass.

In the early 1990s, researchers again examined the question. Two Danish studies considered a typical Danish wind turbine of the period, operating under typical Danish conditions.[1] A German study at Munich's Technical University, by far the most extensive, examined the energy payback of wind turbines from 10 kW to 3 MW in size.[2] The results of the three studies are comparable: medium-sized wind turbines installed in areas with commercially usable wind resources will pay for themselves easily within one year (Table 12.1). At 7-m/s (16-mph) sites such as those on the North Sea coast or in California's mountain passes, turbines will return their energy in three to five months and at sites typical of North America's Great Plains in four to six months. Even at low wind sites, the turbines will pay for themselves in less than one year. As expected, much of the energy used to manufacture the turbine is represented by the rotor and nacelle. But more than one-third of the total energy consumed by the wind turbine is represented by the concrete foundation and steel towers.

According to the German study by Gerd Hagedorn, wind turbines produce 4 to 33 times more energy during their 20-year lifetimes than that used in their construction. Coal plants produce 64 times more energy and nuclear

Table 12.1
Energy Balance or Payback

Wind Turbine Diameter (m)	Wind Turbine Power (kW)	Energy Consumed (MWh)	Energy Generated for Wind Regime (MWh/yr)			Payback Within Various Wind Regimes (months)		
			Average Danish production[a]					
	95	58	210			3.3		
			Roughness Class[b]			*Roughness Class*[b]		
			0	1	2	0	1	2
	150	212	529	395	315	4.8	6.4	8.1
			Average Annual Wind Speed[c]					
			7 m/s 16 mph	5.5 m/s 12 mph	4 m/s 9 mph	7 m/s	5.5 m/s	4 m/s
12.5	45	49	132	94	55	4.5	6.3	10.7
27	225	169	787	533	305	2.6	3.8	6.6
32	300	296	1049	710	411	3.4	5.0	8.6
80	3000	2817	8989	6025	4027	3.8	5.6	8.4

[a] Erik Grum-Schwensen, Wind Stats, Spring 1990.
[b] A. Gydesen et al., Danish Environment Ministry, 1990.
[c] G. Hagedorn and F. Ilmberger, German Ministry for Technology Development (BMFT), Munich, August 1991.

108 times more than that used in their construction. Current photovoltaic technology produces one to three times the energy represented by their materials. When fuel is included, coal and nuclear plants deliver only one-third of the total energy used in their construction and in their fuel supply because fuel consumption dwarfs the amount of energy in the plant's materials.[3]

Number of Homes Served

A common tactic in public policy disputes is to downgrade the benefits and magnify the impacts of a proposed action. Next to energy balance, opponents of wind energy most frequently turn their attention to the amount of energy generated and its significance.

When a utility tries to place a power plant in perspective for a public that seldom understands the difference between a kilowatt and a kilowatt–hour, it often attempts to do so by equating the plant with a point of reference known to all. This could be the number of light bulbs the plant will power or some other measure. Frequently, that measure is the number of homes the plant will serve. For example, Southern California Edison's Liz McDannel reports that the utility uses a rule of thumb in testimony before the state's Public Utility Commission and in public statements that the average house consumes 500 kWh per month or 6000 kWh per year in its service area.

Wind companies do the same. In attempting to explain the importance of a wind turbine array in the energy picture, they will translate the turbines' annual generation into the number of average households that amount of electricity will provide. Thus the 2.7 TWh generated in California by wind turbines each year will provide for the residential electrical needs of 450,000 households. If 2.5 people live in each household, that is enough electricity to meet the needs of 1.125 million people.

Critics, including some electric utilities who refuse to believe that wind turbines can actually perform useful work, immediately challenge the accuracy of such estimates. Admittedly, they are approximations. Engineers use approximations all the time. They prefer approximations, because estimates strip away extraneous information that give a false sense of precision.

The California Energy Commission itself in its comprehensive 1992 Electricity Report succumbs to simplification. The document explains in a footnote that 1 million kilowatt-hours is enough electricity to meet the needs "of about 160 typical California homes," or 6250 kWh per household. Yet the report also says that the state's 10.13 million households consumed 65.321 TWh in 1989, or 6450 kWh per household. It adds that there are 2.8 persons per household.[4] Thus wind generation produces enough electricity to meet the residential needs of 1.17 million Californians.

Californians are the most energy-efficient residents in the United States. With the exception of California, residential electricity consumption has steadily increased in the United States during the past decade. Nationwide, the average household consumed nearly 50% more (9500 kWh per year) electricity than a household in California, according to the Energy Information Agency. In some states, Oregon and Texas, for example, residents use twice as much (13,000 kWh per year) as in California. If California's wind turbines were located in Texas, they would meet the needs of only half as many households.

On the other hand, if California's wind turbines were located in the Netherlands, they would serve twice as many households. California's prodigious use of electricity shocks European energy analysts. They find it unimaginable that Californians consume so much electricity. Their eyes glaze over in amazement when they hear that California leads the nation in energy efficiency. In the Netherlands, the average household consumption actually decreased from 3200 kWh in 1980 to 2800 kWh in 1989.[5]

Like their American counterparts, European utilities frequently use the number of households served as a measure of electricity generation. The Irish state utility estimates that the wind turbines at Bellacorick will generate 17 million kilowatt-hours per year, enough for 4500 Irish households (3800 kWh per household). In Britain, developers estimate the 12 million kilowatt-hours production from one wind farm will serve 3000 homes annually (4000 kWh per household).[6] Danish utilities estimate that the average household uses 5000 kWh per year. Actual consumption is slightly higher, at 5381 kWh per household per year.[7] The European Wind Energy Association estimates that an average four-person household in the European Community consumes 4380 kWh per year.[8] On the European continent, California's turbines could serve the needs of more than 2 million people.

Emissions Offset

Wind energy is a clean source of renewable energy. This simple statement does little to convey its significance. Wind energy is at the forefront of what Worldwatch Institute's Chris Flavin calls the modern approach to battling pollution, an approach that turns away from "end of the pipe" treatments and toward solutions that deal with the pollutants at their source. In some sectors, such as electricity generation, this means switching to zero-emission sources such as wind turbines.[9]

The electricity generated by wind turbines offsets air pollution that otherwise would be generated by conventional power plants. Wind generation offsets the emission of nitrogen oxides (NO_x), sulfur oxides (SO_x), particulates, and carbon dioxide (CO_2) that otherwise would be emitted to generate the same amount of electricity. This is wind energy's principal environment attribute, its environmental *raison d'être.*

In public policy debates about the role of wind energy, it is often necessary to place wind generation in context. How many pollutants, for example, will a wind plant offset? There is no simple answer. The amount of pollutants offset by wind generation depends on the type of power plant or fuel source wind energy would supplant. Natural gas is a relatively clean-burning fuel; oil, less so. Modern coal plants are considerably cleaner than their predecessors, yet they still emit a staggering amount of pollutants. Nuclear plants emit no combustion by-products at the plant site, although there are emissions from the entire fuel cycle of nuclear energy, because of the energy used in refining uranium.

Utilities supply electricity from several sources simultaneously. Thus at any one instant, the electricity a wind turbine offsets may represent a mix of generating sources, including coal, natural gas, and nuclear power. In the United States, the generating mix is still dominated by coal, and many coal-fired power plants have yet to comply with limits on the emission of sulfur oxides. The generating mix also varies from region to region. The midwestern United States, for example, burns more coal than the west. At a national level, then, wind generation may offset more air pollutants, on average, than if the generation were compared to the emissions from a new, cleaner-burning power plant or from a mix of power plants weighted more heavily with cleaner-burning natural gas (Table 12.2). This often creates confusion among planners, when advocates and critics bombard them with conflicting data about wind energy's ability to reduce air pollution.

Planners may also want to consider the amount of pollutants that wind energy offsets relative to specific conventional technologies, such as a proposed coal-fired power plant. Most wind plants in California during the mid-1980s offset natural gas-fired generation from conventional plants. No coal-fired power plants operate in California. Several were envisioned for just outside the state's borders during the 1970s, but none were ever built. Yet the state imports nearly 20% of its electricity from coal and nuclear plants operating in the southwest. Planning agencies at both the state and regional level are still considering out-of-state coal plants as a means of augmenting future generation. State agencies are looking to the plants as a source of cheap electricity. The South Coast Air Quality Management District (SCAQMD) con-

Table 12.2
Average Emissions Factors for Power Generation in the United States

	lb/kWh	g/kWh
NO_x	0.0088	4
SO_x	0.00902	4.1
Particulate	0.000506	0.23
CO_2	2.068	940
CO_2 equivalent	2.42	1100

Source: EPA, Renewable Electric Generation, 1992, p. III-19.

siders them an option for providing electricity to Los Angeles, which would enable the agency to close the polluting plants currently in use. A study for SCAQMD (the agency responsible for reducing air pollution in the Los Angeles basin) recommended: "Another strategy for *exporting air pollution* is to import electricity from the Southwest" (emphasis added).[10]

California utilities are also wont to "repower" some existing natural gas-fired power plants with combined-cycle technology. Although new coal plants emit far fewer emissions than do existing coal plants, new gas-fired combined-cycle plants are cleaner still (Table 12.3). Natural gas is currently the fossil fuel of choice for many utilities in the United States.

However, emissions at the power plant represent only one part of the entire fuel cycle. Meridian Corporation, in a study for the U.S. Department of Energy, found substantial emission of particulates from coal mining, transport, and processing (Table 12.4). These emissions are typically ignored in assessing the impact from a coal-fired power plant or, conversely, the emissions benefit of wind energy. Meridian found 10 times more particulate emissions from the complete fuel cycle than from those of the power plant alone. Another significant impact often overlooked when comparing coal-fired power plants and wind energy is the emission of trace metals contained in coal. Mercury emissions from power plants may become a major health issue during the 1990s. Coal mining and processing also create tons of solid waste. Two-thirds of the solid waste from coal-fired power plants are due to mining and processing, the remainder is ash at the power plant.[11]

Meridian estimates that during its fuel cycle coal emits 200 times the pollutants of photovoltaics, the only renewable technology they considered. By comparison, a Danish study on the energy balance of wind turbines for the Environment Ministry estimates that a coal-fired power plant emits 360 times more SO_x, NO_x, and carbon dioxide than does a wind turbine to generate an equivalent amount of electricity over a 25-year period.[12]

Table 12.3
Emissions Factors for New Power Plants (CEC)

	New Coal		*New Natural Gas Combined Cycle*		*New Oil Combined Cycle*	
	lb/kWh	g/kWh	lb/kWh	g/kWh	lb/kWh	g/kWh
NO_x	0.0055	2.5	0.000162	0.07	0.00041	0.19
SO_x	0.0066	3.0	0.000011	0.00	0.00055	0.25
SO_x without controls	0.0132	6.0				
Particulates	0.00033	0.15	0.000143	0.07	0.00031	0.15
CO_2	2	908	1.07	487	1.42	647

Source: Technology Characterization Report, ER-92, California Energy Comission, Sacramento, CA, November 22, 1992, pp. AII-6, E5-32. Heat rate = 11,000 Btu/kWh for coal; 9000 Btu/kWh for combined cycle. 1979 New Source Performance Standards, except for SO_x without controls. Coal CO_2: Technology Characterization Report, ER-94.

Table 12.4
Emission Factors for Fuel Cycle (U.S. DOE)

	Conventional Coal		*Nuclear*		*Photovoltaics*	
	lb/kWh	g/kWh	lb/kWh	g/kWh	lb/kWh	g/kWh
NO_x	0.00598	2.7	0.000068	0.03	0.000016	0.01
SO_x	0.00594	2.7	0.000058	0.03	0.000046	0.02
Particulates	0.0032	1.5	0.000006	0.00	0.000034	0.02
CO_2	2.116	962	0.01718	7.8	0.01178	5.35
Trace metals	0.000248	0.11				
Solid waste	0.468	213	0.00007	0.03	0.00003	0.01

Source: Meridian Corporation, "Energy System Emissions and Material Requirements," U.S. Department of Energy, Washington, D.C., 1989, pp. 25, 26, 28.

Using the emission factors in Tables 12.2 to 12.4, the amount of emissions offset by wind energy can be calculated for a variety of conditions once the wind resource is known. For example, at a 7-m/s (16-mph) site, a wind turbine delivering nearly 30% of the energy in the wind will generate about 1000 kWh per square meter of rotor area. Thus a wind turbine 25 m (82 ft) in diameter (500 m^2) will generate 500,000 kWh per year. Under these conditions, this wind turbine will offset 500,000 kg (1 million pounds) of CO_2 per year emitted by a new coal-fired power plant (Table 12.5).

If used to offset generation from a new coal-fired power plant, the 2.7 TWh that California's wind plants produce each year would prevent the emission of 5700 million pounds (2600 million kilograms) of carbon dioxide, 30 million pounds (14 million kilograms) of NO_x and SO_x, 9 million pounds (4 million kilograms) of particulates, 670,000 lb (300,000 kg) of trace metals, and 1300 million pounds (580,000 kg) of solid waste.

Water Consumption

To someone in a temperate climate, the water required for conventional power plants seems of little significance. But in arid areas of the world such as in the southwestern United States, such considerations as water, who has it, and how it is used are volatile political issues. Unlike thermal power plants, wind turbines have no need for cooling water.

Conventional power plants use enormous amounts of water for the condensing portion of the thermodynamic cycle. Because a portion of the heat from combustion of fossil fuels is transferred to the exhaust gases, fossil fuel-fired power plants use one-fourth less water than nuclear plants. Nevertheless, additional water is used for washing and processing coal.

Much of the water required for conventional power plants simply passes through the cooling system and the heated water is returned to its source.

Table 12.5
Typical Emission Benefits from Wind Energy

	Per m^2 of Rotor Area at 1000 kWh/m^2		*Per 25-m-diameter Turbine (500 m^2)*	
	kg	lb	kg	lb
Average U.S. Emission Rate				
NO_x	4	8.8	2,000	4,400
SO_x	4	9.0	2,050	4,510
Particulates	0.23	0.5	115	253
CO_2	940	2068	470,000	1,034,000
From New Coal-Fired Power Plants in the United States (Fuel Cycle Includes Mining and Processing)				
NO_x	2.5	5.5	1,250	2,750
SO_x	3	6.6	1,500	3,300
SO_x without controls	6	13.2	3,000	6,600
Particulates	1.5	3.2	750	1,600
CO_2	908	2000	454,000	1,000,000
Trace metals	0.11	0.248	55	124
Solid waste	213	468	106,500	234,000
From New Gas-Fired Combined-Cycle Plants in the United States				
NO_x	0.07	0.162	35	81
SO_x	0.005	0.011	3	6
Particulates	0.07	0.143	35	72
CO_2	487	1071	243,500	535,500
From New Oil-Fired Combined-Cycle Plants in the United States				
NO_x	0.19	0.414	95	207
SO_x	0.25	0.55	125	275
Particulates	0.15	0.319	75	160
CO_2	647	1423	323,500	711,700

In the case of coal cleaning, the waste water will be heavily laden with suspended solids and will need treatment. Although wind turbines and photovoltaic cells need no cooling water to produce electricity, they may use slight amounts of water in arid areas for cleaning. Fluidyne, a small family-owned company, has carved a niche for itself washing wind turbine blades in dusty California (Table 12.6). Meridian estimates that the coal and nuclear fuel cycles use 30 to 40 times more water than do photovoltaic panels, even if the panels were washed more frequently than necessary.

In arid regions, the amount of water used is less critical than the amount of water consumed, that is, the water lost to evaporation from the cooling cycle. As part of considering all the impacts of power plant construction and operation, the California Energy Commission weighs the consumptive uses

Table 12.6
Water Use

	acre-ft/GWh	gal/GWh	gal/kWh	liters/kWh
Coal	3.12	1,016,656	1.0	3.8
Nuclear	4.12	1,342,507	1.3	1.5
Photovoltaics (washing)	0.1	32,585.14	0.03	0.1

Source: Meridian Corporation, "Energy System Emissions and Material Requirements," U.S. Department of Energy, Washington, D.C., 1989, p. 23.

of conventional technologies (Table 12.7). For planning purposes, the CEC assumes that wind power plants and photovoltaic arrays use no water.[13]

Social or External Costs

Not all costs of conventional generation are reflected in the price of electricity. Environmental impacts from conventional sources (for example, air pollution from fossil-fueled plants and radiation exposure from nuclear plants) exact costs from society at large. Environmental and safety regulations internalize some of these costs by increasing the cost of generation. However well-intentioned, regulation has failed to internalize all the social costs of conventional generation. This is particularly true for the costs of pollutant emissions that meet air quality standards. Even though these emissions meet society's accepted limits, they still exact a social or environmental cost, for example through additional sickness and death. At the same time that these costs are not reflected in the price of conventional sources of generation, the air quality benefits of wind-generated electricity are not incorporated into its purchase price.

These external effects, both costs and benefits, cause serious distortions in the energy market. Prices must reflect overall costs, including external effects, to ensure optimal functioning of the market and proper allocation of resources. Where distortions exist, as in electricity generation, intervention is necessary to "level the playing field." If these external effects are not addressed, serious

Table 12.7
Water Consumption by Power Plants (Evaporative Loss)

	acre-feet/Yr/MW	m^3/Yr/MW	gal/kWh	liters/kWh
Nuclear	12.5	15,400	0.62	2.3
Coal	9.9	12,200	0.49	1.9
Oil	8.7	10,700	0.43	1.6
Combined cycle	5.1	6,300	0.25	0.95

Source: CEC Technical Assessment Manual, Vol. 1, Electrical Generation, 1979, p. A-4.

misallocation of resources will continue at a significant social and environmental cost to all, especially to those downwind of conventional power plants.

The concept of external costs, those outside the market's pricing system, is not new. Environmental economists such as Herman Daly have espoused inclusion of these costs in societal decision making since at least the early 1970s. However, concern about global warming and increasing use of integrated resource planning to determine the type of new power plants that will be built have renewed interest in the subject, encouraging a new crop of economists to tackle the thorny issue.

Wind generation offloads conventional plants that otherwise would be used to meet demand. By offsetting conventional plants, wind generation offsets the air pollutants that otherwise would have been emitted. This benefit is rarely quantified, despite several techniques for doing so. One approach examines the number of deaths caused by pollution from conventional plants and places a value on the number of lives saved by cleaner sources. Another examines the costs of increased illnesses, and those of reduced agricultural output, from acid rain for example. A third looks at society's "revealed preference" for emission cleanup through pollution-control regulations. All entail a great degree of uncertainty. What is certain is that the environmental benefits of relatively benign resources such as wind and solar energy are worth more than the zero sum currently placed on them.

To illustrate the conflicting data that analysts must face, consider the number of people killed by conventional power generation. Leonard Hamilton, a physician at Brookhaven National Laboratory, for example, examined the risks of injury and death in conventional fuel cycles. He determined the maximum number of deaths per gigawatt-year of generation possible from the air pollution emitted by conventional power plants[14] (Table 12.8).

Other analysts have tried to analyze the average number of deaths. In a study for the Swiss Federal Energy Office, analyst Andrew Fritzsche found deaths from coal-fired electricity due to both occupational accidents and public health effects to be an order of magnitude fewer than in Hamilton's research: 10 deaths per gigawatt-year from coal (1.1 deaths per terawatt-hour) and 9 for oil (1 death per terawatt-hour).[15] This is the lower limit of estimates of deaths from the fuel cycle for coal and oil. Samuel Morris, a colleague of

Table 12.8
Deaths from Electricity Generation: Upper Bound Air Pollution Effect on Public Health

	Deaths per GW-yr	Deaths per TWh
Coal	220	25
Oil	140	16
Natural gas	150	17

Note: One gigawatt-year, or 1000 MW at 100% capacity for an entire year, is equivalent to 8.76 TWh.
Source: L. D. Hamilton, Brookhaven National Laboratories.

Hamilton, estimates that public health impacts from coal kill a mean of 21 people to as many as 60 per gigawatt-year, or 2.4 to 7 deaths per terawatt-hour.[16] Brookhaven National Laboratory attempted to make sense of all this in a 1992 report[17] (Table 12.9). The Brookhaven report reduced the upper limit for Hamilton's estimate regarding coal to 150 deaths per gigawatt-year, kept the upper limit for oil, but raised the upper limit for Morris. Determining who is right is more than an academic exercise. The outcome will influence public policy, affect the preference for one technology over another, and potentially cost billions of dollars.

Offsetting air pollution is not only socially desirable, it has economic value as well. Federal regulators in the United States, such as the Environmental Protection Agency, estimate that each premature death averted is worth \$1.6 to \$8.5 million, based on how much people are willing to pay for protection.[18] Although some regulations have indirectly valued life far higher, at up to \$65 million per death averted, researchers are more comfortable with values approaching \$5.5 million per life saved. For example, occupational regulations controlling asbestos value one life at \$8.3 million, and those controlling vinyl chloride at \$6.7 million.[19] A British study of social costs uses a value of £2 million (about \$3 million) per death, midway in the EPA range.[20]

If Fritzsche is correct and the toll from occupational and public health risk of coal-fired generation is only 1.1 deaths per terawatt-hour, the monetary cost is substantial: from \$0.002 per kilowatt-hour to \$0.01 per kilowatt-hour (Table 12.10). If, on the other hand, Morris or Hamilton is closer to the target, the human costs of one of the world's principal sources of electricity is enormous. If coal kills a total of only 7 persons per terawatt-hour, the social cost ranges from a low of \$0.011 per kilowatt-hour to an astounding \$0.06 per kilowatt-hour, or as much as the retail price of electricity in many parts of the United States.

These figures reflect only the costs directly attributable to the loss of life. There are other environmental costs as well. For example, there are effects on wildlife and crops. During the late 1980s, the European Community attacked the problem of placing a value on such nonprice factors. The report by the Fraunhofer Institutes' Olav Hohmeyer sent tremors through the European utility industry and aroused public interest in the field. The 1988 report specifically examined the environmental costs of fossil fuel-fired and nuclear-

Table 12.9
Estimated Deaths from Power Generation per Gigawatt-year

	Hamilton				*Morris*			
	Occupational Accidents	Disease	Public	Total	Occupational Accidents	Disease	Public	Total
Coal	0.46	0–4.7	4–150	10–150	0.53–0.93	0.13–8.7	0–320	1–320
Oil	1.63		1.3–130	3–130				
Gas	0.21							
Nuclear	0.35	0.18	0.067	0.5	0.14–0.6	0–0.90	0.2	0.352

Table 12.10
Cost of Deaths from Coal-Fired Electricity Generation

		Dollars/kWh[a]		
	Deaths per TWh	Low	Mid	High
Fritzsche	1.1	0.002	0.006	0.010
Morris, low	2.4	0.004	0.013	0.020
Morris, high	7	0.011	0.39	0.060
Hamilton, revised	17	0.027	0.094	0.146

[a] Low, $1.6 million/life; mid, $5.5 million/life; high, $8.5 million/life.

fueled generation in Germany. Hohmeyer cautions that his results represent only a first cut at approximately the monetary value of these costs and explains at length the difficulties that he encountered. Nevertheless, when considering all subsidies, health and environmental costs, and resource depletion effects, Hohmeyer estimates that fossil-fired generation costs 0.04 to 0.09 DM ($0.02 to $0.05) per kilowatt-hour, and nuclear-fueled plants of 0.1 to 0.2 DM ($0.06 to $0.13) per kilowatt-hour in 1982 currency[21] (Table 12.11).

Table 12.11
Summary of External Effects of Energy Systems

	External Costs			
Effects	DM/kWh		Dollars/kWh	
Fossil Fuels				
Flora	0.0069	0.0104	0.0041	0.0062
Fauna	0.0001	0.0001	0.0001	0.0001
Health	0.0018	0.0460	0.0011	0.0276
Materials	0.0025	0.0044	0.0015	0.0026
Climate	0.0001	0.0001	0.0001	0.0001
Depletion	0.0229	0.0229	0.0137	0.0137
Public services	0.0007	0.0007	0.0004	0.0004
Tax treatment	0.0032	0.0032	0.0019	0.0019
Public R&D	0.0004	0.0004	0.0002	0.0002
	0.0386	0.0882	0.0232	0.0529
Nuclear Power				
Major accidents	0.012	0.12	0.0072	0.0720
Depletion	0.0591	0.0623	0.0355	0.0374
Public services	0.0011	0.0011	0.0007	0.0007
Tax treatment	0.0014	0.0014	0.0008	0.0008
Public R&D	0.0235	0.0235	0.0141	0.0141
	0.0971	0.2083	0.0583	0.1250

Note: 1 DM = $0.6.
Source: Olav Hohmeyer, *Social Costs of Energy Consumption* (Berlin: Springer-Verlag, 1988), pp. 8, 100.

Hohmeyer also estimated the net social and environmental effects of wind and solar energy. At a minimum, says Hohmeyer, wind generation should be worth 0.06 to 0.12 DM ($0.03 to $0.07) per kilowatt-hour in 1982 currency, based on external benefits alone. "Even without the inclusion of all external benefits, and with a deliberate bias against renewable energy sources," he says the net external effects "are of the same order of magnitude" as the retail price of conventional sources[22] (Table 12.12).

Since the report's release, Hohmeyer has won both praise and derision for his methodology. Responding to critics, he reexamined his original work and reported in a 1991 address to the Canadian Wind Energy Association that he might indeed have erred. He said that he understated the external cost of fossil fuels. He revised his estimate to 0.05 to 0.28 DM ($0.03 to $0.17) per killowatt-hour.[23]

Subsequent to Hohmeyer's pioneering effect, several others have examined the subject, including former Congressman Richard Ottinger at Pace University. The Environmental Protection Agency summarized the work of these authors on a comparable basis in 1990 dollars[24] (Table 12.13). No explanation is given for the discrepancy between Hohmeyer's work and that of EPA as reported in its summary. With the exception of the Bonneville Power Administration, which buys power from coal-fired power plants, the studies consistently estimate external costs of fossil fuels at from $0.01 to $0.10 per kilowatt hour. Where considered, the external costs of wind energy were less than $0.001 per kilowatt-hour.

Because of its dependence on hydroelectricity, BPA looked specifically at land use and water pollution externalities. Like nuclear, hydroelectricity emits no air pollutants. But the reservoirs for large hydroelectric projects often inundate prime agricultural or recreational land. BPA estimates the environmental cost of land and water impacts for biomass, geothermal, solar, and municipal solid waste is $0.001 per kilowatt-hour in 1990 dollars; for coal and hydro, $0.002 per kilowatt hour. According to BPA, the environmental cost of wind energy, which is often criticized for its land requirement, is $0.0005 per kilowatt-hour, one-fourth that of coal and hydro.[25]

Table 12.12
Social Costs and Benefits of Wind Energy in Germany

	DM/kWh		Dollars/kWh	
Environmental costs (noise)	−0.0001	−0.0001	−0.0001	−0.0001
Public R&D	−0.0026	−0.0055	−0.0016	−0.0033
Net economic effect	0.0053	0.0094	0.0032	0.0056
Avoided external cost	0.0535	0.1188	0.0321	0.0713
	0.0561	0.1226	0.0337	0.0736
Mean		0.0890		0.0534

Source: Olav Hohmeyer, *Social Costs of Energy Consumption* (Berlin: Springer-Verlag, 1988), p. 104.

Table 12.13
Estimates of Externality Costs (1990 Cents/kWh)

	Range	Hohmeyer	Pace	BPA	Tellus	JBS
Gas turbine	0.1–6	0.6–2.9	0.7–1	0.1	6	1.6–4.1
Oil turbine	0.3–10.3	0.6–2.9	2.6–6.9	0.3	10.3	
Coal	0.6–10	0.6–2.9	2.6–5.9	0.7–1.1	4.5–10	2.8–8.2
Nuclear	0–5.7	0–5.7	3			
Photovoltaics	0–0.4	0–0.2	0–0.4			
Wind	0–0.1	0	0–0.1			
Biomass	0–0.7		0–0.7			
Geothermal	0					
Solid Waste	(3.7)–48.2		2.9	(3.7)–48.2		

Source: EPA, *Renewable Electric Generation,* 1992, p. I-26.

Consulting economist William Marcus of JBS Energy has testified before the California Energy Commission on a theoretical and quantitative approach to estimating the value provided by nonpolluting sources of electricity. In essence, society reveals it preference for clean air through its regulation of pollutants emissions. Thus the costs of the air pollution control devices reveal the value of nonpolluting sources. But Marcus argues, there are costs to society of air pollution from power plants even after pollution control devices have been installed. He contends that even with controls, power plants emit tons of pollutants. These remaining pollutants continue to cause environmental damage, incurring costs to society. The costs of compliance, says Marcus, represent only the minimal value that society places on clean air. By evaluating the cost of controls on nitrogen and sulfur oxides, Marcus was able to estimate the value of reducing air pollution in different regions: the South Coast Air Quality Management District for the Los Angeles basin, the remainder of California, and air basins outside the state. The social costs of major pollutants, says Marcus, could be worth $0.02 to $0.03 per kilowatt-hour.[26]

Because of the difficulty in determining the cost of damage, Hohmeyer's study placed little value on the possible climatic effects caused by fossil fuel–fired plants. Many now believe that these costs may also be significant. There is a near-universal consensus that carbon dioxide emissions, if not already incurring dramatic costs due to climatic change, place society at tremendous risk of doing so in the future. But revealed preference is difficult to apply to carbon dioxide because it is an unregulated pollutant.

Marcus calculated the cost of reforestation as a surrogate for the revealed preference of reducing carbon dioxide emissions. Next to energy conservation, reforestation is the most cost-effective method for limiting carbon dioxide concentrations in the atmosphere. Marcus estimates that the minimal cost of reducing CO_2 emissions alone is $0.012 per kilowatt-hour for a gas-fired plant and $0.02 per kilowatt-hour for a coal-fired plant.

Certainly, there is no universal agreement on the exact monetary value of the external costs from conventional generation. But just as certainly, it is clear that the costs are substantial: from $0.01 to $0.10 per kilowatt-hour. Disregarding these costs, and hence the benefits of wind energy, biases the market toward fossil fuels.

Global Warming

Although known for more than 100 years, the greenhouse effect leapt to the front pages during the hot, dry summer of 1988 when scientists testified on Capitol Hill about global warming. There is a scientific consensus, said James Hansen of NASA's Goddard Institute, that the earth is warming, that the 1980s were the warmest decade of the century, and that this warming is probably the result of human activities. Greenhouse gases are rapidly accumulating in the atmosphere, he and others, such as Stephen Schneider of the National Center for Atmospheric Research, warned. The concentration of carbon dioxide, for example, had increased 25% since the industrial revolution, and that of methane had more than doubled.[27]

Man's effect on the global climate now competes with natural processes that take place over tens of thousands of years. Without controls on carbon dioxide and other greenhouse gases, such as methane, nitrous oxides, and chlorofluorocarbons, the earth could warm from 2 to 6 degrees Celsius by 2100. The Environmental Protection Agency estimates that we could see declines in forest production from global warming within 30 to 80 years. When including the effects from other air pollutants, such as acid rain, we could see changes within one decade. Because of the atmosphere's "thermal inertia," some climatic effects appear inevitable even if we drastically reduce emissions of greenhouse gases now. With "business as usual," sea levels may rise as much as 1 m (3 ft) from global warming by 2100. Demand for electricity, primarily for increased cooling, could grow 4 to 5% by 2055.[28]

Carbon dioxide released by the combustion of fossil fuels and the burning of tropical forests, accounts for about one-half of the greenhouse effect. The United States emits one-fourth of all carbon dioxide worldwide, and power plants produce more than one-third of U.S. carbon dioxide emissions. "By themselves, power plants are the largest single source" of carbon dioxide emissions in the United States, according to Jim MacKenzie of the World Resources Institute.[29] To stabilize the concentration of carbon dioxide in the atmosphere, the Natural Resources Defense Council estimates that emissions must be reduced 50 to 80%.[30]

To reduce its share of worldwide carbon emissions sufficiently, there is growing agreement that the United States must immediately launch an ambitious program to promote both greater efficiency and the development of clean energy sources such as solar and wind. "There are no quick fixes to this problem," according to Worldwatch Institute, "the only safe and cost-effective way to slow global warming is the simultaneous pursuit of renewables and efficiency."[31] NRDC's Dan Lashof reaches a similar conclusion: "There are no silver bullets." Although the potential of energy efficiency is enormous, says Lashof, "the reductions in greenhouse gas emissions required to stabilize climate cannot be obtained from efficiency improvements alone." An effective greenhouse gas strategy must include the use of renewables to replace or retire existing power plants. "Solar electric systems will be critical to avoid new commitments to coal-based power generation," he says.[32]

Amory Lovins, long an advocate of the "soft path," stresses that "renewables are a natural partner with efficient energy use and are both economically and environmentally preferable to depleting nonrenewable energy supplies." The World Resources Institute concurs: "Over the longer term, the United States must begin the inevitable transition to non-fossil energy technologies."[33] In California, the Sierra Club's greenhouse agenda is "centered on the need to wean our economy from its heavy dependence on fossil fuels through increased energy efficiency and the development of renewable energy sources."[34] Wind energy is critical to these strategies because every kilowatt-hour generated by a wind turbine offsets from 1 to 2 pounds of carbon dioxide that would otherwise be emitted.

For the sake of comparison, imagine the equivalent number of trees required to fix the same amount of carbon dioxide as offset by wind generation. Trees consume about 26 lb (12 kg) of CO_2 per year; forests fix some 10,000 lb per acre (11,000 kg/ha) per year.[35] At a 7-m/s (16-mph) site, a typical wind turbine 25 m in diameter should generate about 500,000 kWh per year. If the wind turbine were offsetting natural gas–fired generation, it would require 19,000 trees to accomplish the same effect, or a 50-acre (20-ha) copse of trees. Each terawatt-hour of wind generation can supplant a forest of 100,000 acres (41,000 ha). This should not be construed as indicating that wind turbines are preferable to tree planting in the fight against global warming, only that wind energy's contribution can be significant (Table 12.14).

Table 12.14
Emissions Offset by Trees

Number of 25-m-Diameter Turbines	TWh	*Equivalent Number of Trees*			*Equivalent Forest Area (ha)*		
		Natural	Gas	Coal	Natural	Gas	Coal
1	0.0005	19,200	38,500	50	20	100	41
2,000	1	38,000,000	77,000,000	100,000	41,000	200,000	82,000

Note: Assumes a 7-m/s site.
Source: American Forestry Association, 1989.

Tourism

When renewable advocates want to bolster their case that wind and solar plants stimulate local commerce, they often mention that the novelty of the projects will attract tourists. The Disneyland-like visitor centers they envisioned, such as those common at nuclear power plants in the United States, have yet to materialize. But people are curious about wind turbines and do visit them. It is not uncommon to see a sign on a Danish roadway for "Vindmøllepark," with an arrow directing motorists to the local wind farm. Roads in northern Germany and in the Netherlands are similarly signposted. For an American visitor accustomed to hours of searching on poorly marked back roads to find wind turbines in the United States, the number and prominence of the road signs in northern Europe is a refreshing welcome to anyone interested in renewable energy.

Nearly all utility-sponsored wind plants in Europe include some form of public information center, whether a simple kiosk or full-blown visitor center with staff and exhibits. Privately owned wind projects in North America have never felt the need or responsibility to provide explanations of the technology to the public, although the Kern Wind Energy Association and its members in Tehachapi operate a low-power radio transmitter for this purpose, which broadcasts information to motorists.

A visitor's center was as far from a priority for Cornish farmer Peter Edwards as for his American counterparts. But after he installed Britain's first wind power plant, he had a problem on his hands: gawkers were blocking traffic on the narrow, hedgerow-bordered country lane adjoining his site at Delabole, He acted swiftly to avert an accident, graded a car park, and assembled several of the "porta cabins" used at European construction sites into an instant visitors' center (Fig. 12.1).

It worked. During its first six months, 28,000 visitors paid to pass through the center's turnstile. Many more simply drove up, glanced at the wind turbines, bought a Cornish pasty, and drove away. By mid-1993, after 18 months of operation, 55,000 tourists had called. Edwards estimates that twice as many visited the site but many chose not to go through turnstiles and pay the £1 fee.

Edward's experience may not be unique. The Centre for Alternative Technology draws 70,000 people per year to its site near Machynlleth in mid-Wales and has done so for the past decade. CAT has led Britain in the development of "green" tourism.[36] Across the Channel, German planners estimate that wind plants at Westkuste drew 4000 during its first year of operation and 11,000 the next; a small plant at Cuxhaven attracted 3100 during its two years of operation.[37] In North America, Canada's giant Éole drew 11,000 paying visitors to Cap Chat in 1992, and another 9000 the following year even though the turbine was no longer in operation.

Curiosity about wind turbines arises from a complex mix of interest in green or environmental issues and a general interest in large man-made

Figure 12.1. Tourism at Delabole, Cornwall.

artifacts. A survey by the French ministry of tourism revealed that 70% of those polled wanted to take advantage of touring industrial plants. In France, EDF's power plants lead as attractions. The La Rance tidal power plant in Brittany draws the largest crowds, averaging 350,000 per year. The plant, which uses the renewable energy in the area's tides, attracts as many visitors as all of EDF's nuclear power plants combined. As many as 1 million people visit EDF power plants in France every year.[38]

Visitor centers satisfy this curiosity and educate visitors about wind energy's place in the energy mix (Figure 12.2). They also offer an officially sanctioned venue for individuals to see and hear wind turbines and to judge for themselves whether wind energy's impacts are acceptable. Modern wind turbines certainly fail to rival traditional Dutch windmills in their ability to attract throngs of tourists, but there is no inherent reason why wind turbines will not become as popular someday as Dutch windmills are today (Figure 12.3).

Employment

"Wind power pays for people, not fuel," says Carl Weinberg, former Pacific Gas & Electric Co. executive. And the employment wind energy provides is often highly prized in the rural areas where the turbines are located and where unemployment is often endemic. In Bob Lynette's vision of the future, the jobs

Figure 12.2. School group at Ebeltoft. Schoolchildren on day trip to Ebeltoft harbor to visit municipal wind energy information center.

brought by thousands of wind turbines will one day revitalize the depopulated villages of North America's Great Plains.

To many, working with wind energy is not just another job, but work that makes a difference. Employees can go home at the end of the day feeling that they have made a contribution, small as it may be, to a better world. Dave, a foreman for a wind plant near Tehachapi, gives up two hours of his day to commute to and from the remote site where he works, just for the privilege of being there. "There's not one of us who'd give it up," he says with conviction. "It's a special place . . . to work." Dave is a convert. He once was a critic who was disturbed that wind companies had usurped his favorite off-road trails for their windmills.

In a 1993 study for the California Energy Commission, the American Wind Energy Association's west coast office took the first-ever comprehensive employment survey of California wind plant operators and their service providers (Table 12.15). AWEA found 1250 people working directly with wind energy in California, an average of 460 jobs per terawatt-hour of generation per year. Nearly all jobs in California are related to operating, maintaining, and servicing the existing fleet of wind turbines.[39] The results compare favorably with those for Denmark, where BTM Consult estimates that 400 people are employed in the service sector, an average of 440 jobs per terawatt-hour per year.[40] California's wind industry also indirectly creates more than 4000 jobs. Altogether, California's wind industry employs 5000 to 6000 people. While California has dominated electricity generation over the last decade, Denmark has dominated

Figure 12.3. Tourism at Zaanse Schans. Assimilation of the traditional Dutch windmill has been so successful that tourists fly thousands of miles to pose beneath them at Zaanse Schans.

wind turbine manufacture and sales. BTM Consult's Per Krogsgaard estimates 600 people are employed full-time in Denmark manufacturing about 100 MW of wind turbines per year, an average of 6 jobs per megawatt of annual capacity.[41]

In addition to job creation, the public sector also benefits economically from California's wind power plants. California wind companies pay property taxes to local governments totaling $10 to $13 million annually, not including royalties paid to the U.S. Bureau of Land Management for leases on federal land.[42]

Despite wind energy's demonstrated positive economic impact and its proven ability to generate commercial quantities of electricity, energy planners, environmentalists, and renewable energy advocates alike have long been bewitched by the shimmering beauty of another technology.

Table 12.15
Wind Industry Jobs in California and Denmark

	California		*Denmark*	
	Jobs	Jobs/TWh	Jobs	Jobs/TWh
Manufacturing	0	0	600	(6/MW)
O&M and support	1250	460	400	440
Indirect	4350	1500	—	—

Chapter 12 Endnotes

1. See Erik Grum-Schwensen, "The Real Cost of Wind Turbine Construction," Wind Stats, 3:2, Spring 1990, pp. 1–2; and A. Gydesen, D. Maimann, and P. B. Pedersen, "Renere Teknologi pa Energiomradet," Energigruppen, Fysisk Laboratorium III, Danmarks Tekniske Hoejskole, Miljoeministeriet, Miljoeprojekt 138, Denmark, 1990, pp. 123–127.
2. G. Hagedorn, and F. Ilmberger, "Kumulierter Energieverbrauch fur die Hersetellung von Windkraftanlagen," Forschungsstelle für Energiewirtschaft, Im Auftrage des Bundesministeriums für Forschung und Technologie, Munich, August 1991, pp. 79, 98, 100, 111. In this study, the primary energy used to construct the wind turbine was given in units of kilowatt-hours. However, only 35% of the energy burned in a power plant is converted to useful work. To present the data from this study in a format consistent with that from the other studies, the number of kilowatt-hours consumed has been reduced accordingly.
3. Gerd Hagedorn, "Kumulierter Energieaufwand von Photovoltaik und Windkraftanlagen," Lehrstuhl für Energiewirtschaft und Kraftwerkstechnik, Technische Universität, Munich, 1992, p. 95.
4. *1992 Electricity Supply Report,* Draft Final (Sacramento, CA: California Energy Commission, November 1992), pp. 2–1, 2–3, 2–7.
5. Electricity Supply in 1989, Vereniging van Exploitantenan Electriciteitsbedrijven in Nederland, Arnhem, The Netherlands, Feburary, 1990.
6. "Renouvelables: inventaire et état de l'art," *Systèmes Solaires,* 94/95, 1992, p. 20.
7. Benny Christensen, "A Sustainable Cure," *Windpower Monthly,* March 1992, p. 23; and "Nyt om energi og miljo," *Vestjyllands Energi og Miljokontor, Rinkobing,* 29, Summer 1992, p. 13.
8. Andrew Garrad, "A Plan of Action: Wind Energy in Europe," European Wind Energy Association, summary report, Wind Energy in Europe: Time for Action, Rome, 1991, p. 4.
9. Chris Flavin, Worldwatch Institute, oral comments at the American Wind Energy Association's annual conference, Windpower '93, San Francisco, July 1993.
10. W. W. Wade et al., "Economic Analysis to Examine the South Coast Air Quality Management District's Proposed Fuel Oil and Diesel Phase-Out," Spectrum Economics for SCAQMD, Los Angeles, December 12, 1988.
11. Meridian Corporation, Energy System Emissions and Material Requirements, U.S. Department of Energy, Washington, DC, 1989, pp. 25, 26, 28.
12. Gydesen, Maimann, and Pedersen, "Renere Teknologi pa Energiomradet."
13. Technology Characterization, pp. ER 92, California Energy Commission, Sacramento, CA, November 22, 1991, pp. E6–8, E16–6.
14. L. D. Hamilton, "Health and Environmental Risks of Energy Systems." *Proceedings of an International Symposium on the Risks and Benefits of Energy Systems,* International Atomic Energy Agency, Vienna, 1984, p. 37.
15. Andrew F. Fritzsche, "The Health Risks of Energy Production," *Risk Analysis,* 9:4, 1989, pp. 565–577.
16. Samuel C. Morris, "Health Risks of Coal Energy Technology," in *Health Risks of Energy Technologies,* Curtis C. Travis and Elizabeth L. Etnier, eds. (Boulder, CO: Westview Press, 1983), Chapter 4, p. 128.
17. Michael Rowe, Health Risks in Perspective: Judging Health Risks of Energy Technologies, Brookhaven National Laboratory, Upton, NY, September 18, 1992, pp. 94–100.

18. See the Environmental Protection Agency's Guidelines for Performing Regulatory Impact Analysis, EPA 230-01-84-003, p. 11; and Ann Fisher, L. G. Chestnut, and D. M. Violette, "The Value of Reducing Risks of Death: A Note on New Evidence," *Journal of Policy Analysis and Management,* 8:1, 1989, pp. 88–100; and brief overview of the issue in Steven Waldman, "Putting a Price Tag on Life," *Newsweek,* January 11, 1988.
19. Morrall estimated the cost of various federal regulations. Among some of his findings in million 1990 dollars per premature death averted: seat belt standards, 0.1; rear seat belts, 3.2; asbestos occupational exposure limit, 8.3; occupational exposure limit to coke oven emissions, 63.5. The latter is of some interest to historians of the industrial revolution. The standard is designed to limit cancers first found in chimney sweeps that were identified by Percival Potts in the late nineteenth century. In a survey of the record for carcinogens, Travis estimated that the cost of regulating lead is $1.4 million, and vinyl chloride $6.7 million per death averted. J. F. Morrall III, "A Review of the Record," Regulation, November/December 1986, pp. 25–34. C. C. Travis, S. A. Richter, E. A. C. Crouch, R. Wilson, and E. D. Klema, "Cancer Risk Management," Environmental Science Technology, 21, 1986, pp. 415–420. Cited in Rowe, *Health Risks in Perspective,* pp. 52–54.
20. "Some Externalities Come Home," *RENEW,* 83, May/June 1993, pp. 9–11. For the number of deaths per gigawatt-year the article cites "The Social Costs of Fuel Cycles," by David Pearce, University College London, Centre for Social and Economic Research on the Global Environment, London: Her Majesty's Stationery Office, September 1992. The study used mortality data from Andrew Fritzsche of the Swiss Federal Energy Office.
21. Olav Hohmeyer, *Social Costs of Energy Consumption* (Berlin: Springer-Verlag, 1988), pp. 4–8.
22. Ibid, pp. 7–8.
23. "Hohmeyer Confirms Earlier Findings on Social Costs of Energy," *Wind Energy Weekly,* 11:481, January 20, 1992, p. 1.
24. "Renewable Electric Generation: An Assessment of Air Pollution Prevention Potential" Washington, DC: U.S. Environmental Protection Agency, 1992, p. I–26. EPA summarized the work of Hohmeyer, "*Social Costs of Energy Consumption;* Richard Ottinger, *Environmental Costs of Energy* (New York: Pace University, Bonneville Power Administration, "Estimating Environmental Costs and Benefits for Five Generating Resources," "Generic Coal Study: Quantification and Valuation of Environmental Impacts," and "Environmental Cost and Benefits Case Study: Nuclear Power Plant: Quantification and Economic Valuation of Selected Environmental Impacts/Effects;" Tellus Institute, "Full Cost Economic Dispatch: Recognizing Environmental Externalities in Electric Utility System Operation;" JBS Energy, "Valuing Reductions in Air Emissions and Incorporation into Electric Resource Planning: Theoretical and Qualitative Aspects."
25. Bonneville Power Administration, U.S. Department of Energy, "Application of Environmental Cost Adjustments During Resource Cost-Effectiveness Determinations," Portland, OR, 1991, pp. 5–6, as reported in "Policy Alternatives for Estimating Land and Water Management in California," ER 92, California Energy Commission, Sacramento, CA, September 6, 1991.
26. W. B. Marcus et al., "Valuing Reductions in Air Emissions and Incorporation into Electric Resource Planning: Theoretical and Quantitative Aspects," JBS Energy California Energy Commission Docket 88 ER8, Sacramento, CA, August 25, 1989.

27. For more on the greenhouse effect, read Stephen Schneider, *Global Warming* (San Francisco: Sierra Club Books, 1989).
28. Joel Smith, "The Potential Effects of Global Climate Change on the United States," Forum on Renewable Energy and Climate Change, Washington, DC, June 14, 1989.
29. James MacKenzie, "Breathing Easier: Taking Action on Climate Change, Air Pollution, and Energy Insecurity" World Resources Institute, Washington, DC, 1989, p. 19.
30. Dan Lashof, Natural Resources Defense Council, "Options for Mitigating the Greenhouse Effect: The Need for Solar/Renewable Energy," Forum on Renewable Energy and Climate Change, Washington, DC, June 14, 1989.
31. Lester Brown, Chris Flavin, and S. Postel, Worldwatch Institute, "A Sensible Energy Strategy," *Multinational Monitor,* January/February 1989.
32. Lashof, "Options for Mitigating the Greenhouse Effect."
33. MacKenzie, "Breathing Easier," p. 21.
34. Michael Eaton, John White, and P. Brodie, "The Greenhouse Effect: The Need for California Leadership," Sierra Club California, Sacramento, CA, January 19, 1989.
35. Global ReLeaf Facts, Washington, DC: American Forestry Association, April 1989.
36. D. Sheers, "Report on an Application for 24 Wind Turbine Generators at Mynydd-y-Cemais, Machynlleth," Montgomeryshire District Council, Powys, Wales, April 1991, p. 11.
37. Gottfried Vauk et al., "Bioligisch-ökologische Begleitunterschungen zum Bau und Betrieb von Windkraftanlagen" Norddeutsche Naturschutzakademie, Schneverdingen, Germany, 1990, pp. 96–98.
38. Alain Faujas, "Un Nouveau tourisme: 'le tourisme industriel,'" *Journal Français D' Amerique,* 17–30, September 1993, p. 12.
39. Testimony by the American Wind Energy Association for the California Energy Commission's 1994 biennial report on repowering California's wind industry, May 27, 1993. Prepared by Paul White and Paul Gipe of AWEA's west coast office.
40. The survey found 460 direct jobs per terawatt-hour of annual generation. This is comparable to the number of jobs in the Danish industry for operating and maintaining their fleet of wind turbines. There are approximately 400 people employed directly in the service sector of the Danish industry. Wind turbines generated 900 million kilowatt-hours during 1992, yielding 440 jobs per terawatt-hour of wind generation. Birger Madsen, BTM Consult, 1993, unpublished material.
41. Per Krogsgaard, BTM Consult, personal communication, May 1993.
42. An estimated $5.2 million in secured and unsecured property taxes is generated annually from wind energy developments in Kern Country. J. Fitch, Kern County Assessor's office, Bakersfield, CA, May 1993, personal communication. An estimated $5 million in secured and unsecured property taxes is generated annually from wind energy developments in Riverside County. N. Emmerton, Desert Wind Energy Association, North Palm Springs, CA, personal communication, April 1993; AWEA survey of California wind companies, March 1993.

13

The Emerald City

"No matter how noble the cause, there will always be an opposition." Governor Howard Pyle of Arizona (1951–1955).

Wind has long labored quietly in solar energy's shadow. Photovoltaics' glamorous and futuristic image elicits almost mystical expectations of its potential. Gleaming blue panels of solar cells bespeak high technology. To many, photovoltaics have become synonymous with a future solar economy. In popular parlance and in that of the media, when "solar" is mentioned photovoltaics are what first come to mind.

Some solar advocates have looked upon photovoltaics as a sort of energy panacea, often to the detriment of other solar-electric technologies. These advocates suggest that photovoltaics will provide all our energy answers once they become "so cheap you could paint them on the side of your house" as one skeptical utility engineer describes it. At Wisconsin's Midwest Renewable Energy Fair one summer, the session on photovoltaics resembled a tent revival; a religious fundamentalism coursed through the crowd. Like all evangelists, the speakers whipped their followers into a frenzy by citing the weaknesses and sins of competing technologies, including wind energy. Such proponents harm both wind and solar energy; wind by offering a "better" alternative someday in the hazy future, and solar in general by eerily reminding us of the hype surrounding the "too cheap to meter" days of nuclear power.

More dangerous still is the effect such proselytizing has on the direction of public policy. Photovoltaics' image has been used successfully in the United States to win three times more funding for R&D than wind energy has received. Although certainly no measure of efficacy in bringing a technology to market, federal funding is a measure of congressional support, a testament to the marketing adage "sell to perceptions." A European report examining the balance of government support for various technologies concludes that despite wind energy's economics and near-term promise, which are superior to those of photovoltaics, PVs have a "modern," "high-tech" image that outshines that of wind. Because photovoltaics sell better, they attract more research funds than wind energy, whose high-tech aspects are obscured by the turbines' appearance as simple, old-fashioned machines dependent on the mechanical, therefore outmoded, arts.[1]

This preference for photovoltaics is possibly due to widespread ignorance of wind energy's progress. Or, more likely, wind has fallen victim to the "Emerald City" syndrome. Proponents of a sustainable society look to an idealized vision of the future, to a time when we widely employ solar technologies, to a time when we husband our natural resources wisely, to a time when we treat our fellow man with respect. This is our "Emerald City." It is bright and shiny, a wonder to behold, and a worthy objective. Yet as we approach the gate, the luster dims. The city differs from the one we had envisioned.

We have not only reached the gate of wind energy's Emerald City, but we have gone inside to discover that oil is staining the once-bright towers, some of the turbines are not spinning gleefully like the rest, and others whir and creak. It is not the same pretty sight we saw from a distance. We feel disappointed, cheated. We traveled that long road and for what? The reality is less appealing than the vision we had from afar. Beyond the gates, in the far distance, is another Emerald City, bright and radiant. With that sparkle of reflected sunlight, it looks even better than the one we have just entered. But it is a long way off. We had better leave this less-than-ideal place and strike off now for the next or we may never get there in time. Photovoltaics offer that other Emerald City in the distance, the city always just out of reach.

The intent of recounting this parable is not to justify abandoning the quest for the Emerald City; like the pilgrimages of the Middle Ages, it is a noble journey. It is important, however, that the pilgrims keep it in perspective. Proponents should realize that the luster of photovoltaics may also wane as we near the gate and should be less quick to condemn wind energy for its faults. Because wind energy is now widely used in California and northern Europe, its technological warts are exposed to view for anyone willing to examine them. Photovoltaics, in contrast, are not deployed as extensively, and as a consequence, their net social costs may not be as apparent as those of wind.

According to West Germany's Fraunhofer Institute, photovoltaics may have higher social costs than wind energy, primarily because they exclude land from other uses. The Institute's Olav Hohmeyer, the author of a study on the social costs of energy, argues that wind energy is compatible with land uses such as grazing and growing crops. In contrast, the panels of solar-electric technologies block as much sunlight as possible, thus shading the land beneath the panels. If, as some advocates suggest, the solar cells were installed on rooftops instead of in power plants, Hohmeyer warns that the social costs, rather than being eliminated, would simply be shifted to accidents resulting from inexperienced installers.[2]

The latter can also be said of wind energy. If wind turbines were used in distributed applications such as on farms or in suburban backyards instead of in wind power plants, the social costs of massed arrays would be traded for accidents by unskilled homeowners climbing towers to service their own turbines. There are social costs to both approaches. In a central plant, the costs are more easily discernible than they may be in a dispersed system, but both incur costs just the same. Hohmeyer emphasizes that the social costs of both wind energy and other solar-electric technologies are far less than those of either nuclear or fossil fuels.

There are differences between the impacts of each technology, but there are also similarities. Whereas all wind turbines are audible because of either the blades moving through the air or the whine of their gearboxes and generators, photovoltaic arrays are virtually noiseless except for occasional sounds from a tracker's motor as it hunts the sun, or the hum of the inverters. Solar arrays generally hug the ground, in contrast to wind turbines, which thrust boldly skyward. Nevertheless, like wind power plants, massed solar arrays will be highly visible features on the land from any elevated vantage point. Solar arrays that track the sun will also face the "missing-tooth" phenomenon confronting the wind industry. If a solar tracker malfunctions in an array, the panel it controls will be askew from the rest and like a nonspinning wind turbine will become highly noticeable. Something in the ordered rows of panels will be awry, imperfect. When that occurs, the Emerald City will lose some of its luster and attract criticism.

All solar-electric tecnhnologies share these attributes: they produce electricity cleanly and with a renewable resource. Both wind energy and photovoltaics generate electricity directly. Other technologies, such as commercial solar-thermal systems, generate electricity indirectly by heating a fluid that drives a conventional turbine. Examples of the latter are the Solar One central receiver near Barstow, California and the nearby Luz plants. Collectively, these solar technologies have achieved a remarkable success: the 3.3 TWh they generate yearly serve the residential electrical needs of more than 1 million Californians. Wind provides about four-fifths of the total, and Luz's solar-thermal electric plants provide the remaining one-fourth. At their peak photovoltaics generated 25 million kilowatt-hours annually in California and met the residential electrical needs of about 9000 people (Figure 13.1).

In general, solar-electric technologies have lagged behind those of wind. The CEC's Sam Rashkin critically notes that "solar thermal and PV have never met their performance goals." As a consequence, there has been far more generating capacity installed in California in the form of wind turbines than that of either photovoltaic arrays or solar-thermal electric systems.

By the early 1990s, photovoltaics in power production delivered 120 MW worldwide, mostly in rural, stand-alone applications. Consumer electronics used another 130 MW. And manufacturers were churning out 60 MW more every year. By the late 1980s, there were 10 MW of photovoltaics installed in California, the bulk of it in two central stations built by Arco. Since then, Arco has sold their interests in the plants and the new owners are progressively dismantling them. The solar panels are far more valuable on the remote

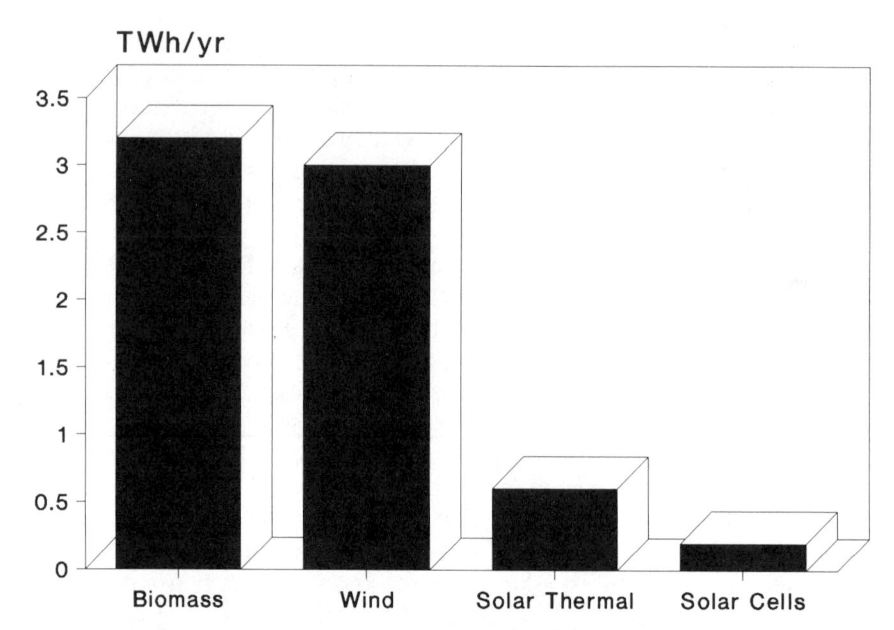

Figure 13.1. Solar energy in California. Wind energy provides more electricity than other forms of solar energy except biomass.

systems market than they are for generating electricity for sale to Pacific Gas & Electric at $0.03 per kilowatt-hour.

The largest solar-electric technology in the world—besides wind, hydro, and biomass—is Luz's parabolic-trough concentrators, Luz's corporate descendants operate 200 MW of capacity on the Mojave Desert north of Los Angeles.

The performance of wind turbines is comparable to that of other solar-electric technologies in California; that is, it delivers about the same amount of electricity relative to installed capacity. However, peak production from solar plants occurs during the time of day when the utility's demand is greatest. There is less of a match between wind generation and demand, except in selected areas such as in Solano County northeast of San Francisco.

Each of these technologies has its advantages and disadvantages. None is a panacea. They all have their place in a mix of renewable technologies that can be used to build a sustainable society. It may be wise to consider Barry Commoner's admonition, "There's no such thing as a free lunch." For any of these technologies to be used, they have to be sited somewhere.

Gorman

On August 17, 1989, more than 200 people packed into a public hearing to shout their disapproval of a proposed wind power plant near Gorman, California.[3] After 5 hours of raucous testimony, the Los Angeles County Planning Commission voted unanimously against the project, ending a bitter dispute between wind energy and one of the state's largest landowners. By the mid-1990s, the scars from the encounter had yet to heal.

Zond's proposal and its dogged pursuit of the project through several public forums damaged not only its own image but that of the entire industry. Unlike U.S. Windpower, which when confronted with overwhelming political opposition to its Rattlesnake Hills project in Washington state, moved its proposal to another site, Zond forged ahead—recklessly, competitors charge—despite the prospect of certain defeat. Emotions ran high on both sides, but especially among residents of Gorman and nearby Frazier Park, who were convinced, through the cunning machinations of the Tejon Ranch Company, that they and their environment would become victims of a marauding corporate Titan.[4] Tejon Ranch, who feared that wind turbines within their viewshed would lower the value of their real-estate holdings, skillfully elevated a dispute between two real-estate developers, itself and Zond, to a debate over the environmental acceptability of wind energy.

There were legitimate environmental concerns about Zond's proposal, especially its potential impact on the California condor. Such concerns could be addressed, and Zond attempted to do so. But Zond's efforts were buried beneath an avalanche of hysteria as their opponents successfully portrayed Zond's project as the greatest environmental calamity to befall Los Angeles County since the automobile. It was a modern version of the Salem witch trials, and several influential members of the environmental community were ready to burn Zond at the stake.

Yet it need not have happened. A lot of blood was needlessly spilled on both sides. Zond could have exercised some foresight. Their own staff warned as early as 1984 that entering Gorman would invite a pitched battle. And the environmental community, especially the Sierra Club, could have censored local activists or disassociated themselves from the activists' antiwind rhetoric. The Sierra Club almost did.

During the fall of 1988, the Sierra Club's Los Angeles Chapter beat back a proposal to oppose the Gorman project outright. Instead, they approved a resolution, calling on Los Angeles County to delay approval of the project until the county established a wind energy ordinance addressing such environmental concerns as regular maintenance, removal of inoperative turbines, and site restoration if the project were eventually abandoned.[5] That, however, was not the end of the matter.

Views within the Sierra Club's Los Angeles Chapter were split. The chapter's conservation chair, the person responsible for determining which issues the club tackles, made an inspection visit of the site at Zond's invitation. Looking over the terrain, she asked a series of leading but not unexpected questions. After hearing Zond's explanation that the dispute revolved around the Tejon Ranch's plans for a massive real-estate development on adjoining property, she turned to them and commented, "but there's nothing wrong with houses." If one of the key local members of the Sierra Club thought that residential development on a wind-swept mountaintop with its attendant roads, utilities, and human impact was preferable to wind energy, it was going to be a long, rocky road for Zond. It was.

Worse was ahead. Two influential club members sought no counsel from Zond, refused similar invitations, and had no plans to seek peace. To Sally and Les Reid, dealing with the enemy was out of the question. The only acceptable solution was either Zond's annihilation or its unconditional surrender. The Reids are not ones to cross. Both are experienced organizers, and at the time, Sally was on the national board of the Sierra Club. Their initial setback at bringing the moral and political force of the club to bear against the Gorman project did little to deter them. They were seasoned infighters and knew club policy well.

Unlike other environmental groups in the United States who depend on grassroots support, the Sierra Club exacts a modicum of discipline on what can be said publicly in its behalf. Launching a legal offensive, for example, requires a lengthy series of approvals. A local activist cannot simply walk into an attorney's office and file a lawsuit on behalf of the Sierra Club. The intent is to avoid situations where one representative of the Sierra Club is saying one thing, such as advocating the use of renewable energy, while another is saying the opposite. That was exactly the situation the Sierra Club faced with Gorman, and the Reids knew it. They also knew how to circumvent it.

The Gorman site lay in Los Angeles County, the jurisdiction of the club's Los Angeles Chapter, but was close to the Kern–Los Angeles County line. The Kern-Kaweah Chapter remained neutral. It was the Los Angeles Chapter's call. However, the Reids knew that if Kern-Kaweah opposed the project, the conflict would move to the state level, where they held greater sway. After an impassioned plea by a self-styled authority on the condor and without a word in Zond's defense, the Kern-Kaweah Chapter jumped into the fray on the side of the Reids and Tejon Ranch. Their decision sent the issue up to Sierra Club California, where the Reids and their allies carried the day. Sierra Club California overruled the Los Angeles Chapter and officially sided with the Reids, unleashing a torrent of authorized abuse on Zond and on wind energy in general. With moral right on their side, no accusation was too outlandish, no tactic too abhorrent in their just cause.

As in every war, there were collateral damage and innocent victims of the fighting. Wind energy became a casualty of the Gorman battle. The effects were still being felt half a decade later. The Sierra Club suffered, too. The controversy reinforced the image that the club never met any change it liked, that even renewable energy development, one of the planks in the club's energy platform, was unacceptable when it stepped from the hypothetical to the real.

There was a certain NIMBY element at work in Gorman, too, says Sierra Club California's energy chair Rich Ferguson, who has tried to prevent just such disputes between renewables and the club's traditional preservation wing from erupting into a public showdown. After all, the wind farm would have been installed in what the Reids consider their backyard. Their opposition may have had less to do with such lofty goals as protecting the condor than with the view from their favorite overlook.

There also may have been more insidious "understandings" at play as well. On the 150th anniversary of the Tejon Ranch's creation, the *Los Angeles Times* quoted Mary Griffin, president of the Kern Audubon Society, as saying: "We think that Tejon will eventually donate a great deal of that land for open space and the public good. They have had a long history of cooperating with the Audubon Society and they've been very nice to our chapter."[6] There is no evidence of a *quid pro quo,* but the implication is clear. Audubon brought the full weight of its national, regional, and local groups to bear against Zond's project. Audubon and the Sierra Club performed yeomen's work by rallying their troops to defend habitat from the forces of darkness. The Tejon Ranch Company was appreciative.

For its part, the Tejon Ranch Company, who aided and abetted local activists financially, thought it had a lot to lose if Zond succeeded. "Who would want to buy a million-dollar home with a view of a wind farm?" they probably wondered. Tejon's stock price at the time was largely determined by speculation in the ranch's real-estate development potential. During the height of the Gorman controversy, which was also the height of the California real-estate boom, Tejon's stock traded at $49 per share. Real-estate values crashed soon after, and by 1992 that same stock was trading at $13 per share, based on its value for ranching and minerals. (The ranch leases land for several oil and gas wells and for a large cement plant which burns liquid hazardous wastes.) The difference represents the stock's speculative value. Financial markets are notoriously edgy. Any hint that an "undesirable" neighbor might move in next door could send the stock price plummeting, costing investors millions. The Company was obligated to protect their interests.

The first step was hiring Cerrell and Associates, high-powered political consultants and advisers to Hollywood stars. Tejon Ranch got their money's worth. The conflict made the pages of every environmental magazine and nearly every major newspaper in the country. Tejon Ranch also hired a bevy of other consultants, who dutifully testified with their sponsor's position fully in mind. One Tejon engineer testified that no one knew whether or not the site was even windy. (Zond dropped six inches of paper in his lap that said otherwise.) Another said that Zond's turbines could not possibly offset as many pollutants as claimed. This "consultant" used as a reference a common conversion table found in any dictionary that neglects to account for the thermal efficiency of conventional power plants. Such a serious error at a large engineering firm would have cost the fellow his job. The audience applauded him as a hero.

The hearing climaxed the whole sad affair. The allies showed no mercy. They massacred Zond, leaving them to nurse $1 million in fruitless expenses while the Reids added another corporate scalp to their collection. But before it was all over, one lone commissioner rose to chastise the audience. He warned that no one should be fooled that this was a victory for the environment. There were other interests at play here, interests who had cynically wrapped themselves in a green flag. And he expected every one of them back clamoring for action when those interests brought their project before the planning commission.

At the height of the Gorman hearing, an old man took the podium. Suddenly the television news crews switched on their Klieg lights. Something was afoot. They had been alerted that a suitably newsworthy "sound bite" was on its way. Tension in the room mounted. The old man proceeded to lovingly describe the beauty of his racing pigeons, their speed and grace, how they had become a part of his family, and then with perfect timing and dramatic flair, pleaded with the planning commission to protect his pigeons from "the Cuisinarts of the air." The arrow went straight home, sending up a roar from the audience. A new image had been created and the cameras flashed it across the country.

Although often credited to staging by Cerrell and Associates, the term was conceived by the Sierra Club. The club's Los Angeles area representative, Bob Hattoy, later bragged to a Washington lobbyist that he coined the infamous expression. Hattoy knew how to turn a phrase. He brought the 1992 Democratic national convention and a television audience of millions to tears with his story of contracting AIDS. Wind energy had made one powerful enemy.

Ironically, one Tejon Pass project is moving ahead—but not in the United States. Costa Rica expects to develop a 20-MW wind plant in the *Tejona* Pass of its Guanacaste province.[7]

Cynicism and Hope

In San Ramon, California, two representatives of the Sierra Club's Mother Lode Chapter attended a meeting at Pacific Gas & Electric. They were there to talk about birds. Like errant schoolchildren, they sat in the back, at the edge of but not quite a part of the proceedings. When called upon to comment as a "public-interest group," they self-righteously demanded to be called "concerned citizens." Everyone there was a concerned citizen, and everyone there represented an interest group. When they shouted that there were too few environmentalists at the meeting, Pacific Gas & Electric's Mary Ilyin blurted out, "we're all environmentalists or we wouldn't be at this meeting." Mary has spent her professional life laboring in the hostile vineyards of utility bureaucracy to elevate renewables to the level of a real alternative. She resented being relegated to the "business," therefore the rape-pillage-and-plunder group, and judged by someone she had never met and who had no idea of her beliefs.

Not long afterward, the same two members of the Mother Lode Chapter encountered a representative of the wind industry at a Washington fund raiser for the Endangered Species Act. "Why are you here?" They demanded to know.

"You're the enemy." They did not know, and had no interest in learning, that he had been invited.

During the mid-1980s, desert activist Howard Wilshire launched a crusade to stop the wind industry from plundering California hillsides. With the scientific credibility of his employer, the U.S. Geological Survey, solidly behind him, he undertook a so-called "study" of wind energy. But at nearly every turn, he was thwarted by the growing productivity of the industry. When confronted with facts that undermined his conclusion, he would conveniently ignore them. After several rejections by environmental magazines for its unsupported conclusions, his paper was finally published in an obscure journal. It is frequently cited by opponents of wind energy. The Reids wielded it well at Gorman, and nuclear advocates in Richland, Washington used it with success at Rattlesnake Hills. As late as the mid-1990s, Wilshire still doubted that California wind turbines were generating significant amounts of electricity.

Such cynicism as these incidents exhibit has reached epidemic proportions among environmentalists. It permits some to bask arrogantly in the moral clarity of the "born again." A lack of introspection allows them to see only black and white, where others see moral ambiguity. Yet, as seen in the Gorman case, the issues are not always as clear-cut as they first appear.

Margo Guda of the Fundashon Antiyano pa Energia notes that today's environmentalists are often newcomers to the field. They are missing the institutional memory of those activists who were hardened in past battles over coal or nuclear power. Guda, whose group advocates renewable energy and conservation on the Caribbean island of Curaço, says that these newcomers are acting morally outraged on issues perceived by the activists of two decades ago as solutions to problems. They have no context, she fumes. Hardly concealing her exasperation, Guda asks where these "activists" were hiding 20 years ago, when so many coal and nuclear power plants were proposed.

Guda and others facing this quandary say that environmentalists must ask themselves some difficult, uncomfortable questions. For example, if wind energy is good in someone else's backyard, is it not also good in mine? Some have asked that question and have renewed their support of wind energy while redoubling their efforts to ensure that sites are developed properly. Others have taken the easy way out and opted for today's most popular energy panacea: efficiency and conservation.

Overreaction to wind energy's failings endangers the technology's principal reason for existence, its environmental benefits. If wind became just another energy technology, its ultimate contribution could be hindered. Some opponents, like the Reids, hope to encourage just that. This conversion of environmental support to hostility by misguided but influential environmental leaders certainly plays into the hands of those who want to quash renewables and begin building new coal and nuclear plants again.

When Yorkshire Water sought a planning permit for their project at Ovenden Moor in the Pennines, Sir Bernard Ingham rose in righteous rage and wrote local papers that "unless we blow up in revolt, we shall find every mountain, hill and cliff colonized by flailing pylons." Ingham, a former aide to British Prime Minister Margaret Thatcher, was no stranger to politics or the politics of energy, having worked for the British nuclear industry.[8] He quickly formed a group, charmingly called Country Guardians, with one mission: to smother Britain's fledgling wind industry in its cradle.

In September 1992, Cliff Groff, a member of the Kennewick City Council in southeastern Washington state, called a wind consultant for help. He had a problem. There was a company, he said, that wanted to put windmills on the Rattlesnake Hills just west of Kennewick. "And?" The consultant's secretary asked. "We don't want them," he replied. "Why?" she queried. "We're 100% for nuclear power." He gave no other reasons. Kennewick is one of the tricities that serve the Hanford Nuclear Reservation. This is a region where the jerseys of the Richland high school football team are emblazoned with an emblem of a mushroom cloud. Groff was faithfully protecting what he considered his constituents' interests. He was also successful: U.S. Windpower eventually withdrew its proposal because of local opposition.

At least Groff is forthright. Like the Tejon Ranch, Ingham is more diabolical, wrapping himself in the green mantle of protecting the countryside from assault. But opponents such as the Reids are the most damaging of all. Because if they succeed too often, if there are too many such "victories" as at Gorman, they will steal one of modern environmentalism's most promising developments: hope.

As professor Robert Paehlke explains in *Environmentalism and the Future of Progressive Politics,* renewable energy, including wind, gives activists a means of describing the kind of future they want, rather than, as too often has been the practice, of fighting the future they want to avoid. It gives environmentalists an opportunity to support a positive alternative. Critics charge, often justly, that environmentalists oppose everything and are seldom for anything. Through the proselytizing work of visionaries such as Amory Lovins, renewables and energy conservation provide a vehicle for expressing environmentalism's positive side.[9]

Lovins' "soft path" showed that there *was* a way out of the energy crisis and many related environmental issues. He provided environmentalism with the intellectual means for breaking free from what Paehlke calls the politics of negativism that too often pits environmentalists against nearly everyone else.[10]

Paehlke describes one of environmentalism's fundamental tenets as originally proposed by Kenneth Boulding in his "spaceship earth" analogy: an economy that seeks to minimize throughputs by reuse, repair, and the use of renewable processes. Thus, says Paehlke, "virtually all environmentalists accept that conservation and renewable resources, especially energy resources," should receive encouragement. Boulding established the ideological test that all technologies must pass before they can receive environmentalism's stamp of approval.[11] When done right, wind energy passes that test.

Neither the Sierra Club nor any other environmental group has ever given wind energy a blank check. They have, as they should, always reserved the right to oppose specific projects or specific practices. Wind turbines are unsuitable in some areas. No one wants to see a wind power plant in Yellowstone National Park or in Montana's Bob Marshall Wilderness. Some types of wind turbines are undesirable because of the noise they make, the oil they spill, or the birds they kill. This entails a responsibility to tell those who build and use them that there is something amiss that needs correction, and why. The corollary, though, is that if wind energy is unsuited in some places, it is suited in others. There must come a realization that a sustainable economy will require renewable power sources and that wind energy will be a part of it. And there is no wind energy without windmills.

The all-too-frequent lip service paid to wind energy by the environmental community eventually begins to sound hollow. "We support wind energy, but . . ." has become a cliché recalling another era in U.S. history when integration was on the front page. "Why, some of my best friends work with windmills, but. . . ."

Environmentalists will have to face an unpleasant fact, warns Michael Grubb. As renewables move from research to full deployment, trade-offs will be necessary. "Sustainable development," he says, "can mean many things, but it will certainly not mean invisible development, and environmental groups will have to accommodate this reality."[12] He goes on to warn that nuclear's prospects could brighten if environmental support for renewables erodes as a result of their large-scale deployment.

Environmentalists have a responsibility to move wind energy forward actively, not to stop it. They must avoid taking the easy route, simply standing by and letting the market and local activists control the direction of public policy. They must stop wringing their hands about the excesses of the industry, and set to work to make an imperfect technology, one built and operated by imperfect people, perform better.

Environmental groups have come to the aid of wind energy in the past. *Public Citizen,* one of Ralph Nader's offspring, came to Zond's aid on the Gorman project. The Sierra Club's North Star Chapter lobbied aggressively to bring wind energy to Minnesota, as has the Izaak Walton League. The Union of Concerned Scientists have become outspoken proponents of renewable energy, wind in particular. In Great Britain both Greenpeace and Friends of the Earth have actively campaigned in support of wind energy. After Sir Bernard Ingham created the Country Guardians, Friends of the Earth countered his move with equal dramatic flourish. Simon Roberts, of Britain's Friends of the Earth, fired a salvo at "yet another whining urban émigré finding his rural idyll disturbed."[13] Friends of the Earth has gone further than any other environmental organization in support of wind energy by urging its 320 local groups in Britain to publicly aid wind projects proposed for their areas. With the exception of UCS, most of the major American environmental groups, the seven sisters, provide only lukewarm support for renewables, focusing their activities on attempts to increase ineffectual federal research.

At the same time, advocates of renewables must also realize that with few exceptions, the people who build wind farms are not environmentalists, nor are they, as a rule, cynical opportunists. Wind companies are businesses. They know that it is environmental restrictions on other fuels that make wind energy attractive. They understand that when the Bonneville Power Administration tries to save salmon runs by reducing hydro flows in the Pacific northwest, BPA creates a need for new sources of energy, preferably renewable energy such as the displaced hydro power. But wind companies respond to the needs of business. If their competitors underbid them by cutting costs for environmental protection, they are forced to do the same. Business is a delicate balancing act and chief executives are always walking a tightrope between the needs of the community, their employees, and the marketplace.

Like others in positions of leadership in the society at large, wind company executives must be sensitized to environmental issues. If wind companies want to wear white hats, and most do, it is insufficient for them simply to comply with the law. They must learn the why, not just the how of meeting environmental standards to ensure that the environment, like safety, becomes an everyday part of doing business. But business schools spend little time introducing their MBAs to Aldo Leopold, Garrett Harden, Fairfield Osborne, and others. Chief executives are quick learners, but often stubborn, and frequently, the antibusiness rhetoric and suspicion of local activists serves only to sour business leaders on the message.

Activists, who seldom work in the private sector and who have little or no appreciation for the challenges that businesses face, must grit their teeth and begin to take an active part in the wind industry. They risk getting their hands dirty, and possibly their consciences as well, but it is only through regular contact with the key players that they can promote their views. By sitting at the table when projects are planned, environmentalists can negotiate conditions to minimize wind energy's impact before the shouting starts. Their constant presence can sensitize executives to environmental concerns and ensure that issues are addressed early in the planning process, not pasted on later.

It is necessary as well to maintain the contact and the pressure after the projects are built and for years afterward. Attention often flags after projects are completed. Wind plants will be in operation for 20 to 30 years—possibly in perpetuity. There is always the risk that significant environmental damage will occur long after the project has been built, for example through overgrazing, widening of roads, failing to control erosion, or repowering. The environment must become an institutionalized priority just as safety has. If it does, the bigger question—whether wind energy can coexist with respect for life and land—can be answered unequivocally. Yes, wind and life, wind energy and the environment, are compatible. They are not mutually exclusive.

Chapter 13 Endnotes

1. Sara Knight, "Balancing the Energy Budget," *Windpower Monthly,* 9:1, January 1993, pp. 15–16.

2. Olav Hohmeyer, *Social Costs of Energy Consumption: External Effects of Electricity Production* (Berlin: Springer-Verlag, 1988), pp. 54–58. Hohmeyer estimates the social costs of solar energy is 0.0044 DM per kilowatt-hour; wind energy is 0.00009 DM per kilowatt-hour.
3. Phyllis Bosley, "California Wind Energy Development: Environmental Support and Opposition," *Energy & Environment,* 1:2, 1990, pp. 171–182.
4. In the course of public debates I have been physically threatened twice: once in Cumberland Gap, Tennessee, by a coal miner who threatened me for supporting passage of the Surface Mining Act, which he felt endangered his livelihood, and once in Gorman by one of the organizers of Save Our Mountains, whose livelihood was not at risk.
5. Resolution of the Conservation Committee, Los Angeles Chapter, Sierra Club, October 21, 1988.
6. Jonathan Gaw, "As It Turns 150, Tejon Weighs Development," *Los Angeles Times,* September 19, 1993, p. 3.
7. "Costa Rica's State Utility Considers 20-MW Wind Farm Site," *Wind Energy Weekly,* August 16, 1993, p. 4.
8. Lawrence Ingrassia, "We'll Refrain from Commenting About the Source of All the Wind," *Wall Street Journal,* August 17, 1993.
9. Robert C. Paehlke, *Environmentalism and the Future of Progressive Politics* (New Haven, CT: Yale University Press, 1989); see Chapter 4, "The Energy Crisis: Limit and Hope," and Chapter 9, "Environmentalism and the Restoration of Progressive Politics?"
10. Ibid., p. 109.
11. Ibid., p. 204.
12. Michael Grubb, "The Cinderella Options: A Study of Modernized Renewable Energy Technologies," Part 2, "Political and Policy Analysis," *Energy Policy,* October 1990, pp. 711–725.
13. Janice Massey, "The Battle for Britain," *Windpower Monthly,* 10:6, June 1994, p. 25.

III

WHERE WIND ENERGY IS HEADED

In previous sections we have seen that wind energy has come of age as a commercial generating technology, and now confronts many of the same issues facing any other energy technology. There is one important difference: wind energy is renewable. Wind can make a difference toward building a sustainable economy. How much difference wind can make, and how soon, are the subjects of Part III.

14

Wind's Future

Those skeptical about wind energy's role in our future energy supply often correctly note that the generation produced by a single small or medium-sized wind turbine looks feeble compared to that produced by a nuclear power plant. If this argument were extended to its logical conclusion, on a cosmic scale, the energy conversion of one wind turbine is infinitesimal alongside that of the universe. These cynics miss the point. On a local, regional, and national level, arrays of multiple wind turbines can make a significant contribution. One wind turbine, or one wind plant, even when very large, will make only a small dent in the electricity consumption of a huge nation such as the United States. The beauty of wind energy is that unlike nuclear power, it is decentralized and modular. Wind generating capacity can be added steadily and progressively, one turbine at a time.

Potential in the United States

Collectively, thousands of wind turbines can make a difference. The resource is truly enormous. Wind energy currently produces 0.1% of the electricity in the United States, mostly in California, but it could produce significantly more.

Even in California, which has seen the world's most extensive wind development, there are ample wind resources for future expansion. According to the California Energy Commission, the state has the potential for 7000 MW of wind generation in prime locations and an additional 6400 MW at less energetic sites: altogether more than six times that currently produced. Battelle's Pacific Northwest laboratory arrived at similar results, based on the wind turbine technology available in the early 1990s. Modest advances in wind technology could enable California to tap 10 times the wind energy currently produced—and meet 10% of the state's electricity consumption—after excluding environmentally sensitive areas from wind development.[1]

But California's resources are overwhelmed by those of other states. Two sites on the Great Plains, one in Minnesota on the eastern edge of the Plains, the other in Montana on the western edge, illustrate the potential.

Buffalo Ridge, Minnesota

For nearly a decade, Paul Helgeson and John Dunlop of Minnesota's Department of Public Service administered a unique statewide resource assessment program in cooperation with local utilities and other state agencies. The program was significant because it discovered a powerful wind resource in a small slice of southwestern Minnesota called Buffalo Ridge, an area previously unidentified in national surveys.

Buffalo Ridge alone could produce as much wind-generated electricity as is currently produced in California. The ridge is actually a low-lying hill 100 km (60 miles) long running southeastward from Sica Hollow, South Dakota to Spirit Lake, Iowa. The "ridge" is more correctly termed a low rise that forms the drainage divide between the Mississippi and Missouri watersheds. The strong winter winds that sweep across the tilled soil of this subtle topographic feature give the ridge a resource comparable to that found in California: an annual average wind speed of 7.4 m/s (17 mph) at 30 m (100 ft) above ground level.[2]

If the windiest sites are found within 1 km of the summit, Buffalo Ridge's best resources encompass 200 km^2 (80 mi^2) of the tristate area. Theoretically, this area could contain upward of 1250 MW of modern 500-kW wind turbines in widely spaced, open arrays. The 2500 turbines this represents could generate 3 TWh annually—as much electricity as produced in California by seven times more turbines in densely packed arrays.

The wind resource becomes less energetic off the crest of the ridge. Still, the potential is staggering. After mapping the region of which Buffalo Ridge is a part, Helgeson found that more than 2000 km^2 (800 mi^2) exhibit the equivalent of a Battelle class 5 wind resource. This one region of the Great Plains could contain upward of 25,000 wind turbines generating 24 TWh annually, slightly less than 10 times California generation, or half of Minnesota's 1990 electricity consumption.[3]

Pacific Northwest

In a study performed for the Northwest Power Planning Council in late 1989, Don Bain estimated that there are sufficient resources in the Pacific northwest to support 18,900 MW of wind-generating capacity, at costs ranging from \$0.075 to \$0.15 per kilowatt-hour. Bain, an energy analyst with the Oregon Department of Energy, found most of these resources on the Blackfeet Indian Reservation east of Glacier National Park in northwestern Montana. The reservation contains 11,150 MW of wind resources, with 7120 MW of that potentially developable at $0.0831 per kilowatt-hour in nominal 1988 dollars.[4] The Council's staff estimates, for comparison, that power from two 600-MW plants burning pulverized coal will cost $0.085 per kilowatt-hour, or slightly more than the cost at the best wind sites.[5] Altogether the 25-m wind turbines Bain envisioned using could generate 34 TWh annually, more than 10 times California's generation and five times that of a well-running 1100-MW nuclear power plant.

There are even better sites in the northwest than those on the Blackfeet reservation, but they cover a less extensive area. Even so, says Bain, there are 16 sites in the Pacific northwest that could supply a total of 12,800 MW of capacity at less than $0.10 per kilowatt-hour in nominal dollars. If the wind resources identified by Bain were ever developed, the wind plants would occupy only 1% of the northwest's total land area.

Battelle Resource Assessment

The work in Minnesota and the northwest barely scratched the surface. Analysts only realized the full potential of wind energy in the United States after Battelle's Pacific Northwest Laboratory examined the question in the early 1990s. As a part of the National Energy Strategy's review of U.S. energy policy, the Department of Energy directed Battelle to assess the nation's wind-electric generation potential. Using Battelle's most recent Wind Resource Atlases, Dennis Elliott and Larry Wendell proceeded to sort regions by their resources and land uses. What they found surprised even wind energy proponents: wind energy could meet more than one-fourth of the nation's electricity consumption, using the technology that was then available and excluding environmentally sensitive areas. With expected advancements in the technology, Battelle found that wind turbines could theoretically generate twice the total consumption of electricity in the United States.

Battelle uses a system of wind power classes to characterize wind resources. The classes provide a shorthand method of expressing wind speed and power density at various heights above the ground (Table 14.1). (This table can be used with the Battelle map of the United States in Appendix D.) These classes represent a range of overlapping conditions. The lower end of class 5 overlaps with the upper range of class 4, and the upper end of class 5 overlaps with the lower end of class 6. As mentioned in Chapter 6, commercial wind plants in California are generating bulk power in areas with class 5 resources.

Realizing that wind energy is incompatible with some existing land uses, Elliott and Wendell considered several different land-use scenarios. Under the "moderate" scenario, Battelle excluded all urban areas and all parks, 50% of forested lands, 30% of agricultural land, 10% of range land, 20% of mixed agricultural and range land, 10% of barren land, and 100% of wetlands.[6] This scenario excludes 50% of windy areas with class 5 or greater resources. The

Table 14.1
Battelle Wind Power Classes

	At 10 m (30 ft)			*At 30 m (100 ft)*			*At 50 m (160 ft)*		
Power Class	Nominal Speed (m/s)	Power Density (W/m²)	Nominal Energy (kWh/m²/yr)	Nominal Speed (m/s)	Power Density (W/m²)	Nominal Energy (kWh/m²/yr)	Nominal Speed (m/s)	Power Density (W/m²)	Nominal Energy (kWh/m²/y)
	3.5	50	440	4.1	80	700	4.4	100	880
1									
	4.4	100	880	5.1	160	1,400	5.5	200	1750
2									
	5.0	150	1310	5.9	240	2,100	6.3	300	2630
3									
	5.5	200	1750	6.5	320	2,810	7.0	400	3500
4									
	6.0	250	2190	7.0	400	3,510	7.5	500	4380
5									
	6.3	300	2630	7.4	480	4,210	8.0	600	5260
6									
	7.0	400	3500	8.2	640	5,610	8.8	800	7010
7									
	9.5	1000	8760	11.1	1600	14,020	11.9	2000	17520

Note: Increase in speed and power with height assumes $\frac{1}{7}$ power Law and Rayleigh distribution.

45,000 km² (17,000 mi²) remaining after the exclusion of environmentally sensitive lands cover less than 0.6% of contiguous U.S. land area.[7]

After these exclusions, Battelle estimates that the United States could produce 700 TWh per year from more than 300,000 MW of wind capacity. This is enough capacity to generate 27% of total 1990 electricity consumption and nearly 10% of the total U.S. energy consumption.[8] The land used to do so would be one-tenth of that occupied by the wind turbines, or 0.06% of the lower 48 states.

During peak development in California during the mid-1980s, manufacturers were essembling 300 to 400 MW per year. At that rate, which seemed frenetic at the time, it would take 1000 years to reach full U.S. potential. With today's 500-kW wind turbines, it would require 600,000 machines. This seems like a staggering number, except when compared to the 8 million cars built in North America every year.

Elliott's team assumed that the turbines would be installed on towers 50 m (160 ft) tall. If 30-m (100 ft) towers were used instead, wind generation would still meet 20% of U.S. electricity consumption. Battelle's assessment also considered a hypothetical array of wind turbines 10 diameters apart downwind by 5 diameters apart across the wind. This 10 by 5 spacing is more open than that found in California, where turbines are far more concentrated. The winds in California's mountain passes are nearly unidirectional, allowing for a much tighter crosswind spacing than would be best on the Great Plains. A typical spacing in California of 8 by 2 diameters is three times denser than that used in Battelle's study (Table 14.2). Battelle's spacing approximates that of Danish arrays, which are less dense than California's.[9]

Table 14.2
Array Spacing and Generation[a] (Rotor Diameter)

Spacing	Cross-Wind	Down-Wind	Relative Density	acres/MW	ha/MW	Million kWh/mi^2	Million kWh/km^2
Open[b]	5	10	1	39	16	41	16
Moderate	3	10	1.7	23	9	68	26
Dense	2	8	3.1	12	5	127	49

[a,b] Spacing used in Battelle PNL study of U.S. wind resource potential.

Array losses and turbulence decrease greatly with more open spacing. Thus the open arrays envisioned by Battelle would enable these turbines to work more efficiently and reliably, and probably with less maintenance, than the turbines operating in California's tight arrays, where the turbines are buffeted by intra-array turbulence. But it is probably unwarranted to go to a less dense packing ratio than Battelle's 10 by 5 diameter arrays.

To err conservatively, Battelle assumed that altogether, array effects would rob 25% of total generation from their hypothetical array. Zond's Victory Garden Phase IV, by comparison, loses slightly less than 10% from all effects, including those of the array. Their much larger Sky River project, which has long transmission distances, loses only 12%. Thus well-sited projects in open arrays should be able to keep total losses to well under 15%. Battelle also used a conservative overall efficiency of only 25%.[10] Modern wind turbines at good sites could capture up to 35% of the energy in the wind. These changes, which are within the reach of current technology, would increase the potential shown in Battelle's survey by more than 50%. Put another way, technology performing better than that assumed in Battelle's assessment could reduce by one-third the number of turbines required to deliver the same amount of electricity.

As enormous as the potential is for class 5 and greater resources, Battelle found even greater untapped resources in midwestern and plains states if wind technology continues to advance such that class 4 resources become economical. With moderate exclusions, 6% of the U.S. land area has class 4 and greater wind resources. This is 10 times the land area with class 5 potential and is sufficient to generate 5500 TWh: more than twice the amount of electricity consumed in the United States and more than two-thirds of total U.S. energy demand in 1990[11] (Figure 14.1).

There are likely to be additional windy areas that have yet to be identified. Since Battelle last updated its survey of wind resources in 1987, researchers have found a large area of class 5 resources on Buffalo Ridge in southwestern Minnesota.[12] In many areas of the United States there is a paucity of wind data, and meterologists now believe there may be other regions where there are class 5 or greater wind resources waiting to be discovered. In a later revision of their 1991 study for the National Energy Strategy, Battelle included more specific data on environmentally sensitive areas that should be excluded from wind development. After the revisions, Battelle found slightly more (1

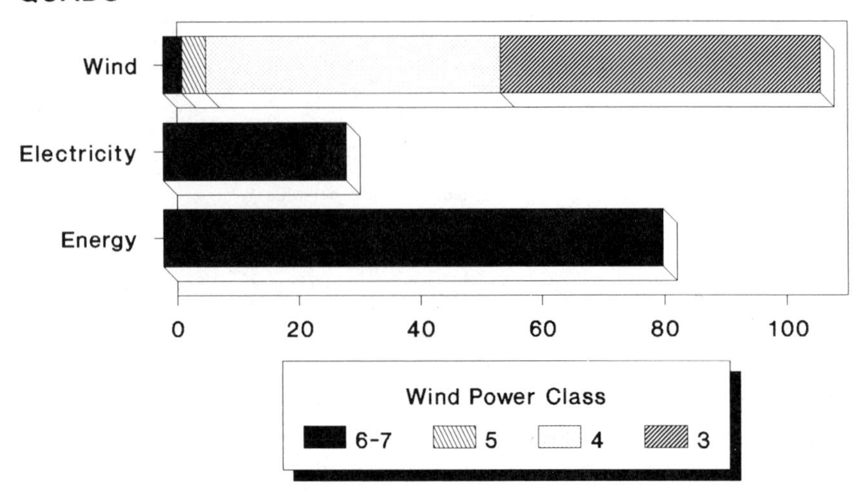

Figure 14.1. U.S. wind energy potential. The potential for wind generation in the lower 48 states at 50 m above ground level with moderate environmental exclusions, in comparison to 1990 electricity and total energy consumption in the United States.

to 2%) available windy land under the moderate exclusion scenario than in the original study.[13]

As expected, the five states with the most class 4 and greater wind resources are found on the Great Plains: North Dakota, followed by South Dakota, Wyoming, Montana, and Minnesota. Battelle also found good sites along the Atlantic seaboard, around the Great Lakes, in the Appalachian Mountains, on the Hawaiian islands, and throughout the Aleutians. The fact that North Dakota has the nation's greatest potential, more than 100,000 MW, and could become a major exporter of wind-generated electricity has led the American Wind Energy Association's Randy Swisher to declare the state "the Saudi Arabia of wind energy."[14]

Potential in Canada and Europe

The United States has no monopoly on the enormous potential of the Dakotas. The Great Plains extend northward beyond the international border. Researchers Michael Grubb and Neils Meyer estimate that total Canadian wind resources may be twice as great as those in the United States, even after excluding more areas than are excluded under Battelle's "moderate" scenario (Table 14.3).

While the wind resource in North America is truly immense, there are sizable resources on other continents as well. Grubb and Meyer estimate that Great Britain has the best wind resources in Europe, with a gross potential

Table 14.3
Onshore Wind Energy Potential in the United States, Canada, and Europe After "Second-Order" Environmental Exclusions

	After Environmental Exclusions (TWh/yr)	1989 Electrical Consumption (TWh)	Potential Contribution (%)
United States[a]	5549	2700	206
Canada	9200	490	1878
Denmark	10	26	38
Netherlands	3	67	4
Norway	12	109	11
Sweden	7	140	5
Finland	10	51	20
Great Britain	40	285	14
European Community	130	1600	8

Source: Michael Grubb and Niels Meyer, *Renewable Energy: Sources for Fuels and Electricity,* "Wind Energy: Resources Systems, and Regional Strategies" (Washington, DC: Island Press, 1993), pp. 186–199.
[a] The U.S. estimate includes only wind resources of class 4 and greater with moderate environmental exclusions.

of 2600 TWh per year. Using the strict siting limitations that can be expected in Europe, where population densities are much greater than those in North America, they find that 20 to 150 TWh could be generated onshore. Excluding the same proportion of sites in Britain as in Denmark, Britain still has the potential of generating 40 TWh per year.[15]

Using this methodology and reviewing wind resources worldwide, Grubb and Meyer believe that wind energy could meet 20% of the world's electricity needs within the next four to five decades, and in certain locations could provide 25 to 40% of a region's electricity without disrupting local utilities.[16]

According to Grubb, a researcher at the Royal Institute of International Affairs, the message is clear: the resource is huge. The question is one of choice. There is more than enough wind energy available to warrant going after it. "Renewable energy is an enigma," says Grubb. "Everyone is in favor of it," but until recently, few have taken it seriously.[17]

Technological advances are always desirable, but it is not the limitations of current technology, nor a lack of wind resources, that will slow wind energy's progress. "The revolution required," says Grubb, "is one of attitudes." The attitude that incremental additions of dispersed generating capacity can collectively meet a large part of our needs is central to that progress. Grubb suggests further that the current debate over the role of renewable energy, with wind at the forefront, represents a revolution in technical thought, a philosophical shift from a paradigm dominated by large, centralized power sources to one reliant on dispersed, intermittent resources.[18] Just how wind energy would integrate with the existing mix of power plants is of particular interest to utilities.

Utility Integration

In the past, utilities have been reluctant to consider wind energy as a potentially significant new source of generation, out of concern about the quality of the power produced, the difficulty of regulating wind's variable generation, and confusion about the true value to the utility provided by wind. Many utilities worried that wind energy, because it is intermittent, was "unreliable." Zond chief executive Ken Karas pokes fun at the latter perception by describing a hypothetical utility engineer who was taught to believe that if "it doesn't burn, doesn't smell, and doesn't glow, it doesn't work."

Historically, utilities have been resistant to change. Until recently, utility executives were adverse to risk, especially after the nuclear fiasco of the 1970s. They are bound by their franchise to provide reliable service, day and night, 365 days per year, under all conditions—with no exceptions. It is a measure of their success that it is newsworthy when electricity does *not* flow from the flip of a switch.

On the surface it appears that a resource which cannot be controlled at will threatens this hard-won reliability. Fortunately, a better understanding of wind technology is slowly overcoming the once bedeviling specter of wind energy's intermittency—what to do when the wind stops blowing. Although "technology has surpassed the institution," says BTM Consults' Per Krogsgaard, the idea lingers. Krogsgaard notes that executives now directing utilities base their perceptions on ideas conceived in the 1940s. According to conventional wisdom, utilities would need the same amount of generating capacity with or without wind power plants, because the wind is intermittent: it is sometimes unavailable when most needed. A reporter for the widely read Washington Post swallowed this myth as late as 1991. "Because it is unavailable when the wind isn't blowing," wrote Thomas Lippman, "wind power will never be more than a supplemental source of electric power, even in high wind areas such as the Great Plains."[19]

Disregarding the fact that no technology, alone, whether coal or natural gas, is more than a supplement to a utility with a diversified mix of generating plants, Lippman ignored findings of utilities in the United States and Europe that wind energy does indeed provide capacity benefits. Although wind turbines may be idle due to a lack of wind at times of a utility's peak demand, there is a statistical probability that they will be available, especially if there are multiple turbines dispersed geographically. In this, wind turbines are no different from conventional power plants. No generating plant operates 100% of the time, and no power plant is 100% dependable during peak loads.

The work of Don Smith, a consultant to Pacific Gas & Electric, as well as that of engineers in Europe, has refuted the notion that wind energy cannot supply secure power. The firm power of wind energy per kilowatt-hour generated is remarkably similar to that of conventional sources. The question then becomes not if there is any capacity value in wind energy, but what its value is in offsetting the construction of conventional power plants. Utilities have traditionally viewed wind energy solely as a fuel saver. Each kilowatt-hour generated by a wind turbine offsets a kilowatt-hour that would otherwise have been generated. But in some cases, the capacity value of wind energy to a utility is equal to that of the fuel it offsets.[20]

Smith, for example, found that an exceptional fit between the wind resource in Solano County and Pacific Gas & Electric's demand justified a credit of nearly 80% of a wind plant's installed capacity, based on a loss of load analysis. This is equivalent to that of a fossil-fired power plant. Using the same analysis, he found a lesser, but not insignificant value of 20% for the Altamont pass.[21] The capacity value for the Altamont is surprisingly similar to the areawide capacity factor.

Utilities also fear that the variable generation of wind turbines may disrupt the utility's ability to regulate power quality, as the percentage of the load carried by wind turbines on a utility system—the penetration level—increases. One study estimated severe load following impacts at 12% penetration. Using actual field data, however, Southern California Edison found that wind plants would not affect its load-following capabilities. SCE discovered that diversity in the response of different wind turbines to changing wind conditions, the spatial diversity of winds over a producing area, and the geographic separation between the San Gorgonio and Tehachapi passes sufficiently dampened fluctuations in collective generation to simplify power regulation.[22]

Wind energy's varying generation can be absorbed by large interconnected networks, such as Britain's, without significant economic penalties, provided that no more than 20% of the generating capacity is derived from the wind.[23] British researcher Michael Grubb asks rhetorically how fluctuations of a few thousand megawatts from wind plants could ever cause problems in England, where the load on the system varies by 10,000 to 15,000 MW every day. Grubb argues that how much wind generation a utility can absorb is not a technical question but an economic one. He suggests that it is only above 20% penetration rates that economic penalties become significant.[24]

The addition of wind energy to an existing utility may incur some costs at higher penetration levels, yet these may be only temporary. Over time, the utility can reoptimize the mix of generating plants to better maximize the benefits of each technology. Reoptimization may not only eliminate costs attributable directly to wind energy, but may also cut costs of other technologies, resulting in a lower systemwide cost of energy.

If integration were a problem severely limiting wind development, it would already be seen in Denmark. Danish utilities have the highest penetration of large-scale wind development of any major utility system in the world. They also collectively operate 100 MW of their own wind capacity.

ELSAM is a consortium of seven regional utilities serving Jutland and the island of Funen. Vestkraft and Nordkraft are two utilities in the consortium. Vestkraft, which serves Jutland's west coast, meets nearly 8% of its generation from wind energy. Nordkraft, which serves the area around Ålborg, meets 7% of its demand with wind.[25]

ELSAM must manage 75% of all Danish wind capacity, yet the large-scale fluctuation of wind generation as the wind ebbs and flows has failed to become the problem ELSAM once feared. Wind generation can vary as much as 6 MW per minute, says ELSAM's Peggy Friis, but peak demand already fluctuates about 10 MW per minute, and the utility can capably handle such a fluctuation.[26]

Wind energy provides 0 to 15% of the load carried on the ELSAM system at any one time and can reach 25% of total demand under extreme conditions on winter mornings when demand is at a minimum. In addition to wind energy, ELSAM also absorbs generation from municipally owned combined heat and power plants. Their peak production occurs during the winter heating season. The high penetration of both sources has at times required ELSAM to balance load and supply through its interconnections with other utilities. As the country continues to install more wind turbines, ELSAM's capability for balancing load and supply will decrease, possibly requiring export of surplus generation to Germany. Currently, 20% of Denmark's conventional generation is produced by new power plants with better than average ability to operate at minimum power. ELSAM will repower another 20% of its fossil-fired capacity during the 1990s. These new power plants will further aid ELSAM in absorbing wind generation economically in Jutland.[27] Effectively, ELSAM is reoptimizing its mix of generating technology to include wind energy.

On small networks, even higher penetration may be possible. On the Aegean island of Andros, a cluster of seven medium-sized wind turbines provides 12% of the diesel-powered grid's overall generation. During periods of low consumption, at night in the off-season, the turbines deliver up to 65% of the island's power without severely affecting voltage and frequency.[28]

Utilities deal with short-term variation in wind plant generation in the same way that they deal with existing variations in their load. The effects of medium-term variations at low penetration rates are easily handled with standard operating practices. Hawaii Electric Renewable Systems, for example, reports that on windy nights, the Kahuku wind plant carries the entire north shore of Oahu. As seen in Denmark, wind energy can meet a substantial portion of demand without disrupting the utility system. Turbines in the Altamont Pass provide 3% of PG&E's energy needs, but on the afternoon of May 14, 1989, they met 7% of the utility's peak demand without causing problems.

At higher penetration rates, more reserve capacity may be needed, or the operation of the entire system may need to be reoptimized, as ELSAM's was, too take better advantage of the renewable resource. Because wind is a dispersed, modular resource, utilities have found it unnecessary to maintain more than normal spinning reserve, since it is unlikely that all of the wind capacity would suddenly be lost at one time.

Rather than disrupting a utility system, wind energy can provide valuable flexibility. "We use nine different sources of energy to produce electricity," said Southern California Edison president Michael Peevey. "This diversity provides our customers with considerable protection against supply disruptions and volatility in world energy markets."[29]

Storage, which may enable better use of intermittent resources, is helpful, but unnecessary. Most utilities already have some form of storage, because storage provides benefits to utility systems with or without wind plants.

Wind plants use the same hardware and are integrated into a utility's system in the same way as other power plants. In the United States, where nearly all wind plants have been built by independent power producers, each operator constructs and operates its own power collection systems. Pad-mounted transformers collect the power from one to several wind turbines and increase the voltage for delivery to a central substation. From the substation, the operator typically delivers the power to the utility.

The one exception to this has been in the Tehachapi Pass. After exhausting local transmission capacity, operators were forced to build a 38-mile (60-km) 230-kv transmission line to a SCE dispatch center near Los Angeles. As with other natural resources, where transportation routes to market determine the economic value of the resource, transmission capacity may limit wind development in areas of western North America. Areas of energetic winds, like the Tehachapi Pass, are seldom near population (load) centers, so transmission lines become critical links for carrying the product—electricity—to market.

Because most wind turbines worldwide drive induction generators, utilities had been concerned that wind plants composed of multiple induction generators would cause voltage flicker on rural distribution lines and consume considerable reactive power. On "stiff" transmission systems such as that in the Altamont Pass with PG&E's nearby Tesla substation, and in the San Gorgonio Pass with SCE's adjacent Devers substation, the reactive power of the wind turbines has not proven a serious problem. In Tehachapi, where wind plants draw reactive power from a switching center 20 miles away, special measures have been taken to ensure stable voltage. Although each turbine is installed with its own power-factor correction capacitors to reduce reactive power consumption, additional capacitance has been installed as part of some intrastation power collection systems in the Altamont and Tehachapi areas.[30]

By the mid-1990s, wind turbines in North America and Europe had operated more than 1 billion hours interconnected with local utilities and had generated nearly 40 TWh of electricity without incident. Despite initial concerns, utilities have adeptly managed the integration of wind energy with conventional sources: clocks still run on time, computers still compute, and electricity continues to flow at the flip of a switch.

Wind energy has truly become another conventional source of generation. According to a survey by the Electric Power Research Institute, many European utilities now view wind energy as a part of their future generation mix.[31] Southern California Edison president Michael Peevey sees it as part of California's, acknowledging that wind energy has "performed much better on the Southern California Edison system than we anticipated."[32] And SCE should know; it uses more wind-generated electricity (nearly 2 TWh per year) than any other utility in the world.

Whether it is through research by PG&E engineers showing that wind turbines provide capacity value in northern California, ELSAM's successful integration of high penetration levels in Jutland, or SCE's long-distance transport of wind generation to its load centers, utilities will play a determining role in the future of wind energy. They are, and will remain, the market for wind-generated electricity. It is partly the utility industry's generation and transmission system that will determine where and how fast the industry grows.

It is in these areas that utilities in North America and Europe are experiencing wrenching changes. Flemming Tranæs, of the Danish windmill owners' association, for example, sees the role of Danish utilities changing from one of distributing power from a central station, to one of collecting power from numerous dispersed generators scattered throughout their service area. The thought that utilities are in the business of collecting and distributing power and not necessarily in the business of generating it is as radical as the utilities' realization during the industry's formative years that they were providing a commodity, not an appliance.[33]

During the late 1970s, the long-distance transport of privately produced, wind-generated electricity in the United States was inconceivable. But "wheeling"—the transport of power over long distances on utility-owned lines—of privately produced power is nearing a reality in several areas of the country. The long-distance transmission of wind-generated power is not far behind. It is not inconceivable, says Arigna Fuel's Michael Layden, that wind-generated electricity produced in Ireland and Britain could someday be transported to the continent by the same undersea link that now carries nuclear power from France to England.

Distributed Generation

While it is still common for utilities in North America and Europe to sell and service electrical appliances, it was unthinkable just a few years ago that utilities would sell, install, and service solar power systems for their customers in the same way that they sell electric hot water heaters. But the concept of distributed generation has quickly gained momentum. Idaho Power began offering packaged solar systems for those living beyond its lines during the early 1990s. PG&E and SCE have toyed with the idea in California. By 1993 Wisconsin Power & Light was paying new homeowners $2500 as an incentive to build their own hybrid power systems and stay off-the-grid.[34]

One European utility examining wind energy at the end of the line is Électricité de France. The French utility finds itself facing a problem of its own making. After years of stimulating consumption and a policy of extending service to all comers, EDF discovered that consumers were placing greater and greater loads on rural lines with low population densities. In areas with fewer than 10 persons per square kilometer (25 per square mile), it costs EDF four to five times the production cost of electricity to serve its customers.[35]

Marc Vergnet of Vergnet S.A., a manufacturer of small wind turbines, notes that winter winds match EDF's winter peak, giving wind energy a high capacity value, whether on or off EDF's lines. Vergnet sees a potentially large market for distributed wind turbines, even in industrialized countries. Half a million people live in remote areas of France, and another 4 million live in zones of low population densities that could eventually be served by distributed forms of generation. EDF spends 8 billion FFR ($1.4 billion) per year for rural electrification, he says. "If only 2 to 3% of that sum would be dedicated to renewables, it would radically change the market overnight.[36]

Mike Bergey of Bergey Windpower agrees. He believes that distributed generation will ultimately account for 10 to 15% of the total wind energy market. The distributed market, says Bergey, will include more than just small turbines.[37] The European Wind Energy Association estimates that there is a potential demand for as many as 15,000 distributed wind turbines per year through the 1990s.[38]

The future of wind energy could have a far different face than it has today. In the Netherlands, for example, EGD (Energiebedrijf voor Groningen en Drenthe) is pursuing the installation of both individual wind turbines and wind plants. Unlike other Dutch utilities, which are limiting wind development to small wind plants, EGD is installing medium-sized turbines for pumping stations, sewage plants, farms, and businesses near the end of EGD's distribution lines. Utilities of the future could well see a mix of uses for wind energy: some wind turbines at the end of their distribution lines; others beyond the line, providing all the power for those off-the-grid; and wind plants generating large quantities of bulk power.

Transportation: A New Market

As utilities wrestle with a fast-changing world, they are hungrily eying new markets. Southern California Edison's Michael Peevey has only to look out his window at the dirty brown haze obscuring the skyline of Los Angeles to see his company's future: electric vehicles.

Peevey introduced his theme in public appearances in 1991 by noting that "transportation causes most of the pollution in southern California." He would then follow by laying down a competitive gauntlet before the oil industry. "Electric vehicles produce less nitrogen oxides, carbon monoxides, and reactive organic gases than gasoline-powered vehicles," he said, "even taking into account emissions from the power plants that generated the electricity to charge electric vehicle batteries." If only 1% of the 200 million vehicles registered in the United States were replaced with electric ones, the country would save 35 million barrels of oil per year and cut the trade deficit at least $1 billion annually," Peevey states.[39]

Cynics scoff that electric vehicles merely shift pollutants from the tailpipe to the smokestack. Not so according to researchers at the University of California at Davis, who note that power plants operate at a far higher thermodynamic efficiency than do auto engines, even after accounting for the losses incurred through transmission of the electricity to consumers. Altogether, electric vehicles can cut automotive emissions of hydrocarbons 99%, carbon monoxide 99.7%, nitrogen oxides 84%, sulfur oxides 98%, and respirable particulates 5% when the electricity is generated by gas-fired power plants. About 40% of California's electricity is produced by burning natural gas. Electric vehicles provide slightly fewer emission benefits outside California, where much of the generation would come from coal-fired power plants.[40]

The utility's marketing program will receive a major push in 1998, when the California Air Resources Board will require that 2% of all new cars sold in the state each year emit *no* pollutants. Currently, only electric vehicles can meet this requirement. By 2003, 10% of all vehicles sold must be zero-emission. These regulations could put almost half a million electric vehicles on California's roads by 2003.[41]

Current electric vehicles consume about $\frac{1}{2}$ kwh per mile traveled. Californians drive 250,000 million vehicle-miles per year. If electric vehicles were used to meet just 10% of that market, the demand for electricity would jump 12.5 TWh per year. This move would increase total California electricity consumption a modest 5%, provide a new source of revenue for the state's utilities (albeit at the expense of oil companies), and create a new market for wind-generated electricity. Ideally, the new demand created by "zero-emission vehicles" would be met by zero-emission sources.

California's existing wind plants could fuel 350,000 vehicles for an entire year. One contemporary wind turbine 40 meters in diameter at a good site would generate enough electricity to fuel 200 vehicles per year (Table 14.4). Transportation opens up a potentially huge new market for wind generation. Electric cars could absorb large amounts of generation that utilities might otherwise have difficulty using. When wind energy is abundant and demand weak, excess generation could be directed into electric vehicles, a form of mobile storage.

Other forms of transportation could also become a potential new market for clean sources of electricity. California is considering linking its major cities in a high-speed rail network patterned after France's TGV. These high-speed trains use electrically driven motors that could be powered by wind plants in the mountains through which the trains would pass.

California's notorious air pollution and Southern California Edison's aggressive pursuit of new markets are pushing electric transportation to the fore in the United States. It is not the first time that SCE has blazed a path other utilities would eventually follow.

Table 14.4
Wind Generation and Electric Vehicles

	Rotor Diameter (m)	Swept Area (m^2)	Capacity (kW)	Site: 7 m/s, 1200 kWh/m^2 (kWh/yr)	Number of Vehicles Fueled/Turbine[a]
Early 1980s	12.5	125	50	150,000	20
Mid-1980s	18	250	100	300,000	40
Late 1980s	25	500	200	600,000	80
Early 1990s	35	1000	400	1,200,000	160
Mid-1990s	40	1250	500	1,500,000	200

[a]At 15,000 miles/vehicle/yr (25,000 km/vehicle/yr).

In the early 1980s, SCE sent shock waves through the utility industry by proclaiming that it would not build any new conventional power plants before the turn of the century, and that the utility would develop renewable sources of energy instead. After the press conference in Hollywood announcing the ground-breaking promise, SCE's chief executive Glenn Gould turned to his son Wayne and, referring to political pressure, said, "When you don't have any options, you don't have any choice." Then, after a pause, smiled and added, "Besides, renewables are the right thing to do."

Repowering California Wind Plants

Europeans seem to agree that expanding the use of wind energy is the right thing to do. During the mid-1990s they were installing 250 to 350 MW of new wind-generating capacity annually. In the United States, on the other hand, new installations fell off dramatically after California wind companies exhausted their remaining standard offer Number 4 contracts. By mid-decade, U.S. wind companies were slowly beginning to develop projects outside California, in Minnesota, Iowa, Washington, Wyoming, Montana, Alberta, Mexico, and elsewhere.

As part of the National Energy Strategy, the national research laboratories working for the U.S. Department of Energy were assigned to project what contribution wind energy would make by the year 2000 at the current pace of development and without any incentives. The resulting interlaboratory report estimated that wind turbines could generate 10 times the electricity now produced in California, or 1% of the nation's consumption.

With modest market incentives, the laboratories reported, the wind industry could generate 20 times the electricity produced in California, or 2% of the nation's supply. Under the same scenario, by 2010, wind generation would increase to more than 100 times that of today, providing 12% of the nation's electricity consumption. The authors note an important caveat: possible future environmental restrictions on fossil fuels, such as on carbon dioxide emissions, could stimulate much faster growth. The potential contribution of wind energy could be greater and could come on line sooner than they projected.[42]

While the potential for expansion outside California is far greater than that within the state, wind companies have not turned their backs on the place where it all began. In fact, one of the best near-term prospects for growth of wind energy in the United States is through the "repowering" of existing sites in California. The industry is poised to begin replacing old turbines with newer, more modern machines. Repowering has been proposed in Denmark as well.

In a study for the California Energy Commission, the American Wind Energy Association's west coast office estimated that repowering California's wind plants by replacing first- and second-generation designs with contemporary machines could lower operations and maintenance costs, reduce the density of turbines on the landscape by more than 50%, and increase annual generation 30%. The study concluded that repowering with modern turbines would make California's wind industry more competitive with other energy sources, preserve existing jobs, and reduce the industry's aesthetic impact on the landscape, thus improving its public acceptance.

California's turbine stock is comprised largely of early, less cost-effective designs. Some of the poorest-performing turbines are 10 years old. These contributed significantly to the decline in the statewide average specific yield during the early 1990s. In 1988, consultant Bob Lynette estimated that 1000 machines in California were so poorly designed or manufactured that they were unsalvageable.[43] Lynette's estimate may have been far too conservative. More than 3000 turbines comprising 230 MW of California wind capacity may be unsalvageable and best suited for the scrap yard (Table 14.5). These first-generation designs are costly to maintain, notoriously unreliable, installed on short towers, and often sited poorly. Many of these turbines are also noisy and unsightly.

The bulk of California's operating capacity is provided by the more reliable first-generation Danish designs and a significant number of second-generation turbines, including 100 to 150-kW Danish machines and U.S. Windpower's 56-100. These turbines are all salvageable, but they have seen more than a decade of heavy use and will need substantial repairs before they can be reused. This fleet, once reconditioned, will have value as an inexpensive source of wind turbines for the Midwest or elsewhere. Used Danish wind turbines in California, for example, have been bought "as is" for $8000 to $25,000 (Table 14.6). After reconditioning, they have been resold to farmers on the Great Plains for $50,000 to $60,000. At a price of $300 to $350 per square meter of rotor area, these reconditioned turbines are a good value compared to new turbines costing up to $500 per square meter.

Repowering began in Palm Springs in 1994, when U.S. Windpower installed 115 of its new 33M VS turbines on a site formerly occupied by ESI-54 and Jacobs machines. Nearby, Bill Adams replaced three inoperative 600-kW Floda

Table 14.5
First-Generation Turbines in California

Model	Orientation[a]	Number of Blades	Nominal Rotor Diameter (m)	Capacity (kW)	Units
Salvageable					
Jacobs, United States	u,p,tv	3	8–9	18–20	630
Carter, United States	d,p	2	10	25	350
Aeroman, Germany	u,a,m	2	12.5	40	320
Enertech, United States	d,p	3	13.5	40–60	550
Bonus, Denmark	u,a,m	3	15–16	65–100	640
Micon, Denmark	u,a,m	3	15–16	65–75	720
Nordtank, Denmark	u,a,m	3	15–16	60–75	1,130
Vestas, Denmark	u,a,m	3	15–17	65–90	1,700
Wincon, Denmark	u,a,m	3	15–16	65	100
Windmatic, Denmark	u,a,m	3	14–17	65–95	340
USW 56-100, United States	d,p	3	17.6	100	4,000
					10,480
Unsalvageable					
Fayette, United States	d,p	3	10–11	75–95	1,400
Storm Master, United States	d,p	3	12	40	420
Century, United States	u,a,tv	3	12	75–100	50
Windtech, United States	d,p	2	15.8	75–80	220
ESI 54, United States	d,p	2	16	50–80	680
Windshark, United States	d,p	2	16.4	80–90	190
Wenco, China	u,a,m	2		100	20
Polenko, The Netherlands	u,a,m	3	19.6	100	30
Bouma, The Netherlands	u,a,m	3	20	200	40
Carter, United States	d,p	2	23	250–300	50
ESI 80, United States	d,p	2	24	250–300	50
					3,150

[a] u, Upwind; d, downwind; p, passive; a, active; m, mechanical; tv, tail vane.

machines on Whitewater Hill with Vestas V39s. This was only the beginning. At the time, U.S. Windpower was negotiating for more existing sites in the San Gorgonio Pass and was hoping soon to begin replacing turbines in the Altamont Pass as well.

If wind companies replace all of California's first- and second-generation turbines, they will liberate more than 12,000 turbines and 1200 MW of generating capacity that can be used elsewhere (Table 14.7). These inexpensive used turbines could spur growth in new markets around the world.

The idea is attractive in Denmark as well. Danish consultant Birger Madsen estimates that 1900 of Denmark's current 3500 wind turbines could be replaced with more modern designs. Repowering Denmark's existing fleet could add substantially to wind generation. Repowering alone would increase generation by 25 to 75% over that of 1993. One alternative considered would add nearly 300 MW and generate an additional 550 million kilowatt-hours per year, by replacing all existing turbines of 130 kW or less with 250-kW machines. A second alternative would add nearly 400 MW and 750 million kilowatt-hours

Table 14.6
Estimated Installed Cost of Rebuilt Wind Turbines from California

Diameter (m)	Area (m^2)	Capacity (kW)	Cost[a] ($)
8	50	18	16,000
10	79	25	26,000
11	95	40	31,000
12.5	123	40	40,000
13.5	143	40	47,000
15	177	65	57,000

[a]At $325/m^2$.

per year, by replacing all the existing turbines 130 kW or less with 500-kW turbines. Madsen estimates that 40% of the replacements could be completed before the year 2000, and 80% before 2010.[44]

Mate for Methane?

To maximize the benefits from our economic and environmental investment in renewable technologies, it is paramount to curb growth in consumption. Otherwise, the repowering of developed sites in California and Denmark and the installation of new wind-generating capacity elsewhere never carries us far down the road to sustainability. We simply run in place. And as existing plants begin nearing retirement, there will be mounting pressure to replace them with coal, natural gas, and in some quarters "modular" nuclear power plants.

Table 14.7
Repowering California Wind Plants

	Units	MW	TWh/yr
Current Fleet			
Unsalvageable	3,078	233	
Salvageable first and second generation	12,509	1,211	
State of the art	1,286	317	
	16,873	1,761	2.8
After Repowering			
Unsalvageable	0	0	
Salvageable first and second generation	0	0	
State of the art	7,012	1,761	
	7,012	1,761	3.6

To build a sustainable future, proponents must ensure that wind energy is used not only to meet growth in demand, but also to substitute for existing fossil fuel–fired power plants. This can only be accomplished if consumption is reduced through conservation and improved energy efficiency. Only then can wind energy and other renewables begin to substantially replace existing power plants.

There is ample room for cutting unnecessary consumption in North America. As seen previously, Californians consume twice as much electricity as does the average European with the same standard of living. Unfortunately, those environmentalists who fear any form of new development have seized on the potential for cutting consumption as the new energy panacea. Rather than using it as a tool to build sustainability, critics of development use conservation as another arrow in their quiver to pierce the need for new power plants of any kind, including wind. If they successfully stymie wind energy's expansion today, there may not be an industry tomorrow, when utilities begin building new power plants.

After listening to testimony on the virtues of conservation and energy efficiency at a hearing in the early 1990s, Richard Bilas, a professor of economics who sits on the California Energy Commission, commented—with no small amount of sarcasm—that "DSM [demand side management] is the nuclear power of the 1990s." Even as an advocate of energy efficiency in the Pacific northwest, where electricity is widely squandered, K. C. Golden of the Washington Environmental Council wearies of struggling with "misty-eyed" activists who believe that "conservation is easy, free, and unlimited."[45]

Amory Lovins, the best known proponent of the soft path in the United States, has long stressed that rather than being enemies of renewables, energy conservation and efficiency programs should be important allies. Eric Heitz of the Energy Foundation says that it is improved efficiency that will ultimately slay the fossil dragon that has kept renewables at bay for so long, that is, reduce total demand to within reasonable reach of renewables.

Michael Grubb of the Royal Institute of International Affairs believes that wind energy, along with other renewables, forms an essential third leg of a new energy triad of renewables, improved efficiency, and natural gas. Neither renewables nor efficiency is sufficient without the other. Natural gas provides the bridge to make the transition from today's dependence on fossil fuels to a future energy supply based on renewables.

After noting that local activists have vigorously opposed some biomass, geothermal, and wind power plants, Worldwatch Institute's Chris Flavin pleads that a "balanced expansion of renewable energy and natural gas seems preferable to the crash development of either." While urging a strategic alliance among environmentalists and the gas industry, Flavin stresses that natural gas should only be considered as a bridge fuel, and that its increased use must not delay the growth of renewables now under way.[46]

"Wind power is a mate for methane," Carl Weinberg, former manager of research and development for Pacific Gas & Electric, told audiences across the United States in 1994. He sees hybrids of wind power plants and gas turbines as providing an attractive alternative to coal. Weinberg believes that by 2004, utilities will begin closing fully functional coal-fired power plants and replacing them with energy savings and renewables, especially wind energy.[47]

But hitching wind energy's star to that of natural gas is a risky strategy. Public policy in the United States tends to swing from one extreme to the other. And utilities throughout North America have turned their attention toward gas—and only gas. "Gas-fired combustion turbines are the crack cocaine of the electric utility industry," according to Ken Stump, a Pacific northwest environmentalist frustrated that nearly all the new power plants now planned will be gas-fired.[48] Somewhat more diplomatically, Angus Duncan warns that "the gas gorilla is rumbling through the Northwest."

Duncan, one of Oregon's delegates to the Northwest Power Planning Council, urges that the region return to the "clean and renewable resource base that we had in the 1930s" when hydro met all demand. "We've gone through our nuclear phase, and now our fossil phase," he says, going on to say that it is time to move on to the region's second renewable phase. The hydro system of the Pacific northwest is particularly well suited for adapting to wind energy, says Ralph Cavanagh of the Natural Resources Defense Council. "I'd be happy to see wind take it all," he says of the new power plants that will be needed.[49] But gas is preferred by regulators and administrators in the northwest.

Most utility commissions have difficulty incorporating into their deliberations future regulatory and fuel risks associated with conventional fuels, such as natural gas. And some utilities, Harvey Sachs found, still have institutional inhibitions against decentralized, intermittent, nonfossil technologies such as wind energy, despite their achievements.[50]

Few regulators in their rate making have reflected the growing concern about global warming and probable limits on carbon dioxide emissions. The Union of Concerned Scientists' Don Aitken contends that it is obvious that the emission costs of carbon dioxide are greater than zero but less than infinity. Therefore, he says, a regulatory policy that determines the cost of carbon dioxide emissions as being equal to zero is "a form of ostrich math for those with their heads in the sand." This explains why NRDC's Cavanagh put U.S. utilities on notice. He demands to know "who will pay" if a flood of new coal and gas plants in the pipeline are eventually saddled with carbon taxes or emission limits. The financial exposure of a small 250-MW coal plant to future carbon taxes, he warns, exceeds the original cost of the plant itself. Deliberately using jargon familiar to attorneys specializing in regulatory law, Cavanagh advises that "these taxes are reasonably foreseeable."[51] Cavanagh's message is clear: if a utility builds a new fossil-fired plant today and taxes or emission limits are imposed, the utility's stockholders will pay.

To avoid such an eventuality, some utilities are beginning to call for minimal carbon taxes as a way of thwarting more onerous measures. The Edison Electric Institute, the political arm of the U.S. electric utility industry, and at least one utility, New England Electric System, have called for a modest carbon tax.[52] They are suggesting the tax to head off a more substantial levy that could be imposed to control global warming gases.

Southern California Edison again took the lead when it became the first electric utility in the United States to pledge a 20% reduction in carbon dioxide by the year 2010. "Despite the uncertainty," said Michael Peevey, "it makes good sense to take actions now to reduce those emissions. We are taking a 'no regrets' approach." Whether or not carbon dioxide emissions "are eventually determined to cause global warming, we won't be sorry we took early action."[53]

Sustained Orderly Development

Thus the dilemma facing wind energy in the United States: With no national policy encouraging use of renewable resources, and only a few progressive utilities that see a need for diversifying their generating mix, how do wind turbine manufacturers continue to drive down costs and improve reliability and performance with only small, sporadic orders?

Utilities in North America will begin a massive building program in the late 1990s to replace aging power plants. There is a clear need for the sustained orderly development of renewables now, says the Union of Concerned Scientists' Aitken, to avoid another boom-and-bust cycle like that of the early 1980s. Worse yet, he warns, when American utilities decide they need new capacity, they will need it "tomorrow," and the nation could ruefully discover that no one any longer builds wind technology. Coal and nuclear plants could proliferate, even though renewables were ready—once—but withered from lack of a market during the mid-1990s.[54]

Aitken argues that the rapidly falling cost of wind energy during the 1980s was due to the sustained development of the technology resulting from California's standard offer contracts. He cites, as his example, PG&E's calculation of the cost of wind energy in the Altamont Pass as wind companies, notably U.S. Windpower, progressively worked off a backlog of utility contracts.

U.S. Windpower, says Aitken, steadily cut costs through manufacturing economies.[55] Following a path blazed by Henry Ford and his Model T, U.S. Windpower has built more than 4000 of its model 56-100 since its introduction in 1983. While other manufacturers introduced larger wind turbines to reduce the cost of energy, USW systematically reduced the cost of building the 56-100. The experience gained from multiyear purchases enabled the manufacturer to lower costs progressively by climbing down the "learning curve."

Aitken's "sustained orderly development" is what the Danes accomplished with their national program. They created a stable market that ensured a reliable stream of orders for domestic manufacturers that enabled producers to advance the technology in response to the "pull" of the market.

Harvey Sachs, of the Center for Global Change, uncovered a parallel in the trucking industry in a study done for the South Coast Air Quality Management District. Every year, U.S. manufacturers build 100,000 heavy trucks comparable in size to a 300-kW wind turbine. That is equivalent, says Sachs, to about 30,000 MW of complex rolling stock per year, at a cost of $300 per kilowatt. It is conceivable, Sacks postulates that if wind turbines were produced in similar volumes, the cost could eventually fall to the same level. Even after including installation, the total costs of new capacity would be less than $500 per kilowatt—half of the price in the mid-1990s. This is well within the range needed to compete against fossil fuels at sites on the Great Plains.[56]

Leveling the Field

One way to redress the imbalance created by utility regulation in the United States is through the creation of a Clean Energy Trust Fund similar to the trust fund that was used to build federal highways. The fund would raise money from taxes on the social costs of conventional resources, and compensate wind and other forms of solar energy for their social benefits. Taxing the emissions from coal, oil, and natural gas power plants could be used to pay a per kilowatt-hour "social cost adjustment" or "environmental benefit payment" to clean sources of electrical generation.

Sierra Club California specifically suggests the approach of "gradually increasing user fees on fossil fuels, based on carbon and pollutant emissions" and recommends that "the funds collected then be used to finance the development of renewable sources."[57]

For the system to work, taxes must be linked with benefit payments. Taxing conventional sources for their pollution alone will be insufficient. Taxes would discourage one activity, in this case pollution, but would not necessarily encourage the preferred activity: the increased use of wind and solar energy. Pollution taxes must be coupled with a mechanism for directly compensating wind and solar sources for their social or nonprice benefits.

One way of handily dealing with the pricing problem in the United States is to broaden the definition of "avoided costs" under PURPA, to include social and environmental costs. If wind generation were compensated by prices similar to those now found in Europe, the market for wind energy would grow dramatically. Similarly, least-cost planning models could be modified to incorporate social and environmental costs over the entire life cycle of the power plants being considered. If these steps are taken, wind energy would find itself competing on the "level playing field" our market economy demands.

Another approach proposed by David Moskovitz, a former member of Maine's Public Service Commission, is "green pricing." Moskovitz suggests that utilities offer their customers the choice of paying a small amount more every month on their utility bill to finance renewable development. The funds could be administered by a "green" board of directors, says Moskovitz, composed of environmentalists who would review proposals for specific projects. In this way, says Moskovitz, environmentalists could influence decisions before they go beyond the boardroom, by imposing strict environmental safeguards on renewable projects the utility proposes building.

Traverse City Power & Light, on Michigan's scenic Leelanau peninsula, will be one of the first utilities in the United States to experiment with the concept. The municipal utility asked its 6000 customers to opt for its "green rate" and pay an average of $90 more per year to install one 500-kW wind turbine on a prominent hill overlooking the town on Grand Traverse Bay. Customers are asked to make a three-year commitment for a rate increase of $0.0158 per kilowatt-hour on rates averaging $0.07 per kilowatt-hour. If more customers enlist than expected, the utility will install two turbines instead of one.

Others seethe at the mere thought of "green pricing." FloWind's Hal Koegler fears that wind energy would be segregated with other renewables in a "green pricing ghetto." "Why not have a check-off for nuclear, coal, and gas plants as well?" he demands. Green pricing implies, say critics, that renewables are uncompetitive with conventional sources. Green pricing also puts renewables development back into the hands of utilities, benefiting them at the expense of independent developers, such as FloWind, that have brought wind energy to its current state of competitiveness.

Monuments to Sustainability

Wind energy in North America will eventually come down from the mountaintops and go forth onto the plains as it has begun to do in Minnesota and Iowa. Unlike the defacto wilderness of ridge crests in California's Sierra Nevada mountains, much of the Great Plains is in agricultural use. Many of the people of the Plains states living on the land also live by the land. Like the farmers on the lowlands of continental Europe, they are accustomed to working with the land and its renewable resources.

In 1995, Congress will begin hearings on what will be a long and agonizing task of reducing farm subsidies. Carl Weinberg believes that wind energy may prove to be a life preserver to midwestern farmers, as it has to farmers in the Netherlands and Germany, as crop subsidies wither on both sides of the Atlantic.

Wind energy could be both a vehicle and a beneficiary of a midwestern rural revival. Bill Grant, an environmental activist with the Izaak Walton League, shares a vision with wind consultant Bob Lynette of one day seeing thousands of wind turbines sprouting on farms in North America's heartland. They envision wind energy breathing life into dying midwestern towns, generating prosperity as well as electricity. Instead of plowing giant furrows in the earth for coal and uranium, they believe we can plant graceful wind turbines that practice Lacarrière's "aeleosynthesis" amid the wheat, sugar beats, soybeans, and corn.

These wind farms of the future could provide economic opportunities for those who inhabit the land not only through royalties to landowners and jobs to the millwrights of a new age, but also to the residents themselves through community investment in locally owned wind cooperatives on the Danish model. Grant sees a place—and a need—for both forms of development.

Gradually, these regions could become self-sufficient in low-cost, locally generated electricity that stimulates commerce, reversing decades of flight from the land by providing residents new occupations. Grant and Lynette then see the region eventually becoming an exporter of wind-generated electricity to less fortunate areas of the continent.

A similar scenario could be acted out in the oil fields of western North America. In a land where the wind once "pumped the water," thousands of unemployed roughnecks search for work after the oil industry "restructuring" of the 1980s. They could find fulfilling work as windsmiths. Both professions require similar skills. While one drills down for a finite fossil fuel, the other climbs skyward for a renewable resource. It would be fitting that the men and women dislocated by the vagaries of oil would find their future as wind catchers and by doing so learn to feel nature's rhythms as their forebears did.

"We have become so successful at controlling nature," says Vice President Al Gore, "that we have lost our connection to it."[58] Wind and solar energy provide an opportunity to reforge that link. Wind energy is not some exotic new technology like nuclear power. Only wind energy's current manifestation is new. We have lived peacefully with the wind before, and we can do so again.

Wind turbines could become as common on the European landscape as windmills once were. Our twenty-first-century pastoral ideal may very well include the hills of England and Wales dotted by sheep grazing among soft-white whirling flowers, while on the continent, farmers tend to their wind turbines that rise above tidy, thriving farms.

The "wonderful, soaring structures" referred to in an editorial by Alice Ledogar, deputy director of New England's Conservation Law Foundation, can become "monuments to sustainability." With their "arms wide open," they welcome the future in "beautiful union" with the environment of which they are a part[59] (Figure 14.2).

Californians have proved that wind energy can work on a grand scale. Europeans have shown how wind energy can be developed harmoniously with its neighbors and the environment. We can learn from both experiences. To UCS's Aitken, "wind [energy's success] has brought all renewables to the threshold of acceptability." As worldwide wind generation pushes past that of a conventional power plant, whether nuclear or fossil fuel–fired, wind can no longer be considered an "alternative" source of energy. Wind energy has indeed come of age. Wind works!

Figure 14.2. Fiberglass flowers. Wind turbines in the setting sun near Tehachapi, California. In the afternoon light the towers become less distinct leaving the white rotors to dance against golden hillsides and blue skies.

Chapter 14 Endnotes

1. D. L. Elliott, L. L. Wendell, and G. L. Gower, "An Assessment of the Available Windy Land Area and Wind Energy Potential in the Contiguous United States," Battelle Pacific Northwest Laboratory, Richland, WA, August 1991, 44, B.2. Scenario 3, moderate environmental exclusions for California and class 4 and greater wind resources.
2. Paul Helgeson, "Documenting the Minnesota Wind Resource," unpublished summary of Minnesota's Wind Resource Assessment Program, Minnesota Department of Public Service, St. Paul, MN, April 1991.
3. Assumes 5 by 10 spacing; 1000 kWh/m^2 in first case; 750 kWh/m^2 in second case; 40-m-diameter 500-kW turbine. This estimate is more conservative than Battelle's, which assumes a 2640-km^2 area. Battelle estimates that the southwestern region of Minnesota could generate 30 TWh, 63% of the state's electricity supply. Elliott, Wendell, and Gower, "An Assessment of the Available Windy Land Area," p. 53.
4. Don Bain, "Wind Resources," Staff Issue Paper 89–40, Northwest Power Planning Council, Portland, OR, October 16,1989.
5. G. Lee, "Will Coal Ride the Region's Resource Rails," *Northwest Energy News,* Northwest Power Planning Council, Portland, OR, January/February 1990.
6. Elliot, Wendell, and Gower, "An Assessment of the Available Windy Land Area," p. 23.
7. Ibid., pp. 29–38.

8. Ibid., pp. 46–48. See p. 44 for expected yields: class 5 wind resources yield 14 million kWh/km^2, class 6 yield 18 million kWh/km^2, and class 7 yield 23 million kWh/km^2.
9. Ibid., p. 54.
10. Ibid., pp. A1–A4.
11. Ibid., pp. 29, 48.
12. Ibid., p. 53.
13. M. N. Schwartz, D. L. Elliott, and G. L. Gower, "Gridded State Maps of Wind Electric Potential," paper presented at the American Wind Energy Association's annual conference, Windpower '92, Seattle, WA, October 19–23, 1992.
14. Elliott, Wendell, and Gower, "An Assessment of the Available Windy Land Area," p. 52. 21,000 average MW, class 5 and greater resource at 30 m, moderate environmental exclusions. California has 1700 average MW under the same conditions.
15. Michael Grubb and Niels Meyer, *Renewable Energy: Sources for Fuels and Electricity,* Chapter 4, "Wind Energy: Resources Systems, and Regional Strategies" (Washington, DC: Island Press, 1993), p. 192.

	Population Density		
	Inhabitants/ mi^2	Inhabitants/ km^2	Second-Order Exclusions
United States, lower 48 states	80	30	$\frac{1}{2}$
Denmark	300	120	$\frac{1}{65}$
United Kingdom	600	230	$\frac{1}{65}$
The Netherlands	950	370	$\frac{1}{150}$

16. Ibid., p. 157.
17. Michael Grubb, "The Cinderella Options: A Study of Modernized Renewable Energy Technologies," Part 1-A, "Technical Assessment", *Energy Policy,* July-August 1990, pp. 525–542.
18. Michael Grubb, "The Cinderella Options: A Study of Modernized Renewable Energy Technologies," Part 2, "Political and Policy Analysis," *Energy Policy,* October 1990, pp. 711–725.
19. Thomas Lippman, "A Breath of Fresh Air for Wind Power," *Washington Post,* November 25, 1991.
20. E. Bossanyi, "System Integration," *Wind Energy Conversion Systems,* Leon Freris, ed. (London: Prentice Hall, 1990), pp. 357–371.
21. Don Smith, "Wind Energy Resource Potential and the Hourly Fit of Wind Energy to Utility Loads in Northern California," *Proceedings of Windpower '90,* the annual American Wind Energy Association conference, Washington, DC, September 24–28, 1990, pp. 47–52.
22. Jay Stock, "Wind Power Generation: One Utility's Perspective" Southern California Edison Co., Rosemead, CA, February 1985.
23. Alexi Clarke, "Wind Energy: Progress and Potential," *Energy Policy,* October 1991.
24. Grubb, "The Cinderella Options" (Part 1-A), pp. 525–542.
25. Peggy Friis and Mogens Held, "Commercial and Experimental Windpower in ELSAM Utility Area, Denmark," paper presented at annual conference of the American Wind Energy Association, Windpower '93, San Francisco, July 1993.
26. Ibid.
27. Ibid.

28. "Wind in Small Grids," *Sustainable Energy News,* 5, June 1994, p. 13.
29. Michael Peevey, president, Southern California Edison Co., address before the American Wind Energy Association's annual conference, Windpower '91, Palm Springs, CA, September 24, 1991.
30. For more information about the technical issues surrounding the integration of wind energy with utilities, read Ervin Bossanyi's chapter in L. L. Freris, ed., *Wind Energy Conversion Systems* (Hemel Hempstead, Hertfordshire, England: Pentice Hall International, 1990), and Carl Weinberg's and Dan Ancona's chapter in David Spera, ed., *Wind Turbine Technology* (New York: ASME, 1994).
31. Jamie Chapman, "European Wind Technology," Electric Power Research Institute, Palo Alto, CA, March 1993, pp. 1–6.
32. Peevey, address before Windpower '91.
33. See David Roe's *Dynamos and Virgins* (New York: Random House, 1984) for a fascinating account of shifting utility perceptions.
34. "Wisconsin Utility Initiatives: WP&L Strengthens Support for Off-Grid Renewables," The Renewable Quarterly, *RENEW,* Madison, WI, Winter 1994, p. 6.
35. "Un entretien avec Marc Vergnet," *Systèmes Solaires,* 94/95, 1993, pp. 18–21.
36. Ibid.
37. Mike Bergey, address before the annual conference of the American Wind Energy Association, Windpower '94, Minneapolis, MN, May 10, 1994.
38. Andrew Garrad, "Wind Energy in Europe: Time for Action!," European Wind Energy Association Strategy Document, August 31, 1990, p. 24.
39. Peevey, address before Windpower '91,
40. Q. Wang, M. DeLuchi, and D. Sperling, "Emission Impacts of Electric Vehicles," University of California, Davis, paper presented at Transportation Reseach Board Annual Meeting, Washington, DC, January 1989. Also reported in "California 2000: Exhausting Clean Air: Major Issues in Managing Air Quality," Assembly Office of Research, Sacramento, CA, October 1989.
41. Peevey, address before Windpower '91.
42. "The Potential of Renewable Energy: An Interlaboratory White Paper," Office of Policy, Planning and Analysis, U.S. Department of Energy, Washington, DC, March 1990, p. F-9.
43. Robert Lynette, "California Wind Farms: Operational Data Collection and Analysis," Solar Energy Research Institute, Golden, CO, March 28, 1988, p. ii.
44. BTM Consult, "Ressourceundersogelse vedrorende aeldre vindmoller i Danmark," Skjern, Denmark, January 1992, and "Rapport fra styregruppe om udskiftning af gamle vindmiller," Danish Energy Ministry, Copenhagen, June 1993, pp. 3.1–3.3.
45. K. C. Golden, Washington Environmental Council, address before the American Wind Energy Association's annual conference, Windpower '92, Seattle, WA, October 19–23, 1992.
46. Chris Flavin, "The Bridge to Clean Energy," *World Watch,* July/August 1992, pp. 10–18.
47. Carl Weinberg, oral comments before Solar '94, annual conference of the American Solar Energy Society, San Jose, CA, June 29, 1994.
48. Ken Stump, Greenpeace, oral comments at a Washington Environmental Council Seminar on renewable energy, Seattle, WA, March, 1993.
49. Ralph Cavanagh, oral address before the annual conference of the American Wind Energy Association, Windpower '92, Seattle, WA, October 1992.
50. Harvey Sachs and Frank Muller, "Technology Policy and Wind Power in the U.S. Utility

Sector," paper presented at the American Wind Energy Association's annual conference, Windpower '92, Seattle, WA, October 1992.

51. Cavanagh, address before Windpower '92.
52. Washington Comment, *Electrical World,* May 1992, p. 26.
53. Peevey, address before Windpower '91.
54. Donald Aitken, "Sustained Orderly Development of *Solar Electric Technologies," Solar Today,* May/June 1992, pp. 20–23.
55. Ibid.
56. Sachs and Muller, "Technology Policy."
57. Michael Eaton, John White, and P. Brodie, "The Greenhouse Effect: The Need for California Leadership," Sierra Club California policy statement, Sacramento, CA, January 19, 1989.
58. Al Gore, *Earth in the Balance: Ecology and the Human Spirit* (New York: Houghton Mifflin, 1992), p. 225.
59. Alice Ledogar, "On Visual Impacts," *Conservation Law Foundation Newsletter,* Boston, Spring 1994.

A

Tables

Conversions

Speed

1 m/s = 2.24 mph
1 mph = 0.446 m/s
1 knot = 1.15 mph
1 mph = 0.870 knot

Length

1 meter = 3.28 feet
1 foot = 0.305 meter
1 kilometer = 0.620 mile
1 mile = 1.61 kilometers

Area

1 square kilometer = 0.386 square mile
1 square kilometer = 1,000,000 square meters
1 square kilometer = 100 hectares
1 square mile = 2.59 square kilometers
1 hectare = 10,000 square meters
1 hectare = 2.47 acres
1 acre = 0.405 hectare
1 acre = 4049 square meters

Volume

1 cubic meter = 35.3 cubic feet
1 cubic foot = 0.028 cubic meter
1 liter = 0.264 gallon
1 gallon = 3.78 liters
1 cubic meter = 1000 liters
1 cubic meter = 264 gallons
1 gallon = 0.0038 cubic meter

Flow Rate

1 liter/second = 0.0044 gallon/minute
1 gallon/minute = 227 liters/second
1 cubic meter/minute = 264 gallons/minute
1 gallon/minute = 0.0038 cubic meter/minute

Weight

1 metric ton = 1.10 tons
1 kilogram = 2.20 pounds
1 pound = 0.454 kilogram

Energy Equivalence of Common Fuels

1 kWh = 3413 Btu
= 3.41 cubic foot of natural gas
= 0.034 gallon of oil
= 0.00017 cord of wood
=
1 therm = 1E+05 Btu
= 100 cubic foot of natural gas
= 1 gallon of oil
= 29.3 kWh
= 0.005 cord of wood
=
1 gallon of oil = 1E+05 Btu
1 cord of wood = 2E+07 Btu
1000 cubic feet of natural gas = 1E+06 Btu

Approximate Primary Energy Offset by Direct Generation of Electricity

$$1 \text{ kWh} = 10{,}000 \text{ Btu}$$
$$= 600 \text{ barrels of oil}$$

The amount of energy needed by a conventional power plant to generate 1 kWh of electricity. Fossil fuel–fired power plants convert only one-third of their fuel's primary energy into electricity. The Energy Information Agency assumes that 10,235 Btu of primary energy is needed in a thermal power plant to offset 1 kWh of hydroelectricity. Cited in D. L. Elliott, L. L. Wendell, and G. L. Gower, "An Assessment of the Available Windy Land Area and Wind Energy Potential in the Contiguous United States," Battelle Pacific Northwest Laboratory, Richland, WA, August 1991, p. 45.

Scale of Equivalent Power (Instantaneous Power)

Typical Wind Turbine Rating by Rotor Diameter (m)	Power (kW)	Equivalent
1.5	0.25	$\frac{1}{3}$ horsepower electric motor, electric drill
2	0.50	$\frac{2}{3}$-horsepower electric motor
3	1	Hair dryer, electric space heater
7	10	Garden tractor
10	25	
18	100	Passenger car engine
25	250	
40	500	Heavy truck engine
50	1,000	Race car engine, small diesel locomotive
100	3,000	Diesel locomotive
	500,000	Coal-fired generator, small nuclear reactor
	1,000,000	Large nuclear reactor

Scale of Equivalent Energy

Energy (kWh)	Approximate Size Wind Turbine to Provide the Same Annual Energy Output by Rotor Diameter[a] (m)	(ft)	Equivalent
1			Auto battery, 100-watt light for 10 hours
10			Electric space heater for 10 hours
50	0.25	1	Average per capita consumption in India
1,000	1.5	5	
3,500	2.5	8	Average residential consumption in northern Europe
4,000	3	10	
6,450	4	13	Average residential consumption in California
12,000	5	16	Average residential consumption in Texas
15,000	6	20	
20,000	7	23	Typical electric home consumption in Oklahoma
75,000	10	33	
250,000	18	60	
500,000	25	80	
750,000	33	110	
1,000,000	36	120	

[a] Class 4 wind resource, 30 m (100 ft) hub height (7 m/s), 30% conversion efficiency.

Air Density and Temperature

$$\text{Air density in kg/m}^3 = \frac{1.01325 \times 10^5 \text{ newtons/m}^2}{2.87 \text{ J/kg} \times (273.15 + T)}$$

where T is the temperature in Celsius (see Figure A.1).

Air Density and Altitude

See Figure A.2. When pressure and temperature are not known, Battelle PNL decreases air density by 1.194×10^{-4} kg/m^3 per meter increase in elevation above sea level. This approximates the U.S. Standard Atmospheric profile for air density [Dennis Elliott et al., *Wind Energy Resource Atlas of the United States* (Golden, CO: Solar Energy Research Institute, October 1986), p. 145].

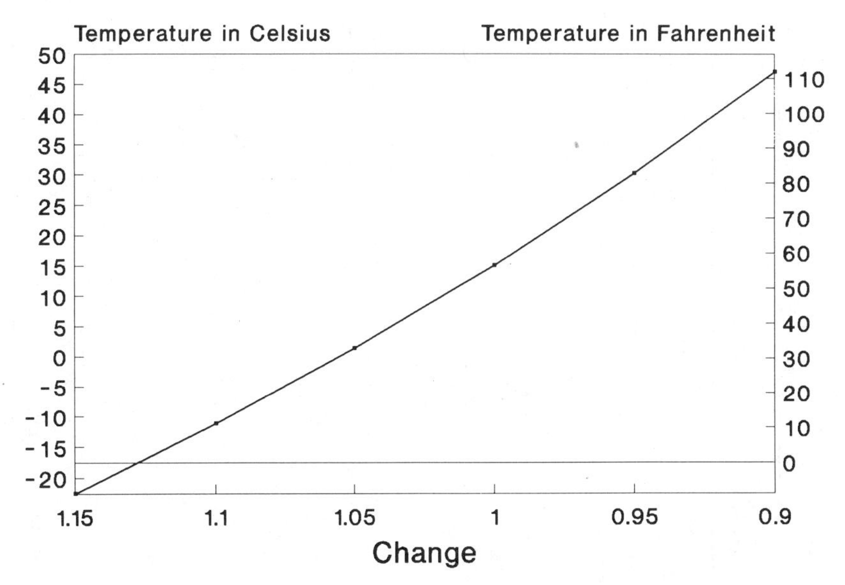

Figure A.1. Change in air density with temperature.

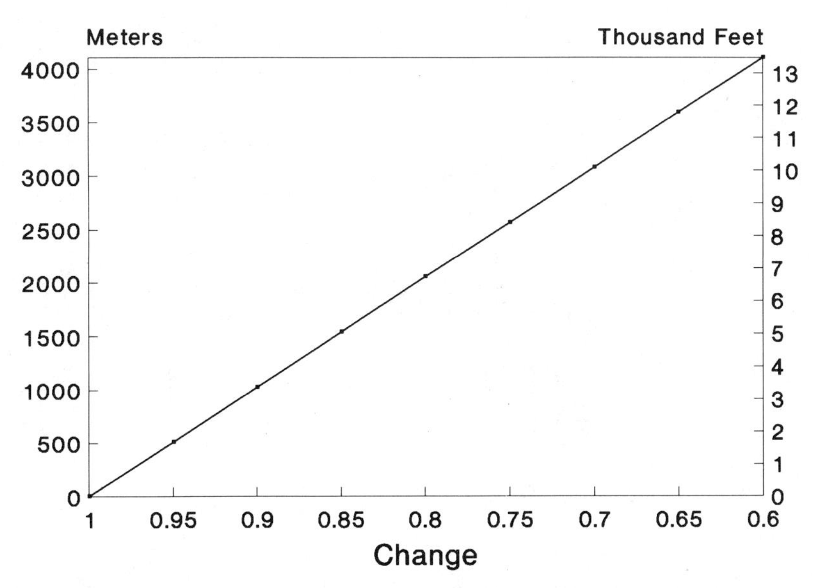

Figure A.2. Change in air density with elevation above sea level.

Rayleigh Wind-Speed Distribution

Wind-Speed Bin		*Probability of Occurrence at Annual Average Wind Speed (m/s) of:*										
m/s	mph	4	4.5	5	5.5	6	6.5	7	7.5	8	8.5	9
4	9.0	0.1790	0.1668	0.1520	0.1371	0.1231	0.1105	0.0992	0.0893	0.0807	0.0731	0.0664
5	11.2	0.1439	0.1471	0.1432	0.1357	0.1264	0.1168	0.1074	0.0985	0.0903	0.0828	0.0761
6	13.4	0.1006	0.1152	0.1217	0.1224	0.1194	0.1142	0.1080	0.1014	0.0947	0.0882	0.0821
7	15.7	0.0620	0.0812	0.0943	0.1019	0.1049	0.1047	0.1023	0.0986	0.0942	0.0893	0.0844
8	17.9	0.0339	0.0519	0.0673	0.0789	0.0864	0.0905	0.0919	0.0914	0.0895	0.0867	0.0834
9	20.2	0.0166	0.0302	0.0444	0.0571	0.0671	0.0742	0.0788	0.0811	0.0817	0.0811	0.0796
10	22.4	0.0072	0.0160	0.0272	0.0387	0.0492	0.0579	0.0645	0.0691	0.0719	0.0733	0.0735
11	24.6	0.0028	0.0078	0.0154	0.0247	0.0343	0.0431	0.0507	0.0567	0.0612	0.0642	0.0660
12	26.9	0.0010	0.0035	0.0082	0.0148	0.0226	0.0307	0.0383	0.0449	0.0503	0.0545	0.0576
13	29.1	0.0003	0.0014	0.0040	0.0084	0.0142	0.0209	0.0278	0.0343	0.0401	0.0450	0.0490
14	31.4	0.0001	0.0005	0.0019	0.0045	0.0085	0.0136	0.0194	0.0253	0.0310	0.0361	0.0406
15	33.6	0.0000	0.0002	0.0008	0.0023	0.0048	0.0085	0.0131	0.0181	0.0233	0.0283	0.0328
16	35.8	0.0000	0.0001	0.0003	0.0011	0.0026	0.0051	0.0085	0.0125	0.0170	0.0215	0.0259
17	38.1	0.0000	0.0000	0.0001	0.0005	0.0014	0.0029	0.0053	0.0084	0.0120	0.0160	0.0200
18	40.3	0.0000	0.0000	0.0000	0.0002	0.0007	0.0016	0.0032	0.0055	0.0083	0.0116	0.0151
19	42.6	0.0000	0.0000	0.0000	0.0001	0.0003	0.0009	0.0019	0.0034	0.0056	0.0082	0.0111
20	44.8	0.0000	0.0000	0.0000	0.0000	0.0001	0.0004	0.0011	0.0021	0.0036	0.0056	0.0080
21	47.0	0.0000	0.0000	0.0000	0.0000	0.0001	0.0002	0.0006	0.0012	0.0023	0.0038	0.0057
22	49.3	0.0000	0.0000	0.0000	0.0000	0.0000	0.0001	0.0003	0.0007	0.0014	0.0025	0.0039
23	51.5	0.0000	0.0000	0.0000	0.0000	0.0000	0.0000	0.0002	0.0004	0.0009	0.0016	0.0026
24	53.8	0.0000	0.0000	0.0000	0.0000	0.0000	0.0000	0.0001	0.0002	0.0005	0.0010	0.0017
25	56.0	0.0000	0.0000	0.0000	0.0000	0.0000	0.0000	0.0000	0.0001	0.0003	0.0006	0.0011
26	58.2	0.0000	0.0000	0.0000	0.0000	0.0000	0.0000	0.0000	0.0001	0.0002	0.0004	0.0007
27	60.5	0.0000	0.0000	0.0000	0.0000	0.0000	0.0000	0.0000	0.0000	0.0001	0.0002	0.0004
28	62.7	0.0000	0.0000	0.0000	0.0000	0.0000	0.0000	0.0000	0.0000	0.0000	0.0001	0.0003

Wind-Speed Bin		*Probability of Occurrence at Annual Average Wind Speed (mph) of:*											
mph	m/s	9	10	11	12	13	14	15	16	17	18	19	20
8	3.6	0.0834	0.0760	0.0686	0.0616	0.0552	0.0496	0.0447	0.0403	0.0365	0.0332	0.0303	0.0277
10	4.5	0.0735	0.0716	0.0678	0.0632	0.0584	0.0537	0.0492	0.0451	0.0414	0.0380	0.0350	0.0323
12	5.4	0.0576	0.0608	0.0612	0.0597	0.0571	0.0540	0.0507	0.0473	0.0441	0.0410	0.0382	0.0355
14	6.2	0.0406	0.0472	0.0509	0.0524	0.0523	0.0512	0.0493	0.0471	0.0447	0.0422	0.0398	0.0374
16	7.1	0.0259	0.0337	0.0394	0.0432	0.0453	0.0460	0.0457	0.0448	0.0434	0.0417	0.0399	0.0380
18	8.0	0.0151	0.0222	0.0285	0.0335	0.0371	0.0394	0.0406	0.0409	0.0406	0.0398	0.0387	0.0374
20	8.9	0.0080	0.0136	0.0194	0.0246	0.0290	0.0323	0.0346	0.0360	0.0367	0.0368	0.0365	0.0358
22	9.8	0.0039	0.0077	0.0123	0.0171	0.0216	0.0254	0.0284	0.0306	0.0321	0.0330	0.0334	0.0334
24	10.7	0.0017	0.0041	0.0074	0.0113	0.0153	0.0191	0.0224	0.0252	0.0273	0.0288	0.0298	0.0304
26	11.6	0.0007	0.0020	0.0042	0.0071	0.0104	0.0139	0.0171	0.0201	0.0225	0.0245	0.0260	0.0271
28	12.5	0.0003	0.0009	0.0022	0.0042	0.0068	0.0097	0.0127	0.0155	0.0181	0.0203	0.0221	0.0236
30	13.4	0.0001	0.0004	0.0011	0.0024	0.0043	0.0065	0.0091	0.0116	0.0141	0.0164	0.0184	0.0201
32	14.3	0.0000	0.0002	0.0005	0.0013	0.0026	0.0042	0.0063	0.0085	0.0108	0.0130	0.0150	0.0168
34	15.2	0.0000	0.0001	0.0002	0.0007	0.0015	0.0027	0.0042	0.0060	0.0080	0.0100	0.0120	0.0138
36	16.1	0.0000	0.0000	0.0001	0.0003	0.0008	0.0016	0.0027	0.0041	0.0058	0.0075	0.0093	0.0111
38	17.0	0.0000	0.0000	0.0000	0.0002	0.0004	0.0009	0.0017	0.0028	0.0041	0.0056	0.0071	0.0088
40	17.9	0.0000	0.0000	0.0000	0.0001	0.0002	0.0005	0.0010	0.0018	0.0028	0.0040	0.0054	0.0068
42	18.8	0.0000	0.0000	0.0000	0.0000	0.0001	0.0003	0.0006	0.0012	0.0019	0.0028	0.0039	0.0052
44	19.6	0.0000	0.0000	0.0000	0.0000	0.0001	0.0002	0.0004	0.0007	0.0012	0.0020	0.0028	0.0039
46	20.5	0.0000	0.0000	0.0000	0.0000	0.0000	0.0001	0.0002	0.0004	0.0008	0.0013	0.0020	0.0028
48	21.4	0.0000	0.0000	0.0000	0.0000	0.0000	0.0000	0.0001	0.0003	0.0005	0.0009	0.0014	0.0020
50	22.3	0.0000	0.0000	0.0000	0.0000	0.0000	0.0000	0.0001	0.0001	0.0003	0.0006	0.0009	0.0014
52	23.2	0.0000	0.0000	0.0000	0.0000	0.0000	0.0000	0.0000	0.0001	0.0002	0.0004	0.0006	0.0010
54	24.1	0.0000	0.0000	0.0000	0.0000	0.0000	0.0000	0.0000	0.0000	0.0001	0.0002	0.0004	0.0007
56	25.0	0.0000	0.0000	0.0000	0.0000	0.0000	0.0000	0.0000	0.0000	0.0001	0.0001	0.0003	0.0005
58	25.9	0.0000	0.0000	0.0000	0.0000	0.0000	0.0000	0.0000	0.0000	0.0000	0.0001	0.0002	0.0003
60	26.8	0.0000	0.0000	0.0000	0.0000	0.0000	0.0000	0.0000	0.0000	0.0000	0.0000	0.0001	0.0002
62	27.7	0.0000	0.0000	0.0000	0.0000	0.0000	0.0000	0.0000	0.0000	0.0000	0.0000	0.0001	0.0001

Estimates of Annual Energy Output

How to Use AEO Tables

The following tables estimate the amount of energy a wind turbine of a given size will produce in a given wind regime using its swept area, a Rayleigh distribution of wind speeds, and an assumed overall conversion efficiency. This assumption is given in the column labeled "Total Effic." and has been derived from a survey of wind turbine product literature. The results are approximations and may not necessarily correspond to estimates by manufacturers for similar conditions.

The first series of tables can be used if the average wind speed is known in either m/s or mph. The second series of tables can be used with Battelle's wind atlas. Each series of tables comprises two parts: one for small wind turbines and the second for medium-sized wind turbines up to 45 m (148 ft) in diameter.

To estimate the potential generation from a 7-m wind turbine installed on the Great Plains where the average annual wind speed, at hub height, is 6 m/s (13.4 mph), follow these steps. Find row *6.0* in the first column. This is the wind speed in m/s. Then move along the row until it intersects with the last column, labeled *7*. This is the wind turbine's rotor diameter in meters. The value where row *6.0* and column *7* intersect, *18*, is the estimated generation in thousands of kilowatt-hours per year. Under these conditions a 7-m turbine will produce approximately 18,000 kWh per year.

Estimated Annual Energy Output at Hub Height in Thousand kWh/yr for Wind Speed in m/s

Wind Turbines 1 to 7 Meters in Diameter

Average Speed		Power		*Rotor Diameter [m (ft)]*							
		Density	Total	1	1.5	2	3	4	5	6	7
m/s	mph	(W/m²)	Effic.[a]	(3.3)	(4.9)	(6.6)	(9.8)	(13.1)	(16.4)	(19.7)	(23.0)
4.0	9.0	75	0.28	0.1	0.3	0.6	1.3	2.3	3.6	5.2	7.1
4.5	10.1	110	0.28	0.2	0.5	0.8	1.9	3.4	5.3	7.6	10
5.0	11.2	150	0.25	0.3	0.6	1.0	2.3	4.1	6.5	9.3	13
5.5	12.3	190	0.25	0.3	0.7	1.3	2.9	5.2	8.2	12	16
6.0	13.4	250	0.21	0.4	0.8	1.4	3.3	5.8	9.0	13	18
6.5	14.6	320	0.19	0.4	0.9	1.7	3.8	6.7	10	15	20
7.0	15.7	400	0.16	0.4	1.0	1.8	4.0	7.0	11	16	22
7.5	16.8	490	0.15	0.5	1.1	2.0	4.6	8.1	13	18	25
8	17.9	600	0.12	0.5	1.1	2.0	4.5	7.9	12	18	24
8.5	19.0	720	0.12	0.6	1.3	2.4	5.3	9.5	15	21	29
9	20.2	850	0.12	0.7	1.6	2.8	6.3	11	18	25	34

[a] Assumed efficiency based on published data.

Wind Turbines 10 to 45 Meters in Diameter

Average Speed		Power	Total	*Rotor Diameter [m (ft)]*														
m/s	mph	Density (W/m^2)	Effic.[a]	10 (33)	11 (36)	12 (39)	13 (43)	14 (46)	15 (49)	16 (52)	17 (56)	18 (59)	19 (62)	20 (66)	21 (69)	22 (72)	23 (75)	25 (82)
4.0	9.0	75	0.25	13	16	19	22	25	29	33	37	42	46	52	57	62	68	80
4.5	10.1	107	0.25	19	23	27	32	37	43	48	55	61	68	76	83	92	100	118
5.0	11.2	146	0.3	31	37	45	52	61	70	80	90	100	110	120	140	150	160	190
5.5	12.3	195	0.3	39	47	56	66	77	90	100	110	130	140	160	170	190	210	250
6.0	13.4	253	0.3	52	62	74	87	100	120	130	150	170	190	210	230	250	270	320
6.5	14.6	321	0.3	66	80	100	110	130	150	170	190	210	240	260	290	320	350	410
7.0	15.7	401	0.28	80	90	110	130	150	170	200	220	250	280	310	340	370	410	480
7.5	16.8	494	0.28	90	110	140	160	190	210	240	270	310	340	380	420	460	500	590
8	17.9	599	0.25	100	120	150	170	200	230	260	300	330	370	410	460	500	550	650
8.5	19.0	718	0.25	120	150	180	210	240	280	320	360	400	450	500	550	600	660	770
9	20.2	853	0.22	130	160	190	220	250	290	330	370	420	460	510	570	620	680	800

Average Speed		Power	Total	*Rotor Diameter [m (ft)]*									
m/s	mph	Density (W/m^2)	Effic.[a]	26 (85)	27 (89)	28 (92)	29 (95)	30 (98)	31 (102)	32 (105)	33 (108)	34 (112)	35 (115)
4.0	9.0	75	0.25	87	94	101	108	116	124	132	140	149	158
4.5	10.1	107	0.25	128	138	148	159	170	183	194	206	219	232
5.0	11.2	146	0.3	210	230	240	260	280	300	320	340	360	380
5.5	12.3	193	0.3	270	290	310	330	350	380	400	430	450	480
6.0	13.4	253	0.3	350	380	400	430	460	500	530	560	600	630
6.5	14.6	321	0.3	450	480	520	560	590	630	680	720	760	810
7.0	15.7	401	0.28	520	560	600	650	690	740	790	840	890	940
7.5	16.8	494	0.28	640	690	740	790	850	910	970	1,030	1,090	1,160
8	17.9	599	0.25	700	750	810	870	930	990	1,060	1,120	1,190	1,260
8.5	19.0	718	0.25	840	900	970	1,040	1,110	1,190	1,270	1,350	1,430	1,520
9	20.2	853	0.22	870	940	1,010	1,080	1,160	1,240	1,320	1,400	1,490	1,580

Average Speed		Power	Total	*Rotor Diameter [m (ft)]*									
m/s	mph	Density (W/m^2)	Effic.[a]	36 (118)	37 (121)	38 (125)	39 (128)	40 (131)	41 (134)	42 (138)	43 (141)	44 (144)	45 (148)
4.0	9.0	75	0.25	167	176	186	196	206	216	227	238	249	261
4.5	10.1	107	0.25	245	259	273	288	303	318	334	350	366	383
5.0	11.2	146	0.3	400	420	450	470	500	520	550	570	600	630
5.5	12.3	195	0.3	510	540	570	600	630	660	690	730	760	790
6.0	13.4	253	0.3	670	710	750	780	830	870	910	950	1,000	1,040
6.5	14.6	321	0.3	860	900	950	1,000	1,060	1,110	1,170	1,220	1,280	1,340
7.0	15.7	401	0.28	1,000	1,050	1,110	1,170	1,230	1,300	1,360	1,420	1,490	1,560
7.5	16.8	494	0.28	1,220	1,290	1,360	1,440	1,510	1,590	1,670	1,750	1,830	1,910
8	17.9	599	0.25	1,340	1,410	1,490	1,570	1,650	1,730	1,820	1,910	2,000	2,090
8.5	19.0	718	0.25	1,600	1,700	1,790	1,880	1,980	2,080	2,180	2,290	2,400	2,510
9	20.2	853	0.22	1,670	1,760	1,860	1,960	2,060	2,160	2,270	2,380	2,490	2,610

[a] Assumed efficiency based on published data.

Estimated Annual Energy Output at Hub Height in Thousand kWh/yr for Wind Speed in mph

Wind Turbines 1 to 7 Meters in Diameter

Average Speed		Power	Total	*Rotor Diameter [m (ft)]*							
mph	m/s	Density (W/m²)	Effic.[a]	1 (3.3)	1.5 (4.9)	2 (6.6)	3 (9.8)	4 (13.1)	5 (16.4)	6 (19.7)	7 (23.0)
9	4.0	76	0.28	0.1	0.3	0.6	1.3	2.3	3.7	5.3	7.2
10	4.5	105	0.28	0.2	0.5	0.8	1.8	3.2	5.0	7.2	10
11	4.9	139	0.25	0.2	0.5	1.0	2.2	3.8	6.0	8.6	12
12	5.4	181	0.25	0.3	0.7	1.2	2.8	5.0	7.8	11	15
13	5.8	230	0.21	0.3	0.7	1.3	3.0	5.3	8.3	12	16
14	6.2	287	0.19	0.4	0.8	1.5	3.4	6.0	9	13	18
15	6.7	353	0.16	0.4	0.9	1.6	3.5	6.2	10	14	19
16	7.1	428	0.15	0.4	1.0	1.8	4.0	7.1	11	16	22
17	7.6	513	0.12	0.4	1.0	1.7	3.8	6.8	11	15	21
18	8.0	610	0.12	0.5	1.1	2.0	4.5	8.1	13	18	25
19	8.5	717	0.12	0.6	1.3	2.4	5.3	9	15	21	29

* Assumed efficiency based on published data.

Wind Turbines 10 and 45 Meters in Diameter

Average Speed		Power	Total	*Rotor Diameter [m (ft)]*															
mph	m/s	Density (W/m²)	Effic.[a]	10 (33)	11 (36)	12 (39)	13 (43)	14 (46)	15 (49)	16 (52)	17 (56)	18 (59)	19 (62)	20 (66)	21 (69)	22 (72)	23 (75)	24 (79)	25 (82)
9	4.0	76	0.25	13	16	19	22	26	29	34	38	42	47	52	58	63	69	75	82
10	4.5	105	0.25	18	22	26	30	35	40	46	52	58	65	72	79	87	95	104	112
11	4.9	139	0.3	29	35	41	49	56	65	70	80	90	100	110	130	140	150	170	180
12	5.4	181	0.3	37	45	54	63	73	80	100	110	120	130	150	160	180	200	210	230
13	5.8	230	0.3	47	57	68	80	90	110	120	140	150	170	190	210	230	250	270	300
14	6.2	287	0.3	59	72	90	100	120	130	150	170	190	210	240	260	290	310	340	370
15	6.7	353	0.28	70	80	100	110	130	150	170	200	220	250	270	300	330	360	390	420
16	7.1	428	0.28	80	100	120	140	160	190	210	240	270	300	330	360	400	440	480	520
17	7.6	513	0.25	90	110	130	150	170	200	230	260	290	320	350	390	430	470	510	550
18	8.0	610	0.25	100	130	150	180	210	240	270	300	340	380	420	460	510	550	600	660
19	8.5	717	0.22	110	130	160	180	210	240	280	310	350	390	430	480	530	570	620	680

Average Speed		Power	Total	*Rotor Diameter [m (ft)]*									
mph	m/s	Density (W/m²)	Effic.[a]	26 (85)	27 (89)	28 (92)	29 (95)	30 (98)	31 (102)	32 (105)	33 (108)	34 (112)	35 (115)
9	4.0	76	0.25	89	96	103	110	118	126	134	143	151	161
10	4.5	105	0.25	122	131	141	151	162	173	184	196	208	220
11	4.9	139	0.3	190	210	230	240	260	280	290	310	330	350
12	5.4	181	0.3	250	270	290	310	340	360	380	410	430	460
13	5.8	230	0.3	320	350	370	400	430	460	490	520	550	580
14	6.2	287	0.3	400	430	460	500	530	570	610	640	680	730
15	6.7	353	0.28	460	500	530	570	610	650	700	740	790	830
16	7.1	428	0.28	560	600	650	690	740	790	840	900	950	1,010
17	7.6	513	0.25	600	640	690	740	790	850	900	960	1,020	1,080
18	8.0	610	0.25	710	760	820	880	940	1,010	1,070	1,140	1,210	1,280
19	8.5	717	0.22	730	790	850	910	980	1,040	1,110	1,180	1,250	1,330

Average Speed		Power	Total	*Rotor Diameter [m (ft)]*									
mph	m/s	Density (W/m²)	Effic.[a]	36 (118)	37 (121)	38 (125)	39 (128)	40 (131)	41 (134)	42 (138)	43 (141)	44 (144)	45 (148)
9	4.0	76	0.25	170	179	189	199	210	220	231	242	254	265
10	4.5	105	0.25	233	246	260	273	288	302	317	332	348	364
11	4.9	139	0.3	370	390	410	440	460	480	510	530	560	580
12	5.4	181	0.3	480	510	540	570	600	630	660	690	720	750
13	5.8	230	0.3	610	650	680	720	760	800	840	880	920	960
14	6.2	287	0.3	770	810	850	900	950	1,000	1,040	1,090	1,150	1,200
15	6.7	353	0.28	880	930	980	1,030	1,090	1,140	1,200	1,260	1,320	1,380
16	7.1	428	0.28	1,070	1,130	1,190	1,250	1,320	1,390	1,450	1,520	1,600	1,670
17	7.6	513	0.25	1,140	1,210	1,280	1,340	1,410	1,480	1,560	1,630	1,710	1,790
18	8.0	610	0.25	1,360	1,440	1,510	1,590	1,680	1,760	1,850	1,940	2,030	2,120
19	8.5	717	0.22	1,410	1,490	1,570	1,650	1,740	1,820	1,910	2,010	2,100	2,200

[a] Assumed efficiency based on published data.

Estimated Annual Energy Output for Battelle Wind Power Classes at 30 m (98 ft) Hub Height in Thousand kWh/yr

Wind Turbines 1 to 7 Meters in Diameter

	Battelle Power Class at 10 m		*Wind Speed and Power at 30-m Hub Height*			*Rotor Diameter [m (ft)]*							
	Power Density[a] (W/m²)	Speed[a] (m/s)	Power Density[a] (W/m²)	Speed[a] (m/s)	Total Effic.[b]	1 (3.3)	1.5 (4.9)	2 (6.6)	3 (9.8)	4 (13.1)	5 (16.4)	6 (19.7)	7 (23.0)
	50	3.5	80	4.1	0.28	0.2	0.3	0.6	1.4	2.5	3.9	5.6	7.6
1													
	100	4.4	160	5.1	0.25	0.3	0.6	1.1	2.5	4.4	6.9	9.9	13
2													
	150	5.0	240	5.9	0.21	0.3	0.8	1.4	3.1	5.6	8.7	12	17
3													
	200	5.5	320	6.5	0.19	0.4	0.9	1.7	3.8	6.7	10	15	21
4													
	250	6.0	400	7.0	0.16	0.4	1.0	1.8	4.0	7.1	11	16	22
5													
	300	6.3	480	7.4	0.15	0.5	1.1	2.0	4.5	7.9	12	18	24
6													
	400	7.0	640	8.2	0.14	0.6	1.4	2.5	5.5	9.9	15	22	30
7													
	1000	9.5	1600	11.1	0.12	1	3	5	12	21	33	48	65

[a] Increase in speed and power with height assumes $\frac{1}{7}$ power law.
[b] Assumed efficiency based on published data.

Wind Turbines 10 to 45 Meters in Diameter

	Battelle Power Class at 10 m		*Wind Speed and Power at 30-m Hub Height*			*Rotor Diameter [m (ft)]*															
	Power Density[a] (W/m²)	Speed[a] (m/s)	Power Density[a] (W/m²)	Speed[a] (m/s)	Total Effic.[b]	10 (33)	11 (36)	12 (39)	13 (43)	14 (46)	15 (49)	16 (52)	17 (56)	18 (59)	19 (62)	20 (66)	21 (69)	22 (72)	23 (75)	24 (79)	25 (82)
	50	3.5	80	4.1	0.25	14	17	20	23	27	31	35	40	45	50	55	61	67	73	79	86
1																					
	100	4.4	160	5.1	0.3	33	40	48	56	65	74	80	100	110	120	130	150	160	170	190	210
2																					
	150	5.0	240	5.9	0.3	50	60	71	84	100	110	130	140	160	180	200	220	240	260	290	310
3																					
	200	5.5	320	6.5	0.3	66	80	95	110	130	150	170	190	210	240	260	290	320	350	380	410
4																					
	250	6.0	400	7.0	0.28	80	90	110	130	150	170	200	220	250	280	310	340	370	410	440	480
5																					
	300	6.3	480	7.4	0.28	90	110	130	160	180	210	240	270	300	330	370	410	450	490	530	580
6																					
	400	7.0	640	8.2	0.25	110	130	160	190	220	250	280	320	360	400	440	490	530	580	630	690
7																					
	1000	9.5	1600	11.1	0.22	240	290	350	410	470	540	620	700	780	870	970	1,100	1,200	1,300	1,400	1,500

Wind Turbines 10 to 45 Meters in Diameter (*Continued*)

	Battelle Power Class at 10 m		*Wind Speed and Power at 30-m Hub Height*			*Rotor Diameter [m (ft)]*									
	Power Density[a] (W/m²)	Speed[a] (m/s)	Power Density[a] (W/m²)	Speed[a] (m/s)	Total Effic.[b]	26 (85)	27 (89)	28 (92)	29 (95)	30 (98)	31 (102)	32 (105)	33 (108)	34 (112)	35 (115)
	50	3.5	80	4.1	0.25	93	100	108	116	124	132	141	150	159	169
1															
	100	4.4	160	5.1	0.3	220	240	260	280	300	320	340	360	380	400
2															
	150	5.0	240	5.9	0.3	340	360	390	420	450	480	510	540	570	610
3															
	200	5.5	320	6.5	0.3	450	480	520	560	590	640	680	720	760	810
4															
	250	6.0	400	7.0	0.28	520	560	600	650	690	740	790	840	890	940
5															
	300	6.3	480	7.4	0.28	630	670	730	780	830	890	950	1,010	1,070	1,130
6															
	400	7.0	640	8.2	0.25	740	800	860	930	990	1,060	1,130	1,200	1,270	1,350
7															
	1000	9.5	1600	11.1	0.22	1,600	1,800	1,900	2,000	2,200	2,300	2,500	2,600	2,800	3,000

	Battelle Power Class at 10 m		*Wind Speed and Power at 30-m Hub Height*			*Rotor Diameter [m (ft)]*									
	Power Density[a] (W/m²)	Speed[a] (m/s)	Power Density[a] (W/m²)	Speed[a] (m/s)	Total Effic.[b]	36 (118)	37 (121)	38 (125)	39 (128)	40 (131)	41 (134)	42 (138)	43 (141)	44 (144)	45 (148)
	50	3.5	80	4.1	0.25	178	189	199	209	220	232	243	255	267	279
1															
	100	4.4	160	5.1	0.3	430	450	480	500	530	560	580	610	640	670
2															
	150	5.0	240	5.9	0.3	640	680	720	750	790	830	870	920	960	1,000
3															
	200	5.5	320	6.5	0.3	860	900	950	1,010	1,060	1,110	1,170	1,220	1,280	1,340
4															
	250	6.0	400	7.0	0.28	1,000	1,060	1,110	1,170	1,230	1,300	1,360	1,430	1,490	1,560
5															
	300	6.3	480	7.4	0.28	1,200	1,270	1,340	1,410	1,480	1,560	1,630	1,710	1,790	1,870
6															
	400	7.0	640	8.2	0.25	1,430	1,510	1,590	1,670	1,760	1,850	1,940	2,040	2,130	2,230
7															
	1000	9.5	1600	11.1	0.22	3,100	3,300	3,500	3,700	3,900	4,100	4,300	4,500	4,700	4,900

[a] Increase in speed and power with height assumes $\frac{1}{7}$ power law.
[b] Assumed efficiency based on published data.

Sound Power Level for Selected Wind Turbines

Turbine	Diameter m	Diameter ft	Area m^2	Capacity kW	Wind Speed (m/s)	Blade Tip Speed (m/s)	Sound Power Level [dB(A)]
Kuriant	11	36	93	15	10	37.7	94
Aerotech	14	45	150		5	48.4	100
WindMatic	15	48	167	55	7	38	98
Bonus	15	50	181	55	10	38	98
Vestas	15	51	186	55	10	40	104
Lagerwey	16	51	191	75	7	68	92
Bouma	16	52	201		10	30.2	96
Aerotech	17	56	227		7	49	94
USW 56-100	18	58	243	100		67	98
Lagerwey	18	59	254	80	7	61	91
Lolland	19	61	275	80	8	39.2	97
Micon	19	62	284	108	5	46	93
Bouma	20	66	314		6	45	99
Nedwind	23	75	415	250	7	52	98
Bonus	24	78	445	150	8	50	94
Micon	24	79	452	250	10	50	100
Nordtank	25	81	475	150	8	49	97
HAT	25	82	491	300	7	69	103
WindMaster	25	82	491	300	7	68	105
Micon	26	85	531	250	8		99
Vestas	27	89	573	225	8	61	97
Vestas	29	95	661	225	8	62	97
Holec	30	98	707	300	8	47	95
Bonus Combi	31	102	755	300	8	50	96
Nordtank	31	102	755	300	8	55	101
Tacke	33	108	855	300	8	60	96
Enercon	33	108	855	330	7	43	98
WindMaster	33	108	855	550	7	48	107
Vestas	35	114	951	400	8	64	98
Nedwind	35	115	962	500	7	70	103
WEG	37	121	1060	400		52	98
Bonus	37	121	1075	450	8	58	97
Nordtank	37	121	1075	500	8	59	100
Mod OA	38	125	1140	200		80	105
Vestas	39	128	1195	500	8	61	101
Enercon	40	131	1257	500	8		99
Nibe A,B	40	131	1257	630	10	71	112
Tacke	43	141	1452	600	8	61	99
NEWECS	45	148	1590	1000	7	75	106
Näsudden	75	246	4418	2000		80	115
Maglarp WTS3	78	256	4778	3000		101	114
WTS4	79	260	4927	3000	7-17	124	120
Mod 2	91	300	6561	2500		86	116
Growian	100	328	7854	3000		83-110	119

Sources: ECN and product literature.

DEWI Magazin
Deutches Windenergie Institut
Eberstrasse 96
2940 Wilhemshaven
Germany
phone: +49 44 21 48 08 0
fax: +49 44 21 48 08 43

Home Power
P.O. Box 520
Ashland, OR 97520-0520
phone: 916 475 3179
fax: 916 475 3179

Independent Energy
620 Central Avenue N
Milaca, MN 56353-1788
phone: 612 983 6892
fax: 612 983 6893

Neue Energie
Interessenverband Windkraft Binnenland
Postgraben 37
49074 Osnabrück
Germany
phone: +49 541 20 15 93
fax: +49 541 25 93 03

RENEW
Network for Alternative Technology and Technology Assessment
Energy and Environment Research Unit, Faculty of Technology
The Open University
Walton Hall
Milton Keynes, Bucks MK7 6AA
United Kingdom
phone: +44 908 65 20 99
fax: +44 908 66 53 658

Systèmes Solaires
146, rue de l'Université
75007 Paris
France
phone: +33 1 44 18 00 80
fax: +33 1 44 18 00 36

WinDirections
British Wind Energy Association
42 Kingsway
London WC2B 6EX
United Kingdom
phone: +44 17 14 04 34 33
fax: +44 17 14 04 34 32

Windenergie Aktuell
Deutsche Gesellschaft für Windenergie
Lutherstrasse 14
30171 Hannover
Germany
phone: +49 511 28 23 63
fax: +49 511 28 23 77

Wind Energy Weekly/Windletter
American Wind Energy Association
122 C Street, NW, 4th Floor
Washington, DC 20001
phone: 202 408 8988
fax: 202 408 8536

Wind Engineering
Multi-Science Publishing
107 Hight St.
Brentwood, Essex CM14 4RX
United Kingdom
phone: +44 277 22 46 32
fax: +44 277 22 34 53

Windmillers' Gazette
P.O. Box 507
Rio Vista, TX 76093

Windpower Monthly/WindStats/Naturlig Energi
Vrinners Hoved
Knebel
Denmark
DK-8420
phone: +45 86 36 54 65
fax: +45 86 36 56 26

Books on Wind Resources and Siting

Battelle Pacific Northwest Laboratory. March 1987. Wind Energy Resource Atlas of the United States. This is the updated version of Battelle's classic work mapping U.S. wind resources. The update incorporates new data collected since the original atlas was published in 1980. DOE/CH 10094-4. Available from the American Wind Energy Association, 122 C Street, NW, 4th Floor, Washington, DC 20001; or the National Technical Information Service, U.S. Department of Commerce, 5285 Port Royal Road, Springfield, VA 22161.

Battelle Pacific Northwest Laboratory. Shaded relief maps of elevation; 950 maps cover the United States and Puerto Rico. Available from the American Wind Energy Association.

Risø National Laboratory. 1989. European Wind Atlas. Published for the European Community. 656 pp.

Risø National Laboratory. 1992. Wind Atlas Analysis and Application Program. (WA^{S}P). Published for the European Community.

Books on Wind Energy

Baker, T. Lindsay. *A Field Guide to American Windmills.* University of Oklahoma Press. Norman, OK. 1985. 528 pp. The definitive history of the American farm windmill.

Eggleston, David, and Stoddard, Forrest. *Wind Turbine Engineering Design.* Van Nostrand Reinhold. New York. 1987. 352 pp. Essential for wind turbine designers. Includes an extensive discussion of aerodynamics and structural dynamics. Not for the faint of heart. The authors warn readers that a working knowledge of differential equations is essential to understanding the text.

Freris, L. L., ed. *Wind Energy Conversion Systems.* Prentice Hall International. Hemel Hempstead, Hertfordshire, England. 1990. 388 pp. An excellent engineering text by the leaders in British wind energy.

Gipe, Paul. *Wind Power for Home and Business: Renewable Energy for the 1990s and Beyond.* Chelsea Green Publishing. White River Junction, VT. 1993. 414 pp. Directed toward homeowners, farmers, and others who want to use small wind turbines. Includes chapters on wind resources, estimating annual energy output, economics, stand-alone power systems, utility interconnection, water pumping, siting, and safety.

Golding, E. W. *The Generation of Electricity by Wind Power.* E.&F.N Spon. London. 1955. 332 pp. Reprinted by John Wiley & Sons in 1976. Still a classic of English language books on wind technology. Recounts British research on wind energy during the early 1950s.

Johnson, Gary L. *Wind Energy Systems.* Prentice Hall, Englewood Cliffs, NJ. 1985. 360 pp. An engineering textbook strong on electrical engineering and the wind resources of Kansas.

Koeppl, Gerald W. *Putnam's Power from the Wind,* 2nd ed. Van Nostrand Reinhold. New York. 1982. 470 pp. The first half of the 2nd edition is a reprint of Putnam's original book. The second half is a thorough look at large wind turbine development programs in the United States and Europe. Useful as a historical record of megawatt wind turbine technology of the late 1970s and early 1980s.

Le Gourières, Dèsirè. *Energie Eolienne: Thèorie, conception, et calcu pratique des installations.* Editions Eyrolles. Paris. 1980. 268 pp. In English as *Wind Plants: Theory and Design.* Pergamon Press. Oxford. 1982. At the time it was one of the first modern technical books on wind turbine design. It includes numerous sketches of small European wind turbines manufactured during the late 1970s, the most thorough documentation of early French wind turbines of any source and a useful discussion of classic water-pumping windmill design.

Molly, Jens-Peter. *Windenergie: Theorie, Anwendung, Messung.* Verlag C. F. Müller. Karslruhe, Germany. 1990. 316 pp. German language treatment of wind technology development through the late 1980s. One of the most thorough and well-illustrated books on modern wind technology in any language. Includes useful German–English and English–German translation of common technical terms.

Park, Jack. *The Wind Power Book.* Cheshire Books. Palo Alto, CA. 1981. pp 253. Introduction to wind turbine design principles suitable for the nonengineer.

Putnam, Palmer Cosslett. *Power from the Wind.* Van Nostrand Reinhold. New York. 1948. 224 pp. Reprinted in 1974. The classic account of constructing the 1.250-MW Smith–Putnam turbine during the early 1940s. Like Golding, many of Putnam's observations still apply in the 1990s.

Righter, Robert. *Wind Energy in America: A History.* University of Oklahoma Press. Norman, OK. 1995. A thought-provoking account of the people and ideas behind the use of wind energy in the United States. The book emphasizes the conflict between centralized and distributed use of wind-generated electricity.

Spera, David, ed. *Wind Turbine Technology: Fundamental Concepts of Wind Turbine Engineering.* American Society of Mechanical Engineers. New York. 1994. 650 pp. Funded by the U.S. Department of Energy, overseen by NASA, *Wind Turbine Technology* was seven years in production. The book is noteworthy for a chapter on NASA/DOE's big turbines by the program's principal proponent, Lou Divone. *Wind Turbine Technology* is packed with photographs, illustrations, and tables.

D

Maps

Battelle PNL Map of U.S. Wind Resources

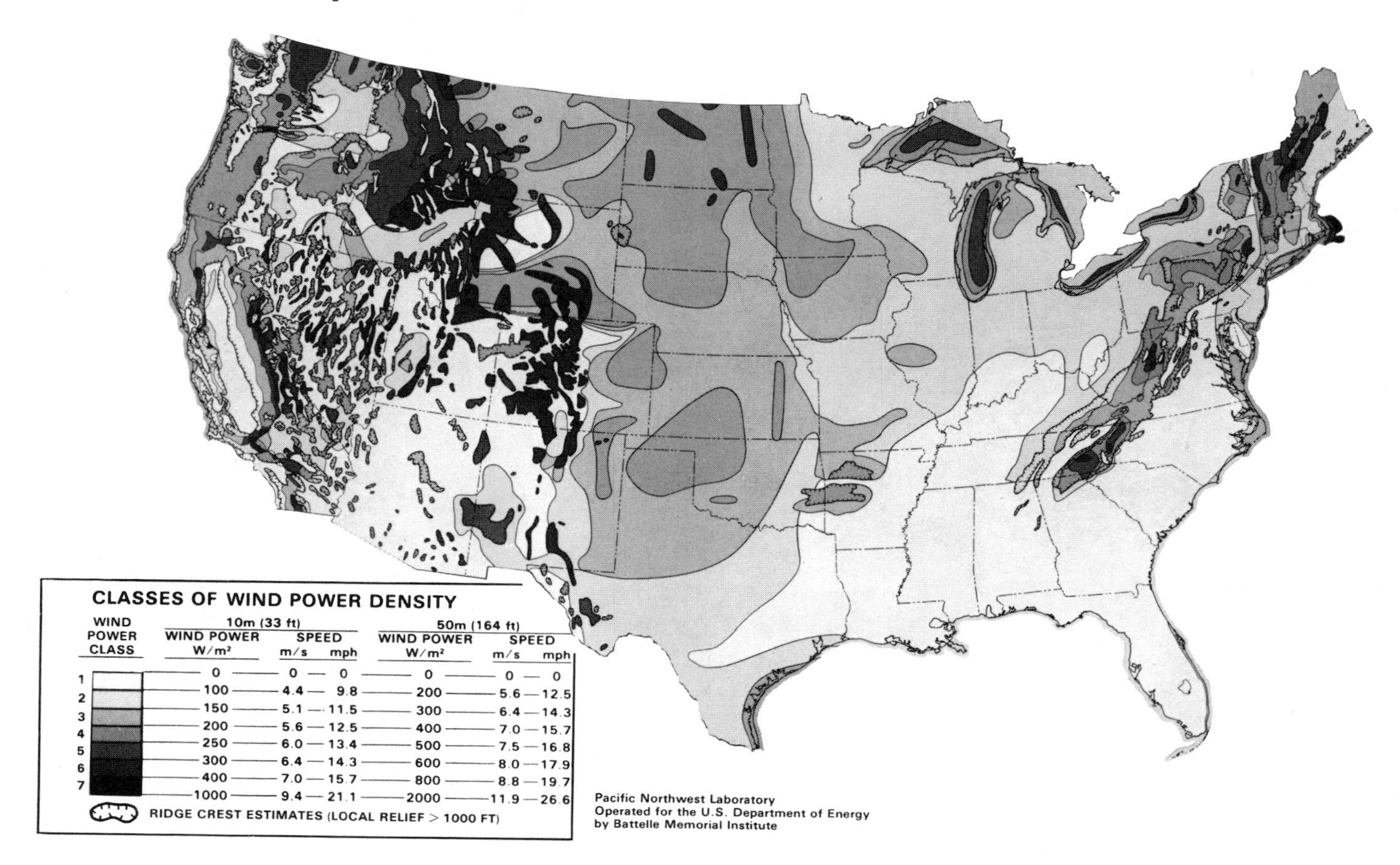

Risø Map of European Wind Resources

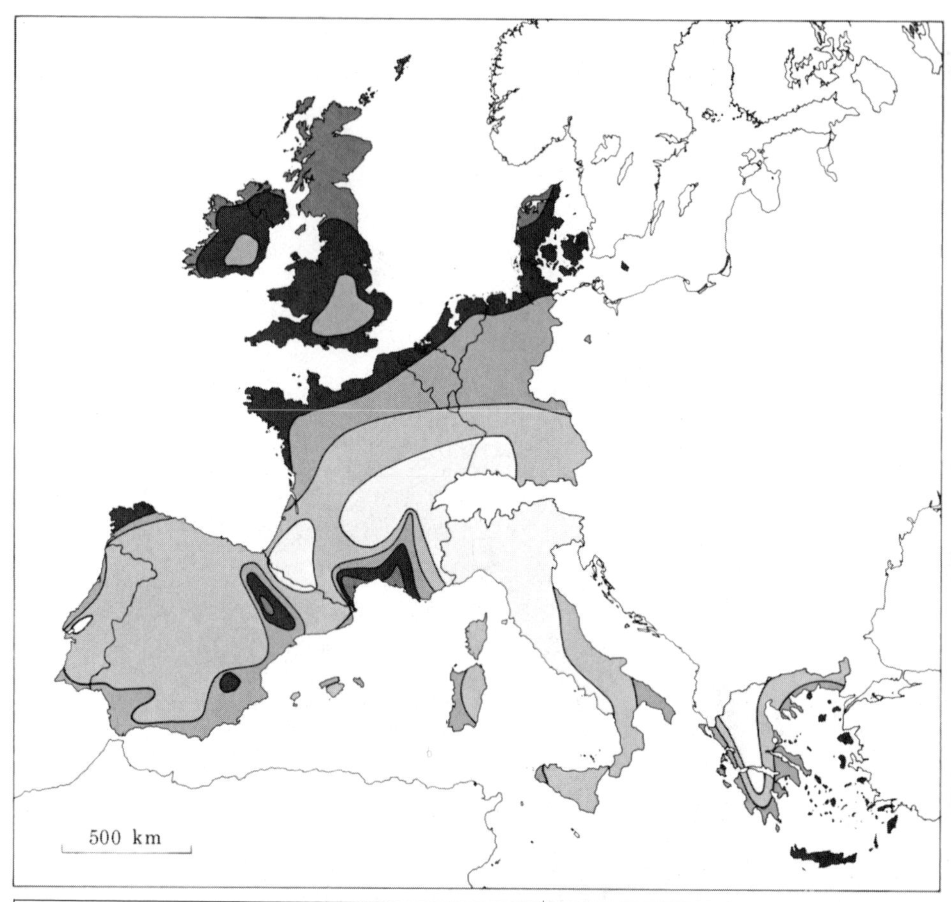

Wind resources[1] at 50 metres above ground level for five different topographic conditions										
	Sheltered terrain[2]		Open plain[3]		At a sea coast[4]		Open sea[5]		Hills and ridges[6]	
	$m\,s^{-1}$	Wm^{-2}	$m\,s^{-1}$	Wm^{-2}	$m\,s^{-1}$	Wm^{-2}	$m\,s^{-1}$	Wm^{-2}	$m\,s^{-1}$	Wm^{-2}
	> 6.0	> 250	> 7.5	> 500	> 8.5	> 700	> 9.0	> 800	> 11.5	> 1800
	5.0-6.0	150-250	6.5-7.5	300-500	7.0-8.5	400-700	8.0-9.0	600-800	10.0-11.5	1200-1800
	4.5-5.0	100-150	5.5-6.5	200-300	6.0-7.0	250-400	7.0-8.0	400-600	8.5-10.0	700-1200
	3.5-4.5	50-100	4.5-5.5	100-200	5.0-6.0	150-250	5.5-7.0	200-400	7.0-8.5	400-700
	< 3.5	< 50	< 4.5	< 100	< 5.0	< 150	< 5.5	< 200	< 7.0	< 400

1. The resources refer to the power present in the wind. A wind turbine can utilize between 20 and 30% of the available resource. The resources are calculated for an air density of 1.23 kg m^{-3}, corresponding to standard sea level pressure and a temperature of 15°C. Air density decreases with height, but up to 1000 m a.s.l. the resulting reduction of the power densities is less than 10%.
2. Urban districts, forest and farm land with many windbreaks (roughness class 3).
3. Open landscapes with few windbreaks (roughness class 1). In general, the most favourable inland sites on level land are found here.
4. The classes pertain to a straight coastline, a uniform wind rose and a land surface with few windbreaks (roughness class 1). Resources will be higher, and closer to open sea values, if winds from the sea occur more frequently, i.e. the wind rose is not uniform and/or the land protrudes into the sea. Conversely, resources will generally be smaller, and closer to land values, if winds from land occur more frequently.
5. More than 10 km offshore (roughness class 0).
6. The classes correspond to 50% overspeeding and were calculated for a site on the summit of a single axisymmetric hill with a height of 400 metres and a base diameter of 4 km. The overspeeding depends on the height, length and specific setting of the hill.

(Courtesy of Risø National Laboratory, Roskilde, Denmark.)

Index